POLYMER TECHNOLOGY

Third Edition

by

D. C. Miles,

F.I.M.
(Consultant)

and

J. H. Briston,

B.Sc., C. Chem., F.R.S.C., F.I.M., F. Inst. Pkg.

Chemical Publishing Co., Inc.
ISBN: 0-8206-0344-9

Printed in the United States of America

CONTENTS

PREFACE TO THE THIRD EDITION

Once again the many advances in polymers and their associated processes have rendered necessary a new edition of this book.

The Material chapters have been updated to include linear low density polyethylene (LLDPE), pearlescent polypropylene, amorphous nylon (Selar PA), poly (ethylene terephthalate) (PET), ethylene/vinyl alcohol copolymers (EVOH), poly (etherimide) (Ultem), and poly (etheretherketone) (PEEK).

The chapters on Processing have also been updated to take account of developments in powder coating, coextrusion, injection and blow molding, solid phase pressure forming, and thermoforming (including blister and skin packaging). New additions include sections on stretch blow molding, multi-layer (coextrusion) bottle blowing, and the decoration of plastics.

The chapter on Testing has been expanded to include water vapor and gas transmission, while the section on optical tests has been enlarged.

As in previous editions, the applicational aspects have also been updated so that the book will continue to serve the needs of the users of polymers, as well as those of chemists, engineers and technicians in the industry.

PREFACE TO THE SECOND EDITION

Since the first edition of this book there has been a tremendous growth in both polymers and processes.

The framework of the original book seemed to meet with the general approval so this has been retained but a great deal of new materials has been added in order to keep it up-to-date. As well as two new complete chapters on Heat Resistant Thermoplastics and Powder Coating there have been substantial additions to most of the other chapters. Chapter 16, for example, contains information on a number of new materials and has been retitled High Performance Thermoplastics.

Additional materials covered include polyimides, aromatic polamides, polysulfone, poly (methyl pentene), parylene, polyphenylene oxide and many others. New processes include the molding of structural foams, coextrusion, powder coating, solid phase forming, form/fill/seal bottle blowing, and film fibers.

As in the first edition a great deal of emphasis has been placed on the applications of materials and processes so that the book will continue to be of interest and help to users of plastics and rubbers as well as to chemists, engineers and technicians in the industry.

PREFACE

The past few years have seen the development of many new polymeric materials, some of them, such as polypropylene and polyisoprene rubber, being the results of completely new polymerization techniques. There have also been corresponding advances in processing techniques. Some of these developments have been the subject of separate textbooks or monographs, while at the other end of the scale there are multi-volume works with each chapter written by specialists.

It seemed to us, however, that there was still room for a book which would cover the most important of these developments in a single volume and incorporate them into the general framework of plastics technology. Originally we were guided to some extent by the syllabus for the plastics technology papers of the Graduateship Examination of the Plastics Institute, since it appeared that there were no single texts available to students. The framework suggested by this approach, namely, Materials, Processes and Testing, remains but with additional material and a slight change of emphasis. These changes should still enable the book to act as a basis for students intending to sit for the Plastics Institute and City and Guilds Examinations as well as the Polymer Technology route to Licenciateship of the Royal Institute of Chemistry.

We have, for instance, included some material on natural, synthetic and modified rubbers following the decision to form a Joint Examining Board for the Plastics Institute and the Institute of the Rubber Industry, at the same time altering the title of the book to *Polymer Technology* from *Plastics Technology*. We also decided to place more emphasis on the applications of plastics, linking these with the properties of the various materials and the conversion processes available. There is an increasing number of people, in a wide variety of industries, who find that they need to know something about the uses of polymers, their advantages and their limitations. Not all of these people have a technical background and we have tried to keep their needs in mind, as well as those of chemists, engineers and technicians.

Finally, for any reader who finds that he needs to know more on any particular topic in this book, we have included suggestions for further reading at the end of each chapter. While these do not include every book written on polymers, they form a fair cross section and can themselves act as stepping stones to more advanced tomes, if desired.

D. C. MILES
J. H. BRISTON

ACKNOWLEDGMENTS

We gladly make acknowledgment here to all those companies and institutions who made available to us items such as technical information on their products, photographs and sketches:

Bakelite Ltd.; Baker Perkins Granbull Ltd.; Bayer Farbenfabriken A.G.; B.I.P. Chemicals Ltd.; B.I.P. Tools Ltd.; David Bridge & Co. Ltd.; Brit. Celanese Ltd.; Brit. Cellophane ltd.; T.H. & J. Daniels Ltd.; The Distillers Co. Ltd.; Farrel-Birmingham Co. Inc.; Fibreglass Ltd.; Fothergill & Harvey Ltd.; Hercules Powder Co.; *Plastics; Plastics International;* Projectile & Engineering Co. Ltd.; Scott Bader and Co. Ltd.; Shaw, Francis and Co. Ltd.; Turner Brothers Asbestos Co. Ltd.

PART I GENERAL

1.

INTRODUCTION

A polymer may most easily be defined as a material which is composed of comparatively simple recurring units. Well-known examples which occur in nature are the polysaccharides such as starch and cellulose, which are made up of very large numbers of glucose units. Another well-known polymer is natural rubber (polyisoprene); its repeating unit is shown below.

$$\left(-CH_2-\underset{\displaystyle CH_3}{\underset{|}{C}}=CH-CH_2- \right)$$

Many other polymers occur in nature but few of them concern us here. As is pointed out in the Preface, our intention in this book is to deal with the technology of certain polymeric materials known as plastics, but for the sake of completeness the subject of rubbers needs to be touched upon.

For a definition of plastics we cannot do better than repeat that given in the British Standards Institute publication, *Glossary of Terms used in the Plastics Industry* (B.S. 1755: 1951) which says, 'Plastics—A wide group of solid composite materials which are largely organic, usually based on synthetic resins or upon modified polymers of natural origin and possessing appreciable mechanical strength. At a suitable stage in their manufacture most plastics can be cast, molded or polymerized directly to shape. Some plastics are rubber-like while some chemically modified forms of rubber are considered to be plastics.'

Rubber is a little harder to define but the following, based on performance characteristics at room temperature, is adequate: 'A rubber is a material capable of being stretched to at least double its original length and then returning, rapidly and forcibly, to substantially its original length. An additional requirement is that it shall be capable of being easily formed in its raw state but after vulcanization it shall be able to retain a given size and form.' The theory and practice of vulcanization are discussed in a later chapter dealing with rubbers but it may be briefly noted here that the effect of vulcanization on a rubber is to reduce its plasticity while effectively maintaining its elasticity. Surface tackiness is decreased, while the strength is increased. In many cases it is possible to produce hard materials, as instanced by the material known as 'ebonite.' The first half of the definition above would include many plastics which are, however, excluded because they cannot be vulcanized.

Plastics themselves can be divided into two main subgroups, namely, thermoplastics and thermosets (or thermo-hardening) materials. Thermoplastics are materials which have the property of softening repeatedly on the application of heat and of hardening again when cooled, while thermosets soften once and then harden irreversibly on the application of sufficient heat (usually referred to as 'curing'). When a thermosetting resin is 'cured' it is insoluble in solvents.

From the definition of rubbers given above, it can be seen that these form an intermediate class of material somewhere between the thermoplastics and the thermosets. Thus, they are thermoplastic during forming operations but are 'cured' c vulcanized after forming to an intermediate elastic or rubbery state and, in some instances, the 'cure' may be prolonged to produce a material closely resembling a thermoset.

The difference in the formation of thermoplastics and thermosetting materials can be explained by reference to the Theory of Functionality, which was first produced by R. H. Kienle. This theory may best be explained using a diagrammatic approach, as follows:

Let the monomer unit be represented as a circle, O, and each functional group (or reactive center) within the monomer be represented by a dash, —.

A monofunctional moner $(f) = 1$ would thus be represented as O—.

Since reaction can take place only between reactive groups or centers the only type of compound which can be formed is of the type shown here, O—O. An example of this type of reaction would be that between a monohydric alcohol with a monocarboxylic acid to give an ester, e.g.,

$$CH_3CH_2OH + CH_3COOH \rightarrow CH_3COO \cdot CH_3CH_2 + H_2O$$

Let us now consider a monomer with functionality $(f) = 2$. Reaction can then continue to give long-chain molecules, in the manner pictured below.

$$—O— + —O— + —O— \rightarrow —O—O—O—$$

An example of this type of reaction is the formation of a linear polyester from a dihydric alcohol and a dibasic acid, e.g.,

$$\underset{\text{Malonic acid}}{n(COOH \cdot CH_2 \cdot COOH)} + \underset{\text{Ethylene glycol}}{n(OH\ CH_2 \cdot CH_2 \cdot OH)} \longrightarrow$$

$$\left[O \cdot CO \cdot CH_2 \cdot O \cdot CO \cdot CH_2 \cdot CH_2\right]_n + 2n \cdot H_2O$$

It should be noted, in passing, that a reactive double bond possesses a functionality of 2 since it can open in such a way that it will connect up with two other groups.

$$\underset{\text{Ethylene}}{CH_2{=}CH_2} \rightarrow {}^*{-}CH_2{-}CH_2{-}{}^*$$

or

$$n \cdot CH_2{=}CH_2 \rightarrow {}^*{-}(CH_2{-}CH_2)_n{-}{}^*$$

The asterisk denotes that the attached bond can react with other reactive groups which may be present. For a treatment of the kinetics governing such reactions one must refer to textbooks on physical chemistry but it should be emphasized that the reaction is not a straight build-up of one chain only. Very many chains are in the process of formation at any instant and each chain competes for monomer with all other chains. Chain formation ceases when the supply of monomer is exhausted or when reaction occurs with a monofunctional molecule, such as a catalyst residue, solvent molecule or an impurity. Monofunctional compounds are also added deliberately to control the chain length (and hence the molecular weight) of the polymer.

$$O— + —O(—O—)_nO— + —O \rightarrow O—O(—O—)_nO—O$$

We have said that only molecules of a functionality of 2 or higher can react to form a polymer but while this is true, it does not necessarily mean that di-functional molecules will always polymerize. Sometimes a functional group is shielded by other groups, and in some cases the formation of a ring compound is favored, either by the geometry and nature of the reacting molecules or by the reaction conditions. If a six-membered ring can be formed, for instance, by intraesterification, then polymerization will rarely occur. Reaction in dilute solution also favors ring formation since the reactant molecules are separated from each other by the solvent molecules, and the chance that the ends of the same molecule will come together is greater than that of their meeting, and reacting, with different molecules. The reaction of methylene diamine and malonic acid, for example, leads to the formation of a six-membered ring as shown below.

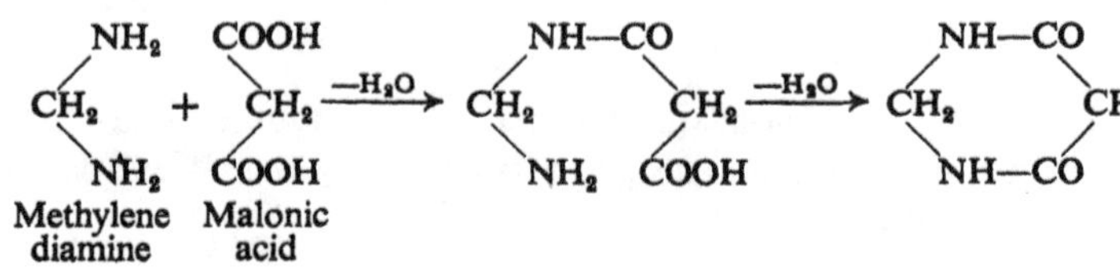

Once the molecule with reactive ends is formed, ring formation is an easy subsequent step. Six-membered rings are the ones most easily obtained because such rings are formed with the least distortion of the bonds connecting the individual atoms. Closer study of these and similar factors governing the formation of polymers is more properly the province of the physical chemist: some excellent texts are available and are given in the reading list at the end of this chapter.

The linear polymers formed by the reaction of difunctional molecules are members of the subgroup of plastics which we have already defined as thermoplastics. The long-chain molecules are held together by relatively weak intermolecular (or Van der Waal) forces, and energy, in the form of heat, will cause the molecules to move relative to each other and the material will then flow. Differences in the softening point of different plastics are due to the relative strength of their intermolecular forces. Polar molecules, such as those of nylon, have greater attraction for each other than nonpolar ones such as polyethylene, and the softening point of the former is, therefore, greater than that of the latter. As is usual in the polymer field this is not the whole story and we find that the softening point of polystyrene is lower than that of polyethylene in spite of the fact that polystyrene, too, is a polar molecule. This is because there are bulky groups attached to the polystyrene molecule, which prevent a close approach of the polymer chains with a consequent lessening of the attractive forces between them. Less energy (i.e. heat) is necessary, therefore, to cause relative movement or slipping of the chains.

We may now take a brief look at the consequences of including a material with a functionality of 3 among our reactants. Using the pictorial representation referred to previously we see that it is possible to build up a complete network. The diagram shown below is, of course, a two-dimensional representation of what is, in reality, a three-dimensional network.

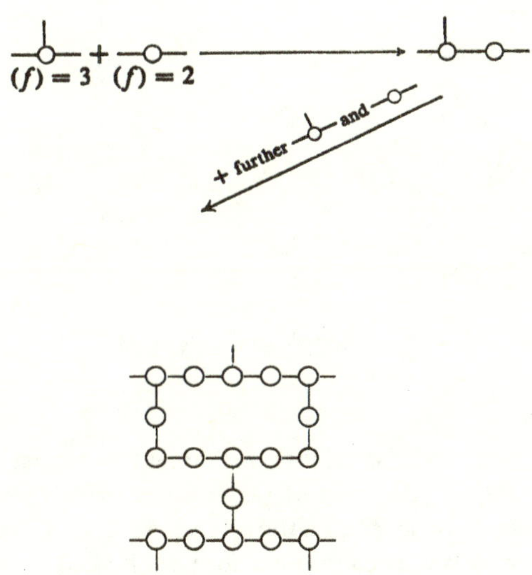

An example of the type of reaction shown above, where the reactants have functionalities of 3 and 2, is given by the formation of phenol-formaldehyde resins.

The three-dimensional type of polymer typified by phenol-formaldehyde exhibits the type of behavior described earlier as thermosetting, the fully cured compound being unable to flow under the action of heat. Each carbon atom in the network is bound by primary valence bonds to

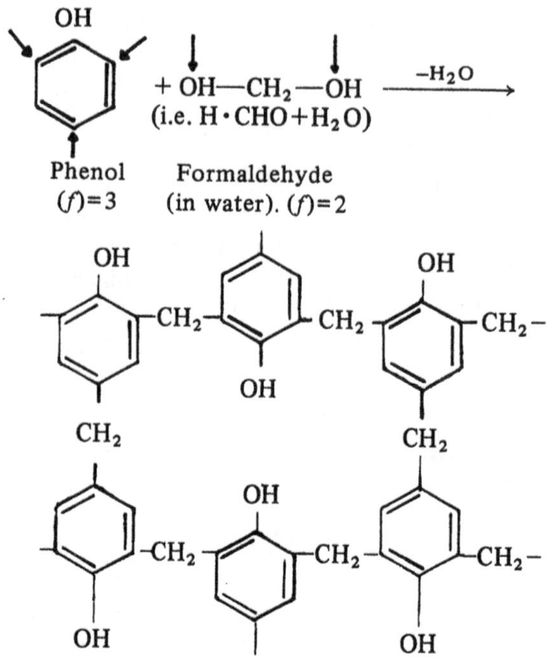

Phenol-formaldehyde

other carbon atoms so that relative movement is almost impossible. In addition, such a structure prevents penetration by solvent molecules and subsequent dispersion so that although solvents may sometimes cause swelling of thermosets they cannot produce a solution.

Intermediate stages are possible in the above and similar reactions, and advantage is taken of this fact in the fabrication of articles from the raw materials. If we consider Stage 1 to be the existence of the starting materials in their monomeric form, and Stage 3 the completely cured compound, then we can visualize a Stage 2 when the full network has not yet been built up and some flow is still possible under the influence of heat. The resin manufacturer stops the reaction at Stage 2 and sells the product in the form of pellets or as a powder. The molder takes this intermediate product and forms it into the required shape, using heat and pressure. The conditions used to promote the flow in the mold are also sufficient to cause the completion of the fully linked network so that the shape, once formed, is permanent.

During the discussion on the case of the reaction between two difunctional materials it was mentioned that polymerization would not always take place because a functional group might be inoperational due to the shielding by other groups. The same arguments apply to reaction between di- and tri-functional groups, an excellent example being the reaction between glycerin and phthalic acid. Glycerin is a tri-functional compound, as can be seen from its formula, namely,

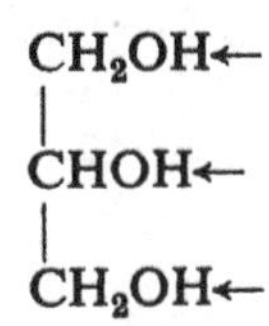

the functional groups being arrowed.

However, the central OH group is, to some extent, shielded by the other two, and temperatures of between 250 and 270°C are necessary to cause reaction of all three OH groups. At lower temperatures (200-240°C) a linear product is obtained, as one would expect from two difunctional compounds.

In passing we may mention compounds having a functionality of 4. A common class of tetrafunctional compound is that of the dienes (compounds having 2 double bonds). The normal reaction of compounds where the 2 double bonds are separated by one single bond (normally referred to as conjugated dienes) is as a difunctional compound but leaving a residual double bond in the molecule.

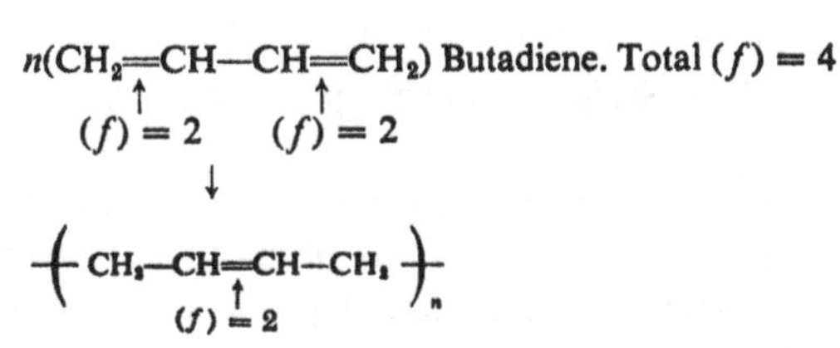

The residual functionality of 2 can be utilized to combine with material such as sulfur to produce three dimensional networks.

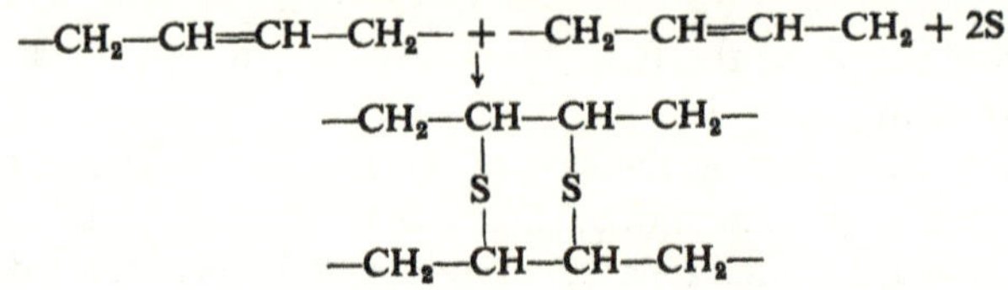

This is the basis of vulcanization of polybutadiene and other rubbers.

The structure of rubbers is intermediate between that of the thermoplastics and thermosets and can best be described as a loosely linked network where the linkages are comparatively few in number (Fig. 1.1).

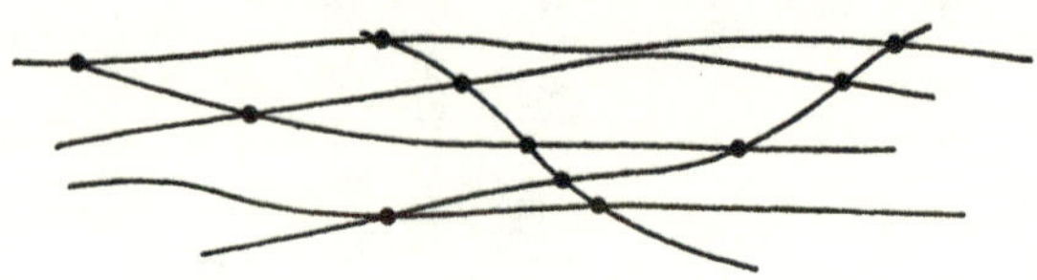

Fig. 1.1

Such a cross-linked structure is formed by vulcanization of rubbers and allows considerable extension to occur, without breaking the chemical bonds, when submitted to a tensional stress. On releasing the stress, the rubber tends to resume its original configuration.

The requirements for fiber formation are, as one would expect, very different and can be summed up as follows: (*a*) the molecules should be linear and of fairly high molecular weight; (*b*) the molecules should possess a fairly high degree of orientation; (*c*) the molecules should have a high degree of linear symmetry, i.e. there should be an absence of bulky side-chains which would prevent orientation and crystal packing; (*d*) the molecules should contain polar groups.

POLYMERIZATION REACTIONS

We have already dealt very briefly with polymerization reactions and have classified them on the basis of the properties of the resulting polymers, i.e. thermoplastic and thermosetting. There is, however, a further classification on the basis of the type of chemical reaction which occurs, namely, addition or condensation.

Addition Polymerization

In addition polymerization, the resultant polymer is obtained by the combination of a number of simple basic substances (monomers) without loss. Addition polymers are usually thermoplastics. The monomers contain a double bond and, with the exception of fluorinated compounds, always contain the methylene group, CH_2. In other words, the basic monomer in addition polymerization is of the general formula $CH_2 = C \cdot R_1R_2$ except, as already stated, in the case of $CF_2 = CF_2$ and $CF_2 = C \cdot ClF$. The explanation appears to lie in the relative sizes of the atoms, that of fluorine being the smallest in size, next to hydrogen, which will combine with carbon. Probably the best known example of addition polymerization is that of ethylene to give polyethylene.

$$n \cdot CH_2{=}CH_2 \rightarrow \left(CH_2{-}CH_2 \right)_n$$

There are four basic methods of carrying out addition polymerization and these have certain advantages and disadvantages which are characteristic of the method, irrespective of the monomer being polymerized. In order to facilitate reference to these methods, when considering the formation of the various addition polymers in greater detail later in the book, a brief description and evaluation of each method is given in this chapter.

(a) Bulk Polymerization. This is the simplest of the four processes since pure monomer only is used, except for the addition of a catalyst when necessary. The mixture is usually stirred in a large reaction vessel, with or without the addition of heat. The chief advantage of this method is obviously the purity of the final product with the concomitant advantages of excellent color and clarity in transparent polymers such as polystyrene. Bulk polymerization may also be considered to be economically attractive, since the raw material costs consist only of the monomer, but the difficulties of manufacture may sometimes balance this. These manufacturing difficulties arise from the fact that addition polymerization reactions are usually exothermic in nature and with large quantities of undiluted reactants there is, thus, an acute problem of heat dispersal. In addition, as the reaction proceeds, there is an increase in viscosity, and stirring becomes first difficult and eventually impossible, if the reaction is allowed to go to completion.

(b) Solution Polymerization. The monomer and catalyst are dissolved in an inert solvent the resultant solution being heated and stirred. In gen-

eral, both heating and stirring problems are eased in solution polymerization, which is also more suitable for continuous processing than is bulk polymerization. There are, however, serious economic problems due to the use of large quantities of expensive solvents which have, therefore, to be recovered and separated. The purity of the resultant polymer is lower than with bulk polymerization because, even with nominally inert solvents, there is frequently some reaction with the active centers of the growing polymer chain, which leads to the existence of solvent fragments in the finished molecule. The polymer is also soluble in the solvent and the removal of the last traces of solvent is rather difficult and expensive. For this reason the solution process is often resorted to for those applications in which the polymer is required in solution form. Surface coatings and impregnating compositions are among the commonest examples where the polymer is applied as a solution, thus rendering evaporation of the original solvent unnecessary.

(c) Emulsion Polymerization. In this process the continuous medium is usually water, and the monomer is dispersed instead of dissolved as in the previous process. The resultant polymer is also insoluble in the medium and is thus precipitated as it is formed. The catalysts used are those which are soluble in the dispersing medium, and emulsifying agents such as sodium or potassium salts of fatty acids are added to stabilize the emulsion. After the reaction is completed, the precipitated product is washed in order to remove emulsifying agents, but these are difficult to remove completely and some contamination always remains. There are, of course, few heat transfer problems in such processes because the heat is mainly carried away by the water. One other factor of interest in this process is that it produces the highest molecular weight product.

(d) Suspension Polymerization. The monomer in this process is suspended in the carrier medium (usually water) together with a monomer-soluble catalyst. When the mixture is stirred, the monomer is broken into small droplets, each of which will contain a little of the catalyst. A stabilizer, such as methyl cellulose, or starch, is also included in the mixture in order to keep the particles separate from each other during the reaction. These stabilizers are easily removed after the reaction by washing or flotation, leaving the polymer in the form of small beads or 'pearls.' This process can be looked upon as a large number of tiny 'bulk' polymerizations without the disadvantage of heat removal problems. The molecular weight of suspension process polymers is also fairly high.

Condensation Polymerization

Condensation polymerization occurs by the interaction of one or more compounds, with the elimination of a simple compound such as water, ammonia, hydrochloric acid, etc. Important examples of this type of polymerization include the nylons (polyamides), polyesters, alkyds, phenol-formaldehyde, etc. Polycondensation reactions can lead to either thermoplastic or thermosetting products, according to the functionality of the reactants. The phenol-formaldehyde reaction already described is an example of a polycondensation reaction (the molecule eliminated in the reaction being water) and the product is thermosetting. Polyamides, on the other hand, are polycondensates with thermoplastic properties.

$$n\left[\underset{(f)=2}{COOH(CH_2)_xCOOH}\right] + n\left[\underset{(f)=2}{NH_2(CH_2)_yNH_2}\right] \rightarrow$$

$$\rightarrow OH\left[CO(CH_2)_xCO\cdot NH(CH_2)_yNH\right]_{n-1}CO(CH_2)_xCO\cdot NH(CH_2)_yNH_2 + (2n-1)H_2O$$

One important difference between polycondensation and addition reactions is the formation of a by-product (water in the two cases mentioned) in the former. If this is not removed during the reaction an equilibrium is set up and the reaction ceases, usually before a useful product is formed.

At a certain stage in a polycondensation reaction, cross-linking will have occurred to such an extent that the molecule is no longer soluble in solvents and the so-called 'gel point' is reached. A considerable amount of work was carried out by W. H. Carothers, in the U.S.A., who derived an equation which could be used to predict the extent of the reaction at which gelation would occur. The Carothers equation is derived as follows:

Let us consider the system (m, n) where m molecules of n functionality react with n molecules of m functionality.

$$\text{Average functionality } (f) = \frac{\text{total number of functional groups}}{\text{total number of molecules}}$$

i.e.,

$$(f) = \frac{2mn}{m+n}$$

Now, let N_0 = number of molecules initially, and N = number of molecules after reaction.

Then $N_0 f$ = number of functional groups at beginning,
$N_0 - N$ = number of molecules lost during the reaction, and
$2(N_0 - N)$ = number of functional groups lost during the reaction.

Now

$$P = \text{extent of reaction} = \frac{2(N_0 - N)}{N_0 f} = \text{fraction of functional groups lost, and}$$

$$\frac{N_0}{N} = \text{degree of polymerization} = \bar{r}n.$$

Substituting, $P = \dfrac{2}{f} - \dfrac{2}{f\bar{r}n}$

At gelation, $\bar{r}n \rightarrow \infty$ and P becomes $= 2/f$

Applying this to the glycerin/phthalic acid reaction mentioned earlier we have:

2 molecules of glycerine of functionality 3.	Total functionality = 6
3 molecules of phthalic acid of functionality 2.	Total functionality = 6
5 molecules	Functionality = 12

$$\text{Average functionality } (f) = \frac{12}{5} = 2.4$$

$$\therefore \quad P = \frac{2}{f} = \frac{2}{2.4} = 83.3 \text{ percent}$$

From this we can see that the theoretical gel point should occur after 83 percent polyesterification. However, actual investigation by Kienle and others has shown the actual gel point to occur at 78.6 percent esterification. The difference between the calculated and actual gel points is due to the assumption that the molecular weight at the gel point is infinite. In actual fact, when the gel point is reached the mass is not one huge molecule, as was once supposed, but is, rather, a large molecule enveloping many smaller units, and in most cases these lower molecular weight polymers can be extracted from the gel.

Curing Techniques

We have seen that cross-linking to form three-dimensional networks can be achieved by reacting two materials, one with a functionality of at least 2 and another with a minimum of 3. Many of these techniques are based on condensation reactions but there are several other techniques which become clear as the reader proceeds through the book. These include the catalytic reaction of amines (epoxies), polyfunctional isocyanates, auto-oxidation and free radicals formed by thermal and radiation energy. The free-radical techniques are discussed here rather than in specific chapters.

(a) Free Radicals from Thermal Energy. The rate at which a chemical decomposes or disassociates into free radicals depends upon the temperature to which it is exposed. Such chemicals, in addition to being used as cross-linking agents, may also be used as initiators in polymerization of double-bond monomers. Included in this group are peroxides and hydroperoxides and while organic peroxides are generally preferred other materials can be used such as azo compounds (azo iso butyro nitrile) and silyl peroxides. It is the thermally unstable –O–O– bond which decomposes to form the free radicals. Such materials may be regarded as derivatives of hydrogen peroxide in which one or both hydrogens are replaced by organic radicals. They are compounds of the general form R·O·O·R' and the thermal decomposition of these peroxides is often complex but essentially where free radicals are required for polymerization the first step is considered to be the homolytic cleavage of the –O–O– bond to give free radicals R·O· and ·O·O·R'. These free radicals so formed attach and add on to the double bond of the monomer. A dimer is thus formed which is still activated and attacks another bond thus starting a chain reaction. This is of course another method of producing three dimensional structure as in the case of styrenated polyester. One of the most popular cross-linking agents for this system is benzoyl peroxide. The free radicals are produced when this chemical is heated to above 90°C.

$$\mathrm{Ph \cdot CO \cdot O \cdot O \cdot OC\,Ph \longrightarrow 2\,Ph \cdot CO \cdot O}$$

$$\mathrm{Ph \cdot CO \cdot O \longrightarrow Ph + CO_2}$$

As mentioned previously, the free radicals so formed attack the unsaturated double bond and this provides a reactive molecule which is capable of further reaction.

$$\mathrm{Ph \cdot CO \cdot O + CH_2CHR \longrightarrow Ph \cdot CO \cdot O \cdot CH_2CHR}$$

While this initial reaction is not necessarily regarded as being fast, once the initiation has begun the further reaction is very rapid. The decomposition rate of most organic peroxides follows first order kinetics and can be expressed by half-life time. It can also be described as the time required to decompose one half of the peroxide originally present in a dilute solution at a given temperature. For example, the half-life of benzoyl peroxide at 100°C is 24 minutes.

(b) Free Radicals from Radiation Energy. A technology which has advanced fairly rapidly over recent years is that of radiation curing. The chemistry of this technique is based on a combination of modern photochemistry and the free radical polymerization which is discussed in the previous section. The main difference is that the less efficient thermal energy is replaced by radiation energy. This latter energy form is either produced by ultraviolet or electron beam techniques. At present, due to the lower installation costs of the former process, it is the preferred method. The basic chemistry of the resultant free radical process is essentially the same for all radiation processes.

This technique is regarded as relatively new but the origin of the cross-linking of photosensitive polymers can be traced back well over 100 years. In fact, the chemistry of photodimerization goes back to 1867 when Fritsche worked on the dimerization of anthracene. This dimerization gave rise to cyclobutane derivatives which formed valuable reactions for the cross-linking of suitably substituted photochemical polymers.

The upsurge of the radiation curing of polymers is attributed to the necessity to save energy and the very stringent air pollution regulations which are coming into force throughout many countries of the world. At present these resins are not cheap in comparison to the more conventional coating types but they have advantages other than the ecological ones viz faster production rates with nonclogging of screens in certain printing applications.

Radiation Curing Components

The simplest system is based upon a polyfunctional acrylate monomer, uv reactive polymer, photoinitiator and a source of radiation energy. The uv reactive polymer is unsaturated and imparts the bulk of the requisite properties such as hardness, chemical resistance, and flexibility. Such a combination of materials still has limited application and can only be used on a commercial basis for clear top coats. To provide more durable coatings with a wider spectrum of uses it is essential to compound these resin systems with pigments, etc., along the same lines as used with the more conventional paints and printing inks.

Radiation Curing Components

1. Polyfunctional Monomer 2. u/v Reactive Polymer 3. Photoinitiator	$\xrightarrow[\text{Energy source}]{\text{Radiation}}$	Cross-linked clear coating

(i) Polyfunctional Acrylates. We have already seen that the basic characteristics are obtained from the uv unsaturated polymer which are in general high viscosity materials. To make them more usable polyfunctional acrylates are added which not only make the overall system more mobile but increase the degree of cross-linking. They can therefore be referred to as reactive diluents analogous to the use of styrene, etc., in unsaturated polyester chemistry. Another common practice with the addition of the polyfunctional acrylates is the addition of monofunctional monomers which not only controls the viscosity but also the degree of cross-linking. Typical monofunctional monomers are 2-ethyl hexyl or hydroxy ethyl acrylate and polyfunctional types are:

Pentaerythritol Triacrylate (PETA)

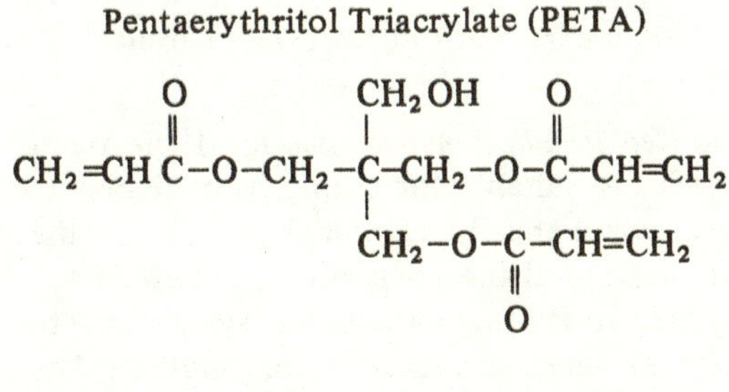

Highest degree of acrylic unsaturation—high degree of gloss.

Trimethylolpropane Triacrylate (TMPTA)

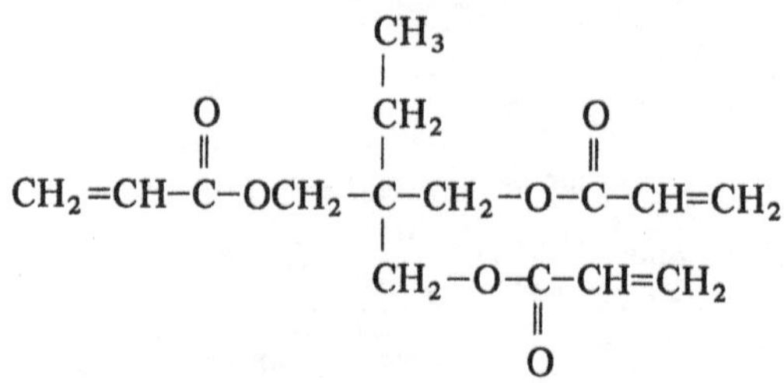

High degree of flexibility—low viscosity.

1,6 Hexanediol Diacrylate (HDODA)

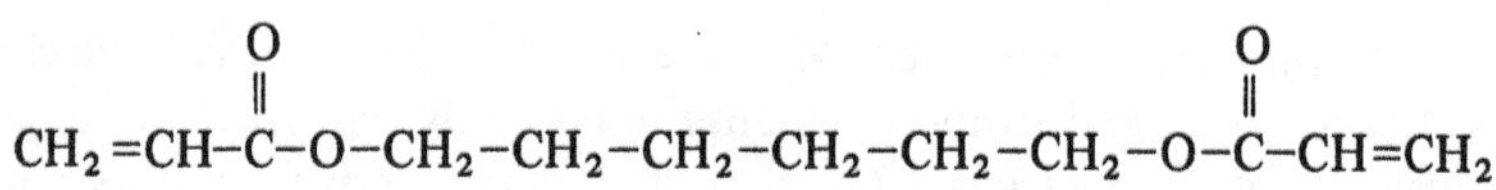

High degree of flexibility—very low viscosity—
improves elasticity and adhesion.

Neopentyl glycol diacrylate

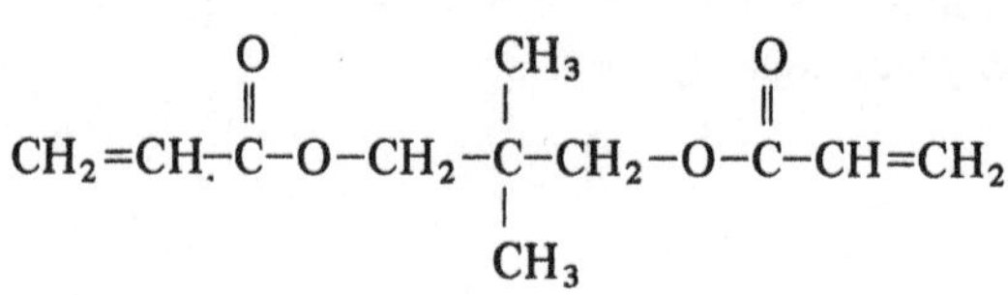

High degree of toughness and hardness.

(ii) Radiation Reaction Polymer Backbone. There are many unsaturated polymers which are suitable for curing by radiation techniques. At present the most commercially exploited groups for the acrylic-based polymers are polyester, urethanes, polyethers, and epoxies.

It is probably true to say that a uv curable system may be based on any conventional polymer which has reactive unsaturation. The order of reactivity is acrylate > methacrylate > vinyl with the acrylate type being preferred due to its fast curing performance.

(iii) Photoinitiators. The peroxides used in thermal free radical polymerization such as the unsaturated polyesters mentioned earlier are probably better known to most readers than the photoinitiators which are the counterparts for uv cure. It is obvious that the light source, not the heat, provides the activation force.

The photoinitiators are stable in the dark but when exposed to intense irradiation from a uv light source they undergo reactions which lead to the formation of radicals. It involves either a photochemically induced homolytic fragmentation of the excited photoinitiator into radicals or electron excitation of the photoinitiator from a ground state singlet to a triplet followed by electron transfer to a monomer with the generation of free radicals. The free radicals so formed are extremely reactive towards the polymerization of acrylic monomers.

The materials which are generally found suitable for uv curing have absorption in the 300–430 nanometer uv range. Typical of this type are aromatic ketones such as benzophenone and its derivatives or some chlorinated aromatic compounds.

Photoinitiator Fragmentation

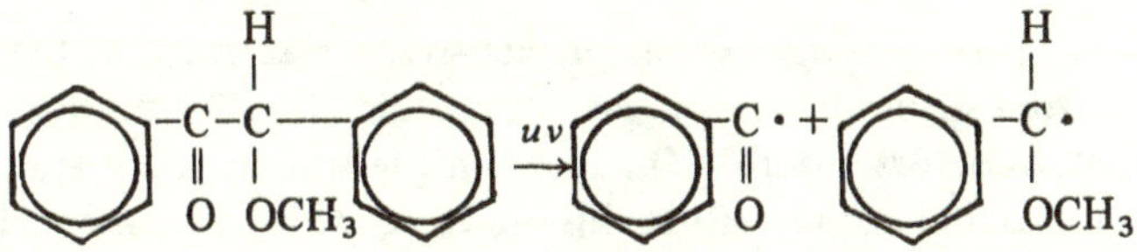

Benzoin methyl ether — Benzaldehyde radical — Methoxytoluyl radical

Mechanism of Radiation Cure

The basic polymer for this type of curing can be represented as in Fig.1 by having two terminal unsaturation sites (a difunctional material). This polymer in turn is then dissolved in a reactive diluent such as an acrylate along with the photoinitiator. The reactive acrylate takes part in the cross-linking reaction or a method analagous to that of melamine-formaldehyde in thermal systems.

On exposure to the uv light, the photoinitiator generates free radicals which in turn begins the polymerization reaction. The components of the resultant solution are converted to a solid cross-linked polymer film almost instantaneously. The resultant film is tack free and because no heat is involved it can be handled immediately. Very little shrinkage occurs and since solvents are not used the thickness of the cured film is virtually the same as that of the coating originally applied.

Mechanism of uv cure

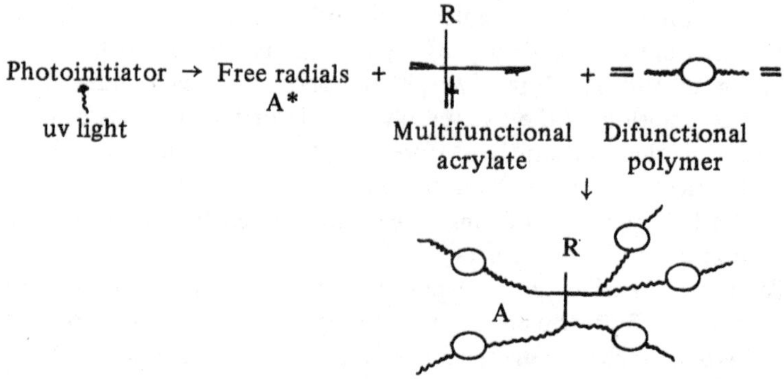

Radiation Equipment

(i) UV Cure Systems. It has previously been stated that the spectral range of interest for uv curing is the near visible 300–430 nm region where most photoinitiators absorb. The medium pressure mercury arc lamp offers the maximum output within this preferred range. Commercially available lamps have an intensity of 200 watts per inch and they can be used to cure coatings of up to 10 mils thickness. Below the 300 nm range the shorter uv rays will only penetrate and cure the very top layer of the coating.

Other equipment is essential besides the lamp and this includes an elliptical reflector, radiation shield, air or water cooling system, power source, and control equipment. In practice the lamp is mounted diagonally across the processing conveyor line at a distance of 2–4 inches from the moving substrate. At normal processing speeds the heat generated by the uv lamp does not exceed 70°C, nevertheless line breakdown can lead to inferior products or even fires. However most commercial units today have automatic shuttering devices which block off the light when production is halted.

(ii) Electron Beam (EB) Cure Systems. The unsaturated monomer is also capable of being cross-linked by means of a beam or curtain of elec-

trons from an EB generator. The mechanism is that of free radical inducement. This type of curing system does not require photoinitiators because of the high level of energy of the electrons (10^5 electron volts as against the 10^3 electron volts from the uv source). The EB systems have an added advantage in that they can also cure pigmented films up to thicknesses of 10 mils.

COPOLYMERIZATION

Addition polymerization can also take place with two or more different monomers, the resulting copolymer having different properties from a mechanical mixture of the polymers formed from the individual monomers. Copolymers are often valuable additions to the growing range of polymers and it is important, therefore, to understand the concept of copolymerization from the beginning. The difference between a mechanical mixture of polymers and a copolymer is shown pictorially below (Fig. 1.2 and Fig. 1.3):

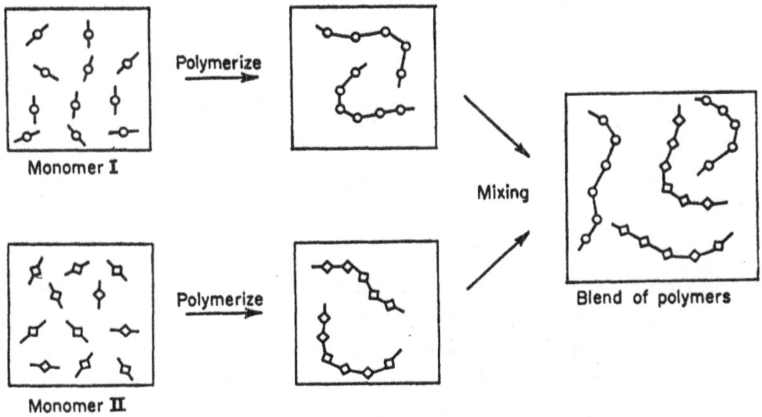

Fig. 1.2 Mixing

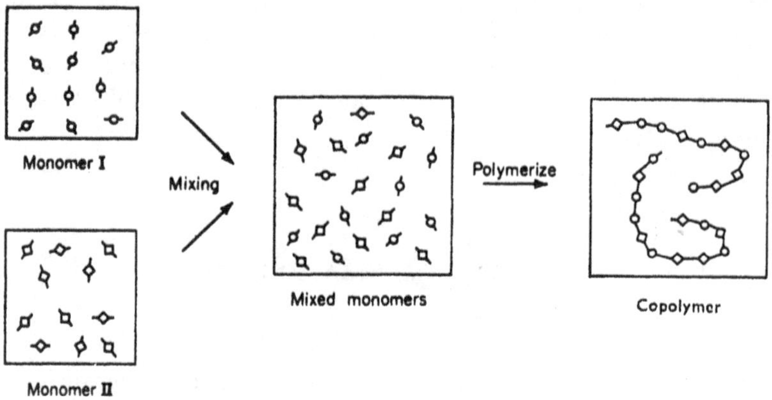

Fig. 1.3 Copolymerization

Many monomers are known which will not polymerize with themselves, even under extreme conditions of temperature and pressure but will copolymerize with other compounds. For example, maleic anhydride will not polymerize with itself, but with SO_2 a long-chain compound is formed:

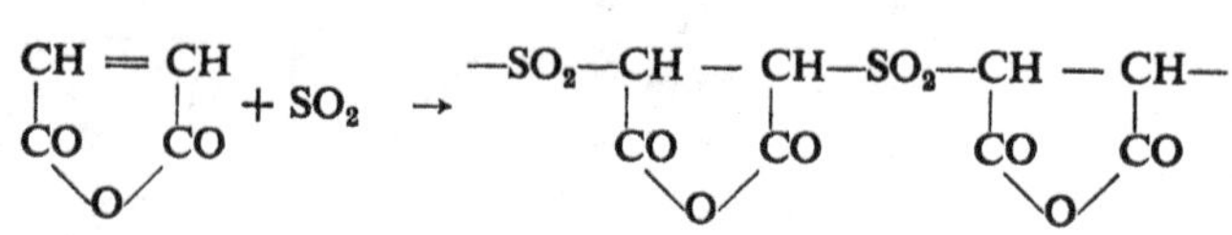

Copolymers can also be formed by grafting side chains on to an already completed polymer chain. One of the earliest uses for grafting was to add reactive groups, capable of bonding dyestuffs, on to synthetic fibers such as acrylonitrile, which are difficult to dye in the normal way. The side chains are grafted on by reaction at certain activated spots in the molecule. These activated spots may be already present in the molecules, as with compounds containing residual double bonds, etc., or may be initiated by bombardment with high-energy radiation such as gamma rays.

CRYSTALLINE MELTING POINT

When a low molecular weight crystalline material, in the molten state, is cooled a sudden change occurs at its melting point as the material goes from the liquid state to a crystalline solid. The change is one from disorder to order and is accompanied by sudden changes in physical properties such as heat capacity and specific volume. Although similar changes occur when a molten thermoplastic is cooled they are not so sudden. If specific volume, for example, is plotted against temperature for a low molecular weight material and an amorphous thermoplastic the results would be as shown in Fig. 1.4 and the upper line in Fig. 1.5. A crystalline thermoplastic would behave as shown in the lower curve of Fig. 1.5, where the change is distinct but occurs over a range of temperature. The temperature at which these changes start to occur is known as the crystalline melting point and is denoted by the symbol Tm.

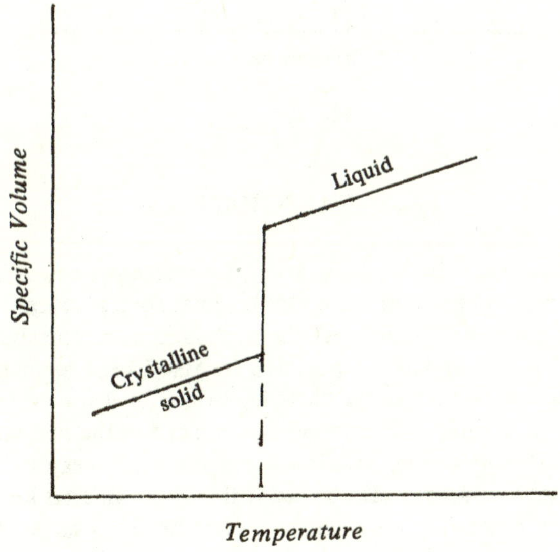

Fig. 1.4

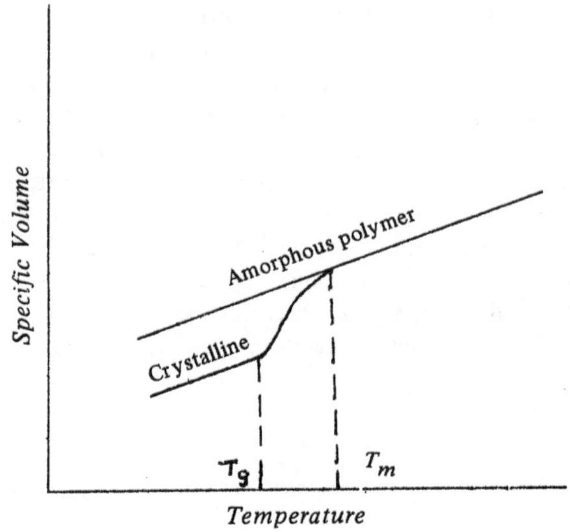

Fig. 1.5

GLASS TRANSITION POINT

In an amorphous, solid thermoplastic, the molecules continue to move relative to each other because the intermolecular forces are low. The same applies to the amorphous portions of a crystalline thermoplastic, although the molecules in the crystalline portions are fixed because of the greater intermolecular forces that occur there (see Chapter 9 for more detail on crystallinity). If cooling is continued below Tm the thermal energy is reduced to a point where even rotation about the bonds ceases. The chains in the amorphous areas will also be set at this stage which is known as the glass transition point, Tg. In general, thermoplastics above their glass transition points are rubbery in nature while below it they are hard and brittle.

MOLECULAR WEIGHTS

The chemical constitution of ethylene gas and that of the polyethylene used for washing-up bowls is exactly the same, namely, two atoms of hydrogen to every one of carbon, hence it is obvious that the size of the polymer chain will have a great influence on the physical properties of the plastic. Between the two extremes mentioned above, there lies a range of liquid to semisolid waxes, the tendency towards hardness and solidity increasing as the number of $-CH_2-$ units in the molecule increases. It is important, therefore, to have some knowledge of the molecular weight of a polymer. The problem is not, however, as simple as it looks in cases where the range of molecular weights is fairly wide. In inorganic chemistry molecular weights of substances are usually constant. Sulfuric acid, for instance, in whatever way it is made, has a molecular weight of 98 (H_2 = 2 : S = 32 : O_4 = 64) but polyethylene not only varies from sample to sample but also within a given sample. Only the 'average' molecular weight can, therefore, be quoted, and even this is the subject of some difficulty. There are two common methods of calculating average molecular weights, namely weight average and number average. These are defined in the following manner:

$$\text{Number average molecular weight} = \frac{\Sigma n_i M_i}{\Sigma n_i} = M_n$$

$$\text{Weight average molecular weight} = \frac{\Sigma w_i M_i}{\Sigma w_i} = M_w$$

Where n_i = number of molecules of molecular weight M_i
w_i = weight fraction of material having molecular weight
$M_i = n_i M_i$

Where all the molecules are the same size, both these expressions give the same result. When, however, there is a size distribution the weight average molecular weight is greater than the number average molecular weight. The ratio M_w/M_n is, therefore, a measure of the molecular weight distribution, being = 1 when all the molecules are the same size, and increasing with increasing molecular size distribution. A simple example will suffice to show the possible differences between the two values.

Exaggerating the problem for the sake of argument we may consider a sample containing 10 molecules of molecular weight = 1000 each and ten molecules of molecular weight = 10,000 each.

The total weight of the 20 molecules will be (10 × 1000) + (10 × 10,000) = 110,000. The number average of the complete sample will, therefore, be 110,000/20 = 5500. The weight average will be

$$\frac{(10{,}000 \times 1000) + (100{,}000 \times 10{,}000)}{110{,}000} = 9092$$

If, however, both sets of ten molecules had the same molecular weight (= say 5000) then

$$Mn = \frac{(10 \times 5000) + (10 \times 5000)}{20} = 5000$$

$$Mw = \frac{(50{,}000 \times 5000) + (50{,}000 \times 5000)}{100{,}000} = 5000$$

Different methods of determining molecular weight give differing averages. Thus, among the common techniques for measuring molecular weight, the following give number average molecular weights, namely, boiling-point elevation, freezing-point depression, osmotic pressure, and analysis of end-groups. Light scattering and sedimentation velocity methods, however, give the weight average molecular weight. Molecular weight may also be determined by measurement of viscosities of dilute solutions of the polymer in question but the picture here is not so clear-cut. The so-called 'intrinsic viscosity' of a polymer solution is not directly proportional to the molecular weight but is proportional to M^a. The value of a has been found to lie between 0.5 and 1.0 for different polymer/solvent systems, the lower figure being associated with poor solvents. If $a = 1$, viscometric methods give the weight average molecular weight but for other values of a the so-called 'viscosity average' lies between the number average and weight average.

Molecular Weight Determinations

Of the classical methods for determining molecular weights, those based on the elevation of boiling point, depression of freezing point or the reduction of vapor pressure of a solution are of limited importance because the concentrations necessary to produce an appreciable reading are so high that nonideal solutions are obtained to which the normal equations do not apply.

Viscosity Measurements

Solution viscosity measurements can be made easily and rapidly and are, therefore, widely used for the determination of molecular weight. There are many types of apparatus available for the measurement of viscosity but the most convenient for laboratory use is that based on the movement of the liquid through a capillary tube. From the time required for a fixed quantity of liquid to pass through a capillary tube of known radius and under a fixed pressure, the viscosity of the liquid can be calculated by Poiseuille's Law. A series of readings at various concentrations is carried out and the relative viscosity η_r is calculated for each concentration

$$\left(\eta_r = \frac{\eta \text{solution}}{\eta \text{solvent}}\right)$$

The 'specific viscosity' is then calculated ($\eta_{sp} = \eta_r - 1$) and finally the 'reduced viscosity' where $\eta_{red} = \eta_{sp}/\text{conc.}$ The reduced viscosity still varies with the concentration of the solution so that the various values for reduced viscosity are plotted against concentration and the value at zero concentration is found by extrapolation. This value is known as the 'intrinsic viscosity' $[\eta]$ and, in general, it is found to be proportional to M^a, i.e.

$$[\eta] = KM^a$$

where K, and a, are constants varying with different polymer/solvent combinations. The values for K, and a, must be found by calibration against some other method of molecular weight determination.

End-Group Analysis

This is a useful method for determining the molecular weight of polyesters, polyamides and other linear polymers which contain a reactive group at one or both ends of the polymer chain. In a polyester, for example, which is formed from a hydroxy acid, there would be a carboxylic group at one end and a hydroxy group at the other.

$$n\left[OH(CH_2)_xCOOH\right] \rightarrow OH(CH_2)_xCOO\left[(CH_2)_xCOO\right]_{n-2}(CH_2)_xCOOH$$

In such a case the 'acid number' (a measure of the number of reactive acid end-groups) is measured by titration with a standard solution of a base. The degree of polymerization then

$$= \frac{\text{original acid number}}{\text{final acid number}} = \frac{\text{number of acid groups originally present}}{\text{number of acid groups in polymer}}$$

Osmotic Pressure

The van't Hoff Law governing the relationship between molecular weight and osmotic pressure of dilute solutions does not hold for solutions of high polymers and has to be modified.

If

π = osmotic pressure,
T = absolute temperature,
R = the gas constant,
c = concentration,
M = molecular weight of solute,
B = constant

then the modified relationship is

$$\pi = \left(\frac{RT}{M}\right)c + [Bc^2]$$

where the expression in square brackets is the modifying factor. B itself is an empirical constant varying with the polymer/solvent combination used.

If the osmotic pressure is measured at several different concentrations (at constant temperature) and π/c plotted against c, the intercept at zero concentration gives the value of RT/M from which M can be calculated. If the molecular weights are not too high a straight line should be obtained, the slope of which will be equal to B. Satisfactory results are obtained with this method only in the case of very dilute solutions (below one per-cent).

Light Scattering

This method depends on the fact that the turbidity of a polymer solution is related to the molecular weight of the dissolved polymer. Turbidity can be determined experimentally by measuring the loss in intensity of a beam of light when it is passed through a cell containing the polymer solution. It can also be determined by measuring the intensity of the light which is scattered by the solution at a known angle, usually 90°C.

Sedimentation

In a dilute solution of a polymer, there is no tendency for the individual molecules to settle out since they are kept in a constant state of motion by collisions with molecules of the solvent. The energy imparted by these collisions is far greater than the gravitational force acting on the polymer molecules, hence no general downward movement occurs. If, however, the gravitational force is increased by thousands of times, by whirling the polymer solution in an ultracentrifuge, sedimentation can occur. Measurements are made either of the rate of sedimentation or of the position of the solvent/solution interface when equilibrium is reached. Both measurements give an indication of the relationship between the forces acting on the polymer molecules, namely impact energy from the solvent molecules, and gravitational forces, and from them it is possible to calculate the weight of the polymer molecules.

HISTORICAL DEVELOPMENT OF THE PLASTICS INDUSTRY

A brief note of the historical development of a particular polymer is given at the beginning of the appropriate chapter and it is intended, here, to give some slight framework into which these subsequent notes can be fitted. Events have moved so rapidly in the last 100 years that we cannot do more than indicate a few of the landmarks along the way but a fuller picture is available, and details of two books covering the centenary of the plastics industry are given at the end of this chapter.

The first man-made plastic was cellulose nitrate, or celluloid, and credit for its discovery, in 1862, goes to Alexander Parkes. He was not, in fact, the first to discover the chemical entity cellulose nitrate, but was the first to create a workable plastics material from it. Parkes felt sure that cellulose nitrate was the material the electrical industry had been searching for as a substitute for natural insulators such as gutta-percha and shellac, which were rapidly becoming scarce. Parkes's great contribution to the plastics industry consists of the discovery that camphor was an efficient plasticizer for cellulose nitrate, giving moldable compositions. The measure of Parkes's success is that after 100 years, no one has yet discovered a better plasticizer than camphor for cellulose nitrate. Parkes failed to commercialize his discovery and seven years later, in the U.S.A., John Wesley Hyatt discovered the use of camphor independently and took out patents on the composition which he called 'celluloid'.

To celluloid then, the honor of being the first man-made plastic, but the raw material, cellulose, was a natural material and we must look elsewhere for the holder of the title of the 'first synthetic plastic'. This title is held by phenol-formaldehyde resin (or Bakelite) which was developed independently by Leo Baekeland (after whom the resin was named) and James Swinburne (later Sir James Swinburne, Bart.). The original patents are due to Leo Baekeland, it being a matter of history that James Swinburne's applications for patents were filed one day later than Baekeland's. The latter were however, rather more extensive. This brings us up to 1907, just 45 years after the official birth of cellulose nitrate. In the interim period other plastics based on natural products had been developed, notably, casein and cellulose acetate. The former can be said to date from 1895, when Spitteler discovered the hardening action which formaldehyde had on casein, while cellulose acetate really began its useful history in 1894, when Cross and Bevan took out a patent on the preparation of cellulose acetate, using zinc chloride as the condensing agent. Other workers had obtained cellulose acetate by various methods but Cross and Bevan were the first to realize the full significance of its discovery. Cross and Bevan (in collaboration with Beadle) are even more renowned, however, for their work on the process by which Viscose rayon or artificial silk is produced.

Compared with today, then, the rate of progress was painfully slow, and at the end of 60 years from Parkes' announcement of cellulose nitrate there were still only four commercially established plastics, namely, celluloid, cellulose acetate, casein, and phenol-formaldehyde. The pace however began to quicken, during the 1930's, and materials such as polystyrene, PVC, the acrylics and urea-formaldehyde were produced on an industrial scale. All of these had been known earlier only as laboratory curiosities and their potentialities had been unrecognized. In addition to these rediscoveries, there were two completely new plastics developed during the 1930's, namely, nylon and polyethylene. The contrast between growth during the first 67 years to 1929, and the subsequent decade, is tremendous and needs some explanation. There would seem to be little doubt that the great surge in the industrial development of plastics during the 1930's was mainly due to the elucidation of the true nature of plastics, for which credit is due to the German chemist Staudinger. It was Staudinger who first proposed the theory that plastics (and rubbers) were composed of giant molecules and he was the first to use the term 'macromolecule'. In addition to his theoretical contributions to polymer chemistry Staudinger was also responsible for the discovery of the relationship between viscosity of dilute polymer solutions and the molecular weight of the dissolved polymer. Staudinger's name is also linked with that of polystyrene, for it was he who first recognized the true nature of the reaction which occurred when styrene was heated and suggested that the structure of the resultant polystyrene was a long chain of styrene molecules joined together via the vinyl groups.

Another great name connected with the elucidation of the structure of polymeric materials is that of the American chemist Carothers. Although perhaps better known for his discovery of nylon, Carothers was also responsible for a great deal of fundamental research into the nature and formation of high polymers and, indeed, the discovery of nylon arose directly from these researches. Carothers's first paper on polymerization pointed out that all high polymers, whether naturally occurring or synthetic, consisted of recurring structural units. He also introduced the concepts of addition polymers and condensation polymers, where addition polymers consisted of a recurring structural unit identical with that of the monomer, and condensation polymers had recurrent structural units different from those of the monomer. The latter are formed by reactions which occur with elimination of simpler molecules such as water and ammonia.

The outbreak of the Second World War in 1939 increased the demands for plastics of all types, but in addition to the intensified production and use of the existing range of plastics other completely new ones were developed. In the U.S.A., for instance, both polytetrafluorethylene and the silicones were discovered during the war years and plants were in production for these materials by 1943 and in 1944 respectively, while in the U.K., Whinfield and Dickson were responsible for the discovery of the polyester fiber polyethylene terephthalate, in 1941. Workers in Germany were also active, of course, during this period, and between 1939 and 1941 Dr. Otto Bayer discovered and developed useful polymers (known as polyurethanes) from the reaction between di-isocyanates and hydroxyl compounds.

The pressures of war also exerted their influence on the development of synthetic rubber. During the First World War, Germany had produced a synthetic rubber, known as Buna methyl rubber, which though not a great success, undoubtedly laid the foundation for the tremendous industry that grew up in time for the Second World War. The 1930's were again the turning point and a successful product was obtained in Germany by the copolymerization of butadiene and styrene. In the U.S.A., work before the war had concentrated on the production of neoprene, a synthetic rubber based on chloroprene, first mentioned in 1931. Neoprene was yet another of the discoveries of Carothers and was developed, not as a rubber substitute but as a synthetic material for special uses, based on its properties such as resistance to oil, aging and heat. The entry of the U.S.A. into the war, however, highlighted the need for a more general purpose rubber and plants were soon erected for butadiene/styrene rubbers similar to those produced in Germany.

The years since the war have also seen immense strides in the growth of established plastics and the development of new ones. One of the developments after the war was that of Professor Karl Ziegler in Germany who succeeded in polymerizing ethylene, not by the usual method of applying extremely high pressures and fairly high temperatures but by using temperatures and pressures very little over normal atmospheric. This was in the early 1950's and was achieved by using special catalysts, the resulting material being found to differ from the usual polyethylene by possessing a higher density, greater stiffness and a higher softening point. Closely following on this came Professor Natta's work in Italy on the polymerization of propylene. Previous polymerization of propylene had produced only soft, rubbery materials of no particular value but Professor Natta's contribution to polymer science was the achievement of an ordered arrangement of molecules in the polymer by means of directed or 'stereo-specific' catalysis. This ordered (or 'isotactic') polypropylene proved to have a stiffness comparable to that of the new high-density polyethylenes together with a harder and glossier surface and a higher melting point. The discovery of stereo-specific catalysis has meant that it is now possible to tailor-make polymers and exercise some control of the properties of the finished polymer.

One of the results of stereo-specific catalysis has been the attainment of that long-sought goal, the synthesis of natural rubber. It had long been known that natural rubber was a polymer of isoprene but attempts to produce, by polymerization of isoprene, a material having the molecular structure of natural rubber had always failed. Rubber contains a series of double bonds in the polymer chain, together with one methyl grouping per isoprene unit. The methyl groups in natural rubber are all on the same side of the molecule, but prior to the discovery of stereo-specific catalysis it was not possible to order the position of the methyl groups during polymerization.

Two other families of plastics of great interest, which have appeared on the scene since the war, are the polycarbonates, developed independently in Germany and the U.S.A. (by Bayer and the General Electric Company, respectively) and the polyacetals. The first commercial plant for polyacetals was brought on stream in the U.S. by Du Pont in 1959 to be followed in 1962 by the Celanese Corporation (also in the U.S.A.) and in 1963 by Farbwerke Hoechst in Germany. Both polycarbonates and polyacetals are important because they are tough materials of high temperature resistance and rigidity and are thus capable of displacing metals for some engineering applications.

We have treated the growth of the plastics industry, so far, only in terms of polymeric materials (with a brief discursion into the subject of the theoretical understanding of polymer structures) but the full story should include some mention of processing equipment.

The early extruders were based very largely on those used for applying gutta-percha or rubber insulation to copper conductors. The first machines were manually operated ram types but these gradually evolved into mechanically operated rams while, still later, the screw-operated extruder was developed. At first these extruders were fed with hot strips of material and were steam-heated, and it was not until the 1930's that electrically heated extruders, capable of converting cold granules of thermoplastics into pipe, rod, etc., were built.

Injection molding machines also have a long history, stemming from the machine used for the die-casting of type metal invented in 1836. The first injection-molding machine for plastics was patented by John Hyatt, in 1878, for covering metal buckles with celluloid but because of the inflammability of celluloid the injection-molding process received very little attention until the advent of cellulose acetate molding powders in 1919. Special machines were then designed to handle this material, but the first successful commercial injection molding machine was produced by Eckert and Ziegler in 1926 and was exhibited at the British Industries Fair in 1930.

Other techniques for shaping thermoplastics that have been fully developed only since the Second World War are blow molding and vacuum forming. The immediate outlet for blow molding was, of course, the manufacture of bottles, but the versatility of this method of fabrication is already finding other outlets, especially for awkwardly shaped articles formerly made by welding two or more components together. Vacuum forming, too, has filled a gap in processing techniques and has made it possible to produce articles with extremely thin walls, thus making an economic proposition of many projects formerly thought to be too expensive.

In theromsetting materials there has not been quite the revolution in processing techniques that has occurred with the thermoplastics but there have still been very definite changes.

In the main, the trend has been towards greater automation of compression molding techniques (particulary automatic feed and ejection), larger machines capable of taking greater numbers of molds, and faster cycling (by the development of pelleting and preheating).

SUGGESTED FURTHER READING

(a) Polymer Chemistry and Technology

Textbook of Polymer Chemistry. (1984 3rd ed.) F. W. Billmeyer (New York: John Wiley and Sons Inc.).

Fundamental Principles of Polymeric Materials. (1982) S. L. Rosen (New York: Wiley—Interscience).

Principles of Polymer Science (1981 2nd ed.) F. Rodriguez (New York: McGraw-Hill).

Modern Plastics Technology. (1975) Raymond B. Seymour (Prentice Hall Co.).

Encyclopaedia of Polymer Science & Technology. (1976) Eds. H. F. Mark and N. M. Bikales (New York: Wiley).

Block & Graft Copolymerisation. (1975) Edited by R. J. Ceresa (New York: Wiley).

Polymer Blends & Composites. (1976) J. A. Manson, L. Sperling (New York: Plenum Press).

Synthetic Paper from Fibers & Films. (1975) M. G. Halpern (New Jersey: Noyes Data Corp.).

Co-ordination & Polymerization. (1975) Edited by James C. W. Chien (New York: Academic Press).

Adhesion and the Formulation of Adhesives. (1976) William C. Wake (England: Applied Science Publishers Ltd.).

Emulsion Polymerization. (1976) Edited by Irja Perma & John L. Gordon (USA: American Chemical Society).

Handbook of Adhesives 2nd Ed. (1977) Edited by Irving Skeist (New York: Van Nostrand).

(b) Applications

Plastic Films. (1989 3rd ed.) John H. Briston (London: Longman Scientific and Technical)

Plastics in Contact with Food. (1974) J. H. Briston and L. L. Katan (London: Food Trade Press Ltd.).

Rigid and Semi-Rigid Plastic Containers. (1994) J. H. Briston (London: Longman).

2.

RAW MATERIALS

GENERAL

In addition to the developments in new products and techniques, outlined in the previous chapter, there has been a gradual evolution in the raw materials used for the manufacture of plastics and rubbers. The original plastic, cellulose nitrate, was based on cellulose, usually in the form of wood-pulp, while the phenol and formaldehyde needed for the early phenol-formaldehyde resins were obtained from coal. Today, the emphasis has shifted very largely to petroleum and natural gas, especially in the U.S.A., where the larger part of plastics production is based on these materials.

In general, then, we may classify the main raw materials for plastics and rubbers into three groups:

(a) Natural products (cellulose, casein, etc.)
(b) Coal
(c) Petroleum and natural gas.

The members of the first group are, on the whole, being replaced continuously, but the last two groups were both formed earlier in the earth's history and are irreplaceable.

The number of polymers and their intermediates is so great as to render it impossible to describe the manufacture of all the necessary raw materials in one chapter. Brief notes on the most important are given, therefore, supplemented by diagrams showing the main routes by which polymers can be obtained from various raw materials.

NATURAL PRODUCTS

Nature has conveniently provided for our use many excellent materials with long-chain structures eminently suitable for modification to useful polymers. The manufacture of cellulose nitrate is, of course, the earliest example of such a modification, other examples being cellulose acetates (also from cellulose) and casein plastics from milk.

Cellulose

Cellulose is a carbohydrate and occurs as the base material of all vegetable matter although not always in a form readily available for the manufacture of industrially useful plastics. Cotton linters and wood are both commercial sources of cellulose and were in early use for the manufacture of cellulose nitrate.

Wood was also extensively used at one time for the production (by destructive distillation) of methyl alcohol and formaldehyde, both of them very useful intermediates in the manufacture of plastics. Little use is now made of this method as other cheaper and more plentiful sources have since been found.

Materials Containing Starch or Pentoses

Starch is another long-chain carbohydrate but its utlization by the plas-

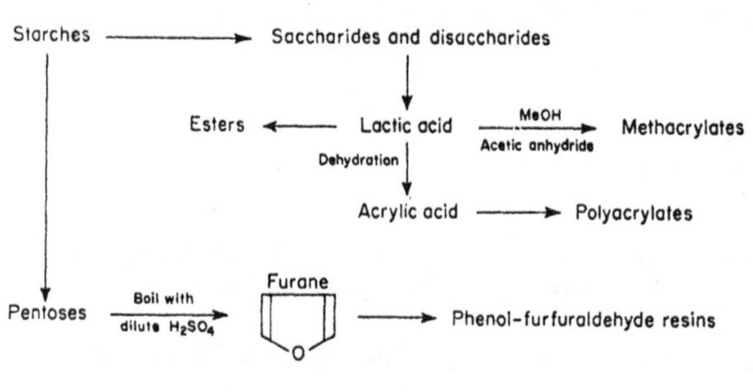

Fig. 2.1

tics industry has been small, although it has been used as an adhesive for paper and as a size for fibers. There are theoretical methods for converting starch and certain pentoses into polymeric materials but they are not always practical and little commercial use has been made of them. A typical series of conversions is shown in Fig. 2.1.

Pentoses may be obtained from many agricultural residues such as corn-waste (oat husks), furfuraldehyde being economically obtained in this way. Furfuraldehyde reacts with phenol to form resins in a manner similar to that of the better-known reaction of phenol with formaldehyde.

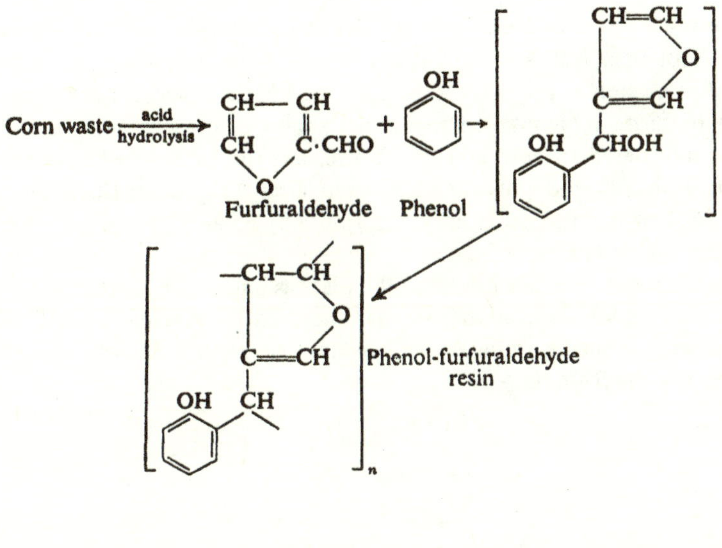

Corn-waste and other agricultural residues, such as bagasse and straw waste, can also be made to provide an alternative route to nylon 6,6. The conventional method utilizes phenol as the starting point while the alternative route from agricultural residues utilizes furfural.

Ethyl alcohol is also readily obtainable from a variety of agricultural materials (a fact well-known in many parts of the world, including the Scottish Highlands and the mountains of Kentucky!) among them being potatoes, corn, barley, cane sugar, beet sugar and molasses. Ethyl alcohol from molasses provided the ethylene for the original manufacture of polyethylene, and butadiene for synthetic rubbers, while an early method for producing a synthetic rubber in Russia utilized ethyl alcohol obtained from potatoes. The alcohol was passed over a mixed dehydration/dehydrogenation catalyst (alumina and zinc oxide) at a temperature of 400°C. The reactants were then cooled and the uncondensed gas, containing most of the butadiene, was washed with turpentine or petrol. The butadiene removed in this way was finally separated by distillation.

Foodstuff shortages, and also the advent of the petrochemical industry, have led to large-scale production of alcohol by this method becoming obsolete.

Materials Containing Proteins, Fats, or Oils

Ground Nuts and Cashew Nuts. The value of ground nuts as a source of food protein is well known, but at one time ground nuts were expected to be of even greater importance as a source of fibers which have properties akin to those of wool. In the initial stages of development such fibers were considered to have a bright future, but the subsequent development of acrylonitrile and polyester fibers rendered the protein fibers obsolete. The protein is obtained from ground nuts by alkaline extraction followed by acid precipitation.

Cashew nuts are of interest to the plastics industry by virtue of the fact that their outer shell contains a dark-colored oil referred to as C.N.S.L. from which can be obtained an unsaturated phenol known as cardanol which has the formula

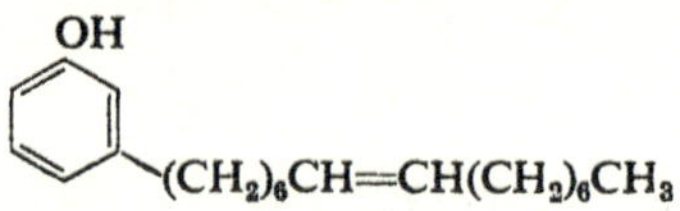

Cardanol will condense with formaldehyde to produce resins similar to phenol-formaldehyde resins but with somewhat inferior physical properties.

Zein. This material is a protein which occurs in maize. It is soluble in aqueous alcohols and can, therefore, be readily spun into fibers. The fibers are hardened in formaldehyde and then exhibit excellent resistance to solvents and greases. Unfortunately the extraction process is rather costly, which has militated against wider use of zein as a raw material.

Casein. This is the chief protein extracted from cow's milk and is composed of α-amino acids. It may be precipitated from skimmed milk by either an acid or by the enzyme, rennet, the latter giving the protein in the most suitable form for plastics production. After precipitation the casein is first washed, pressed, and dried. After extrusion into the required form, the material is hardened by immersion in formalin.

The main sources of casein are France, Argentina, and New Zealand.

Castor Oil. This material has assumed a particular importance in recent years as a source of one of the nylons (nylon 11 or 'Rilsan'). The castor oil is reacted with methanol, alcoholysis occurring to give methyl ricinoleate. Particulars of the manufacturing process are given in the chapter on polyamides, but briefly the sequence of operations is pyrolysis of the methyl ricinoleate to give methyl undecylenate and hydrolysis of this to undecylenic acid. The acid is then reacted with ammonia to give the amino acid, 11-amino-undecanoic acid, which is then made to undergo a polycondensation reaction to give nylon 11.

Miscellaneous Natural Products

Shellac. Shellac is the protective covering produced by the scale insects that attach themselves to certain trees found mainly in Southern Asia. The material so excreted is not a definite polymer but consists largely of esters of aleuritic acid. The refined material has a long history of use as a spirit lacquer for wood and has also been extensively used for the manufacture of gramophone records, as a bonding agent for mica, and as an insulating varnish.

Natural Rubber. Certain trees and other plants (of which the tree *Hevea braziliensis* is the best-known) contain a milky white latex which can be tapped by making an incision in the bark. After collection this latex can be coagulated, washed and rolled into the crepe sheets of rubber with which we are all familiar.

Coal

Coal was the first source of the raw materials used in the manufacture

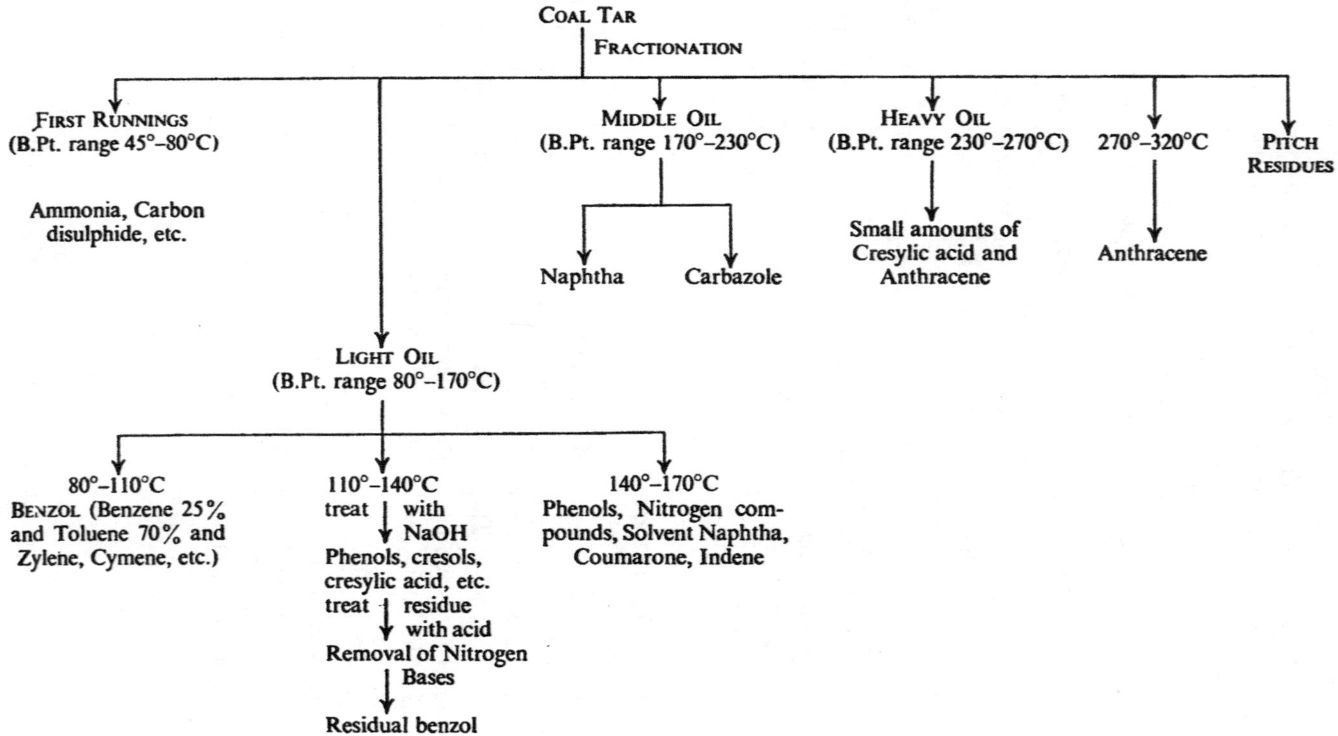
COAL TAR
FRACTIONATION
FIRST RUNNINGS
(B.Pt. range 45°–80°C)
Ammonia, Carbon
disulphide, etc.
MIDDLE OIL
(B.Pt. range 170°–230°C)
Naphtha
Carbazole
HEAVY OIL
(B.Pt. range 230°–270°C)
Small amounts of
Cresylic acid and
Anthracene
270°–320°C
Anthracene
PITCH
RESIDUES
LIGHT OIL
(B.Pt. range 80°–170°C)
80°–110°C
BENZOL (Benzene 25%
and Toluene 70% and
Zylene, Cymene, etc.)
110°–140°C
treat with
NaOH
Phenols, cresols,
cresylic acid, etc.
treat residue
with acid
Removal of Nitrogen
Bases
Residual benzol
140°–170°C
Phenols, Nitrogen com-
pounds, Solvent Naphtha,
Coumarone, Indene

Fig. 2.2

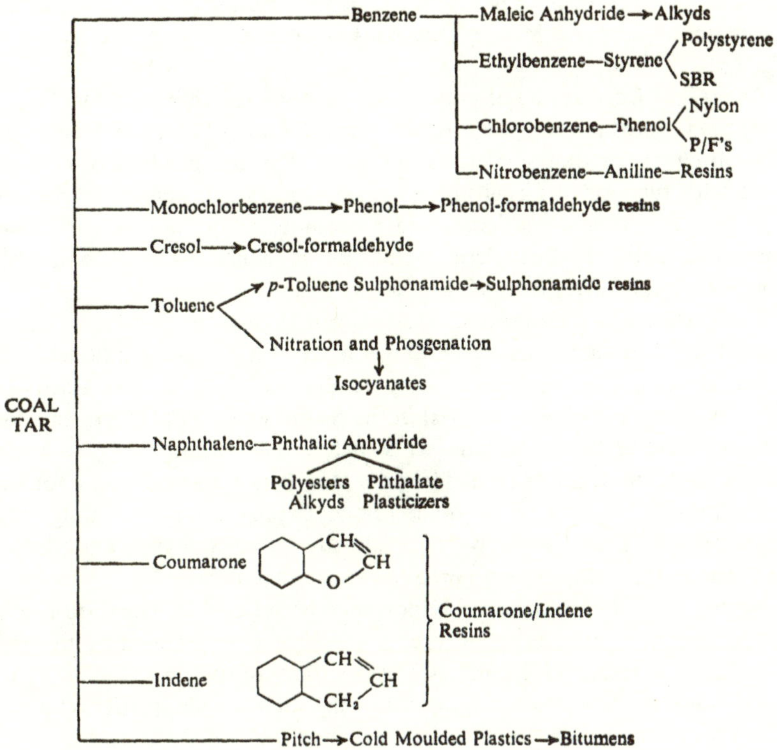

Fig. 2.3 Polymers, intermediates, and raw materials from coal tar fractions

of synthetic polymers but it has now been superseded as the main source by petroleum and natural gas.

The chief products obtained when coal is destructively distilled are gas, ammonia, coal tar and a residue of coke. The most important of these as a source of raw materials is the coal tar fraction. Since each ton of coal produces only about ten gallons of coal tar it is easy to see why demands from the plastics industry soon outstripped the supply.

Coal Tar. Coal tar is a very complicated mixture of chemicals, and the first step in separating the constituents is to subject the coal tar to fractional distillation. Six main fractions are obtained, all of which are still mixtures of several components, as can be seen from Fig. 2.2. The plastics raw materials obtainable from the constituents of coal tar are shown in Fig. 2.3.

Coal Gas. Coal gas is, of course, best-known as a source of fuel but its constituent gases are also used for the manufacture of many substances of prime importance to the plastics industry. The composition of coal gas varies with the grade of coal but, in general, it contains about 50 percent hydrogen, 30 percent methane, and 8 percent carbon monoxide with an appreciable amount of ethylene. It has never been a main source of ethylene, however, a better one being coke-oven gas.

The methane can be reacted with oxygen at high temperatures, and in contact with an aluminum oxide/nickel catalyst it produces methanol and higher alcohols, the methanol being further oxidized to formaldehyde, which is a very important material in the manufacture of plastics. In addition to its use in thermosetting resins such as phenol-formaldehyde, melamine-formaldehyde, urea-formaldehyde, etc., it is a starting point for the production of a family of polymers known as polyacetals (including polyformaldehyde itself), while we have already noted its use as a hardening agent for certain natural polymers such as casein and zein.

Ammonia. Ammonia is of importance because of its conversion into urea and various amines. The urea is used for the manufacture of urea-formaldehyde resins while amines have a variety of uses in the polymer industry, among the most important being their employment as curing agents for epoxy resins.

$$2NH_3 + CO_2 \xrightarrow{\text{Pressure}} \underset{\text{Urea}}{NH_2\cdot CO\cdot NH_2} \xrightarrow{\text{H}\cdot\text{CHO}} \text{U/F resins}$$

Coke. Apart from its widespread use as a fuel, the importance of coke lies in its conversion to calcium carbide with subsequent reaction with water to produce acetylene. For many years coke was the main source of acetylene, but with the development of the petroleum cracking industry its global importance has greatly diminished although at the present time the carbide process is still the only one employed in the U.K.

Other chemicals obtained directly or indirectly from coke are shown in Fig. 2.4.

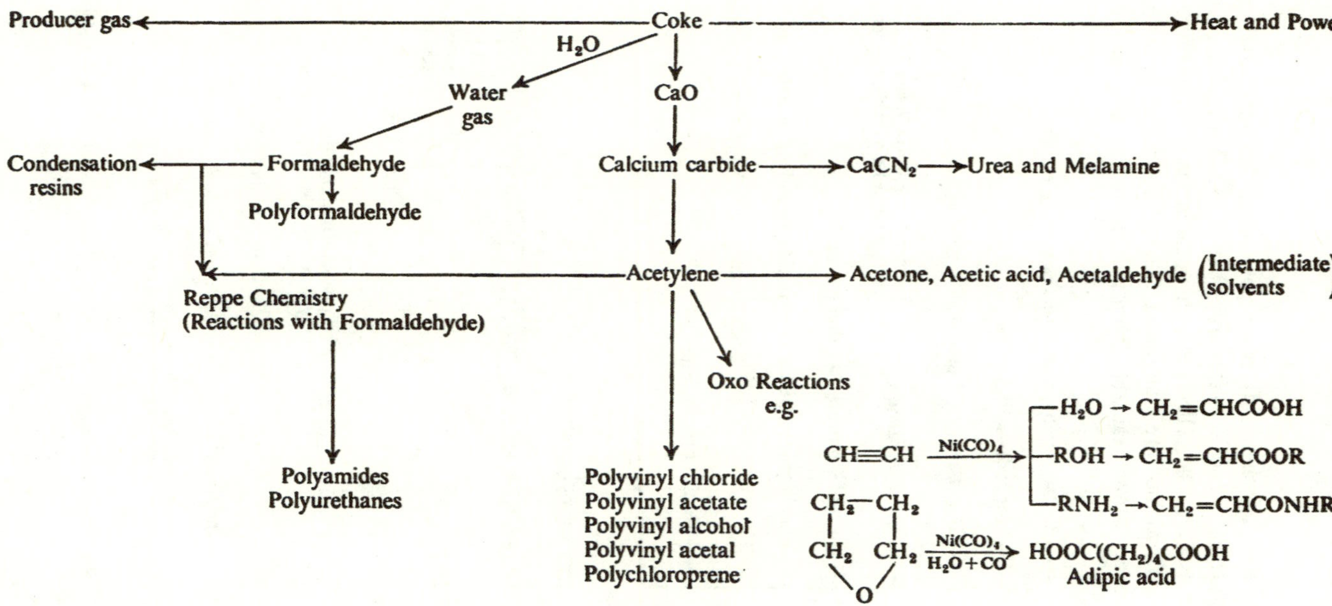

Fig. 2.4 Polymers, intermediates, and raw materials from coke

Petroleum

The production of chemicals from petroleum has reached immense proportions since its inception after the First World War and even an elementary account of the so-called petrochemical industry would fill a book on its own. Some idea of the many materials of importance to the plastics industry which are now obtainable from petroleum can be seen by reference to Figs. 2.5, 2.6, and 2.7.

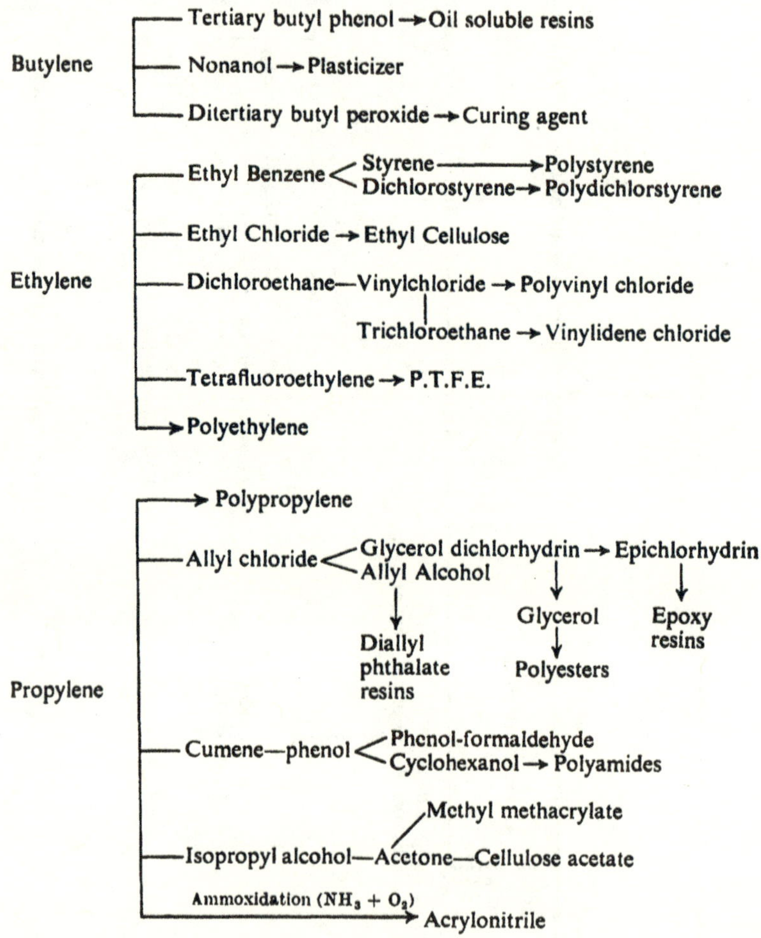

Fig. 2.5 Plastics from petroleum–polymers, intermediates, and ancillary materials

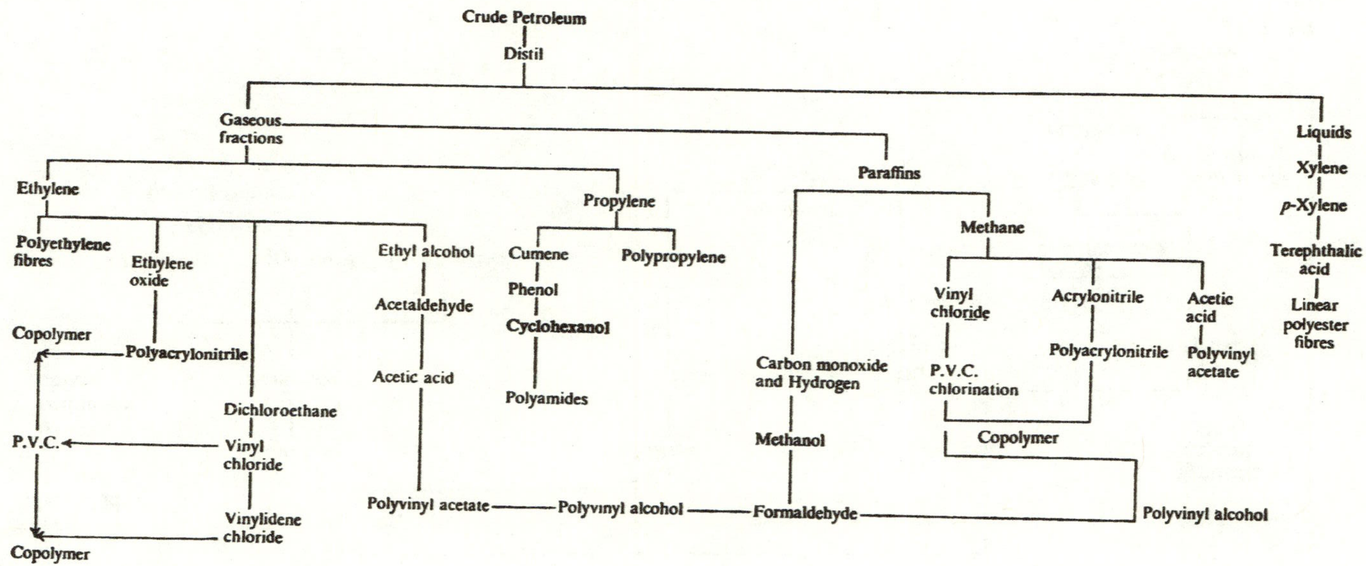

Fig. 2.6 Synthetic fibers and intermediates from petroleum

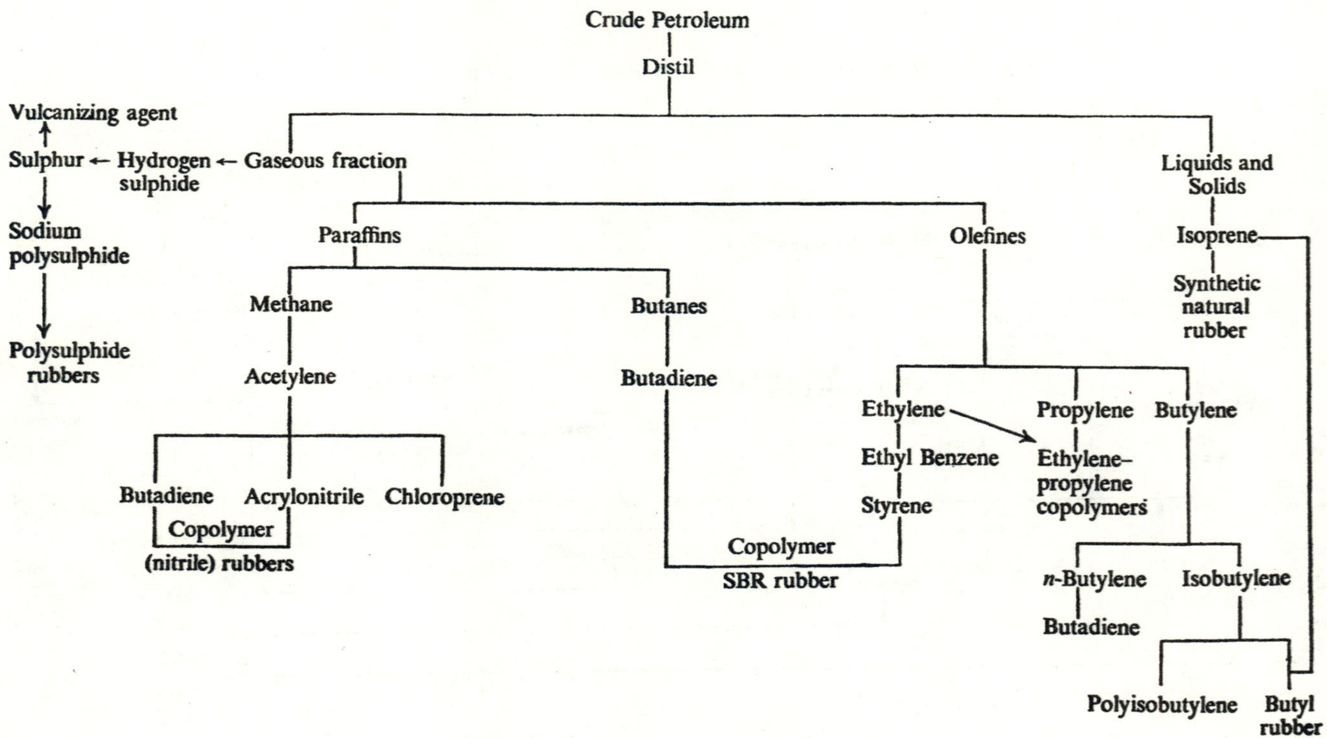

Fig. 2.7 Synthetic rubbers and ancillary materials from petroleum

There are three main methods by which petroleum fractions can be utilized in the production of chemical intermediates of importance to the polymer industries, and these may be summarized as follows:

1. The most important process is the cracking of various hydrocarbon feed stocks to give high yields of unsaturated gases such as ethylene, propylene, butylene, acetylene, the particular feedstock used being dictated by availability and the type of end-product required. In the U.K. the usual feedstock employed is a light naphtha distillate and at least 95 percent of the present U.K. capacity for ethylene is based on naphtha. Other feedstocks which may be used are saturated refining gases such as propane and butane.

The feedstock is vaporized, diluted with steam, and heated rapidly to the cracking temperature. It is held at this temperature for a very short time, then rapidly cooled by quenching to prevent further reaction. The various products are then separated by fractionation.

2. Chemical intermediates are also formed during certain refining operations designed to convert naphtha feedstocks to gasolines with high octane ratings. In addition to this basic function, which is fundamentally a dehydrogenation reaction, converting cyclohexane and similar cyclo-paraffins to aromatic compounds, such as benzene, there are minor reactions such as isomerization and dehydrocyclization of paraffin. *n*-Hexane, for example, is isomerized to 2-methylpentane or dehydrolyzed to benzene.

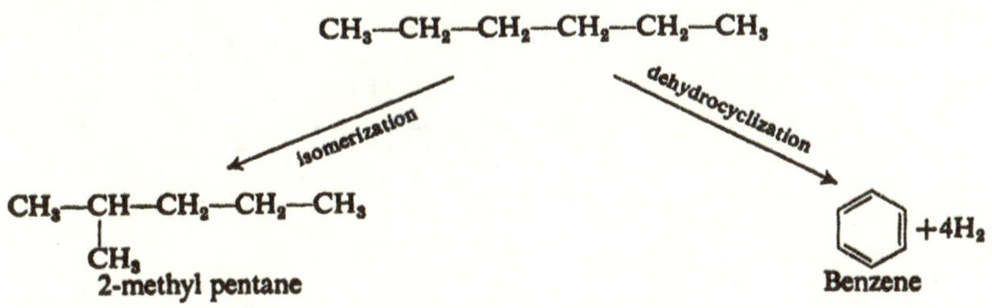

3. The third process used is the isolation of certain individual hydrocarbons from crude oil fractions which are subsequently chemically reacted to give the desired intermediate. Butane, for instance, is dehydrogenated to give butadiene, while cyclohexane is oxidized to cyclohexanone.

At present there is an overwhelming dependence on nonrenewable resources, particularly petroleum. As limitations on the availability of nonrenewable resources increase, so the production of polymers from renewable resources will become more attractive. Attention is being increasingly directed towards a wide range of renewable resources and for convenience

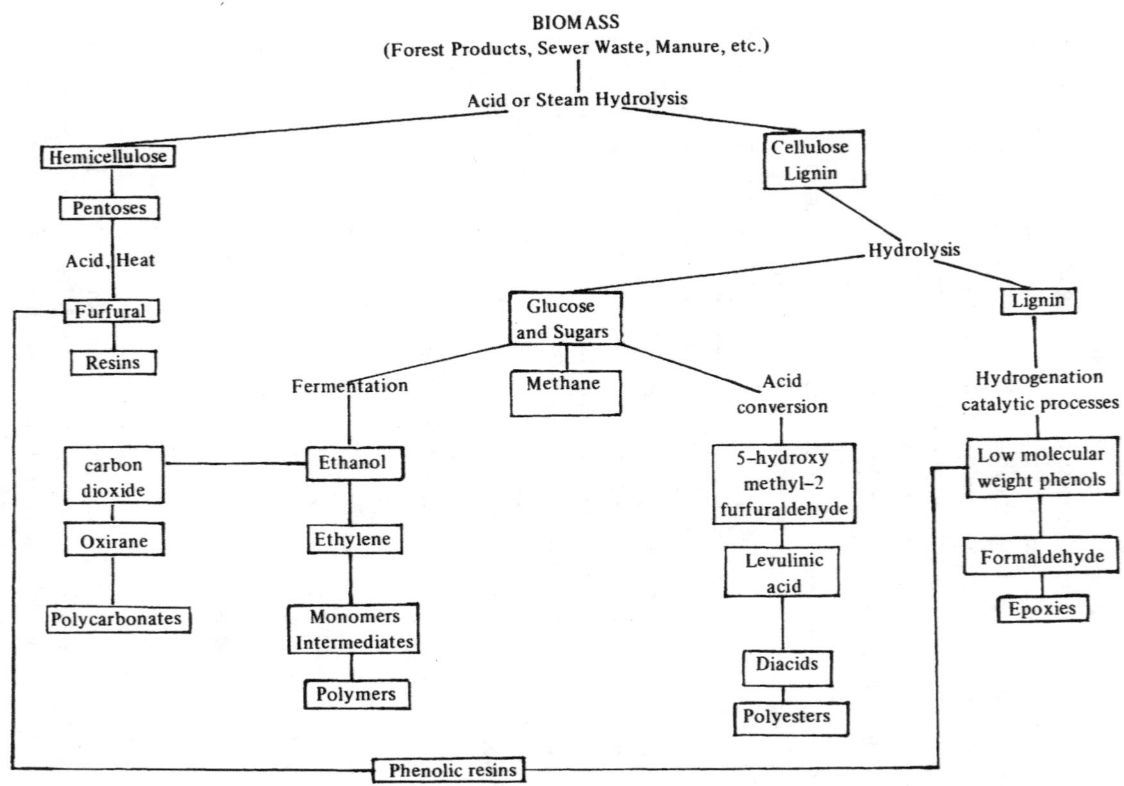
BIOMASS
(Forest Products, Sewer Waste, Manure, etc.)
Acid or Steam Hydrolysis
Hemicellulose
Cellulose
Lignin
Pentoses
Acid, Heat
Furfural
Resins
Hydrolysis
Glucose
and Sugars
Lignin
Fermentation
Methane
Acid
conversion
Hydrogenation
catalytic processes
carbon
dioxide
Ethanol
5–hydroxy
methyl–2
furfuraldehyde
Low molecular
weight phenols
Oxirane
Ethylene
Levulinic
acid
Formaldehyde
Epoxies
Polycarbonates
Monomers
Intermediates
Diacids
Polymers
Polyesters
Phenolic resins

Fig. 2.8

the name 'biomass' has been given to renewable resources with a biological basis (e.g., trees, seaweed, field crops, sewer waste, etc.).

One important monomer is ethylene which can be derived from ethanol, via the fermentation of sugars from plant juices, cellulose, and starch. Methanol can be made from agricultural residues or from municipal solid waste and is important because it is the source of formaldehyde. The major barriers at present are economic rather than technical but this will change as the nonrenewable resources become scarce. The conversion of biomass into feedstocks and polymers of interest is outlined in Fig. 2.8.

PART II MATERIALS

Section A Thermosets

3.

PHENOPLASTS

It is probably true to say that plastics based on the reaction between phenol and formaldehyde have been used for far longer than any others with the exception of the 'mother of plastics', cellulose nitrate. They were the first truly synthetic materials to be used in producing plastics and since their initial production, around 1912, they have developed quickly and have reached industrial importance for a diverse number of applications in many countries. Although their consumption has not grown so rapidly over the past ten years as many of the thermoplastic polymers, they still remain one of the most important polymers of the plastics industry.

The history of the phenolic plastics dates back to the 1870's, when Baeyer studied the reaction of phenol with certain aldehydes and discarded the tar-like products because he was looking for pure and well-defined products. Little interest, therefore, was shown in this group of materials until the beginning of the twentieth century, when they were reinvestigated and their potential as a shellac substitute and a molding composition was realized. Many scientists had been interested in the reaction of phenols with various aldehydes but no great success was achieved until the classical work of L. H. Baekeland, which was made known in a series of papers and patents. His initial work was mainly carried out in the period 1905-1910 and its significance lay in his discoveries of how to control the phenol-aldehyde reaction by adjusting the ratio of the two reactants. The reaction could thus be controlled and stopped at an intermediate stage.

His work was the basis of the phenolic resins manufactured today for such uses as adhesives, laminating resins, molding compositions, castings, surface coatings, foams, and ion-exchange resins. The development of these polymers did not proceed speedily for it was some 20 years before oil-soluble phenolics, suitable for surface coatings, were introduced. This new application was followed by the introduction, in 1935, of the possible ion-exchange properties of phenolic resins. The popularity of these materials for this application was stimulated initially by the purification of sea water in war time and it has subsequently led to a fairly extensive commercial development.

Very little new development has occurred with this family of resins over the past 25 years although the introduction of the Xylok resins around the middle of the 1960's is considered a breakthrough in the technology of achieving high heat-resistant, phenolic materials.

RAW MATERIALS

Phenol (C_6H_5OH)

In its pure form phenol consists of white crystals that melt at about 41°C. For the manufacture of synthetic resins the phenol is normally used in the form of 90 percent phenol and ten percent orthocresol or in the ratio of 80/20. The phenol can be obtained either in the natural state or by synthetic means. The so-called natural phenol is obtained from coal tar by distillation, extraction of the middle oil cut by caustic soda, then acidification and finally distilling. The basic raw material for synthetic production is benzene, which may be converted to phenol by the Sulfonation, Raschig, or Cumene Processes.

(a) Sulfonation Process. Benzene is first sulfonated with sulfuric acid, then fused with caustic soda to form sodium phenate and sodium sulfite. The resultant mixture is then treated with further sulfuric acid, when phenol is liberated.

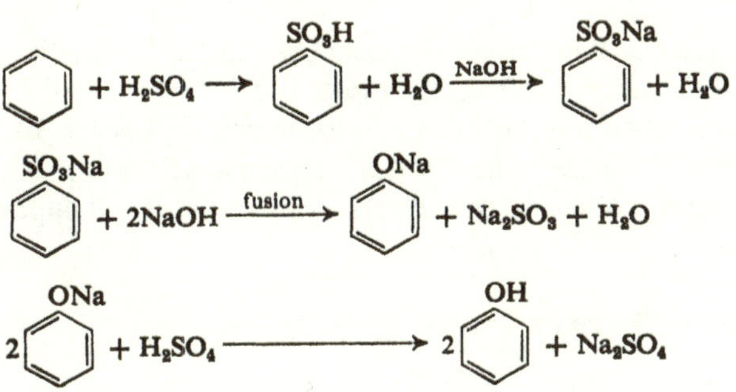

(b) Raschig Process. Benzene, hydrogen chloride, and air are passed over a heated copper-iron catalyst at 200-230°C. The resultant product is chlorobenzene and water in the gaseous state, which is then passed over a silica catalyst at about 500°C. The water hydrolyzes the chlorobenzene to phenol and hydrochloric acid, which material can then be used for further chlorination of the benzene.

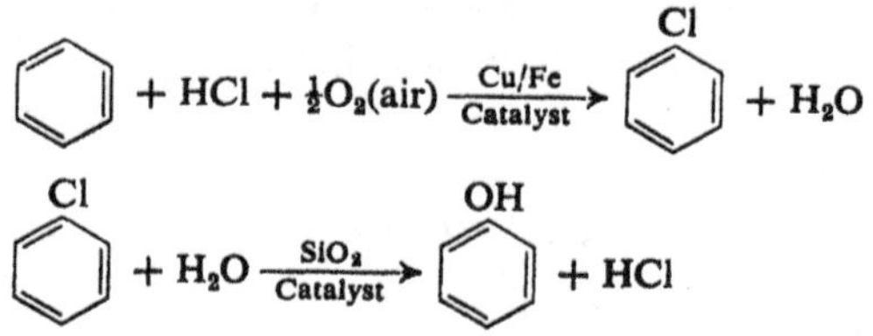

(c) Cumene Process. The catalytic reaction of benzene with propylene yields isopropyl benzene (Cumene) which oxidizes to the hydroperoxide which is then broken down by acid to give phenol and acetone.

Formaldehyde (HCHO)

Formaldehyde is a gas which, on chilling, will condense to give a liquid boiling at −19°C. This aliphatic aldehyde is prepared by the controlled catalytic oxidation of methyl alcohol. The procedure normally adopted is to pass a mixture of the alcoholic vapor and air over a heated catalyst, usually copper, the probable reaction being one of dehydrogenation:

$$CH_3OH \rightleftharpoons H_2 + \underset{(CH_2O)}{HCHO}$$

Vapors of water and formaldehyde condense to yield formalin. There are several grades marketed although for resinification purposes it is normal to use solutions containing between 35-40 percent by weight of formaldehyde. Impurities in the formalin include formic acid and methanol, the latter material is to some extent advantageous in that it will impart stability during storage.

Hexamethylene Tetramine (Hexamine-$(CH_2)_6N_4$*)*

This material is a derivative of formaldehyde and is obtained by passing ammonia gas into 30 percent formalin at room temperature.

$$6CH_2O + 4NH_3 \rightarrow N_4(CH_2)_6 + 6H_2O$$

RESINIFICATION

Liquid or solid phenol-formaldehyde resins can be manufactured from the reaction of phenols with certain aldehydes. The process consists essentially of heating the reactants until condensation takes place, an effect which can be speeded up by the addition of catalysts. The catalysts may be either acids or bases, and depending on which type is used so will the end product differ. However, the same basic reaction takes place in both cases, which is the straightforward addition reaction between phenol and formaldehyde to give phenol alcohols. Thereafter the phenolic resins can be broadly classified either as one-stage or as two-stage resins. If the molar ratio of formaldehyde to phenol is 1 : 1 or greater (i.e. molar excess of formaldehyde) and an alkaline catalyst such as ammonia or caustic soda is used, then a one-stage resin is formed, generally called a resol. The resins so formed have reactive methylol ($-CH_2OH$) groups available in the molecule, and, on heating, the resol molecules condense together forming larger molecules without the addition of hardening agents:

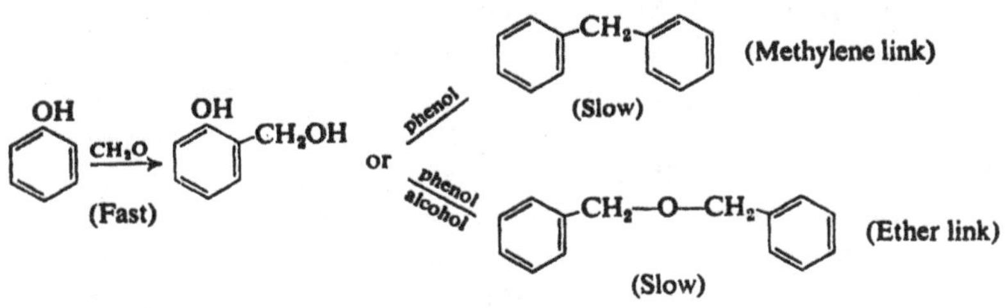

If a molar excess of phenol and an acid catalyst are used then a two-stage resin called a novolak is produced. These are resins which have no reactive methylol groups in the molecule and, on heating, are incapable of condensing with other novolak molecules without the addition of hardening agents. Therefore, to cross-link such materials to form insoluble and infusible resins it is usual to mix, during the compounding stages, materials capable of liberating further formaldehyde, e.g., hexamethylene tetramine or paraformaldehyde:

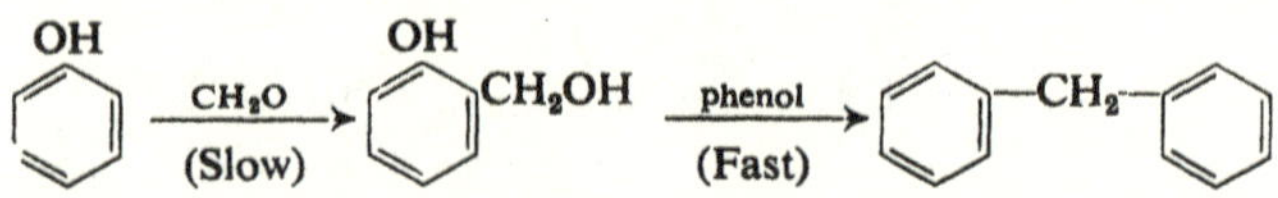

The final curing or heat hardening is considered to be a further condensation process. The following represents the building-up of one only of the simplest types of macromolecule and is shown in two dimensions.

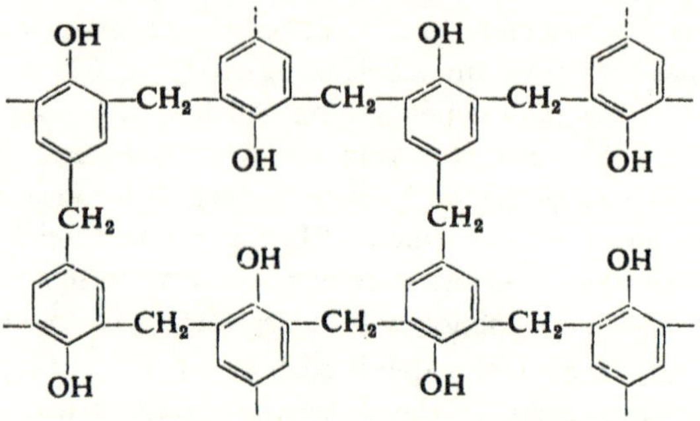

VARIATIONS OF BASIC CONSTITUENTS

Certain variations have been made with the phenolic family of plastics. In particular, furfuraldehyde can replace the formaldehyde, and it has been found that not only does the aldehyde group take part in the reaction but the furan ring also takes part, thereby providing a cross-linked molecular chain. The overall reaction is very similar to the phenol formaldehyde resin reaction although slower. To compensate for the slow speed of reaction it is normally carried out in a strongly alkaline medium. A range of materials with differing properties can be obtained by varying the molar ratio of phenol to furfural. The furfural can be obtained relatively cheaply from certain waste products of agricultural industry such as oat husks (*see* Chapter 2). The resins obtained by reacting phenol with furfural have excellent impact resistance as well as very good chemical resistance. They cure very quickly at high temperatures and possess good flow characteristics. One of their main uses is for modifying conventional phenolic resins, an addition which enables higher loadings of filler to be used with these resins. Other applications include a limited use as temperature-resistant surface coatings. This family is discussed more fully at the end of this chapter.

Cresol may be used to replace phenol in this condensation reaction to give resins suitable for dissolving in alcoholic solvents for coatings having a quick drying time. The meta isomer of cresol is normally used for resinification.

If phenol is replaced by resorcinol then the reaction takes place at a considerably increased rate. This is only to be expected and is due to the fact that the resorcinol possesses two activated hydroxyl groups as opposed to the one of phenol. The resorcinol-formaldehyde resins are generally prepared by using a molar excess of resorcinol, and cure is effected by adding further aldehyde, usually in the form of paraformaldehyde. These resins cure at relatively low temperatures but unfortunately they are rather more expensive than the phenolic type and subsequently they are used in conjunction with the latter materials. They have similar physical properties to the phenolics but give superior properties for certain specific applications such as in the field of adhesives. Resorcinol resins are used in the preparation of plywood; they possess excellent resistance to fungi and mold attack, also to the affect of boiling water. The fact that they may be cured at low temperatures makes them extremely attractive for bonding difficult types of applications such as arches and other curved structures.

There are other types of resins which cannot be classified as phenoplasts but as we have described the manufacture of formaldehyde in this chapter it is convenient to briefly mention them. Formaldehyde can be reacted with cyclohexanone and methyl cyclohexanone. Such resins are the CRF and MS2 types made by Laporte Industries U.K. They can be used as overcoating varnishes, tackifiers, and have a host of printing applications. Another type is the condensation product of high-boiling, aromatic solvents with formaldehyde. This type of resin is designated Kenflex (Kenrich Corp.).

A final variation of some interest is that of replacing phenol with a dark-colored oil known as cashew nutshell liquid, which contains an unsaturated phenol called cardanol. Cardanol condenses with formaldehyde in a similar manner to phenol and may be cured either by heating or by adding paraformaldehyde. The end-product has similar properties to the other phenolic resins except they are somewhat more rubbery. They have found limited use in brake linings, adhesives, and flexible insulating varnishes. Greater use of these cashew nutshell resins has been somewhat limited by their relatively low tensile strength.

MANUFACTURE

The reaction of phenol with formaldehyde is of the condensation type and is therefore normally carried out in resin kettles fitted for vacuum distillation (Fig. 3.1). Several metals may be used in the construction of the resin kettle, the most popular being mild steel, although stainless steel, copper, iron, and nickel have been used.

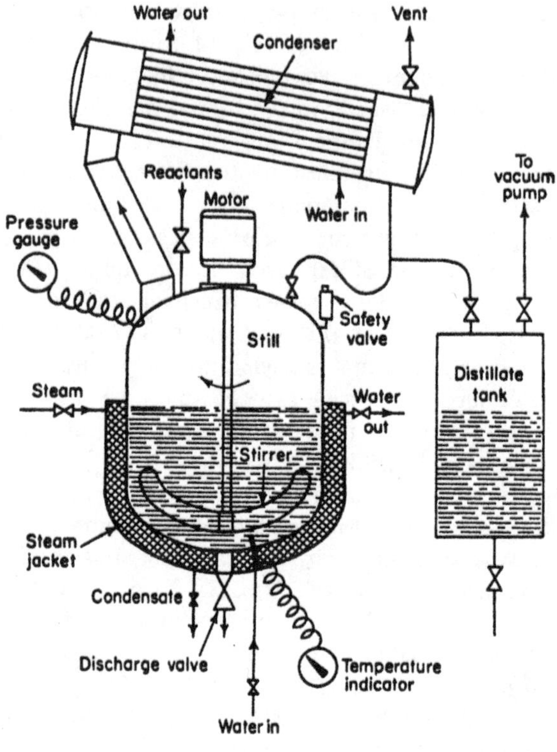

Fig. 3.1 Reaction vessel for manufacture of phenol-formaldehyde resins

Where color is important, as with the novolaks, it is essential to use stainless steel or nickel although for resols, where color is not so important, mild steel, iron, or copper can be used. For the aminoplasts, which are described in the next chapter, the reaction kettle is very similar, and it is even possible to use glass-lined kettles.

The course of the reaction may be followed by measuring the amount of free formaldehyde or by viscosity measurements. On completion of the condensation reaction, which is determined by suitable tests, the resin is then ready for discharging or blending as necessary.

Molding Powders

A novolak is prepared using the formulation details described previously and while the resin is still hot it is run into large shallow trays and allowed to cool. On cooling, the resin is ground to a fine powder and is then compounded with curing agent (hexamine), filler (woodflour), mold lubricant (stearates) and pigments. The properties of a molding powder are largely determined by the type of filler, which can constitute about 50 percent of the powder. Woodflour is normally used as a general-purpose filler but other materials are employed to give improved shock resistance (chopped fabric) electrical properties (mica) and water resistance (silica and mica). These ingredients are premixed and then compounded either on rolls or in a mixer with which a mill is associated. Both types of equipment are steam heated so that the product partly cures during compounding. The final curing is, of course, effected during the actual molding operation. After compounding, the product is ground to the correct particle size, blended, and tested.

Laminating Resins

These are based on resols, the manufacturing details of which have been described previously. The condensation reaction is carried out until a specific degree of polymerization has been reached, the resin being then run out and cooled if a solid resin is required, or a solvent added to the hot resol if a varnish-type material is required.

Casting Resins

These are prepared in a manner similar to resols although in this specific case a large excess of formaldehyde is used, with sodium hydroxide as the catalyst, and the reaction is usually carried out at about 65°C. This encourages the formation of the hydrophilic phenol alcohols, of low molecular weight. The alkaline catalyst is then neutralized with lactic acid, the resultant product is vacuum distilled, a plasticizer such as glycerol is added, and the mix is run into molds and slowly cured at temperatures around 75°C. Oil solubility can normally be achieved by using substituted phenols such as *p*-tertiary butyl phenol or using higher aldehydes like acetaldehyde.

Coatings and Varnishes

Resol resins are used for making varnishes suitable for coil impregnation, but resins of the novolak and resol type can be obtained. These are soluble in conventional paint-drying oils and thereby provide varnishes with excellent water resistance.

Adhesives

The resins normally used for adhesive applications are the liquid resol types with an alkaline catalyst. They may be mixed with certain thermoplastics by a variety of methods to give adhesives suitable for bonding metal to metal. The phenolic plywood glues consist of resols and can either be cured by heat alone at 135°C or by adding small amounts of paraformaldehyde at temperatures as low as 115°C. The method of manufacture is similar as for the aminoplasts described in Chapter 4.

PROCESSING

Molding Powders

Phenol formaldehyde molding powders may be processed by the conventional techniques of compression and transfer molding (*see* Chapter 24). The temperature range is normally between 150-180°C at pressures around 1 ton/in.^{2} or even higher. The molding powders are then fed to the mold either as cold powders or cold or preheated preforms. Phenolic molding powders of the wood-flour type can also be extruded using specialized techniques.

Adhesives

Phenolic resins are used extensively in the production of plywood and comply with B.S. 1203, type WBP (*see* Chapter 4). An interesting technique used in the production of these wood laminates is that of impregnating a thin tissue paper with an alcoholic solution of a resol, placing it between wood veneers, and hot pressing it at 150°C. The actual wood veneers themselves can be impregnated and then compressed during the hot bonding process to produce a denser product which is commonly known as 'compreg'. This material is nearly isotropic and possesses exceptional mechanical properties.

Laminating/Impregnating Resins

Alcoholic solutions of resols can be used for impregnating fabric and

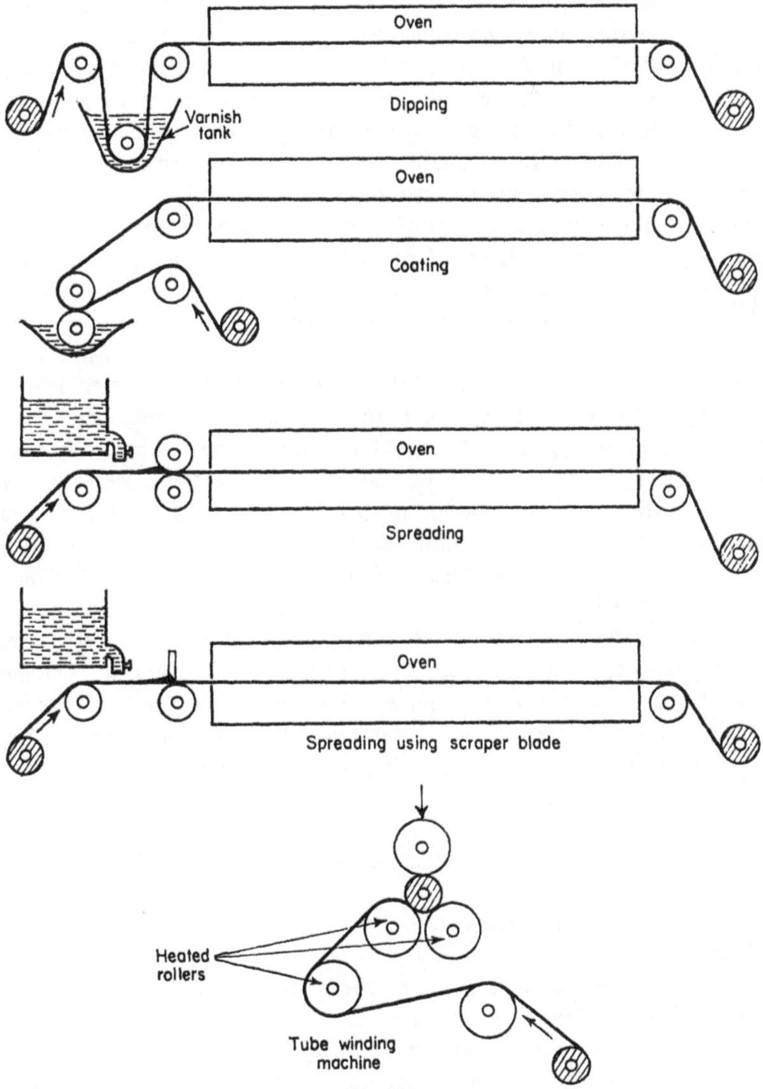

Fig. 3.2

paper to give laminated boards. Various methods of impregnation are used, including 'premix' technique, dipping, coating, or spreading. In the premix process the actual phenolic resin is added to the paper base during the papermaking process. The dipping process is carried out by passing the filler material through a bath containing the varnish, and then on to a drying oven. Coating, as can be seen from Fig. 3.2, is carried out by passing the filler material round one roll and picking the varnish up from an adjacent roll. For spreading methods the resin is introduced to the upper surface of the filler material, which is then passed between the rollers or under a scraper blade. This results in the resin being spread evenly over the surface and removes any excess varnish.

The so-formed treated base must now be consolidated into the finished laminate. For molded shapes, a mold is obviously required and can be either of the open flash or open plunger type. The base/impregnated materials are cut to appropriate shapes and molded at temperatures around 150-180°C with pressures up to 3 ton/in.^{2} Boards or sheets can also be prepared by pressing the laminated structure between platens in multi-daylight presses under the above temperatures and pressure conditions. Tubes and cylinders can also be produced by winding the treated base on a mandrel in a machine which can provide the necessary heat and pressure. The process consists of passing the impregnated base material in roll form over preheating rollers between the end frame to main heated rollers. The mandrel is driven through a universal coupling and rotates between the main rollers with which it is in contact. The pressure required is applied by a third roller bearing on the opposite side of the mandrel. The resin is partially cured when it comes in contact with the heated rollers and is finally consolidated by the pressure roller as the material is wound on the mandrel. Main roll temperatures are of the order of 160°C and the final curing is effected in an oven when the mandrel is finally removed.

PROPERTIES

Appearance and Form of Phenolic Resins

The resins may be transparent or nearly translucent, ranging in color from very pale amber to dark brown, and they may exist either in a solid or liquid form depending upon the molar ratio of the initial reactants and the type of catalyst used. Novolaks are of the solid type and are brittle rosin-like substances which may be crushed to a very fine powder for use in molding compositions or dissolved in an organic solvent for certain bonding applications. The liquid resins of the resol type vary in viscosity from that of water to something approaching that of thick treacle.

The phenol novolak has a specific gravity of approximately 1.25, the cast resin (unfilled) 1.34 and wood-flour or α-cellulose filled moldings 1.35 to 1.4. The physical properties will vary according to the fillers used (Table 3.1). The cured resins are practically insoluble in common solvents, hot oils, aqueous acids, and salts, but are, however, slightly swollen by alcohols, water, and dilute acids and alkalis. They decompose in concentrated oxidizing acids and hot alkalis.

APPLICATIONS

The applications of phenolic resins for molding powders may be classified according to their final end-use, i.e. general purpose, heat and impact resistance or electrical grades. They are low-cost materials, possessing good dimensional stability, heat resistance and also resistance to cold flow. The resultant molded articles are normally black or brown in color which has to some extent limited their wider applications. The general-purpose compounds are usually wood-flour filled and possess a popular balance of properties/price for a wide variety of applications. Typical applications include telephone hand-sets, steering wheels, and fascia panels although the former application is being replaced quite extensively by thermoplastics. The impact resistant grades contain chopped fabric and can be used for pump impellers, washing machine agitators, lavatory seats, and electrical switches. Heat-resistant grades containing asbestos fillers are used in aircraft and missile applications. The electrical-resistant grades are used for automobile ignition parts, coil formers, etc. Phenolic molding powders can be used in electrical equipment where insulation requirements are not too stringent, such as plugs, switches, switch-gear cases, and covers.

The ability of phenolic resins to impregnate and bond such materials as wood, paper and certain fabrics has led to their widespread use for various laminates, not only decorative but also for industrial purposes. Their use for impregnating a wood veneer has been limited to a certain extent by color, and they have a colored glue line which is not always desirable. Industrial laminates are used as gear wheels, bushings, and dollies, and the somewhat thinner ones on which a layer of copper has been deposited, are now employed in the manufacture of printed circuits. These liquid phenolic resins may also be used for the manufacture of grinding wheels by wetting an abrasive grit and then rolling it into a powdered phenolic resin

Table 3.1

PHYSICAL PROPERTIES OF PHENOLIC RESINS

	Cast resin	*Molding resins*			*Laminating resins*		
	(unfilled)	Wood-flour filled	Chopped cotton fabric	Mineral filled	Paper	Fabric	Asbestos
Specific gravity	1.34	1.35–1.4	1.34–1.38	1.9–2.0	1.24–1.38	1.34–1.38	1.6–1.8
Tensile strength (lb/in.^{2})	4–10,000	5–8000	5–8000	3–8000	7–20,000	8–20,000	6–12,000
Flexural strength (lb/in.^{2})	7–12,000	8–12,000	8–12,000	8–12,000	10–30,000	12–30,000	12–20,000
Shear strength (lb/in.^{2})	6–8000	8–10,000	10–15,000	4–15,000	5–12,000	5–12,000	4–8000
Compressive strength (lb/in.^{2})	10–25,000	15–35,000	20–32,000	20–32,000	20–40,000	25–40,000	20–40,000
Impact strength (ft lb/in. notch)	0.2–0.6	0.1–0.5	0.3–3.0	0.25–1.5	0.3–1.5	1–4	0.5–1.5
Specific heat	0.4	0.35	0.34	0.3	0.35	0.35	0.3
Coefficient of thermal expansion/(°C)	3–8	3–6	2–6	2–4	2–3	2–3	2–3

Table 3.1 *(cont'd)*

	Cast resin	*Molding resins*			*Laminating resins*		
	(unfilled)	Wood-flour filled	Chopped cotton fabric	Mineral filled	Paper	Fabric	Asbestos
Dielectric constant (10^3 c/s)	4 plus	5–15	5–10	5–10	4–8	6–12	10–18
Dielectric strength (V/mil)	200–300	100–300	100–250	100–250	250–750	100–500	50–100
Volume resistivity (ohm/cm)	1×10^{12}–10^{14}	1×10^{9}–10^{12}	1×10^{8}–10^{12}	1×10^{9}–10^{12}	1×10^{10}–10^{12}	1×10^{10}–10^{12}	1×10^{9}–10^{10}
Chemical resistance							
Weak acids	*R*	*FR*	*FR*	*R*	*FR*	*FR*	*R*
Strong acids	Decomposed by oxidizing acids resistant to reducing organic acids						
Weak alkalis	*R*	*R*	*R*	*R*	*R*	*R*	*R*
Strong alkalis	*A*	*A*	*A*	*A*	*A*	*A*	*A*
Solvents	*R*	*R*	*R*	*R*	*R*	*R*	*R*
Key							
R–resistant							
FR–fairly resistant							
A–attacked							

which forms a putty which can be pressed into a mold to give a preform, and finally cured.

As we have seen, casting resins are of the liquid type and prepared by using a large molar excess of formaldehyde and an alkaline catalyst. The resultant resin is normally neutralized by adding lactic acid, and is plasticized with glycerol. They may be prepared in a wide range of colors but are to some extent being replaced by other plastic materials such as the polyester resins and polymethyl methacrylate for certain applications. It is for this reason that cast phenolic resins have never developed as was originally hoped, but they still continue to find uses in such applications as billiard balls, umbrella handles, furniture handles, pipe and pen stems as well as cutlery handles, etc.

A comparatively recent development with this group of materials is that of the foamed phenolics. The type of resin suitable for foaming would be the casting type and the foaming action is normally effected by the chemical liberation of gases from blowing agents. The usual blowing agents include the metals that react with acids to liberate hydrogen or certain nitrogen-containing compounds. The curing reaction is exothermic, which facilitates the rapid foaming desirable in order to obtain forms of fine cell size. Lightweight foams with a density less than 1 lb/ft^3 may be produced, although for structural purposes higher density foams up to 20 lb/ft^3 are normally required. The main uses for phenolic foams have been in the field of insulation for buildings, and transport vehicles. The main attractions of these foams are that they may be foamed *in situ* and possess good chemical, fire and heat resistance. For insulation purposes they may also be supplied in slab form. They have low weight per unit volume and are comparatively resilient. Other applications include packaging, buoyancy, handicraft, and decorative materials for stage and movie cinema scenery.

An even later development along this line is that of the syntactic foams, which may be prepared from microscopic hollow spheres of phenolic resins containing nitrogen, and subsequently mixed with phenolic, polyester, or epoxy resins. The resultant product is a putty-like mass which can be trowelled, in a similar manner to cement, into cavities, and finally cured. The end product has very high strength-to-weight ratio. Applications include fillers for acoustic tiles, concrete, and plaster.

FURANES

These resins have been around for 25 years for use essentially as corrosion-resistant mortars. There were many early unsuccessful attempts to utilize these resins in glass-reinforced laminates. They were used, however, as heat-resistant coatings for polyester and epoxy laminates especially in ballistic missile applications. The Quaker Oats Company in the U.S.A. has been responsible for much of the development of these resins.

It has been shown previously in Chapter 2 on Raw Materials that furane resins can be obtained from a variety of naturally occurring materials. The main furane resins involve furfuryl alcohol as the sole polymeric constituent although there are other possibilities with combinations of urea or phenol-formaldehyde.

Furfuryl alcohol can be resinified in the presence of acid catalysts. The system thus formed is neutralized and the water of condensation formed is removed by vacuum distillation to give a stable resin.

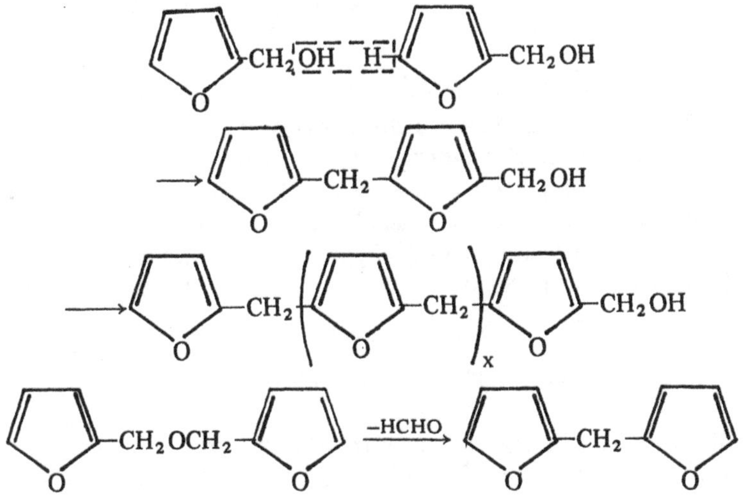

This resin can be further recatalyzed by acid media (e.g., phosphoric acid, *p*-toluene sulfonic acid, etc.). The resultant product is cross-linked to form extremely chemically resistant materials. These cured resins are dark in color with a high degree of heat and chemical resistance.

Several theories of cross-linking have been postulated but one accepted suggestion is that the loss of unsaturation during cross-linking indicates the reaction is essentially a double bond polymerization.

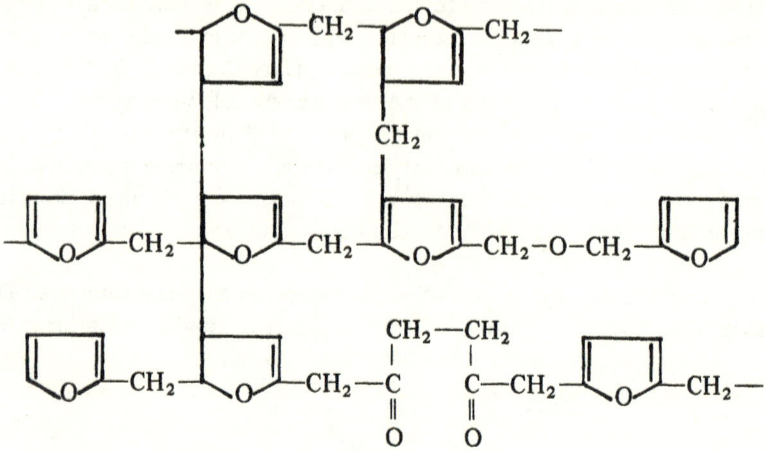

The main field of applications for these resins are as foundry core binders, carbon-graphite binders, and chemically resistant mortars. Phenol furfural and urea-formaldehyde-furfuryl alcohol have been used for adhesive applications, e.g., in plywood. The Quaker Oats Company has developed a series of resins which can now be used in conjunction with glass laminates. Typical uses for such laminates are fume ducts, chemically resistant tanks, scrubbers, and effluent tanks.

PHENOL-ARALKYL

All self-respecting organic chemistry students should be aware of the classical work carried out by Friedel and Crafts. This reaction involves the introduction of an alkyl or acyl group into an aromatic ring in the presence of a catalyst such as aluminum chloride.

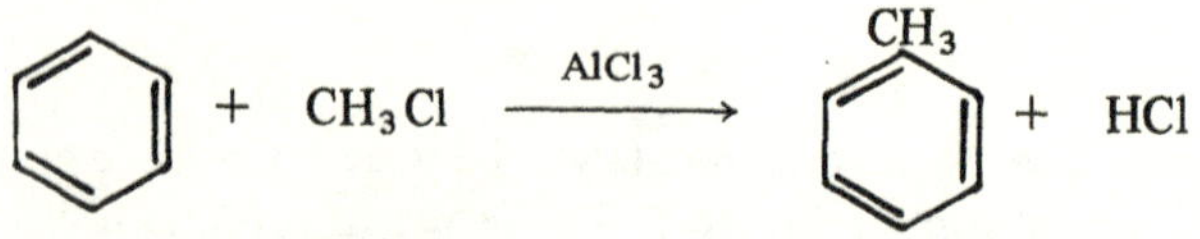

Benzene Methyl Chloride Toluene

This work was followed by their attempts to heat benzyl chloride in the presence of aluminum chloride to form resinous materials. The resultant product was an insoluble infusible mass.

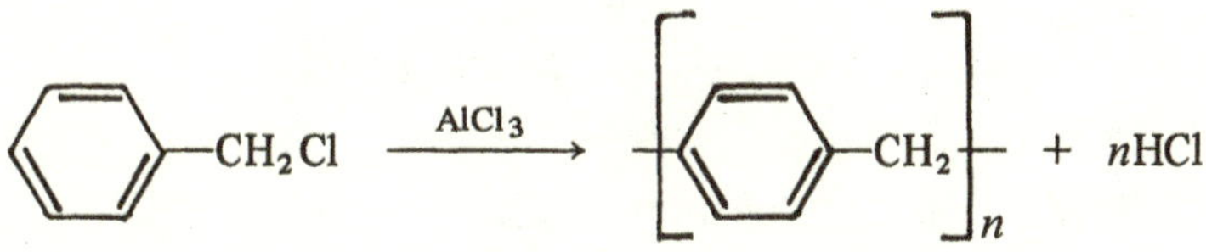

It is clear from our theories of functionality that the resultant product is not thermosetting because it cannot cross-link. Later it was proved that the end-product was simply a highly branched molecule.

Many years later when the interest to develop polymers with heat resistance in excess of 200°C for space exploration became necessary work was revived both in the U.K. and the U.S.A. Work began around the 1950's but it was not until 1964 that a solid breakthrough was achieved. Originally the concept was to produce true thermosets by using compounds having two or three chloromethyl groups.

A typical reaction is as follows:

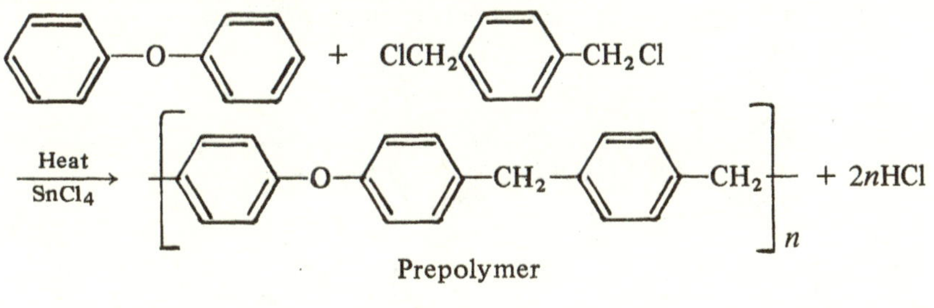

A prepolymer is formed by condensing diphenyl ether with dichloro-*p*-xylene in the presence of stannic chloride. Cross-linking is achieved by adding more of the latter reactant. In 1964 both chemists at the Royal Aircraft Establishment, Farnborough and Albright and Wilson discovered the aralkyl ethers. They found these materials reacted in a similar manner to the chloromethyl compounds but the by-products were alcohols. This type of reaction is illustrated as follows by reacting naphthalene with dimethoxy-*p*-xylene with the elimination of methanol.

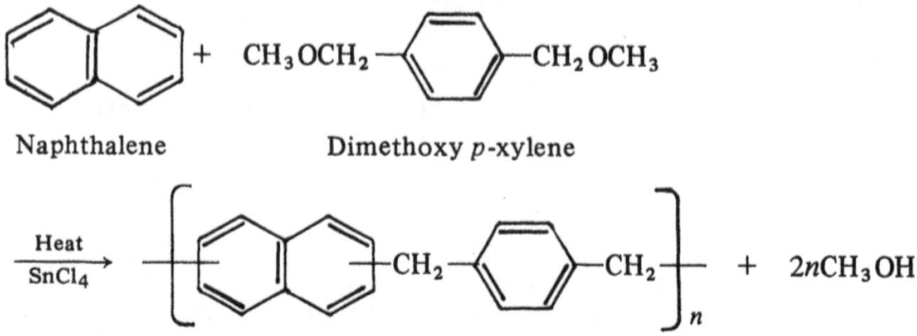

Naphthalene Dimethoxy *p*-xylene

Prepolymer Methanol

A prepolymer is also formed and the cross-linking is achieved by the addition of further aralkyl ether and heating in the presence of ferric chloride. Curing using this type of process is relatively slow and the degree of cross-linking is not high. Therefore applications for these type of resins have been limited to requirements of just high heat stability without high mechanical strength.

Albright & Wilson, Ltd, concentrated their efforts on the cross-linking reaction and found that by condensing the aralkyl ethers with phenolic compounds a prepolymer was formed which could be cross-linked with a variety of techniques. A simple example is shown as follows:

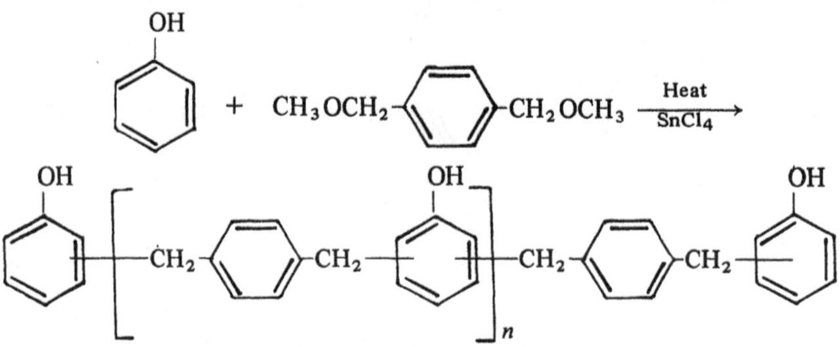

Prepolymer $+ (2n+2)CH_3OH$

These materials are now sold by this company using the registered trademark of Xylok resins.

There are two main methods of achieving cross-linking.

(1) Hexamine. The prepolymer is heated with hexamine which attacks

the vacant ortho and para position on the phenolic nuclei to form methylene or dimethylene amino linkages. Such a reaction is accompanied by the release of ammonia.

(2) Epoxide. The prepolymer is heated with an epoxide and a suitable accelerator. This chemical grouping reacts with the phenolic hydroxyl to form cross-links without any volatile products being evolved.

Commercial resins include a base resin supplied as a 100% solid prepolymer without curing agent, solid prepolymer without curing agent, a one-pack laminating varnish containing curing agent in a ketonic solvent such as methyl ethyl ketone, and a one part varnish, which cures by epoxy addition.

The applications for hexamine cured resins are for electrical insulation both for laminates and moldings. Asbestos filled Xylok resins have resulted in commutators with greatly improved lives. Graphite filled resins are used for the moldings in hot water pumps.

Resins cured by epoxides have higher heat resistance and thermal mechanical strength than hexamine cured types. (See Table 3.2). The electrical industry is the obvious, largest potential user. Typical uses include high performance printed circuit boards; the easy processing characteristics

Table 3.2

PHYSICAL PROPERTIES OF THE GLASS CLOTH-REINFORCED FRIEDEL-CRAFTS LAMINATE

Properties	*Value*	*Test method*
Specific gravity	1.80	
Tensile strength at room temperature		
lbf in.$^{-2}$	58,600	BS 2782, 301C
MN m^{-2}	404	
Tensile strength at 200°C		
lbf in.$^{-2}$	40,800	BS 2782, 301C
MN m^{-2}	282	
Crushing strength		
lbf in.$^{-2}$	69,500	BS2782, 303B
MN m^{-2}	480	
Izod impact strength		
ft lb 0.5 in	10.7	BS 2782, 306A
Flammability oxygen index	32	ASTM D2863-70
Water absorption (%)	0.18	ASTM D570

of these resins should prove useful in structural materials (in the aircraft industry), heavy electrical insulation, and components for chemical plants.

SUGGESTED FURTHER READING

Developments with Thermosetting Plastics. (1975) Edited by A. Whelan and J. A. Brydson (London: Applied Science Publishers Limited).

Thermosetting Materials and Processing. (1994) Edited by J. F. Monk (London: Longman).

4.

AMINOPLASTS

The reaction between urea and formaldehyde has been known since Hölzer first isolated methylene urea in 1884. Synthetic production of urea was achieved some fifty years earlier, in 1828, when Wöhler, preparing ammonium cyanate, unexpectedly obtained urea. Commercial interest was not aroused in the reactions between urea and formaldehyde until a patent was granted to a Czechoslovak scientist named Hans John, in 1918. Other patent claims followed from two Austrians, Ripper and Pollak, and the Germans, Goldschmidt and Neuss. The patent issued to Pollak served as an ideal basis for the commercial development of 'organic glass'. His efforts were directed mainly to the casting of liquid resins whereas Goldschmidt and Neuss saw the potentialities of the resinous products as hot-press molding compounds.

The chemical, melamine (2,4,6-triamino-1,3,5 triazine) was first prepared shortly after the discovery of urea, in 1843, by Liebig. It was first neglected as a commercial material and it was not until a century later that Henkel, in 1835, patented its resinous condensations with formaldehyde. Melamine when reacted with formaldehyde produces resins similar in appearance to those of urea but superior in respect to water resistance.

Thiourea, another chemical used on a somewhat smaller scale for producing aminoplasts was manufactured in this country, in 1924, by British Cyanide Ltd. (later to be known as British Industrial Plastics Ltd.). E. C. Rossiter studied the condensation reactions of this material and, in particular, those of thiourea and urea with formaldehyde. This work was followed, in 1926, by the introduction of the mixed urea-thiourea resin molding powders. Such resins have never formed the basis of a satisfactory molding material, owing to their brittleness and relatively poor water and heat resistance.

The next few years saw consolidation of the early developmental work on the aminoplasts and large-scale production of urea formaldehyde molding powders began in 1938; extension to wood glues followed soon after. The use of such resins has now become firmly established for such applications as molding powders, adhesives, textile treatments, and surface coatings.

Another member of the family which has attracted considerable interest over recent years is benzoguanamine-formaldehyde resin. Benzoguanamine is a triazine, closely resembling melamine although the resins produced from it are more water resistant than the melamine condensates. At present benzoguanamine is more expensive than melamine and this prevents its wider use.

In recent years we have seen the development of aminoplasts in the field of injection molding. The ability to injection mold these plastics have resulted in a stimulated interest.

RAW MATERIALS

Urea (NH_2CONH_2) is a white crystalline solid melting at 132.6°C and having the empirical formula CON_2H_4. It can be produced by various methods, of which probably the most important is the reaction under pressure of carbon dioxide with ammonia in silver-lined autoclaves:

$$CO_2 + 2NH_3 \xrightarrow[70\text{–}230\ \text{atm}]{135\text{–}195°C} CO(NH_2)_2 + H_2O$$

There has been a more recent trend to stainless steel which is passivated by incorporating oxygen in the reactant stream. This allows higher operating conditions with improved degrees of conversion thereby reducing production costs.

Urea may also be synthesized by the acid hydrolysis of cyanamide. Calcium cyanamide is first extracted with water and the soluble lime is precipitated with sulfuric acid:

$$CO_2 + 2NH_3 \xrightarrow[70\text{–}230\ \text{atm}]{135\text{–}195°C} CO(NH_2)_2 + H_2O$$

Melamine ($C_3H_6N_6$) is a white crystalline solid, sparingly soluble in water, and melts just above 350°C. It can be produced on a commercial scale by heating dicyandiamide under pressure with ammonia or by fusion with guanidine salts. Chemically it is the triamine of cyanuric acid

($C_3H_3O_3N_3$) and may also be considered as the trimer of cyanamide. It is generally represented by a triazine structure:

$$\underset{\text{Calcium cyanamide}}{CaCN_2} + H_2SO_4 \longrightarrow CaSO_4 + \underset{\text{Cyanamide}}{NH_2CN} \xrightarrow{H_2O} \underset{\text{Urea (carbamide structure)}}{NH_2CONH_2}$$

The reaction between lime and coke has also formed the basis of the commercial production of melamine as follows:

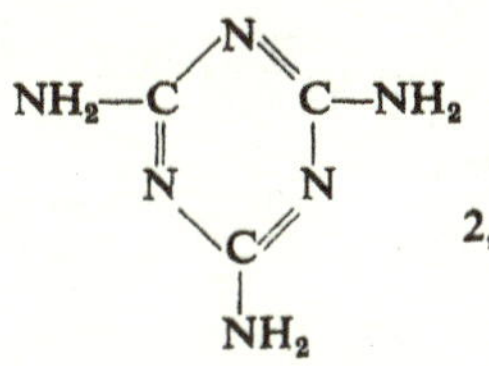

2,4,6-triamino-1,3,5 triazine

However, during the last ten years there has been a very considerable move towards manufacturing melamine from urea:

$$6CO(NH_2)_2 \longrightarrow C_3H_6N_6 + 3CO_2 + 6NH_3$$

This process is used in the U.S.A. and by many manufacturers in Europe, although British Oxygen Co. in the U.K. are still using the dicyandiamide route.

Thiourea ($CS(NH_2)_2$) in the pure state is a colorless crystalline compound melting at 182°C, and is soluble in water. It can be manufactured from ammonium thiocyanate or by the action of hydrogen sulfide on a solution of cyanamide obtained by extraction of calcium cyanamide.

$$NH_2CN + H_2S \longrightarrow CS(NH_2)_2$$

Benzoguanamine ($C_3H_4N_5C_6H_5$) is a white crystalline solid which melts at 227°C and is sparingly soluble in water. The guanamines can be prepared by heating dicyandiamide with a nitrile and a catalyst such as metallic sodium or sodium hydroxide.

Formaldehyde (CH_2O) is discussed fully in Chapter 3.

RESINIFICATION

The basis of the chemistry of the interaction of urea and formaldehyde is the formation of monomethylol and dimethylol urea. Monomethylol urea is obtained after mixing equimolar proportions of urea and formaldehyde to give a neutral solution. Dimethylol urea is prepared in a similar manner using 2 mols of formaldehyde per mol of urea. On a commercial scale the formalin (35–40 percent w/v formaldehyde in water) is adjusted to a pH of 7.5 using sodium or potassium hydroxide. The reactions are as follows:

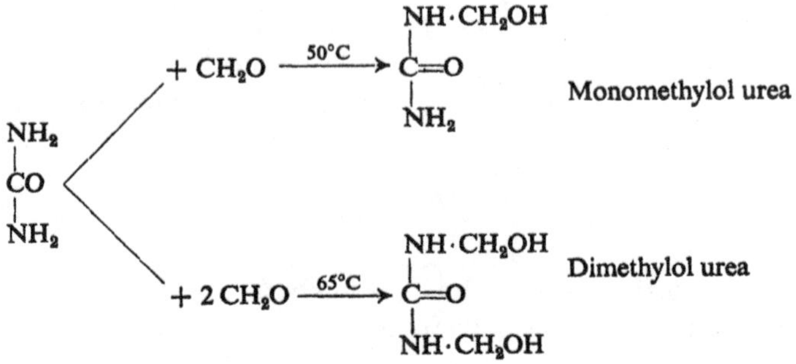

The methylol compounds so formed are relatively unstable and condense in the presence of a catalyst or on heating to produce methylene derivatives which ultimately become cross-linked and infusible products.

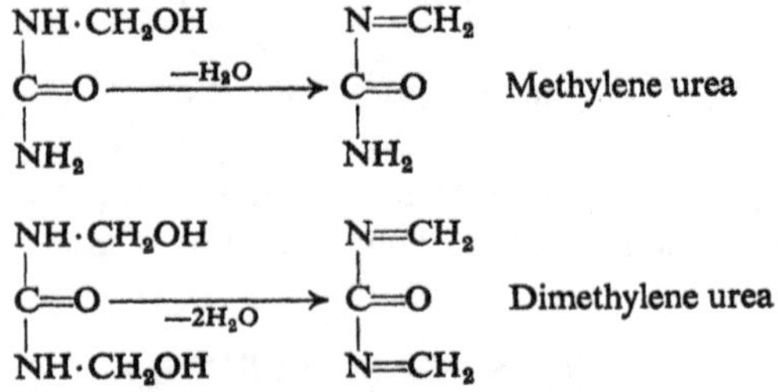

The functionality of monomethylol urea can be considered as 2 and that of formaldehyde as 2. Therefore, as discussed previously, the system will not cross-link. Dimethylol urea, formed from two mols of formaldehyde has a functionality of 4 and cross-links to give thermosetting resins of the following possible structure:

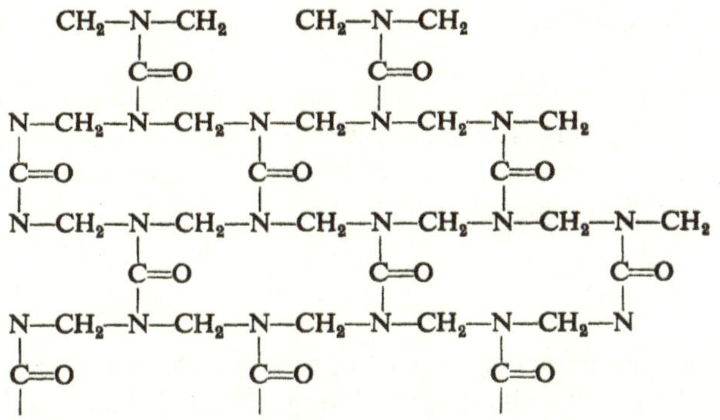

The condensation reactions between melamine and formaldehyde resemble those with urea but compared with the two methylol derivatives of urea, 1 mol of melamine can combine with 1-6 mols of formaldehyde. A similar reaction is also obtained with benzoguanamine and formaldehyde. The resultant resins can be cured by adding acid catalysts such as phthalic anhydride.

MANUFACTURE

The base resins are normally manufactured in a reaction kettle similar to that for phenol/formaldehyde resins described in Chapter 3.

Molding Powders

The base resin, either urea or melamine, is produced in a fairly low state of condensation. Urea-formaldehyde molecular ratios of about 1 : 1.5 are preferred, while melamine-formaldehyde molecular ratios vary from 1 : 1.5 to 1 : 3.0. The condensation is carried out under faintly alkaline conditions, and although the actual process of resin manufacture is quite simple, rigid control of the pH is essential. There are two general methods used for converting the resin into a molding powder. One, which is termed the wet process, consists of kneading the filler (wood-flour or bleached wood pulp) with the aqueous resin until the filler becomes thoroughly impregnated. The resultant mix is then broken up into fine crumbs and dried to the required plasticity. At this stage the material is in a form similar to coarse desiccated coconut and is placed in a ball-mill together with catalyst (acidic, e.g., hexamine), lubricants (stearates) and pigments. It is then ground to a fine powder and finally sieved to remove any coarse particles. The resultant molding powder may be too bulky for certain applications and can be densified by granulating either on heated mixing rolls or in a Banbury mixer.

The other method, termed the dry process, merely consists of mixing the condensed resin with fillers, catalyst, lubricants, and pigments on heated mixing rolls. At the correct temperature the mix resinifies and can then be sheeted off and finally granulated.

Adhesives

These resins are produced by reacting urea and formaldehyde in a molecular ratio of 1 : 2. The initial pH of the reaction medium is usually neutral or slightly alkaline to allow the formation of the methylol ureas, and then raised to a mildly acid medium (pH = 5) to ensure a steady rate of condensation. The reaction can be followed by viscosity measurements and the tendency of the resin to be hydrophobic. At a predetermined stage the process is stopped by neutralization and the resin is then either dehydrated under vacuum, to give a viscous product of 1-10 poises and 60-70 percent solids content, or spray dried to give a resin powder. The resin syrup can then be stabilized by the addition of weak bases or by protective colloids such as starch. The liquid product has a shelf-life of between 3-9 months whereas the spray dried powder is usable for periods of up to two years.

Laminating Resins

For this application the resin is taken to a fairly advanced state of condensation. Urea-formaldehyde resins may require an acid catalyst to effect hardening. Melamine formaldehyde laminating resins are more popular and are obtained in the form of a dry powder by spray-drying the resultant condensation resin. The subsequent powders have a relatively long shelf-life and can be used as and when required by dissolving in water or water/alcohol mixtures, and heating to 60°C to give a solids content of 40-50 percent.

The resin can be applied to fabric or paper by direct immersion or by roller coating. The resultant sheets must be cut into appropriate sizes and the laminates can then be built up from numerous sheets. The resultant laminate can be cured by pressing at 120–150°C under a pressure of 500–2000 lb/in.^{2}. In the preparation of many of the well-known decorative laminates the printed pattern layer is impregnated either with a urea-melamine cocondensate or, as is more common, a phenolic resin, and a protective translucent overlay sheet, treated with melamine resin, superimposed.

Surface Coatings

Such resins are usually prepared by reacting urea and formalin together with an excess of a monohydric alcohol under acidic conditions. The water of condensation so formed is distilled off. The most popular alcohol is butanol, and the resultant resin can be blended with equal parts of a glycerol/phthalic anhydride alkyd resin or a hydroxyl containing acrylic.

This reaction is generally referred to as etherification and sometimes these products are loosely referred to as "blocked" amino resins. On heating to about 150°C the alcohol splits off and the resin undergoes self-condensation and coreaction with the pendant hydroxyl groups of the alkyd or acrylic resin.

Etherification or alkylation with the lower alcohols such as methanol, generally leads to water soluble resins which have found considerable use in textile finishing but unfortunately are incompatible with hydrocarbon solvents.

One member of this family which has been widely used over the past years is the hexamethyl ether of hexamethylol melamine (HMEHMM). The structure is as follows:

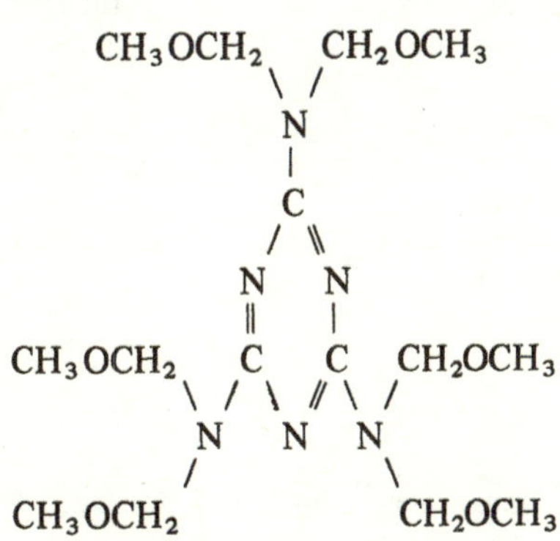

Typical commercial products of this family are Cymel 300 and 301 manufactured by American Cyanamid. A general process would be to dissolve one mole of melamine in excess formalin (9 mol CH_2O) at about 60°C with the pH being kept at alkaline (9.0-9.5). On cooling the hexamethylol melamine separates out and is then filtered, washed with water, and even dried at temperatures not in excess of 40°C. Excess methanol is then added to the resultant dried hexamethylol melamine under acid conditions until dissolution is complete when the solution is neutralized and then filtered with the excess methanol being removed by vacuum distillation. The resultant pure ether is a crystalline solid, colorless, with a melting point temperature of 55°C.

This ether has limited solubility in water but dissolves readily in water containing methanol. It has the ability to react with groups containing a reactive hydrogen atom and its wide range of solubility and polymer compatability renders it a very versatile cross-linking agent for many potential surface coating polymers. If no catalysts are used then the cure times are lengthy but by the addition of strong acid catalysts such as *p*-toluene sulfonic acid the cycles are reduced considerably. This resin can also be used to react with epoxide groups where they react with hydroxyl groups. The resultant cured films have good flexibility with a high degree of resistance to alkalies.

This etherified melamine has been widely used as the cross-linking agent for water soluble alkyds and acrylics. These water-based coating resins are now finding use in the automotive industry.

$$\underset{\text{Alkyd or acrylic}}{R{-}CH_2{-}CH_2{-}OH} + \underset{\text{Amino resin}}{R'{-}\underset{H}{N}{-}CH_2OH} \longrightarrow$$

$$R{-}CH_2{-}CH_2{-}O{-}CH_2{-}\underset{H}{N}{-}R' \xrightarrow{\text{Heat}}$$

$$R{-}CH_2{-}CH_2{-}\underset{H}{N}{-}R' + CH_2{=}O$$

Textile Finishing

Many years ago it became apparent that the tendency of cellulosic fabrics to crease when folded could be overcome by treatment with amino resins. These resins also impart a relatively high degree of resistance to shrinkage on washing. It is also better to first desize and scour the fabric (cotton) before impregnating. This allows the amino resin, which should also be of comparatively low molecular weight, to penetrate the cellulose fibers. Simply, the process consists of padding the fabric through an aqueous solution of amino resin containing an acid or acid liberating catalyst such as magnesium chloride. Excess resin is removed by passing through mangle rolls, which also ensures further penetration. The fabric can then be dried on a stenter with hot air and then passed through an oven at 140–160°C to cure the resin. Washing with alkaline can be used to neutralize the acid catalyst but this is not always done.

In concluding this section some different types of amino resins are discussed.

Ethylene Urea

This is made from reacting urea with ethylene diamine

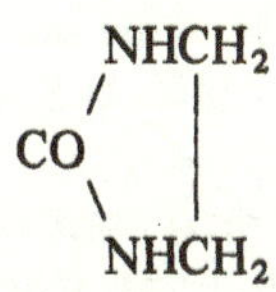

It will react with formaldehyde to form a dimethylol derivative which has a m.pt of 133°C and is readily soluble in water. This resin has been found to give good crease resistance and there is evidence to suggest that the dimethylol ethylene (DMEU) cross-links with adjacent cellulose molecules.

Propylene Urea

This is made by reacting urea and 1, 3 diamine propane.

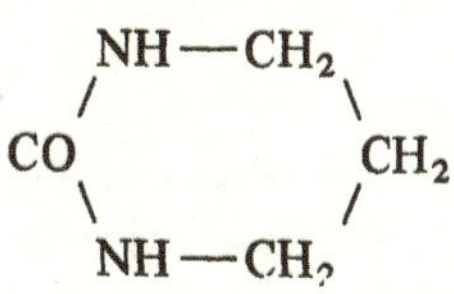

This product also reacts with formaldehyde to form the dimethylol derivative which is more resistant to laundering than the DMEU but more expensive.

Dihydroxyethylene Urea

This is made by reacting glyoxal with urea, although it is more common to form the dimethylol derivative by coreacting urea, glyoxal, and formal-

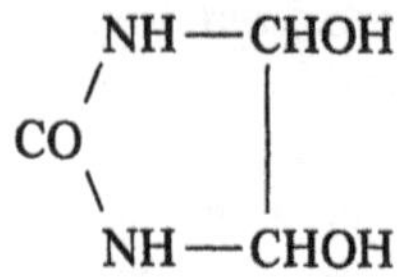

dehyde in 1 : 1 : 2 mol ratio yet without isolating the (DEU). This is probably the most widely used of the cyclic urea finishes and the cross-linking occurs not only by the methylol groups but also via the ring hydroxyls. This product is noted for its absence of formaldehyde odor when finishing fabrics and curing.

The dimethylol dihydroxy ethylene urea (DMDHEU) has been found to be suitable in durable press processes. One of the best known types of processes is the Koratron Process. This was introduced in the early 1960's by the Koret Company of California, U.S.A. The process simply consists of impregnating fabrics (blends of cotton or rayon with polyester) with a solution of amino resin and a metal salt catalyst. The resultant impregnated fabric is then dried carefully to avoid resin fixation. The fabric is then made into garments which are hot pressed and finally baked for 10–20 minutes at 145–155°C. This process is ideal for placing creases and pleats where desired. It must be made clear that in this process no cross-linking must take place on the stored fabric and the reaction must go to completion in the baking oven. DMDHEU has been found to be very suitable for this process because it is capable of existing in the unreacted state for very long periods in the presence of metallic salts such as zinc nitrate or magnesium chloride.

Curing or Hardening Agents

Although certain urea formaldehyde resins will cure solely by the action of heat it is essential on an industrial scale to add materials which will speed up this process. For such resins it is essential that setting-up will not take place in storage and, therefore, the effect should preferably be carried out by the action of heat on a curing agent. Many different materials have been used for molding powders, from the inorganic material zinc sulfate to organic materials such as chloroacetamide. Other materials used include dichlorhydrin, triethyl phosphate and chloro acetone.

Hardeners are nearly always used for urea-formaldehyde resins in adhesive/laminating applications. They are usually acidic or acid-liberating substances. No alkaline hardeners, as far as present knowledge goes, have been used for curing urea-formaldehyde wood adhesives although certain such materials have been used for fixing urea condensates in the treatment of textiles. Ammonium salts are very popular hardening agents and these will react with the free formaldehyde present in the resin to give hexamine and free acid. The reaction is somewhat complex and may briefly be represented as follows:

$$4NH_4Cl + 6CH_2O \longrightarrow N_4(CH_2)_6 + 6H_2O + 4HCl$$

There are other materials that can be used satisfactorily, including phosphoric acid, hydrogen peroxide, ethylene thiocyanate, and glyceryl phthalates.

PROCESSING TECHNIQUES

Molding Powders

Aminoplasts can be molded by the conventional thermosetting molding techniques of compression and transfer molding described in Chapter 24. The thermosetting molding powders can be molded in the temperature range 135-170°C, using pressures of 1-3 ton/in.^{2}, depending on the complexity of the mold. Mold design is virtually the same as for the phenoplasts and, as with the latter materials, care must be exercised in allowing for the bulk factor of the molding powder. It has been found in practice that the UFs are somewhat more critical to mold than the MFs due to the fact that they have less flow and a tendency to set and cure faster in the mold.

Adhesives/Laminates

Urea-formaldehyde resins may be used for gluing wood by a variety of techniques and have now found a major outlet in the hot-press bonding of plywood. This process consists essentially of spreading the adhesive/hardener mixture on to the base wood veneers at a controlled rate and building up the laminate by placing the veneers on top of each other with the grain crossed each time. The resultant plywood laminate is then consolidated by pressing at 30-200 lb/in.^{2} at temperatures between 95-110°C. The glue-line of the plywood is virtually colorless, which is an advantage over the phenolic adhesives, it does not support mold growth, and it will comply with certain sections of B.S. 1203 and 1204 (*see* 'Properties').

In certain constructional applications it is not always convenient to apply heat to the structure and techniques have now been developed whereby cold-setting urea-formaldehyde adhesives can be used, pressure being applied by jigs and clamps.

Another interesting technique is one in which the hardener is applied to the face of one adherend, resin to the face of the other; when the two faces are mated the resin is hardened and the adherends are satisfactorily bonded. This is attractive because the limitations on pot-life are reduced.

An industry which has grown to considerable magnitude over the past few years is based on the manufacture of certain constructional materials from wood-waste and urea-formaldehyde adhesives. The resultant product is more commonly known as 'chipboard.' The boards are made by mixing 5–10 percent of the resin with wood chips and pressing at 200–1000 lb/in.^{2} at a temperature of 100–110°C.

The use of melamine and urea-formaldehyde resins as laminating materials has already been discussed but we should mention here that the resultant laminate is normally pressed between 2000–3000 lb/in.^{2} at temperatures between 140–160°C.

PROPERTIES

The base resins are usually clear water-white syrups or white powders. Correctly cured moldings are odorless and tasteless which makes them eminently suitable for domestic articles. A specification is available covering such moldings, viz. B.S. 1322: 1956, 'Aminoplastic Moulding Materials.' The synthetic resin adhesives are covered by B.S. 1203: 1956, 'Synthetic Resin Adhesives for Plywood (phenol and urea-formaldehyde)', and B.S. 1204: 1956 'Synthetic Resin Adhesives (phenol and urea-formaldehyde) for Constructional Work in Wood'. Urea-formaldehyde resins are less resistant to water than the melamine and phenol types. B.S. 1203 specifies three types of adhesives–Type MR, Type WBP and Type BR. It has been established in practice that the phenolic resins meet the specification for Type WBP, melamine/urea resins for Type BR and urea-formaldehyde resins for Type MR.

Aminoplasts are relatively unaffected by common organic solvents, oils, greases, cold water and cold dilute acids and alkalis. Urea resins are very slightly swollen by water and polyhydric alcohols whereas the melamines show practically no swelling. Both resins are attacked by acids and alkalis.

Typical properties of urea and melamine formaldehyde molding powders (α-cellulose filled) are given in Table 4.1.

Table 4.1

TYPICAL PROPERTIES OF UF AND MF MOLDING POWDERS

Property	*U/F*	*M/F*
S.G. at 15°C	1.48–1.52	1.47–1.52
Tensile strength (lb/in.^{2})	5000–7000	7000–10,000
Elongation at Break (%)	0.6	0.7
Impact strength ft lb/in notch	0.26	0.3
Compressive strength (lb/in.^{2})	25–35,000	25–40,000
Flexural strength (lb/in.^{2})	10–16,000	10–16,000
Thermal conductivity (cal/s/cm^2/°C/cm $\times 10^{-4}$)	7–10	7–10
Heat distortion temperature (°F)	270–280	380
Volume resistivity (ohm/cm)	10^{12}–10^{13}	10^{12}–10^{14}
Dielectric strength (V/mil)	300–400	300–400
Dielectric constant 10^6 c/s	6–7	7–8
Chemical resistance		
Strong acids	Decomposed	Decomposed
Weak acids	Resistant	Resistant
Strong alkalis	Attacked	Attacked
Weak alkalis	Slightly attacked	Resistant
Solvents	Resistant	Resistant

APPLICATIONS

Although urea-formaldehyde resins were first used as molding powders it is not this application which absorbs the greater part of the resins used today. Many of the end-uses which were once exclusively applied to urea-formaldehyde resins are now being replaced by thermoplastic materials. The reason is not always due to material performance, but with thermoplastics the injection molding techniques allow for rapid cycles and therefore higher production rates. However, urea/formaldehyde moldings powders are still used extensively for the production of electrical accessories due to the fact that in addition to having the required mechanical strength and rigidity they can also be molded round metallic inserts. Urea-formaldehyde molding powders are one of the lowest priced nonflammable and nontracking molding materials at present available. Urea-formaldehyde molding powders are cheaper than melamine and are, therefore, generally favored for nearly all electrical accessories except for lampholders, for which MFs are used because of their greater resistance to embrittlement on prolonged exposure to high temperatures.

Urea-formaldehyde is used quite extensively for closures in the cosmetic industry because of its wide color range, its high degree of resistance to solvents, greases and oils, and its surface hardness. Such properties also make it eminently suitable for use in domestic goods such as hair dryer and mixer housings, clock cases, lavatory seats, and even buttons.

The application for which the melamine-formaldehyde molding powders are best known, and which accounts for probably the largest single outlet, is that of fine-quality tableware. The moldings are tasteless, odorless, nontoxic and will withstand staining from coffee, tea, and fruit juices. For this application the molding powders are filled with α-cellulose. Other uses include cutlery handles, aerosol dispensers and domestic kitchen-utensil housings.

It is probably true to say that the product life of the aminoplasts has been revitalized with the development of the technology for the injection molding of thermosets. This new technique has so speeded up the molding process that these older materials can now compete with the newer thermoplastics on an equal footing.

Suggestions have been made in the literature that cellulose-filled molding powders based on a cocondensation product of melamine-benzoguanamine and formaldehyde give moldings with better water resistance as well as resistance to stain by tea and coffee than melamine-formaldehyde. It has been mentioned previously that benzoguanamine is more expensive than melamine and thus these cocondensates are more costly molding materials.

The amino resins have now gained wide acceptance in the wood, furniture, and allied industries. They are used largely in the production of plywood and 'chipboard', and their use in general constructional work is of some importance. The melamines are preferred where better strength and water-resistant characteristics are required.

The main outlet for these resins in the laminating field is in the production of decorative paper-based laminates. For this particular application the melamines are the preferred materials because of their good water- and heat-resistance. Little description is necessary for the wide application of such laminates for surfacing wood, plaster and 'hardboard.' Melamine formaldehyde resins are also used in conjunction with glass-mat and cloth, cotton, asbestos, and certain synthetic fibers. Such laminates in general possess high mechanical strength, good heat-resistance, and good electrical insulating properties; for these reasons they are used in printed circuit panels and electrical switchgear.

The amino resins have also been used quite widely in the textile industry, originally to improve the crease-resistance of cellulose fabrics. The now well-known 'drip-dry' fabrics owe their existence largely to such resins. Fabrics can also be made partially waterproof by treatment with certain amino resins, although the silicone treatment is, in general, now to be preferred.

The tanning process of leather can be greatly accelerated by treatment with solutions of methylol melamines or other low condensed amino resins.

Both urea and melamine-formaldehyde resins have been used to overcome the well-known disadvantage of paper, that of losing its strength when wet. It has been found that treatment with melamine resins makes it longer-lasting though more expensive.

In conjunction with alkyds and acrylics they provide coatings which can be baked to give highly durable and mar-resistant finishes, indoor applications of which include coatings for kitchen and refrigerator cabinets. Melamine/alkyds and acrylics have also been used in automobile finishes, being applied by spray techniques and subsequently stoved (cured) at temperatures around 130°C. Butylated benzoguanamine resins are also used for stoving enamels and when plasticized with oil modified alkyds they cure more slowly than urea-formaldehydes and melamine-formaldehydes, though they give films with better gloss and alkali resistance.

Mention should also be made of another application of urea-formaldehyde, a foamed material which is made by vigorously mixing an aqueous solution of a detergent, an acid catalyst, with a urea-formaldehyde condensate. After foaming, the material is discharged from the mixer, where it sets rapidly, and is then allowed to dry out slowly. Densities around 0.8 lb/ft^3 can be obtained. Prefoamed slabs are used as thermal insulants for refrigerators and houses. A disadvantage of this material is the difficulty of handling it due to its relatively low strength. This is partly overcome in buildings where it can be foamed *in situ* between walls or under floors to improve thermal insulation.

UF foams were originally developed in Germany around the 1940's as heat insulating materials, using the trade name "IPORKA." It was not until some ten years later that this application came to the forefront of the insulation industry.

Not only can this foam be used in new brick houses but it can also be used in older types by drilling holes in the bricks and filling the cavities by injecting foam from a gun.

These foams are cheap, nonflammable and hydrophobic with low densities (0.5 lb/ft^3) although more dense products can be made by using more concentrated resins solutions and different hardeners.

Other uses for these foams are holding blocks for floral arrangements, artificial snow for film and television scenery, also arrester beds at the end of airway runs.

SUGGESTED FURTHER READING

Developments with Thermosetting Plastics. (1975) Edited by A. Whelen and J. A. Brydson (London: Applied Science Publishers Limited).

Thermosetting Materials and Processing. (1994) Edited by J. F. Monk (London: Longman).

5.

POLYESTERS

The name polyester is used to describe a diverse group of materials formed as a result of a condensation reaction between a polyhydric alcohol and a polybasic acid. The history of these materials can be traced back to as early as 1847, when a Swedish scientist, J. J. Berzelius, obtained a white amorphous solid of a resinous nature on heating glycerol with tartaric acid. The reaction was subsequently modified later, in England, by replacing the tartaric acid with phthalic anhydride (acid). This particular work was carried out by Watson Smith who, in 1902, obtained a wide range of resinous products from such reactions, which he named 'glyptal.' The name was in use until made virtually obsolete by the introduction of 'alkyd,' which is derived from a combination of the words ALcohol-aCID. The term 'alkyd' was originally coined by R. H. Kienle who carried out many classical polymerization studies on polyesters and is now used specifically to refer to polyesters used for surface coatings and molding powders.

A great advance was considered to have been made in 1937 when Carlton Ellis, in America, discovered that by adding unsaturated monomers to polyesters the setting (curing) times was substantially reduced. This made the polyester resins attractive for use on a commercial basis, in addition to stimulating further interest.

The industry founded on polyester resins has been relatively slow in growing although, from the period 1919–1926, there was considerable industrial development of these resins in varnishes and paints. The use of polyesters for purposes other than surface coatings can be considered to have started at the outbreak of the Second World War when the first industrial uses of polyesters as laminating resins were seen in the U.S.A. Developments were even later in the U.K., for it was not until 1946 that polyester resins were first offered to specialized industries such as the aircraft industry. Alkyd molding materials were first produced in the U.S.A. in 1948 and then later in the U.K. (1952).

Since the early 1960's we have seen the introduction of several new types of polyesters. The allyl and alkyd resins have developed in the molding field and the use of water soluble and air drying alkyds have also become well established. In recent years we have seen the introduction of aromatic polyesters for heat temperature applications and in the 1970's the vinyl esters were introduced into the field of laminating and casting. Other polyesters were also developed that had a linear structure. These thermoplastic polyesters are dealt with in Chapter 15.

CLASSIFICATION

Polyesters can be broadly classified into unsaturated, saturated, and the more recent vinyl ester polymers. These broad divisions can then be further subdivided as follows:

1. Unsaturated

(a). Laminating and Casting Resins. Probably the oldest member of this group is the type based on the reaction product of dibasic acids and dihydric alcohols. The polyester units so formed must be capable of copolymerizing with a vinyl-type monomer, thereby yielding a vinyl-polyester copolymer or simply a cured polyester with a thermoset structure. Included in this group are also the allyl resins based on diallyl phthalate and diallyl isophthalate. These have found wide usage casting as well as laminating resins.

The newest members of this group are the resins based on vinyl ester. These form a range of thermosetting resins which find wide use in fabricating reinforced plastic structures. While these materials may be regarded as unsaturated, the unsaturation does not occur along the polymer backbone and is terminal. The structure of these resins is discussed more fully later but basically it is as follows:

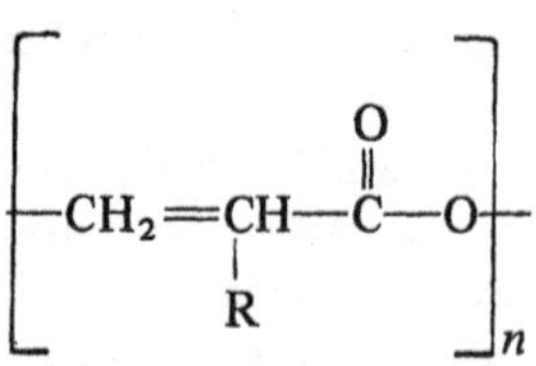

(b) Surface Coatings. The alkyds are the oldest member of this particular group and probably the oldest commercially available type of polyester. These

resins are, in general, of the same type as (a) although the 'glyptal' (surface coatings) types are modified with oils or fatty acids.

Considerable work has been carried out to develop solventless wood lacquers in attempts to replace nitrocellulose. Single coats give a very high gloss surface but they suffer from air inhibition and remain permanently soft and tacky. This tackiness can be overcome by the addition of wax but the result is a dull finish. Wax-free polyester resins have now been developed and an early technique was employed to modify the polyester with a glyceryl a-allyl ether. An interesting product which has found some use in this application is the tetra allyl ether of sorbitol.

In recent years work has been carried out for developing water soluble alkyds. The approach was to build acid groups in excess by the use of polyfunctional acids such a trimellitic anhydride and then to neutralize the residual acid groups with an amine thereby rendering them water soluble.

2. *Saturated*

(a) Fibers and Films. These are based in the reaction of terephthalic acid with ethylene glycol and are linear, high molecular weight polymers which do not undergo any cross-linking reactions. They are dealt with in Chapter 15.

(b) Thermoplastic Molding Resins. The poly(ethylene terephthalate) (PET) polymers described in 2(a) have expanded their uses into applications other than films and fibers, in particular in stretch blow molding. Other polyesters, such as poly(tetramethylene terephthalate)—P.T.M.T.—have been used in injection molding. These are also dealt with in Chapter 15. There are also aromatic polyesters and these are described in the chapter on Heat Resistant Thermoplastics (Chapter 17).

(c) Molding Materials. Alkyd is also used to describe a group of thermosetting molding materials based on the reaction of a dihydric alcohol with an unsaturated acid such as maleic or fumaric in place of the more conventional phthalic acid (anhydride). A vinyl-type monomer is also necessary to effect speedy cross-linking and curing. These materials, when suitably modified with certian fillers, are used as molding powders for compression and transfer-molding techniques similar to those used for the pheno- and amino plasts.

The resins discussed in 1(a) for casting and laminating can also be used in conjunction with chopped glass fiber for dough molding compounds (DMC) as can be the resins based on allyl derivatives and vinyl esters.

(d) Polyester/Polyurethanes. Certain polyesters having a high hydroxyl content can be reacted with various isocyanates to form polyurethanes, which are used extensively as foams, elastomers, surface coatings and adhesives.

Such polyesters are high viscous liquids or waxy solids at room temperature.

Polyesters formed from the ring opening of lactones such as the polycaprolactone from Union Carbide may be used as flexibilizers or the high molecular weight types as modifiers for high density polyethylene.

(e) Plasticizers. These can be polyesters which are completely saturated, normally referred to as polymeric plasticizers. A description of them is given later in this book, under the relevant section on plasticizers.

RAW MATERIALS

The many base raw materials used for manufacturing polyester resin can at times be confusing and, therefore, the preparation of the more important constituents is described.

Saturated Alcohols (Glycols)

(a) Ethylene glycol $[CH_2)_2(OH)_2]$, is the first member of the glycol series which contains no double bond unsaturation. It is derived from ethylene via ethylene dichlorhydrin and ethylene oxide, or straight from ethylene oxide. The ethylene itself is obtained by the catalytic cracking of petroleum.

(i)

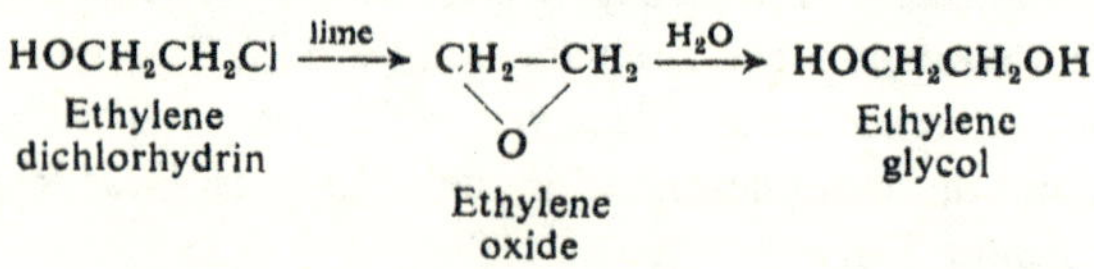

(ii)

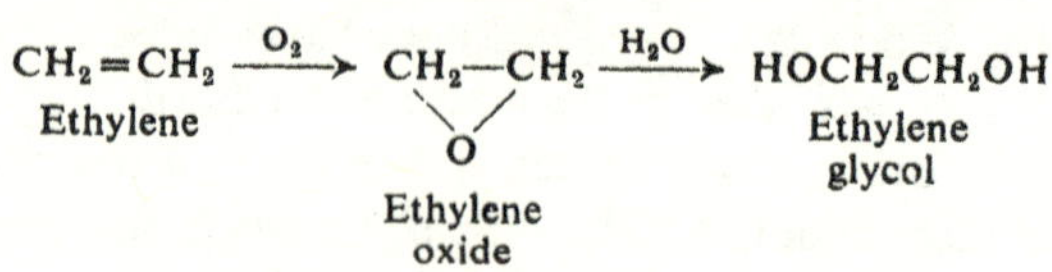

(b) Propylene glycol $[CH_3CHOHCH_2OH]$. Two isomers exist for this material although only 1, 2-propanediol is commercially available. Propylene glycol is obtained from propylene via a similar set of reactions as for ethylene glycol. Propylene is a further product of the catalytic cracking of petroleum.

(c) Diethylene glycol $[HOCH_2CH_2OCH_2CH_2OH]$ is produced either as a by-product in the manufacture of ethylene glycol or by reacting the latter material with ethylene oxide.

(d) Glycerol $[OHCH_2CHOHCH_2OH]$ is used in the production of polyester

resins employed as surface coatings. Traditionally, glycerol has been produced from vegetable oils as a by-product during saponification:

$$\text{vegetable oil} \xrightarrow[H_2O]{NaOH} \text{glycerol} + \text{soap}$$

An entirely different method is also used and is based on molasses, as follows:

$$\underset{\text{(molasses)}}{\text{sugar}} \xrightarrow[\substack{\text{high pressure} \\ \text{catalyst}}]{\text{hydrogen}} \text{glycerol} + \text{other polyols}$$

The most common method, however, is to use propylene which is still the preferred commercial method. The propylene is first converted to allyl chloride, then to glycerol dichlorhydrin, and finally to glycerol.

There are many other saturated alcohols used in the manufacture of polyester resins although space does not permit a lengthy description. These include dipropylene glycol, mannitol, sorbitol, and 1, 4 butanediol.

Mannitol and sorbitol are related to the sugar family but apart from these the remainder are products of the ever-growing petrochemicals industry. The largest two feedstocks are propylene and ethylene, previously discussed.

A reaction which can provide many of the raw materials for polyester production is an aldol-type condensation. Materials such as neopentyl glycol, trimethylol propane, hexanetriol, penta erythritol, and many more can be produced by this reaction. The actual reaction involves aldol-type condensations between preparations of formaldehyde and other low molecular weight aldehydes. The formaledehyde takes part in the condensation and also reduces the 'end' aldehyde group.

$$2CH_2O + CH_3 \cdot CH_2 \cdot CHO \xrightarrow{\text{alkali}} (HO \cdot CH_2)_2C \cdot Me \cdot CHO \cdot$$

$$(HO \cdot CH_2)_2CMe \cdot CHO + CH_2O + MeOH \longrightarrow$$

$$CMe(CH_2 \cdot OH)_3 + H \cdot CO_2Me.$$

Another interesting reaction is in the manufacture of 1–4 butane diol. This involves the hydrogeneration of 1, 4–butyne diol using a nickel-copper manganese catalyst. The 1, 4–butane diol is first obtained from the reaction between acetylene and formaldehyde.

With such a multitude of raw materials the polymer chemist has a wide range of functionalities and molecular weights to enable him to build in many different characteristics into the polyester resins. Commercial materials with functionalities (f) varying from 2 up to 6 are used. The following are some of the alcohols which are used showing their functionality.

(i) $HOH_2C{-}CH_2OH$

$f = 2$

Ethylene glycol

(ii)

$$\begin{array}{c} H \\ | \\ HC{-}OH \\ | \\ HC{-}OH \\ | \\ HC{-}OH \\ | \\ H \end{array}$$

$f = 3$

Glycerol

(iii)

$$\begin{array}{c} CH_2OH \\ | \\ HOH_2C{-}C{-}CH_2OH \\ | \\ CH_2OH \end{array}$$

$f = 4$

Pentaerythritol

(iv) $HOCH_2(CHOH)_4CH_2OH$

$f = 6$

Sorbitol

Unsaturated Alcohols

Allyl alcohol [$CH_2 = CHCH_2OH$] is prepared by the oxidation of propylene to acrolein, followed by reduction or by the isomerization of propylene oxide.

(i)

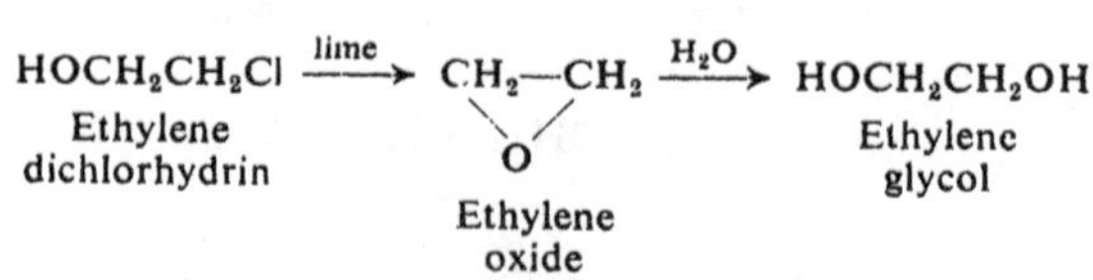

(ii)

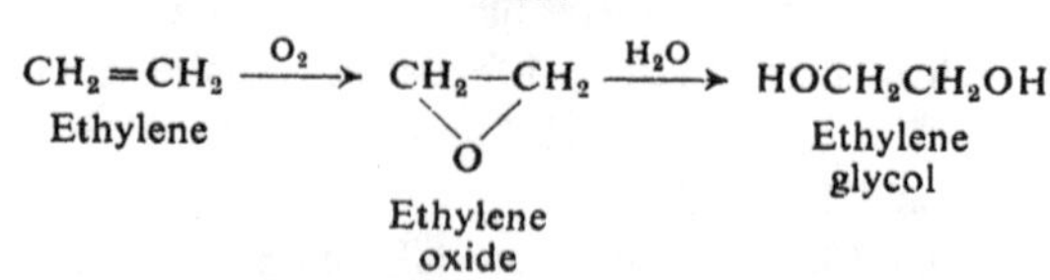

Other saturated alcohols used in polyester production include 2 butene—1,4 diol, and 2,5 dimethyl 3 hexyne, 2,5 diol and 3,6 dimethyl 4, octyne—3,6 diol.

Saturated Acids

(a) Phthalic acid [$C_6H_4(COOH)_2$] is normally produced and used in the form of the anhydride by the controlled oxidation of naphthalene or *o*-xylene.

(b) Adipic acid [$HOOC(CH_2)_4COOH$] is also an important raw material for the manufacture of polyamides, and its production is covered in Chapter 12, along with sebacic acid.

(c) Sebacic acid [$HOOC(CH_2)_8COOH$] is normally prepared from castor oil by treatment with hot caustic soda. The castor oil is 80 to 90 percent the glyceride of ricinoleic acid:

$$\left[CH_3(CH_2)_5CHOHCH_2CH{=}CH(CH_2)_7COO\right]_3C_3H_5 + 6NaOH + 2H_2O \xrightarrow{\text{Heat}}$$

$$\underset{\text{2-Octanol}}{3CH_3(CH_2)_5\underset{\underset{OH}{|}}{C}HCH_3} + 3NaOOC(CH_2)_8COONa + \underset{\text{Glycerol}}{C_3H_5(OH)_3} + 3H_2$$

$$3NaOOC(CH_2)_8COONa \xrightarrow{HCl} \underset{\text{Sebacic acid}}{HOOC(CH_2)_8COOH}$$

(d) Azelaic acid [$HOOC(CH_2)_7COOH$] is made by the ozonolysis of oleic acid:

$$CH_3(CH_2)_7CH{=}CH(CH_2)_7COOH \xrightarrow[\text{Solvent}]{O_2,\ O_3,\ H_2O}$$

$$\underset{\text{Pelargonic acid}}{CH_3(CH_2)_7COOH} + \underset{\text{Azelaic acid}}{HOOC(CH_2)_7COOH}$$

Like the polyfunctional alcohols it is also possible to use polycarboxylic acids such as

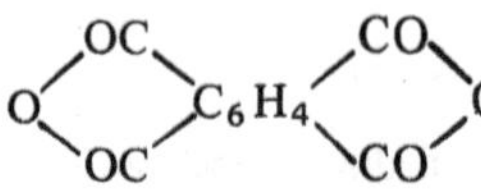 1,2:4,5

$f = 4$
Pyromellitic anhydride

(i)

$C_6H_3(COOH)_3$

$f = 3$
Benzene tricarboxylic acid

(ii)

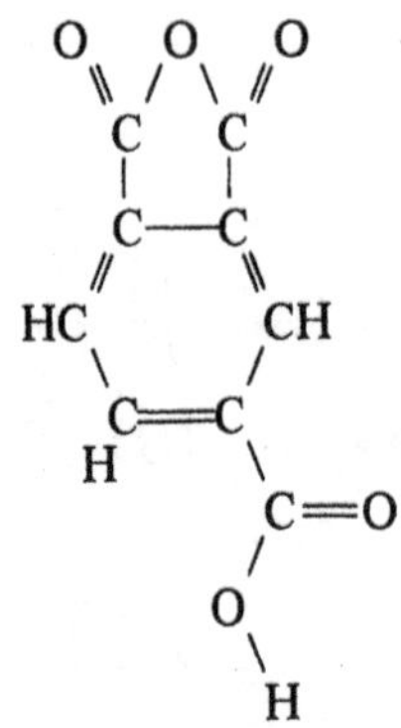

Trimellitic anhydride
$f = 3$
(iii)

Unsaturated Acids

The most commonly used acids are the *cis-trans* isomers maleic acid and fumaric acid. Maleic acid is used in the form of its anhydride.

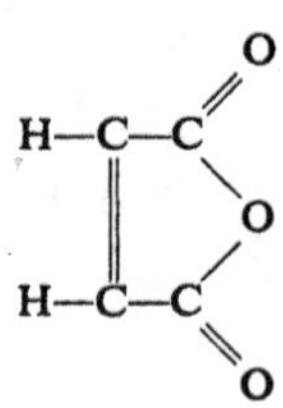

Maleic anhydride
(i)

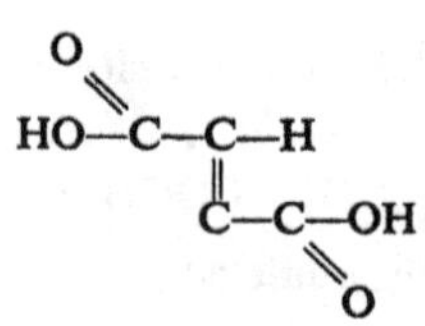

Fumaric acid
(ii)

Maleic anhydride is produced commercially by the catalytic oxidation of benzene with a vanadium catalyst or by the vapor phase oxidation of butylene. Fumaric acid may be produced either by heating maleic acid or as a by-product in the manufacture of maleic anhydride. Both these unsaturated acids are relatively cheap and are the most popular of this group mainly due to price and the fact that more expensive acids do not offer significant advantages in properties. One exception, however, is chlorendic anhydride (hexachloro endomethylene tetra hydro phthalic acid), which is prepared by a Diels-Alder reaction between maleic anhydride and hexachlorocyclopentadiene. This product possesses double bonds which are not reactive under the usual conditions for preparing polyesters and they are therefore used in conjuction with maleic anhydride or fumaric acid in order to achieve cross-linking. Such a material is used when a high degree of flame resistance is required. The chlorendic anhydride in actual fact rapidly absorbs water from the air becoming "chlorendic acid" or more commonly referred to as HET acid.

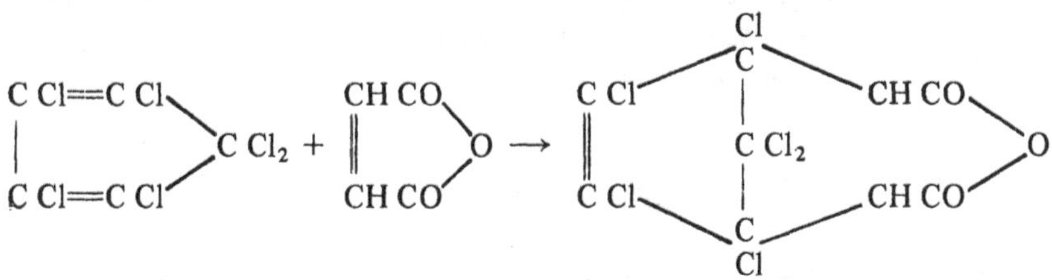

Chemistry of the Production of Polyesters

There is considerable literature and patents associated with this subject but the following illustrates the most common reactions available.

(a) Acid and glycol.

$$HO_2C \cdot R \cdot CO_2H + HOR'OH \rightleftharpoons HO_2C \cdot R \cdot CO_2 \cdot R'OH + H_2O$$

(b) Ester interchange.

$$MeO \cdot OC \cdot R \cdot CO \cdot OMe + HO \cdot R' \cdot OH \rightleftharpoons$$
$$MeO \cdot OC \cdot R \cdot O \cdot R'OH + MeOH$$

(c) Acid chloride and glycol.

$$Cl \cdot OC \cdot R \cdot COCl + HO \cdot R' \cdot OH \longrightarrow$$
$$Cl \cdot OC \cdot R \cdot CO_2 \cdot R'OH + HCl$$

The first two reactions are reversible and continued reaction only occurs if the water or alcohol of condensation is removed completely. The ester interchange process is used mainly for the preparation of saturated polyesters of very high molecular weight suitable for fibers, film, or moldings.

Low molecular weight polyesters used for polyurethane technology and flexibilizing cycloaliphatic epoxies are produced by the simple direct esterification process. These are regarded as diols and include such materials as Desmophen 2000 (from Bayer Chemicals, Germany) and Niax Polyols (Union Carbide, U.S.A.).

CROSS-LINKING MONOMERS

It has been stated previously that for certain applications of polyester resins it is essential to convert the so-called linear polyester structure into the three-dimensional one commonly associated with thermosetting resins. This is normally achieved by adding a polymerizable vinyl type monomer; the more popular ones are listed in Table 5.1.

RESIN PREPARATION

Normally, all polyesters are prepared by using the same basic type of condensation reaction carried out in a resin kettle similar to the one described for phenol-formaldehyde in Chapter 3. The main differrence is that the esterification reaction is carried out under a blanket of inert gas, such as nitrogen or carbon dioxide, whereas this precaution is uncesessary for pheno- and aminoplasts.

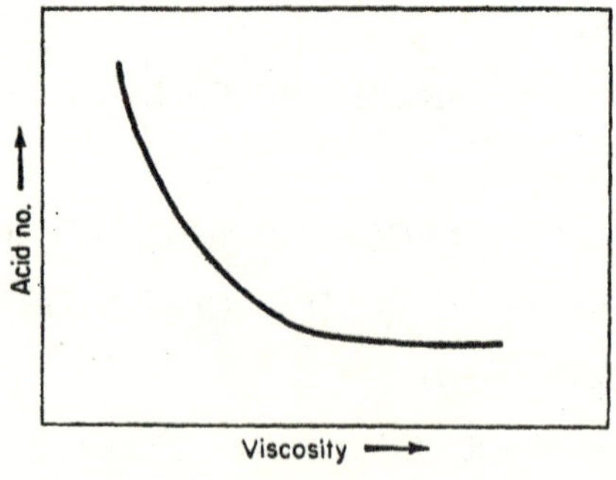

Fig. 5.1

Table 5.1

REACTIVE MONOMERS FOR POLYESTERS

Monomer	*Formula*	*Characteristics*
Styrene	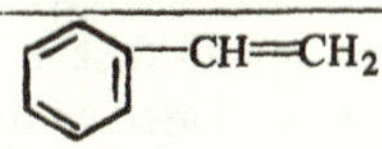	Relatively cheap, reduces viscosity, compatible
Methyl methacrylate	$CH_2{=}C(CH_3){-}CO{-}OCH_3$	Good weathering
Vinyl toluene	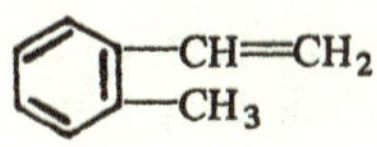	Similar in chemical behavior to styrene–low volatility
α-Methyl styrene	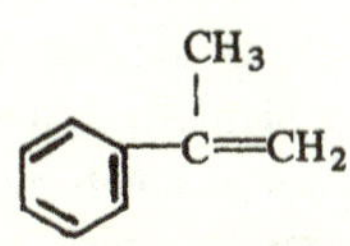	Reduces reactivity
Divinyl benzene	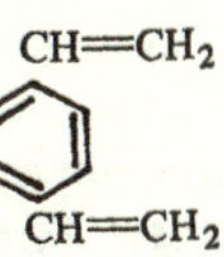	Difficult to blend with the polyester–reduces volatility
Dichlorostyrene	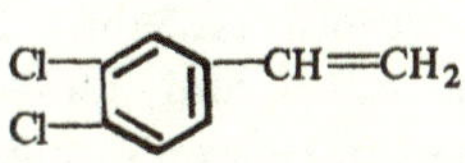	Good heat and fire resistance
Diallyl phthalate	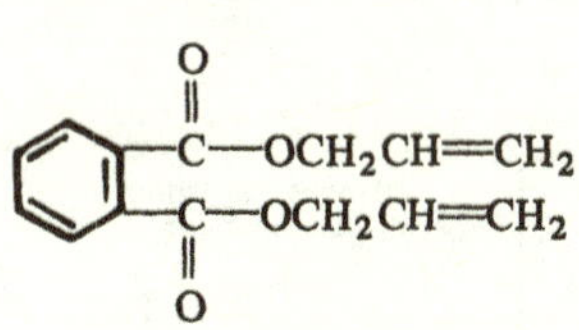	Slow-curing–flexible–low volatility
Triallyl cyanurate	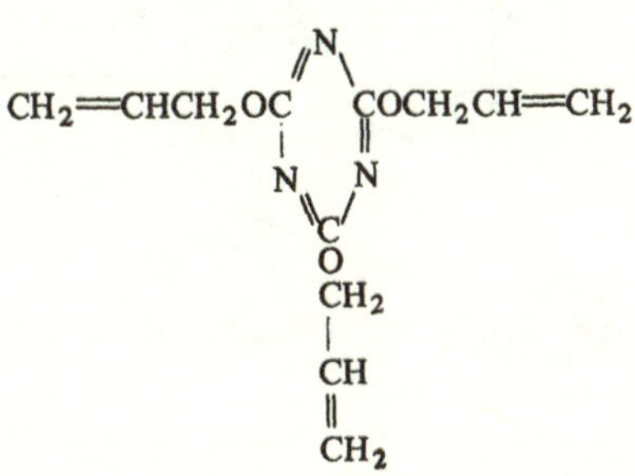	Heat resistance

The alcoholic and acidic constituents are first charged into the resin kettle and the temperature raised slowly to 140 °C, with continuous stirring. The oxygen present in the vessel is then flushed out by the inert gas.

Sometimes esterification catalysts are introduced at the beginning of the reaction which brings about a considerable reduction in the reaction time. Suitable catalysts are *p*-toluene sulfonic acid or sulfuric acid. The reactants are refluxed and an inert solvent such as xylene is added to distill off azeotropically the water of reaction. The temperature is then slowly raised to 190 °C until the acid value of the resin reaches 38–45. The acid value is an indication of the amount of free unreacted acid remaining, and is expressed as the number of milligrams of potassium hydroxide required to neutralize one gram of sample. The vicosity increases throughout the reaction and the reaction is halted at a predetermined acid value. The relationship between viscosity and acid value can be seen in Fig. 5.1.

On reaching the required acid value the reactants are cooled to below 150 °C, when the resin is ready for mixing with the reactive monomer. These resins, when dissolved in this monomer are highly reactive, and the cross-linking reaction can be started off by a free radical mechanism, and will gel. This can, however, be stopped by adding chemicals which prevent the free radicals from reacting. Such materials are quinone, hydroquinone, catechol, and tertiary butyl catechol.

Thus, before adding the resin from the kettle to another kettle containing the reactive monomer it is essential first to dissolve an inhibitor in the monomer. When the inhibitor has been well dispersed the polyester resin is added and stirred thoroughly. Finally, the resin is ready for discharging into containers. A flow diagram of the process can be seen in Fig. 5.2.

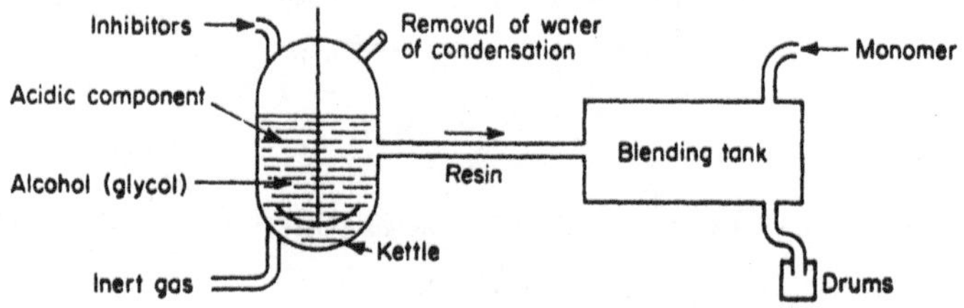

Fig. 5.2 Flow diagram for production of a polyester resin

The following formulation has been found suitable for the production of polyester resins for use in laminating and casting applications:

Materials		*Quantity (parts)*
Propylene glycol		160
Maleic anhydride		115
Phthalic anhydride		85
Xylene		20
Hydroquinone	inhibitors	0.5
Quinone	inhibitors	0.005
Styrene monomer		150

POLYMER STRUCTURE AND CHEMICAL TECHNIQUES

Laminating and Casting Resins

The typical procedure for the preparation of a polyester resin has been described earlier. The chemical reaction mechanism of the typical formulation quoted can now be looked at more closely.

The liquid resin obtained by such a reaction can be converted into the solid state by a further chemical reaction termed copolymerization. This latter term is normally used more with thermoplastics than with the thermoset type of resins. The copolymerization is effected by dissolving the resin in a reactive solvent which reacts with the unsaturated groups to form a three-dimensional polymer. The cross-linking that occurs is activated by a free radical mechanism which causes opening of the double bonds. Materials used for this reaction are sometimes called catalysts, although in the strictest sense the term is incorrect because they do, in actual fact, take part in the reaction and do not remain unchanged structurally.

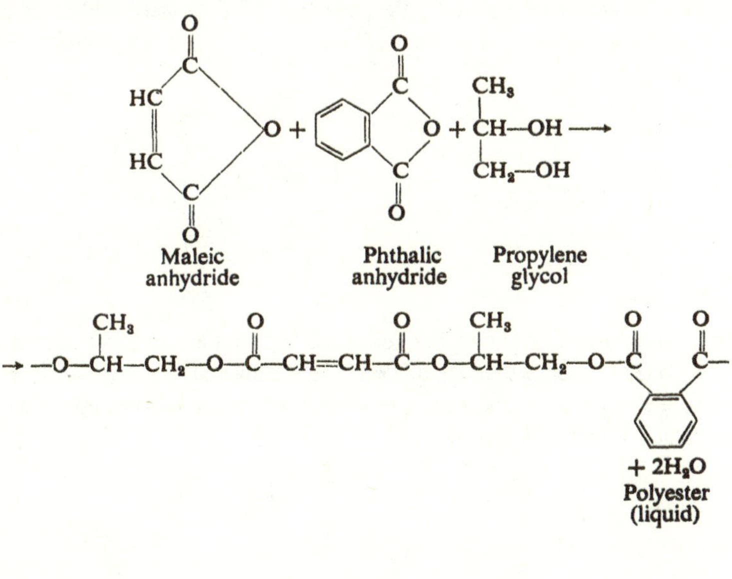

They are now more generally referred to as inhibitors and are usually organic peroxides. Two very popular materials are benzoyl peroxide and methyl ethyl ketone peroxide. The speed of dissociation of these peroxides can be accelerated by the use of certain metallic salts or amines. The most popular accelerator for general use is cobalt naphthenate, which is normally used in conjunction with methyl ethyl ketone peroxide.

The polyesters can be cured either by hot or cold techniques, and for the former, benzoyl peroxide or tertiary butyl hydroperoxide are preferred while, for the cold-cure, methyl ethyl ketone peroxide or cyclohexanone peroxide are more commonly used.

The overall curing action can be schematically represented as follows, where M represents the maleic residue of the previous reaction described, P the phthalic residue, G the gylcol, and X the reactive monomer (such as styrene):

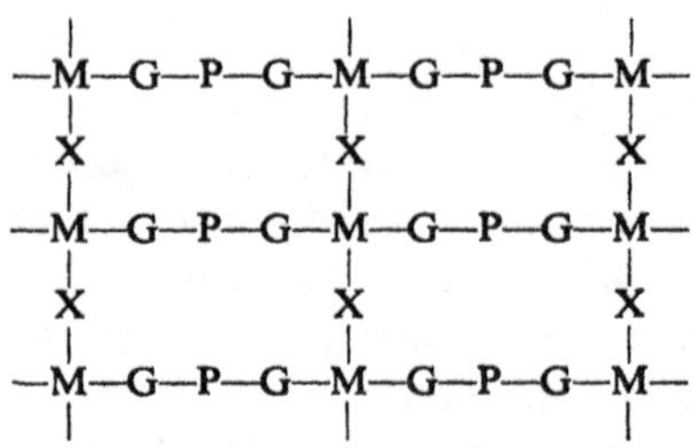

(cured polyester resin)

Vinyl Esters

A typical example of the molecular structure of a typical vinyl ester can be achieved by reacting two mols of methacrylic acid with a polyepoxide. The structure is as follows:

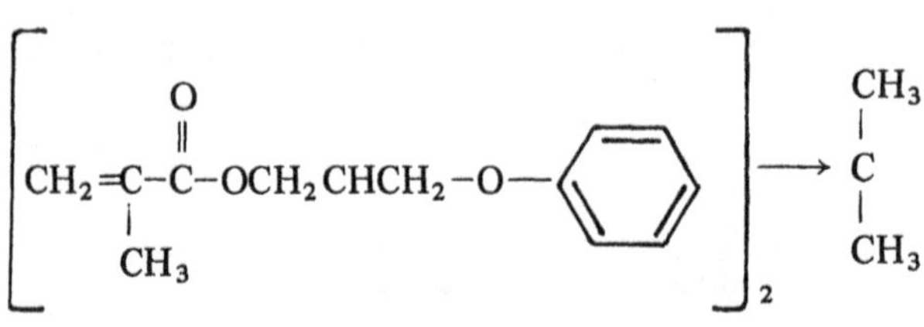

The Derakane resins (Dow Chemical Co., U.S.A.) are the best known of this family and in practice they are diluted, like the unsaturated polyesters, with styrene to about 45–50% and curing is carried out using cobalt naphthenate, peroxides and dimenthyl aniline.

Let us examine the molecular structure of a vinyl ester in comparison to a bisphenol *A*-fumaric acid polyester and a isophthalic acid polyester. It can be seen that chemical attack on all these types of resins occurs through hydrolysis of ester groups or the splitting of carbon to carbon double bonds (unsaturation). Such attack could be achieved by oxidation, halogenation, or by exposure to electron beam. As can be seen from the schematic structure of the cured bisphenol *A*-fumaric acid polyesters and isophthalic polyesters, ester linkages (O=C−O) occur throughout the molecular chain, making the resin very susceptible to attack by hydrolysis. In addition to these ester groups there are carbon to carbon (−C=C−) double bonds randomly spaced along the two nonvinyl polyesters. These double bonds are also very susceptible to attack but the vinyl ester resins have double bonds at the end of the molecular chain which react completely on polymerization, giving a more chemically resistant structure.

(a) Vinyl ester.

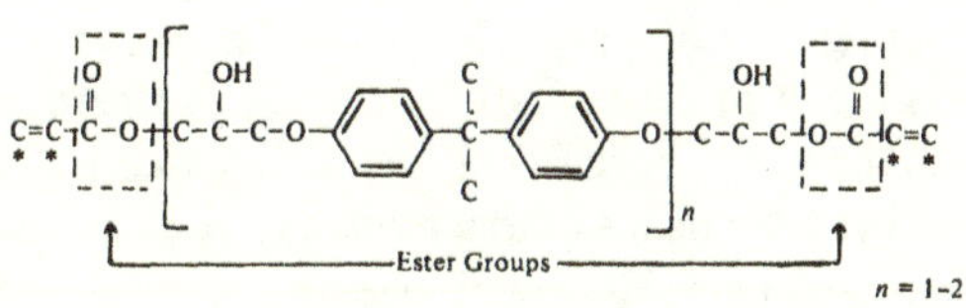

(b) Bisphenol A-fumaric acid polyester.

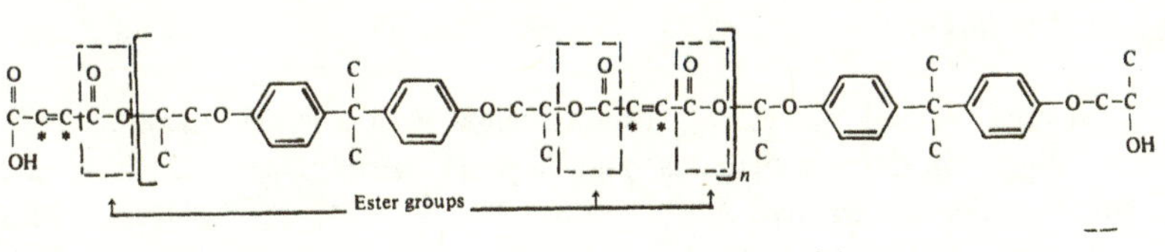

(c) Isophtalic polyester.

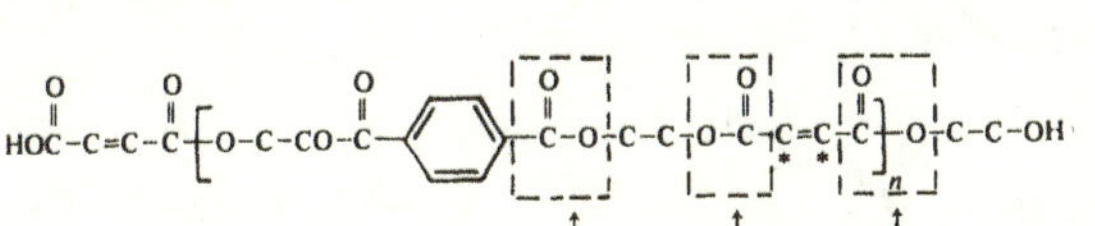

** denotes reactive sites*

It has also been established that the number and arrangement of polar groups on vinyl ester resins make them less polar than polyester resins. They are thus less prone to the effects of water when cured by organic peroxides such as benzoyl peroxide. Any water present in peroxide cured polyesters affects the quality of the finished product. The water is absorbed at the polar sites but with the vinyl esters absorption does not occur and the cure is unaffected.

The unsaturation on the vinyl esters is at the end and beginning of the chains and this makes the carbon-to-carbon double bonds exceptionally active. The Derakane resins therefore cure rapidly with superior physical characteristics.

The secondary hydroxyl groups on the vinyl ester molecule also have a beneficial effect on the quality of finished laminates. The hydroxyl groups react with similar groups on the surface of the glass fibers, resulting in good surface wetting and good adhesion to the fibers.

It is also believed that because the cured vinyl ester resins contain only terminal cross-linking, the entire length of the molecular chain is not restricted and can therefore elongate under stress which makes it suitable to absorb mechanical or thermal shocks. The more conventional polyesters with the cross-linking taking place along the entire chain length are far more rigid and brittle. With these polyesters it is also necessary to use a reactive monomer but with the vinyl esters no monomer is needed and the resin cures by itself. This is another reason why vinyl ester resins of low monomer content have better corrosion resistance than the polyacid/alcohol esters.

POLYMER STRUCTURE AND CURING TECHNIQUE

Surface Coatings

Polyesters formed by reacting dibasic acids with dihydric alcohols consist of long-chain linear molecules, and they are thermoplastic. If, however, the functionality (*f*) is above 2 : 2, i.e. 3 : 2 as it would be if glycerol were reacted with phthalic anhydride, then more complex resins are produced, which are essentially thermosetting and are referred to as alkyds. This latter reaction can be represented diagrammatically as follows, 'G' representing glycerol unit and 'P' phthalic acid.

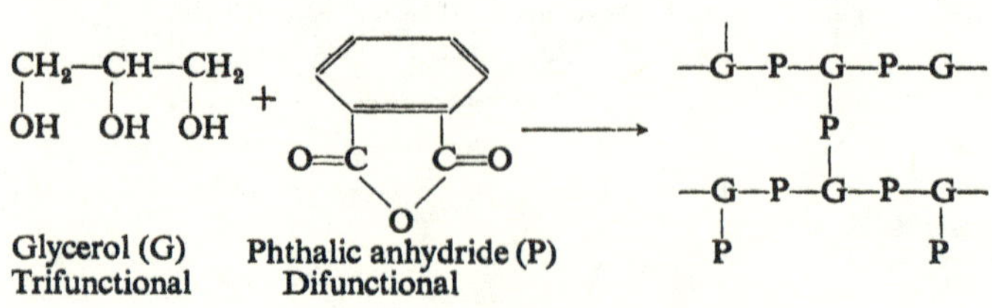

A further alteration to this structure can be achieved by oil modification, which is of immense importance to the paint industry. Oil-modified alkyd resins of the air-drying type cure by virtue of the oxidation of the drying oil added. The heat-cured types, in which the oil modification may or may not be oxidizing, cure by heat alone or can be cocondensed with amino resins. Non-drying oils are coconut and castor oil, and the drying oils include linseed oil and dehydrated castor oil. These reactions can be illustrated as follows:

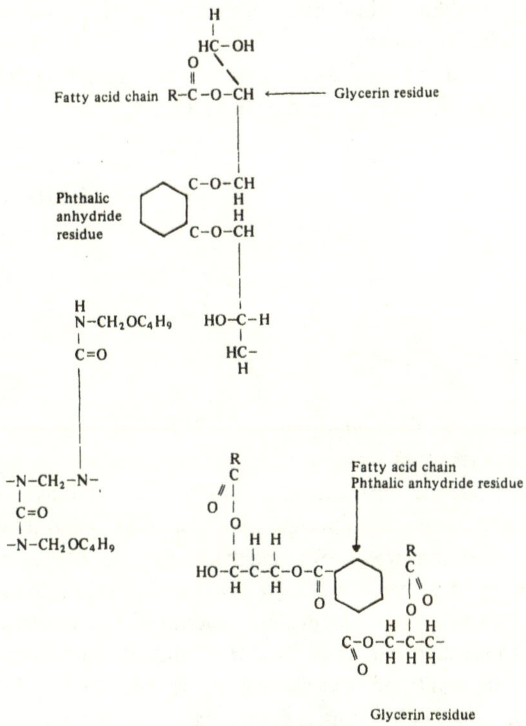

Butylated urea resin

Residual–OH groups or alkyd resin react with alkoxy groups of U–F or M–F resins to liberate alcohol and form an ether linkage.

The resultant modified surface coating resins are conveniently classified according to their oil length, which is approximately the ration of oil to resin.

Short oil alkyd . . .	20-40 percent oil
Medium oil alkyd . . .	41-60 percent oil
Long oil alkyd . . .	greater than 60 percent

Water Soluble Alkyds

Polyfunctionality of certain acids makes possible the synthesis of a wide variety of polymers useful in formulating paints and a wide variety of coatings. Particularly interesting is trimellitic anhydride (TMA) where the third functional carboxyl group is used as the solubilizing site.

The simplest water soluble resin prepared from TMA consists of polymer chains in which the third carboxyl group of the TMS molecule is left unreacted. This grouping is then neutralized by an amine or similar base which renders the polymer soluble in water.

A typical formulation for a baking enamel is as follows indicating the use of TMA as made by Amoco Chemical Corporation, U.S.A.

Materials	*Mol Ratio*
Trimellitic anhydride (TMA)	3
Neopentyl glycol (NPG)	8
Isophthalic acid (IPA)	2
Tall oil fatty acid (TOFA)	3

The alkyl resin is made by first charging the NPG, IPA, and approximately half the TMA into a reaction kettle similar to the one described on page 58. The reaction is then purged with nitrogen and partial condensation is begun. The ingedients are then heated to around 240 °C and the speed depends upon the ability to remove the water of condensation. The contents are maintained at a temperature of 95–98 °C until the acid number is less than ten. The TOFA is then charged when the temperature reaches 230 °C and reacted until the acid number is less than ten. Cool to 200 °C and add the remaining TMA. Hold this temperature until an acid number of 42–44. Cool to 160 °C and then reduce to 80 percent NVM using a suitable solvent such as butoxyethanol and adjust to a pH of 8.0–8.5 with dimethylamino ethanol.

A suitable baking formulation can be obtained by adding water to give a viscosity of 100–120 s No. 4 Ford Cup. Then the alkyd is mixed with a suitable amino resin, e.g., Cymel 301, in a ration 85/15 using titanium dioxide in a pigment/binder ratio 0.98/1.0. The resultant paint must then be ground in a pebble mill or similar machinery until the correct degree of dispersion has been achieved. The resultant product then gives satisfactory results when baked 30 minutes at 150 °C.

As previously discussed, solvent soluble air-dry alkyds were the first cross-linking coatings and are still used widely as exterior house paints as well as industrial coatings. The early attempts to incorporate water-solubilizing functionality into industrial drying-oil based resins resulted in coatings which took hours to dry and thus were unacceptable. Added to this was poor physical characteristics such as low hardness and poor outdoor weatherability. The use of linseed fatty acids and trimethylol propane as opposed to linseed oil with its natural glycerin content was found to yield air-dry alkyds with excellent hardness and flexibility and reasonable drying times.

Miscellaneous Polyesters for Surface Coatings

A technique which has commanded considerable interest in the coating industry is the electrophoretic technique. By means of electrical charges coatings can be applied to metal and both water soluble and solvent-soluble systems can be used. A water soluble formulation has already been discussed and the following is one suitable for both water and solvent techniques in electrophoretic coatings.

Material	*Mol ratio*
Tall oil fatty acid (TOFA)	6
Styrene allyl alcohol copolymer (SAAC)	2
Trimellitic anhydride (TMA)	2

The TOFA and SAAC is charged to a reaction kettle and then purged with nitrogen. The temperature is raised to 230 °C and maintained for four hours until an acid number of less than ten is reached. It is then cooled to 165 °C, the TMA is added, and the temperature is raised to 200 °C until an acid value of 38 is reached. The resin is then cooled to below 150 °C and is reduced to 80 percent NVM with butoxy ethanol and filtered while warm. The pH can be adjusted to 7.5 with triethylamine if it is required for water solubilization and then cured by a suitable amino resin.

Oil-free polyester coatings can also be made and they provide useful materials for coil coating and gloss enamels by baking techniques. A suitable formulation can be using either neopentyl or propylene gylcol with adipic and isophthalic acid as well as trimellitic anhydride. A hydroxyl excess of 15 percent is achieved and the polyester resin can be cured with amino resins in a ration of 4 : 1.

One other interesting group, which was touched on earlier, is the air drying polyesters which are wax free. One big problem in the early developments was air inhibition and consequently the surface remained tacky. One method which overcame this effect was to modify the polyester with glyceryl α-allyl ether or it is possible to replace the styrene with the adipic ester of glyceryl α, α' diallyl ether.

$$\begin{array}{ccc} CH_2CHCH_2OCH_2 & & CH_2OCH_2CHCH_2 \\ | & & | \\ & CHOOC(CH_2)_4COOCH & \\ | & & | \\ CH_2CHCH_2OCH_2 & & CH_2OCH_2CHCH_2 \end{array}$$

The liquid allyl ethers may be therefore used as reactive solvents in conjunction with conventional alkyds, forming solventless systems. The allyl ether groups may also be incorporated into water soluble alkyds and has provided a basis for a range of water thinnable air-drying and stoving media.

These allyl ethers dry by oxidation in air but in comparison with drying oils the polymerization mechanism involved is more efficient. Some of the simpler products yield films which tend to embrittle but films have the great attraction of a reduced tendency to yellow in the absence of light.

Other allyl ethers suitable for use are α-allyl glycerol, pentaerythritol diallyl ether and sorbitol tetra allyl ether.

A typical formulation for an air drying system is as follows:

Material	*Mol ratio*
Sorbitol tetra-allyl ether	5
Diethylene glycol	2
Adipic acid	2
Trimellitic anhydride	3
(Excess hydroxyl–8 percent)	

A water soluble type would be

Material	*Mol ratio*
Pentaerythritol diallyl ether	8
Adipic acid	2
Trimellitic anhydride	3
(Excess hydroxyl–23 percent)	

Molding Powders

It is now evident that glyptals (alkyds) are thermosetting resins, but unfortunately the time required to effect a final cure for certain applications is far too long to be practical. To obtain resins suitable for molding powders it has therefore been found necessary to replace the phthalic acid with an unsaturated material such as maleic acid. The resultant unsaturated polyester can then be made to cross-link rapidly via the ethylene linkages at temperatures in the region of 150 °C or even lower by using initiators such as benzoyl peroxide, as already mentioned. The curing action is also effected with an unsaturated monomer, such as styrene, vinyl toluene, or diallyl phthalate.

$$\underset{\text{Maleic acid}}{HOOCCH{=}CHCOOH} + \underset{\text{Ethylene glycol}}{OHCH_2{-}CH_2OH} + HOOCCH{=}CHCOOH \rightarrow$$

$$\underset{\text{Ethylene glycol maleate}}{{-}OOCCH{=}CH{-}COO{-}CH_2{-}CH_2{-}OOC{-}CH{=}CHCOO{-}}$$

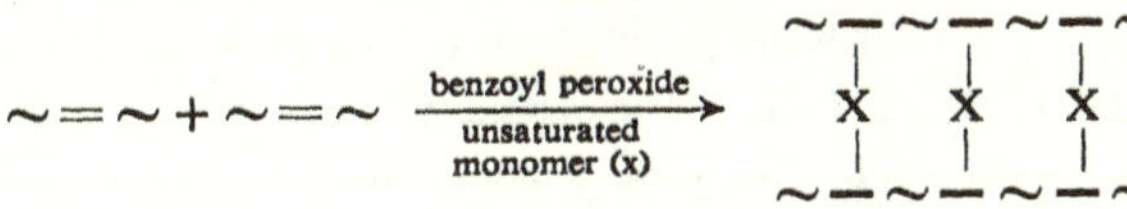

Molding powders have also been produced by using an unsaturated alcohol instead of the more conventional saturated type. For example, phthalic anhydride may be esterified with allyl alcohol to produce the ester, diallyl phthalate, which is cross-linked by using an organic peroxide and opening up the double bonds in the vinyl groupings.

This group of resins is referred to as the "allyl" family. The esters formed from mono or dibasic acids are available as low viscosity monomers and thermoplastic prepolymers. The allyl monomers and prepolymers find use as cross-linking agents for unsaturated polyester resins and in molding compounds. Typical monomers are triallyl cyanurate, diallyl chlorendate and diallyl maleate.

Typical commercial products include diethylene glycol bis (allyl carbonate), it is designated CR-39 Allyl Diglycol made by PPG Industries Inc., (U.S.A.). It can be prepared in the following manner:

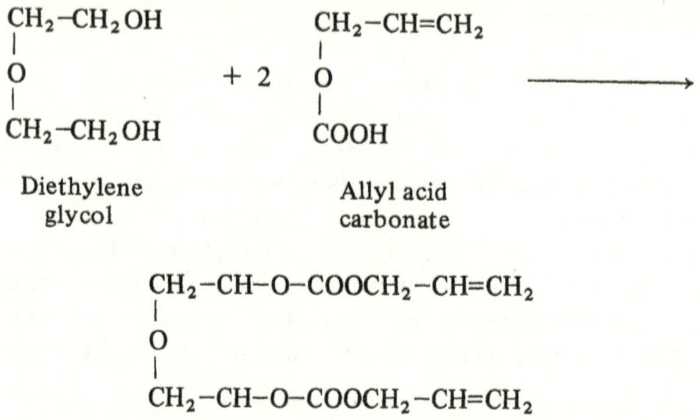

Other materials are Dapon 35 which is the prepolymer derived from diallyl orthophthalate and Dapon M is the isophthalate analogue (both manufactured by FMC Corp., U.S.A.). They are solid, linear, internally cyclized thermoplastics with unreacted allylic groups spaced at regular intervals along the polymer backbone.

The U.K. took another approach in attempts to produce low pressure resins. They utilized mixed esters in which the unsaturation was obtained either from acrylic acid ($CH_2 : CH\ CO_2H$) or methacrylic acid ($CH_2 : CCH_3\ CO_2H$). Such products included a dimethacrylate glycol phthalate using *n*-butyl methacrylate as the comonomer. These resins were relatively slow and difficult to make which restricted their exploitation.

Polyester/Polyurethanes. No elaboration is needed for this type of polyester because it is the standard reaction of a dicarboxylic acid with a dihydric alcohol. However, the structure of the polycaprolactones is more interesting and is described in a little more detail.

Polycaprolactones. It has been shown previously that polyesters can be made by the ring opening of lactones. A series of such resins have been made by the ring opening of ε-caprolactone. The polymers so formed are thermoplastic crystalline materials. These materials are manufactured by Union Carbide Corporation, U.S.A.

On such product is polycaprolactone diol (m.w. 10,000) which is designated PCL 300 with the following structure.

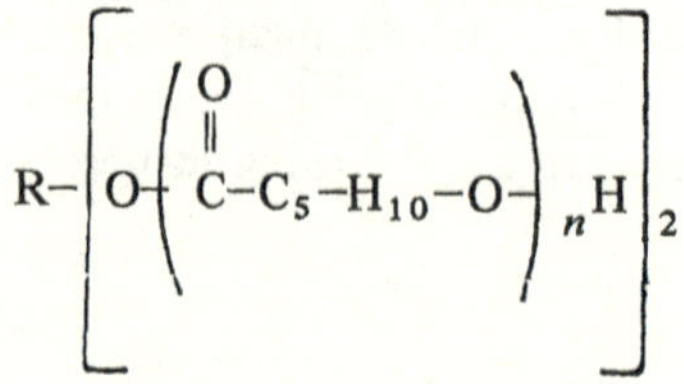

Another in this family with the same basic structural formula but with a higher molecular weight is PCL 700.

$$-\left((CH_2)_5-\overset{\displaystyle O}{\overset{\|}{C}}-O- \right)_n$$

PROCESSING TECHNIQUES

Surface Coatings

The processing techniques as applied to polyesters for surface coatings need very little further elaboration. Very nearly all gloss or enamel paints are today based upon alkyds, not white lead as was customary, and most readers will be fully conversant with the technique of applying them, as well as with the basic details of such mechanized techniques as spraying. Certain specialized techniques are employed in spraying alkyd paints and this process can be carried out either by air atomizing or airless spray equipment. A recent advance in this field is the introduction of a surface coating based upon a polyester resin, suitable for use with wood, which cures in the cold by oxidation or air inhibition.

Laminating and Casting Resins

In general, all thermosetting resin laminates are produced by impregnating the base material with a synthetic resin and then consolidating by applying heat and/or pressure. The degree of impregnation affects the characteristics of the final laminate, which is dependent upon such variables as the type of the base material and resin solvent (if any), resin characteristics (e.g., solids content and viscosity), or impregnation time. The production techniques can be conveniently classified as either (*a*) high pressure or (*b*) low pressure. The fact that no by-products are evolved on curing polyester resins, as with epoxy resins, means that the low-pressure process is more commonly used. The high-pressure process is used with polyesters for dough or gunk moldings, which is described later. The casting techniques are very similar for polyesters as are used for epoxy resins, which are described in detail in Chapter 6.

Molding Powders

The materials are processed by the conventional thermosetting molding techniques such as compression and transfer molding. Specific details of these processes are given in the relevant chapter.

LOW-PRESSURE PROCESS

The molding of polyester resins with glass fibers by this process is made simpler by virtue of the facts that no volatiles are evolved or curing needed, and that the reaction can be effected without heat. Pressures from zero to 500 lb/in.^{2} are all that is normally required to consolidate these low-pressure (contact) resins. The term reinforced plastics' has now become virtually synonymous with low-pressure or contact laminates.

1. Wet Lay-up Technique

A mold is first constructed out of wood, plaster of paris, metal, or epoxy resin reinforced and cheapened with aluminum powder. The surface of the mold must first be smoothed, polished and then treated with a release agent which may be a hard wax or a film former such as polyvinyl alcohol. A 'gel' coat is then applied, which is usually a polyester or an epoxy resin allowed to harden to a tacky state. At this stage the first layer of fibrous reinforcement is applied, the various types of reinforcement are described in Chapter 23. It is also possible to use thixotropic agents if vertical drainage is required. Further resin is then applied to other layers of reinforcement and worked into the interstices of the fibers, usually with a paint brush. This practice is repeated several times until the required thickness is obtained (Fig. 5.3). The resultant laminated surface may then be cured by placing it in a rubber bag and applying

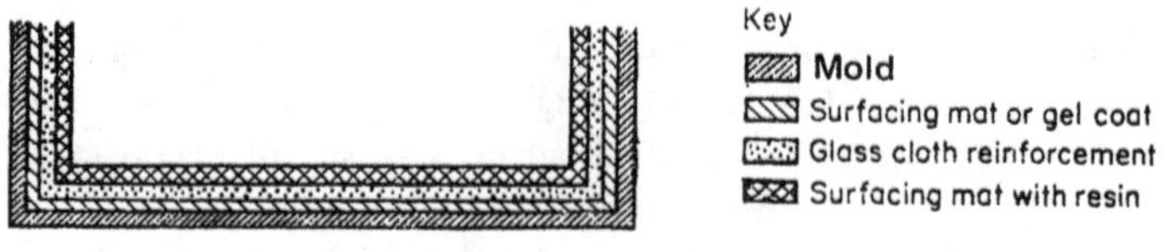

Fig. 5.3 Build-up of fiberglass reinforced laminate

a vacuum (the vacuum bag technique is described more fully later). This process is one of the most popular methods used for reinforced plastics because there is virtually no limit to the size or complexity of the final molding. Mold costs are relatively low, but the main disadvantage is that it is essentially a hand method and is therefore dependent on the skill of the operator. The process is extremely popular for manufacturing shell structures such as automobile bodies and parts, boats and chemical tanks.

2. *Matched Metal Molding*

This is essentially a preforming technique used where large runs of small components are required. The productivity rate is comparable to that of the conventional molding techniques used for plastics. Basically, the method consists of producing a 'preform' of the component using chopped glass fibers and impregnating by the following methods:

(a) Plenum Chamber Technique. In this method the fiber reinforcement takes the form of chopped glass fibers ½-2 in. in length which are passed into a chamber, and vacuum applied. Resin is then sprayed on to the fibers and the impregnated fibers thus assume the shape of the preform. The preform is finally withdrawn and cured.

(b) Direct Fiber Method. The resin binder and the glass fibers are sprayed on to a preform screen through which suction is applied. This method has the advantage of being suitable for producing large preforms, although the method is largely dependent upon the skill of the operator.

(c) Aqua Slurry Technique. This method has basically the same techniques as that in the phenolic pulp preform process. By this method the chopped fibers are fed into a tank of water and agitated continuously with a stirrer. The preform screen is then raised slowly out of the water and the water is drained through a vent, leaving the preform behind. Resin in then poured over the preform and, finally, the mold is closed and pressure applied.

3. *Spray Molding*

The two methods described above both suffer from certain disadvantages. The preform method requires a high initial capital outlay and the wet lay-up technique requires a large amount of labor, which is also costly.

The spray molding technique is relatively simple and consists of mixing two systems in separate containers. The first container is filled with the resin/promoter and the second with the resin/catalyst. The glass fibers are contained separately, having been previously cut into convenient lengths. The resin system is applied by means of a twin-head spray-gun to a cutting device which cuts the glass to the required lengths. In the actual process the mold is first coated with a release agent and then a gel coat is applied by hand or spray-gun. On curing the gel coat, the resin and the glass fibers are applied simultaneously, and the final laminate is consolidated by rolling.

4. Bag Molding

The processes based on this technique are, in general, modifications of the wet lay-up methods previously described. It is possible by this method to obtain moldings with even thicknesses and high glass-fiber/resin ratios. Hot curing techniques can also be applied, thus reducing the cycle times.

(a) Vacuum Bag. Basically this method consists of applying atmospheric pressure to the uncured laminated structure by exhausting the air between the lay-up and a flexible bag. This process allows for pressures up to one atmosphere (14.7 lb/in.^{2}) to be used which, for nomal work, is adequate.

(b) Autoclave. This process is used when higher pressures are required than are obtainable with method *(a)*. Steam may also be used instead of air. The method is applied on only a limited scale for polyester glass reinforcement but more frequently with phenolic glass structures which require both heat and pressure for satisfactory molding.

(c) Pressure-bag Molding. Laminate build-up is first carried out in a female mold, which is then protected by a film of cellulose or polyvinyl alcohol and placed in a rubber bag. The rubber bag is then forced by pressure against the mold. Pressures between 20-50 lb/in.^{2} can be used; this enhances the mechanical strength properties of the molding. Tooling costs are low.

Premix Molding

This method is commonly referred to as 'dough molding' and consists of premixing the polyester resin, inorganic filler and fiber reinforcement into a dough or fibrous mass. The resultant product is then suitable for molding under heat and pressure in a matched mold, thus allowing rapid production. It has been said that in terms of strength these molding materials bridge the gap between conventional molding powders and the polyester/glass laminates, and as such they warrant a wide range of end-uses in their own right without superseding the use of the above-mentioned types. The dough has a low glass content (5-20 percent), is nontacky in the uncured state and retains its shape when deformed. Advantages of this molding material over other polyester molding compositions are: Articles are cheaper to produce; it allows complicated parts to be made easily and shorter press cycles (as low as 15 s have been reported as giving successful moldings); there is less waste from off-cuts, etc. The main disadvantage is that the resultant moldings have lower mechanical strength than conventional polyester laminates, although flexural strengths of the order of 20,000 lb/in.^{2} have been obtained.

Tube and Rod

Glass reinforced tubing may be prepared by a modification of the wet lay-up process. This is carried out either by impregnating glass rovings, mechanically wound on a heated cylindrical mandrel, or by placing glass mat inside the surface of the mold and spinning, thus impregnating by centrifugal force.

The production of glass reinforced rod is carried out by several methods of which the most popular is continuous drawing of reinforced resin through a die which is heated by a high-frequency electrical technique to give a very rapid cure.

Injection Molding

Another interesting method has been developed which consists of forcing the poyester resin into a matched mold containing the reinforcing medium. The principle is that the incoming resin forces air bubbles out of the mold, thus overcoming one of the main difficulties in producing laminates. Multiplicity of molds allows rapid production without the necessity of using hydraulic presses and associated equipment.

Casting Resins

Polyester resins can be readily cast by a variety of techniques which are fully described in Chapter 6.

PROPERTIES

By varying the nature and combinations of the base raw materials, filler reinforcement, and cure time, it is possible to obtain polyester resins exhibiting a wide range of properties. The base resins may be tailored to give certain specific characteristics, for instance, resistance to solvents, alkalis, acids, or bleaching agents may be required, or possibly high heat distortion temperatures flexural or compressive strengths. These are all possible by careful formulation. Certain factors should be borne in mind when formulating; for example, the heat distortion temperature is dependent on the relative amount of unsaturation in the polyester chain, whereas flexibility is dependent on the unsaturated points and on the presence of long chains of carbon atoms. Chemical resistance has been found to depend largely upon the initial choice of glycols and dibasic acids.

It is considered that a discussion of the characteristics of polyester reinforced laminates warrants most attention because, apart from casting and encapsulation, very few other applications exist for the unfilled resin. As mentioned earlier the reinforcing medium is generally a member of the glass-fiber family (*see* Chapter 24). A laminate prepared with these materials would exhibit many of the following properties:

High strength/weight ratio
High impact strength
Dimensional stability
Good electrical properties
Good weather and chemical resistance
Good heat resistance
Ease of fabrication
Good light transmission
Good surface hardness

If an inorganic filler is added, the laminate becomes opaque, but shrinkage on cure is reduced, as also is the cost, as well as heat and fire resistance.

The excellent strength of a glass-fiber reinforced resin is shown in Table 5.2., which lists the properties of cured polyester resin and glass-fiber reinforced laminates.

In general, cured polyester resins will resist most inorganic and organic acids with the exception of strong oxidizing acids. They are not attacked by weak alkalis, although they are attacked by strong alkalies, which cause hydrolysis.

Alkyd moldings are resistant to water, salt solutions, detergents and common organic solvents. In general, they are resistant to acids but have poor resistance to alkalis. They are very similar in this respect to general purpose phenolic molding materials and in direct contrast with the aminoplast moldings which resist alkalis but not acids. Some general properties of alkyd molding materials are listed in Table 5.3.

Table 5.2

PROPERTIES OF CURED POLYESTER RESINS AND GLASS-FIBER REINFORCED LAMINATES

	Cast resin	*Laminate* 1*	*Laminate* 2†
Flexural strength (lb/in.^{2})	15,000–20,000	18,000–26,000	55,000–65,000
Young's modulus in flexure (lb/in.^{2} $\times 10^6$)	0.5–0.8	1.2–1.5	2.0–3.0
Tensile strength (lb/in.^{2})	4000–5000	15,000–25,000	35,000–50,000
Compressive strength (lb/in.^{2})	15,000–22,000	18,000–25,000	30,000–50,000
Impact strength Izod (ft lb/in. notch)	0.2–0.4	10–20	20–30
Specific gravity	1.2–1.3	1.5–1.6	1.6–1.7
Heat distortion temp. (°C)	60–145	–	–
(°F)	140–293	–	–
Specific heat (cal/°C/g)	0.5–0.6	–	–
Thermal conductivity (cal/sec/cm/°C/cm^2)	5–6 $\times 10^{-4}$	2–3 $\times 10^{-5}$	3–4 $\times 10^{-5}$
Coefficient of linear expansion (/°C $\times 10^{-6}$)	80–90	12–20	10–15
Power factor at 23°C (73°F) at 800 c/s	0.01–0.02	0.011	0.001–0.01
at 1 Mc/s	0.015–0.03	0.23	0.01–0.015
Permittivity at 23°C (73°F)			
at 700 c/s	2.8–4.0	1.48	4.0–5.5
at 1 Mc/s	3.0–4.5	1.87	3.0–5.5
Volume resistivity (ohm cm)	10^{15}–10^{16}	10^{14}–10^{15}	10^{14}–10^{15}
Surface resistivity (ohm)	1–9 $\times 10^{13}$	1–9 $\times 10^{13}$	1–9 $\times 10^{13}$
Breakdown voltage (volt/mil)	–	150–200	150–200

Table by courtesy of British Resin Products Limited

*Laminate prepared from chopped strand mat glass fiber with 70 percent resin content.

†Laminate prepared from woven glass fiber satin weave with 30 percent resin content.

Table 5.3

GENERAL PROPERTIES OF ALKYD MOLDING MATERIALS

Molding properties	*Standard cure*	*Fast cure*
Shelf life	Good	Fair
Pelleting properties	Fair	Good
Bulk factor	3–4.3	2.0–2.2
Cup time for blister free time	60–75 s (c.f. Melamine–formaldehyde 70–100 s)	15–30 s
Hot strength	Poor	Fair
Resistance to insert cracking	Good	Poor
Specific gravity	1.6–1.8	2.0–2.2
Impact strength (ft lb/in. notch)	0.1–0.2	0.1–0.2
Electrical properties	Good	Good
Electric strength	250–350 V/mil	250–350 V/mil

APPLICATIONS

The uses of glass reinforced polyester resins are now widely seen in such applications as vehicles and boats. Boat-building is one of the most successful applications of reinforced polyester resins. Boats made almost entirely by the wet lay-up process are becoming widely used throughout the world, their main attraction being their resistance to salt water, sunlight, termites, and rot. While the initial cost may show no apparent saving over wood and other conventional materials, maintenance costs are lower, repairs are easier, and the boats have a longer life. Extensive use has also been made of polyester laminates in storage tanks for oil, water and certain chemicals. Pipes made from polyester/glass have been produced but no large market has yet developed. Use has been made of glass reinforced rod for fishing tackle, etc., where the high strength-to-weight ratio is fully utilized.

A wide variety of household appliances and other articles are now made from polyester laminates. Furniture, luggage, ornaments, sinks, trays and handles, washing machines, and toys are among the extensive range. In office and industry similar applications are found, including machine housings, guards and cabinets, although many of these are now being replaced by thermoplastics.

Another use for polyester resins, one of growing importance, is roof sheeting. For such applications these laminates have the advantage of ease of handling, good strength and weathering properties, a wide range of colors and a high degree of transparency.

Table 5.4

TYPICAL PROPERTIES OF A COMMERCIAL STRUCTURAL POLYESTER FOAM

Property	Method of testing	"Valox®" FV-600 resin (metric)	"Valox®" FV-650 resin (metric)
Specific gravity, 73°F (22.8°C)	–	1.2	1.2
Heat deflection temperature,	ASTM-D648		
°F @ 66 psi (°C @ 46,404.6 kg/m²)		405 (207.2)	420 (215.3)
°F @ 264 psi (°C @ 185,618.4 kg/m²)		340 (170.9)	360 (182.0)
Thermal conductivity, Btu/h/ft²/°F/in. (cal/h/cm²/°C/g-cm)	ASTM-C177	1.10 (1.36)	1.10 (1.36)
UL listing**	UL Standard 94	94 V-O	94 HB
Tensile strength, psi (kg/m² X 10³) @ 73°F (22.8°C)	ASTM-D638	9910 (6,967)	9,920 (6974)
@200°F (93.2°C)	–	6350 (4465)	6,350 (4465)
Elongation at break, percent @ 73°F (22.8°C)	ASTM-D-638	1.3	2.3
Tensile modulus, psi (kg/m² X 10³) @ 73°F (22.8°C) [Tangent]	ASTM-D638	1,028,000 (722,790)	8,200,000 (576,540)
Flexural strength, psi (kg/m² X 10³) @ 73°F (22.8°C)	ASTM-D790	17,900 (12,585)	18,500 (13,007)
@200°F (93.2°C)	–	8400 (5906)	9,200 (6468)
Flexural modulus, psi (kg/m² X 10³) @ 73°F (22.8°C)	ASTM-D790	1,000,000 (703,100)	900,000 (632,790)
@200°F (93.2°C)	–	390,000 (27,420)	370,000 (26,014)
Flexural stiffness-to-weight ratio, @ 73°F (22.8°C)	–	160	150
Compressive strength [10 percent deformation] psi (kg/m² X 10³)	ASTM-D695	11,300 (7945)	11,000 (7734)
Creep [300 h 73°F (22.8°C) @ 1000 psi] (703,100 kg/m²) percent strain	ASTM-D674	0.12	0.12
Charpy impact, unnotched ft.-lb/in. (kg/m/cm)	ASTM-D256	5.3 (.733)	9.9 (1.37)
Gardner impact, 5/8 in. (1.58 cm) Nose: 1" (2.54 cm) Base lb/in. (g/cm)	–	11 (1964.6)	22 (3929.2)
Dielectric strength volts/mil			
perpendicular	ASTM-D149	152	152
parallel	–	200	200
Mold shrinkage, in/in X 10^{-3} (cm/cm X 10^{-3})	ASTM-D1299	–	–
flow direction		4–6	4–6
perpendicular flow direction		6–10	6–10

**This rating is not intended to reflect hazards presented by this or any other material under actual fire conditions.

*¼ Nominal Wall

*All ASTM general test methods are followed, except specimen size with four sided skin is varied to accommodate Structural Foam Processing.

*All properties on this chart are based on the chemical foam process.

Typical applications of alkyd molding powders include valveholders, insulating sleeves, parts for television assemblies, capacitor and commutator parts. They have also found limited uses as oven-door housings and housings for business machines.

In general alkyd molding materials are used in applications that require good arc and track resistance, insulation resistance, nonflammability with a combination of good mechanical and thermal stability. Such applications include automotive ignition systems such as distributor caps, coil tops, rotors, motor controls, circuit breakers, and a host of other electrical components.

We have already mentioned the polymer of diethylene glycol bis (allyl carbonate) and this has been found to possess excellent, exceptional, optical properties. It is therefore evident that applications have been found for this material as coatings where such characteristics are required.

SUGGESTED FURTHER READING

Unsaturated Polyester Technology. (1976) Ed. by Paul F. Bruins (New York, London, Paris: Gordon and Breach Science Publishers).

6.

EPOXY RESINS

The word 'epoxy' is now in everyday use although its derivation may not be obvious to everyone. It comes from the Greek prefix meaning 'over' or 'between' and the English suffix for oxygen. Thus, an epoxy material may be regarded as an 'oxygen between compound.'

Epoxy resins are thermosetting viscous liquids or brittle solids, of which the most widely used are those derived from the condensation reaction between epichlorhydrin and bisphenol A. Commercial exploitation was first initiated about the middle 1930's by I. G. Farben and this work was later followed independently by P. Castan, in Switzerland, and S. O. Greenlee, in the U.S.A., around 1938. Developments were slowed down by the Second World War and the relatively high price and scarcity of epichlorhydrin. However, at the end of the war epichlorhydrin, which was then obtained as an intermediate from the production of glycerol, became commercially available and led to wider uses for epoxy resins.

In 1962 Union Carbide Corporation announced that they would be manufacturing thermoplastic epoxide resins which could be molded by conventional thermoplastic techniques. They designated these materials phenoxy and they differ from the glycidyl ether types mentioned previously only in their molecular weight which is considerably higher.

Other developments in this field include the introduction of cycloaliphatic and brominated epoxy resins. An important new family of epoxy resins is based on the nitrogen-containing heterocyclic hydantoin ring.

RAW MATERIALS

Epichlorhydrin

Epichlorhydrin is a colorless liquid with the characteristic odor of many of the chlorinated hydrocarbon solvents such as chloroform. It has the following structure:

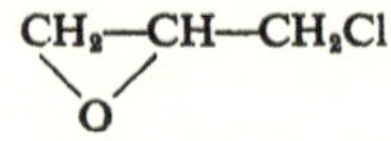

It has been known since 1854, when Berthelot reported its synthesis by saponification of the dichlorohydrins obtained by hydrogen chloride acting on glycerol. It is now produced on a chemical scale by the chlorination of propylene and is, therefore, a petroleum derivative. The resultant allyl chloride is reacted with hypochlorous acid to produce dichlorohydrin, which then undergoes what may be regarded as a 'stripping' process. It is exposed to sodium hydroxide at elevated temperatures, one hydrogen and one chlorine atom being stripped off. The overall reaction is as follows:

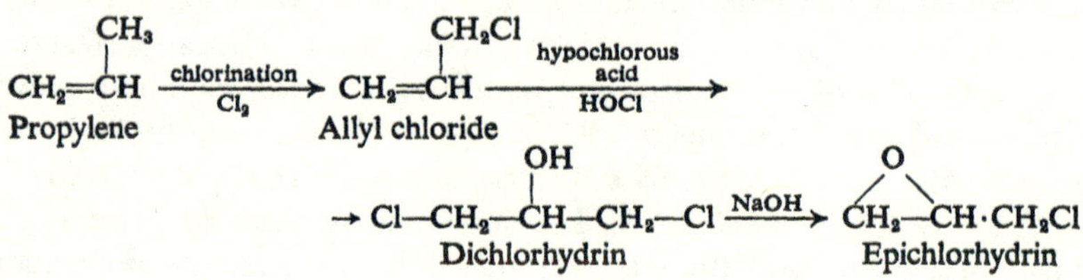

Bisphenol A

It is not uncommon to see other terminology used for this material and it is often referred to as diphenylol propane or bis (4-hydroxyphenyl) dimethylmethane.

Bisphenol A is normally prepared by combining acetone with phenol as follows:

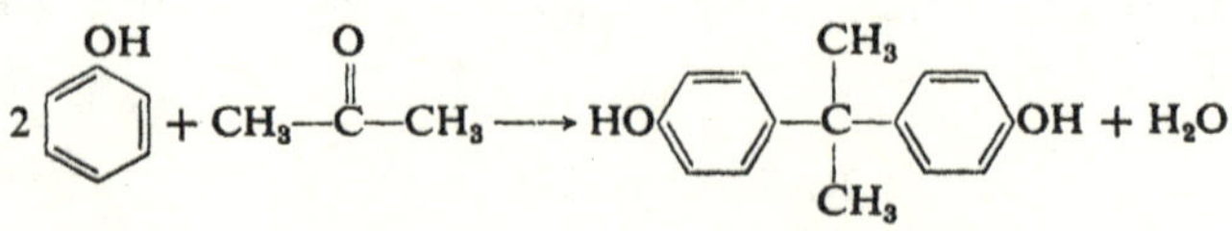

Phenol Acetone Bisphenol A

RESIN PRODUCTION

The most widely used epoxy resins at present are those based on the condensation reaction of epichlorhydrin and bisphenol A, using an alkaline catalyst such as sodium hydroxide. The general reaction may then be written as follows:

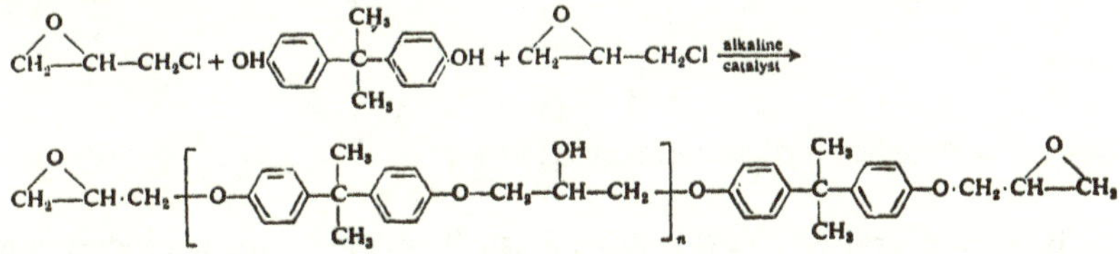

When n = 0 the molecular weight is 340 and the epoxy is in the form of a high-viscosity resin but when n = 10 the molecular weight is in the region of 3000 and the resin is a high-melting solid. These resins may be regarded as glycidyl ethers of bisphenol.

A typical manufacturing procedure is to mix the reactants at 50°C in a conventional resin kettle and slowly raise the temperature to 115°C over a period of 30 minutes, in the meantime slowly adding an alkaline catalyst such as sodium hydroxide. The reaction is then carried out for a further 30 minutes and the resulting product washed free of alkali and salt with hot water. Finally, the temperature is raised to 130°C and the remaining water removed by decanting or vacuum distillation. The resulting resins vary from viscous liquids to solids and the molecular weight is controlled by the epichlorhydrin/bisphenol ratios. Low ratios give high molecular weight solids, and high ratios normally produce liquid resins.

Epoxy Novolak Resins

These resins are produced by condensing the hydroxyl groups of a phenolic resin (in place of the more conventional bisphenol A) with epichlorhydrin. They have an average functionality greater than 3, thereby exhibiting a higher degree of cross-linking and superior thermal stability to the bisphenol-based epoxides.

The epoxy novolaks exhibit excellent reactivity towards the normal curing agents used for epoxy resins and in practice they cure more rapidly with primary amines than do the conventional epoxides.

The following structure is typical of an epoxidized novolak:

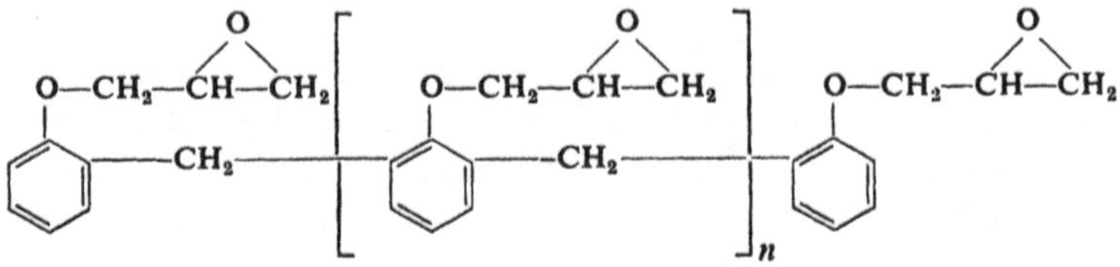

Cyclic and Acyclic Aliphatic Epoxy Resins

These comparatively new resins are usually based on the epoxidation of unsaturated compounds. The cyclo type does not contain the phenolic type structure associated with conventional epoxy resins. Typical base materials are vinylcyclohexene dioxide and dicyclopentadiene dioxide and, due to the fact that the distance between the reactive groups is shorter than with the bisphenol type, they are more highly cross-linked. Therefore by using the same curing agents and comparing the heat distortion temperatures it is found that the temperatures for the newer types are somewhat higher.

The acyclo types are similar to the cyclo epoxides except that their basic structural units are long chains whereas the other resins have a ring structure. They can be broadly divided into two main types: epoxidized polyolefins and epoxidized oils. The epoxidized polyolefins can be prepared by oxidizing the polyolefin with peracetic acid and/or other per acids.

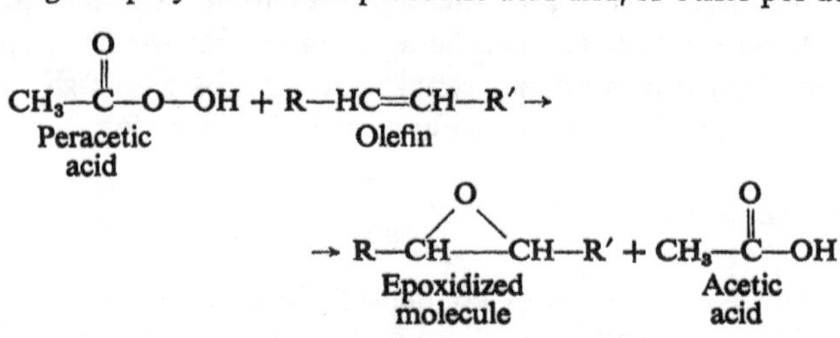

Another method of production is the condensation of an acid such as tetrahydrophthalic with epichlorhydrin, followed by dehydrohalogenation.

They are capable of being cured with the peroxide-type catalyst in a similar manner to polyesters. However, in practice they are cured by techniques more akin to the glycidyl epoxies. Whereas these glycidyl resins are particularly reactive to bases and especially to polyamines, the cycloaliphatic ones are reactive to acidic hardeners. The anhydrides of dibasic acids such as hexahydrophthalic anhydride (HHPA) constitute a very important class of anhydrides for these cyclic epoxies.

The following are some of the commercial resins available:

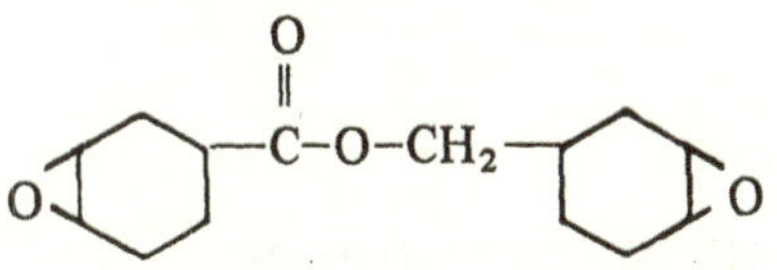

3,4-Epoxycyclohexylmethyl-
3,4-Epoxycyclohexane carboxylate

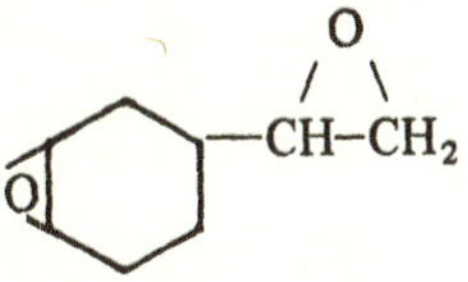

Vinyl cyclohexene dioxide

Phenoxy Resins

These resins differ from the conventional epoxies in that they are, and behave as, thermoplastics. The resins are termed phenoxies and may be classed as a polyhydroxyether.

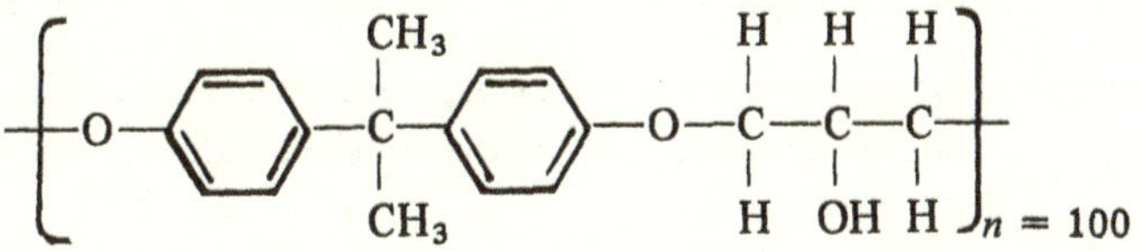

The outstanding physical characteristic of these new materials is that of ductility. The fact that their mechanical strength is not temperature dependent is also extremely attractive for many applications.

Phenoxies are strongly polar materials and as such have poor resistance to polar solvents such as ketones. Their amorphous structure also infers poor solvent resistance whereas the ether structure is indicative of chemical stability both to acids and alkalis.

The phenoxy resins may be processed on conventional thermoplastic machinery and can be injection molded, blow molded, and extruded in addition to being used for adhesives and coatings, as is later illustrated.

Compared to the more conventional epoxies these materials have much higher molecular weights. A typical example, Phenoxy Resin PKHH (Union Carbide), has a molecular weight of around 30,000 as compared to 300–4000 for the uncross-linked epoxies. These resins do not have terminal reactive epoxy groups and are therefore thermally stable materials with a long shelf life as is expected of thermoplastics. The phenoxies are useful as adhesives but unlike the epoxies they do not require any further chemical conversion. No catalysts, curing agents, or hardeners are needed to make them useful although plasticizers help flexibilize the resultant adhesive film. Typical properties of the Phenoxy Resin PKHH are given in Table 6.1.

Table 6.1

TYPICAL PROPERTIES OF PHENOXY RESIN PKHH

Specific gravity	1.18
Viscosity of 40% solids in MEK, Brookfield RVF, 20 rpm No. 5 spindle	5500–7700 cps
Reduced viscosity (0.2 g./100 ml. dimethylformamide)	0.4–0.6
Ultimate tensile strength	630–670 kg/cm^2
Ultimate tensile elongation	50–100%
Softening temperature	100°C
Permeability (1 mil free film at 25°C)	
water vapor (24 h/100 in.^{2})	3.5 g/mil
oxygen (24 h/100 in.^{2})	5–8 cc/mil
carbon dioxide (24 h/100 in.^{2})	15–30 cc/mil
Bulking value	1178 g/dm^3

(Phenoxy is tradename of Union Carbide Corp.).

From the structure it can be seen that the phenoxy resins contain secondary hydroxyl groups (6%) and therefore these resins can be cross-linked through these groups with polyisocyanates, amino and phenolic resins. The resultant coating has improved solvent and heat resistance. The polyisocyanates give satisfactory cure at room temperature whereas the other resins require elevated temperatures (150–175°C) although they do provide stable one-package systems at room temperature. Phenoxy coatings cross-linked with melamine-formaldehyde resins have been granted FDA approval.

These resins are compatible with many amino and phenolic resins, ketone resins, rosin, and coumarone indene. They are incompatible with acrylics, rosin esters, and vinyl resins although they can be plasticized with tri phenyl phosphite.

The hydroxyl groups on the phenoxy groups not only can be cross-linked but they are suitable for esterification.

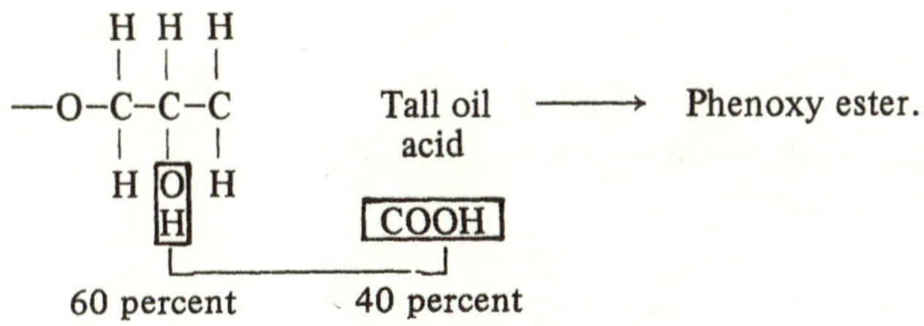

Phenoxy esters are useful materials for metal priming applications. They have better water and alkali resistance than the corresponding epoxy esters made from solid low molecular weight epoxy resins. Conventional formulating and compounding techniques may be used. Typical primers include iron oxide (automotive) wash and zinc rich types. This latter type has outperformed all other primers tested in corrosive atmospheres.

The phenoxy resins when pigmented exhibit high gloss, excellent resistance to salt spray, humidity, and water. As clear coatings they are tough, abrasion resistant, and either the unmodified or cross-linked types are suitable for wood or floor coatings. Coatings reacted with isocyanates or cata lyzed melamine-formaldehyde exhibit excellent resistance to cigarett burns and alcohol spotting.

CURING TECHNIQUES

A wide variety of curing agents may be used to cure epoxy resins and there are many ways in which they can be classified. They can be conveniently subdivided under room temperature of high-temperature curing agents or by the grouping to which the chemical curing agent belongs. However, for simplicity it is intended to deal with the subject as follows.

1. Catalytically

(a) Epoxy-Epoxy Reaction. The epoxy resins are fairly reactive and because of this the group may be readily opened by tertiary amines. The actual opening is believed to take place as follows:

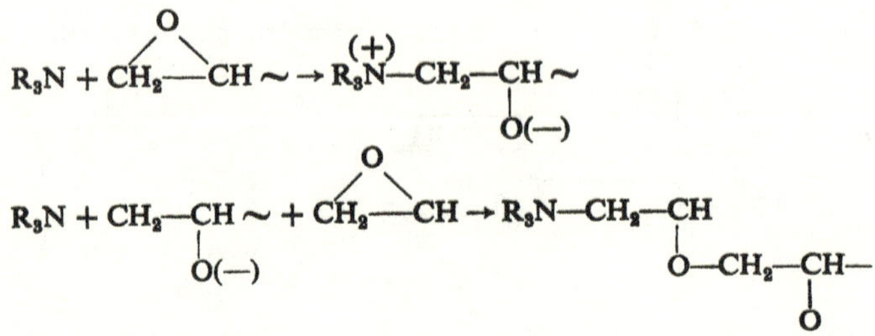

This reaction may continue and can result in a very long-chain system, although the chain may be terminated by blocking off the final activated epoxy.

(b) Epoxy-Hydroxyl Reaction. Epoxy rings react with hydroxyl groups that have been introduced into the resin by curing agents or certain modifiers. The reaction takes place as follows:

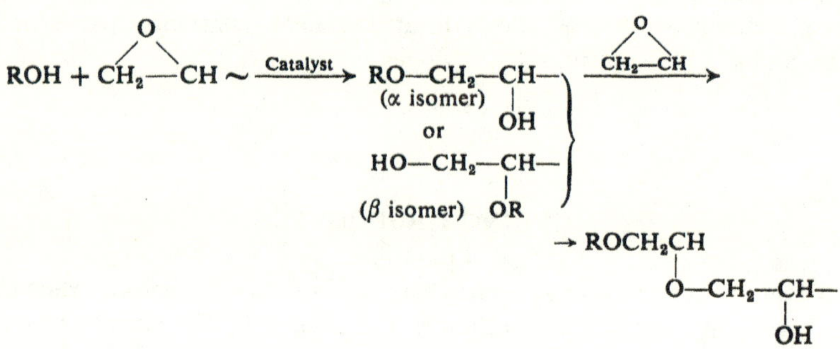

Two isomers are possible, and both provide a new hydroxyl group from which further polymerization may take place. A structure very similar to that shown for the epoxy-epoxy reaction may result and the reaction may continue until a high degree of cross-linking is obtained.

2. *Cross-Linking Agents*

Some commercial curing agents are listed in Table 6.2.

Table 6.2

COMMERCIAL CURING AGENTS

Materials	*Amount used parts per 100 resin*	*Pot life*	*Comments*	*Uses*
Diethylenetriamine (DTA)	8–11	45 min at 20°C	Liquid, volatile, and toxic. The amount used is also very critical	General-purpose room-temperature curing agent
Triethylene tetramine (TETA)	Very similar to diethylenetriamine			
Pipiridine	6	90 h at 40°C	Liquid, toxic, and volatile	Longer pot life and is used for castings
Triethylamine	10	7 h at 20°C	Catalytic	Adhesives
m-phenylenediamine (MPDA)	14	2½ h at 50°C	Solid with low toxicity	Chemical and heat resistant laminates
p,p′ diaminodiphenyl methane (DDM)	27	8 h at 20°C	Solid-toxic	General purpose laminates
p,p′ diaminodiphenyl sulfone	30	2 h at 20°C	Solid	Produces a cured resin with high heat distortion, point
BF_3 monoethylamine	5	3 weeks at 20°C	Solid–has a delayed action effect	Heat resistant resins
Phthalic anhydride	30–45	14 h at 100°C	Solid–produces a low exotherm	Coatings
Hexahydrophthalic anhydride (HHPA)	80	ditto	Solid	Castings
Chlorendic anhydride (HET ACID)	100	½ h at 110°C	Flame resistant	Flame resistant laminates

(a) Amines. Each amine group is difunctional and the following reactions are therefore possible.

(i)

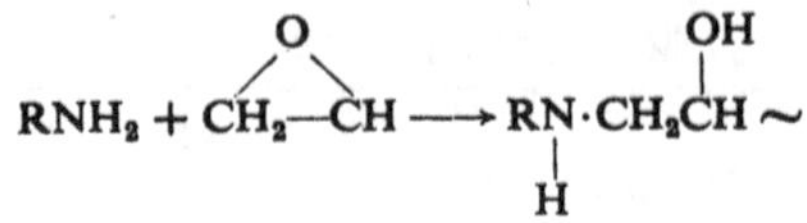

(ii)

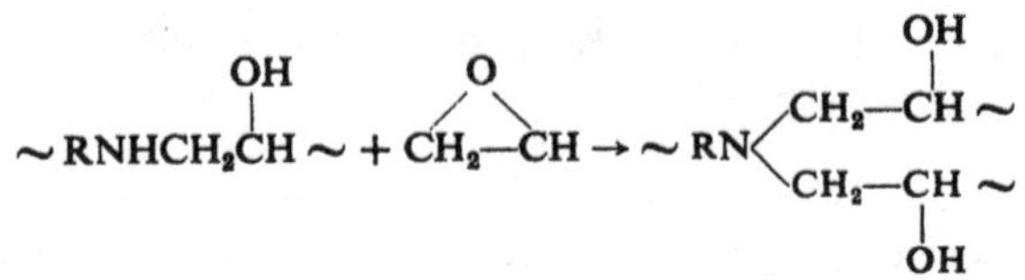

(iii) The hydroxyl groups may then start hydroxy/epoxy reactions.

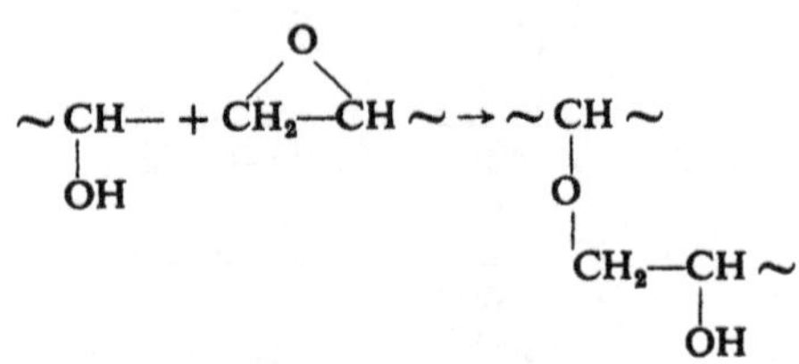

The aromatic amines are much less reactive than aliphatic amines and usually require elevated temperatures to achieve reasonable levels of cure.

Before leaving this section, mention must be made of dicyandiamide in view of its wide use in powder coating systems. The mechanism uses an amine curing agent:

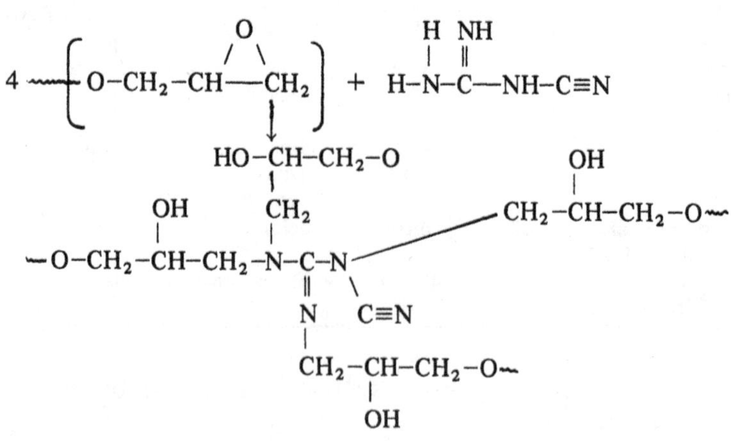

Dicyandiamide has poor solubility and a high melting point, thus it provides a poor rate of cure. Consequently the industry is moving more towards substituted derivatives which have lower melting points with improved compatibility. They have the general formula:

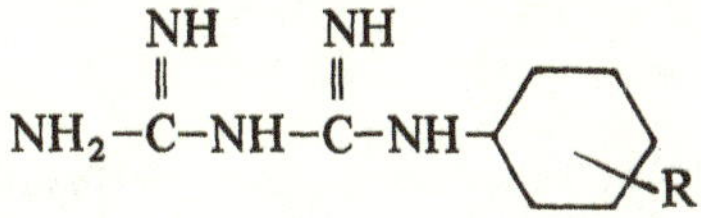

(b) Organic Acids. Organic acids are used mainly to esterify high-molecular weight epoxides suitable for use in surface coatings.

(i) The initial step is the esterification of the epoxy and hydroxyl-groups with the liberation of water.

(ii) The next step is the etherification reaction of the epoxy with the hydroxyl group.

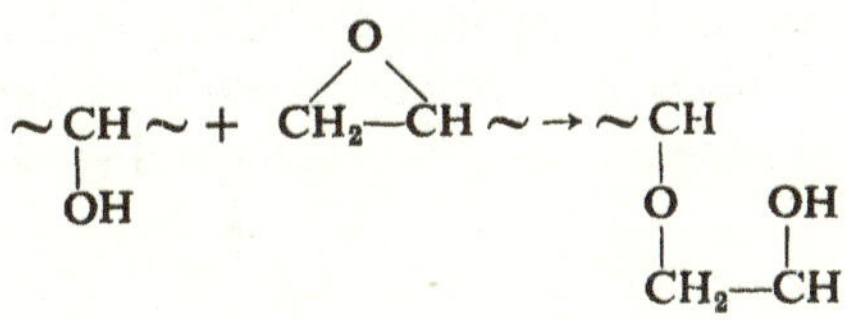

(iii) There is then a final hydration of epoxy groups by liberated H_2O.

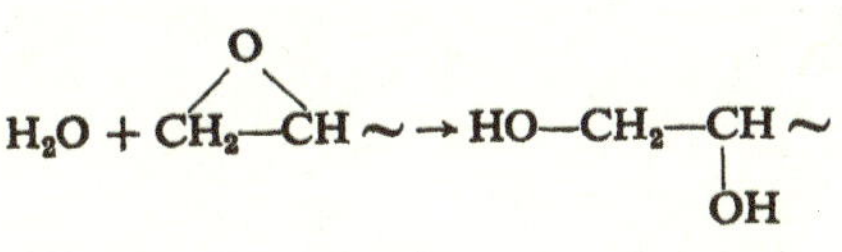

(c) Acid Anhydrides. Acid anhydrides react similarly to organic acids although the reaction is somewhat more complex due to the absence of water in the structure and the necessity for activating the anhydride structure.

The mechanism of curing for this group of chemicals is described in detail in view of the importance for curing cycloaliphatic epoxides. Two typical hardeners of this type are methyl tetrahydrophthalic anhydride (MTHPA) and hexahydrophthalic anhydride (HHPA). Chemically they can be represented as follows:

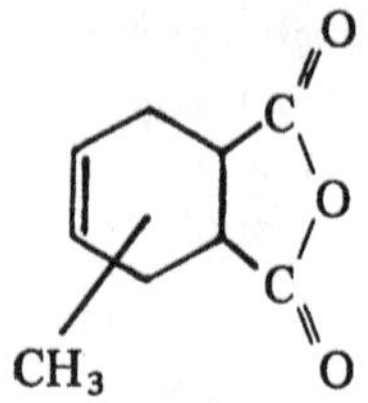

Methyltetrahydrophthalic anhydride

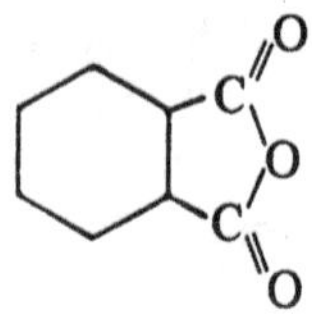

Hexahydrophthalic anhydride

It has been shown that the cycloaliphatic epoxides, made from a totally different method of manufacture than glycidyl ether types, contain very few hydroxyl groups. Therefore for the cycloaliphatic epoxide/anhydride system it is common practice to add hydroxyl rich materials. Even though it is possible to write a balanced reaction for a pure epoxide with an anhydride, in practice no reaction occurs. By adding hydroxyl rich materials the active hydrogen allows the epoxide/anhydride curing to take place in a stepwise process and adding anhydride to hydroxyl yields ester.

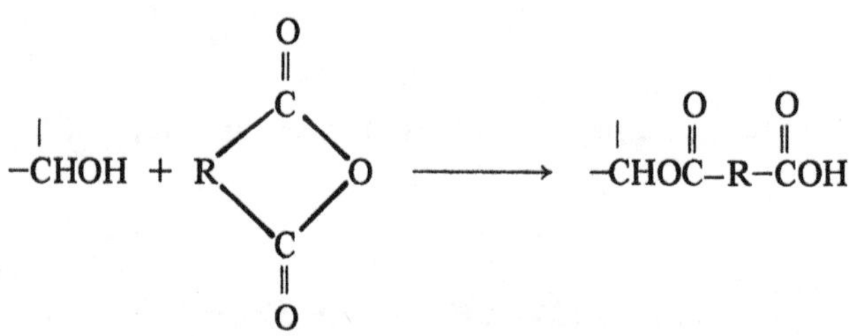

This is then reacted with epoxide groupings to form a hydroxyester.

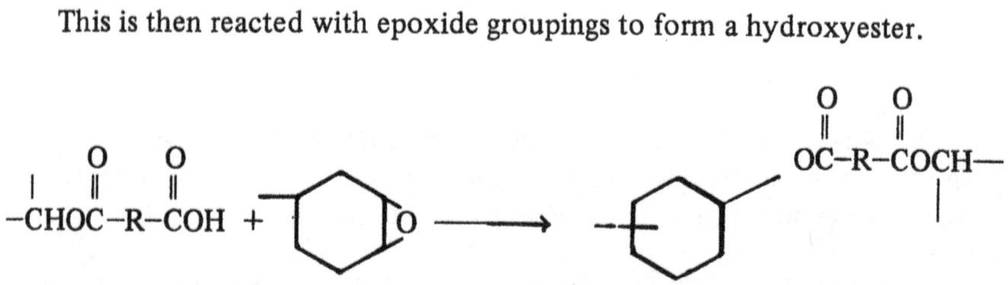

This reaction then repeats until either all the epoxide or the anhydride is consumed. The overall effect is the reaction of epoxide with anhydride to yield a diester with modification to give hydroxyls. The hydroxyl content remains constant throughout the cure. This process is not as straightforward as has been put forward because there is a complication. Under acid conditions the hydroxyl groups react with epoxide to produce an hydroxy-ether.

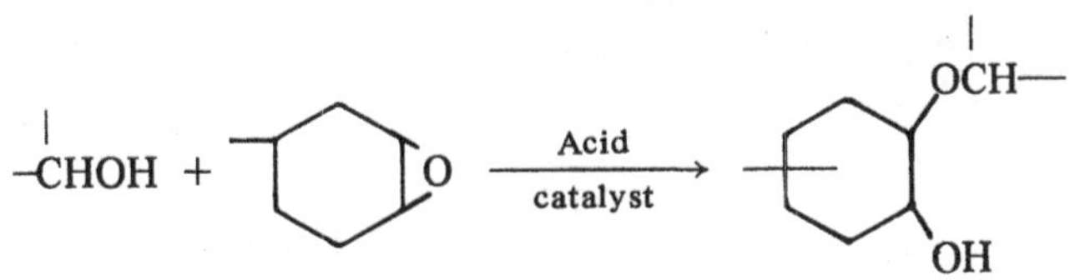

Clearly it can be seen that the important stage of the reaction of a bisphenol epoxy resin is the opening of the anhydride ring. This can be achieved in several ways:

(i) The hydroxyl group in the resin molecule used without a catalyst results in esterification and etherification processes while
(ii) tertiary amines favor esterification and
(iii) boron trifluoride complexes promote etherification.

The etherification process is generally undesirable because of the unreacted anhydride left in the cured film. Several catalysts have been suggested for use with this type of curing and benzyldimethylamine BDMA) is a commonly used tertiary amine for cycloaliphatic epoxide/anhydride system. The following are three catalysts which are finding considerable usage:

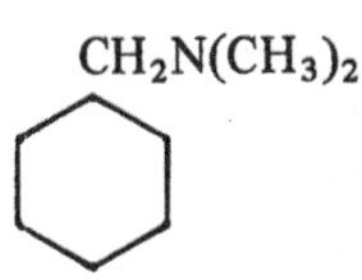

(Benzyldimethylamine)

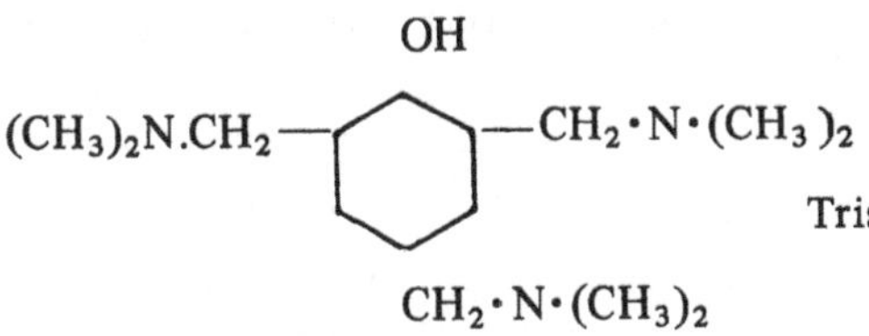

Tris(dimethylaminomethyl)phenol

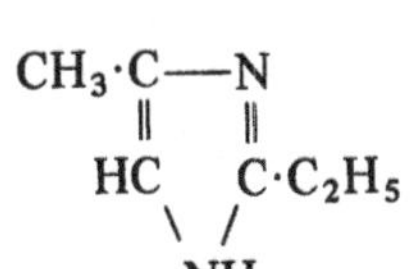

2 Ethyl 4 methylimidazole

Normally, between ½ to 2½ parts per 100 of the base resin are used.

(d) Boron Trifluoride Complexes. The complexes are frequently employed to delay curing and thus give the operator more time in which to work when fabricating various compounds. They are normally amines which are chemically associated with the boron trifluoride molecule such as BF_3-monoethylamine.

$$\mathrm{RNH: BF_3} \rightleftharpoons \overset{\oplus}{\mathrm{H}} + \overset{\ominus}{\mathrm{RN}}\mathrm{: BF_3}$$

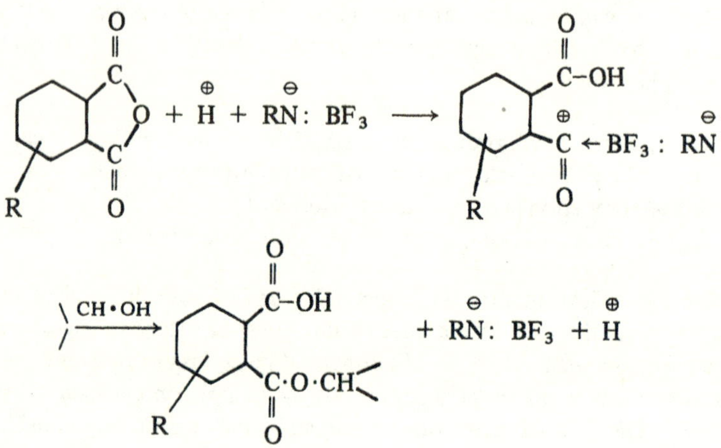

(e) Resinous Curing Agents. Many resins such as the polyamides, polysulfides, and phenolics react with and cure epoxides, but it must be remembered that the polysulfides are not practical curing agents and must be used in conjunction with an aliphatic amine.

(i) Polyamides.
These can be regarded as polyamino amides which are produced by the reaction of polyamines with dimerized or trimerized vegetable oils. They serve a dual function both as curing agent and as flexibilizer, discussed later in the chapter.

(ii) Polysulfides.
These are described under the section on Flexibilizers.

(iii) Phenolics.
These resins react with epoxies by an etherification process. The cross-linking occurs by the reaction with the epoxy group and the phenolic hydroxyl or the methoxy side groups of the phenolic resin.

(iv) Amino Resins.
It has been shown in Chapter 4 that melamine when reacted with formaldehyde in the correct proportions forms hexmethoxy melamine which in turn cross-links alkyd resins at elevated temperatures. It is also possible to cross-link epoxy resins with this modified melamine as follows:

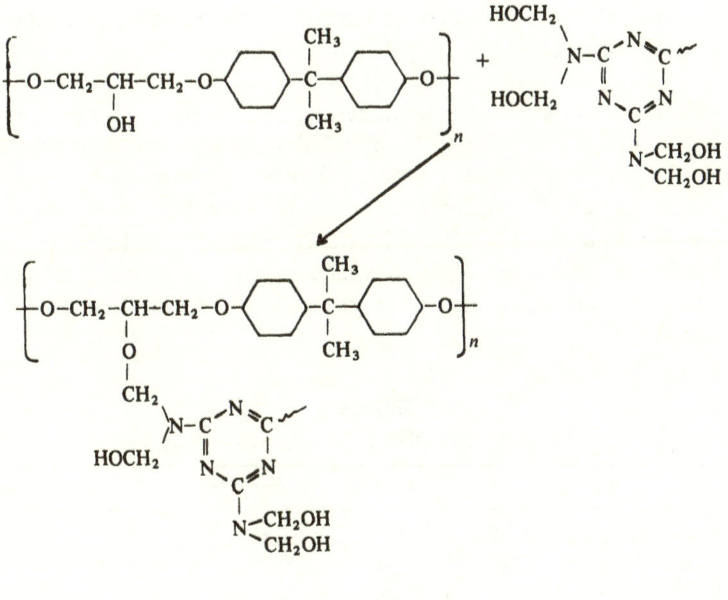

COMPOUNDING INGREDIENTS

Diluents

Diluents are normally incorporated into epoxy resins to reduce the viscosity and permit a higher loading of filler. Various types are used and they may be classified as follows:

(a) Materials that are reactive with epoxies. These include epichlorhydrin, allyl glycidyl ether, and phenyl glycidyl ether.

(b) Nonreactive diluents. These are normally aromatic hydrocarbons such as xylene or toluene. They do not produce a permanent effect, hence the method is not popular.

(c) Polymerizable Monomers. Monomers such as styrene and diallyl phthalate have been used with some success.

FLEXIBILIZERS

Flexibilizers, as the name implies, impart flexibility to epoxy resins and in many cases lower the viscosity of the resin. The following polymers are used for this purpose.

(a) Polyamides. These polyamide resins, trade name Versamid, are supplied by the Chemical Division of General Mills, and are based on vegetable dimers. They are low molecular weight polymers. They produce a two-fold effect, namely flexibilizing the resin and at the same time acting as a non-toxic cross-linking agent. Table 6.3 lists the properties obtained from an epoxy resin and an epoxy/polyamide blend using 65 parts by weight epoxy resin, 35 parts polyamide, 20 parts quartz flour filler, and curing for 60 hours at 25°C.

Table 6.3
PROPERTIES OF EPOXIES AND EPOXY/POLYAMIDE

Test	*Epoxy/Polyamide*	*Epoxy resin*
Impact strength (ft lb/in notch)	1.5	0.5
Heat deformation temp. (°C)	30	56
Compressive strength (lb/in.^{2} $\times$ 10^3)	8	15
Dielectric constant (1 Mc/s)	2.8	3
Power factor (1 Mc/s)	0.015	0.001

(b) Polysulfides. Polysulfides are rubber-like materials and are viscous liquids. They are compatible in all proportions with epoxy resins and reduce the resin viscosity considerably. The curing reaction with epoxy resins is by addition and takes the following form:

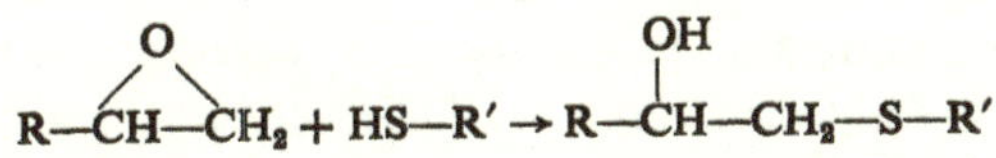

It must be emphasized once again that it is still necessary to add the normal amount of curing agent to complete the cure. The time for conversion varies according to the properties of the components and the temperature of cure, and may vary from a few minutes to days. Table 6.4 lists some of the properties obtainable using 100 parts by weight epoxy resin, 50 parts polysulfide resin, 20 parts curing agent, and curing for 60 hours at room temperature. Also included for comparison are the properties without a modifying resin.

Table 6.4

PROPERTIES OF EPOXIES AND EPOXYL POLYAMIDE

Test	*Epoxy/Modifying resin*	*Epoxy*
Impact strength (ft lb/in. notch)	6	1.5
Heat deformation temp. (°C)	23	56
Compressive strength ($lb/in.^2 \times 10^3$)	2.3	15
Dielectric constant at (Mc/s)	3.8	3.8
Dielectric strength	450	450
Power factor at (1 Mc/s)	0.006	0.004
Volume resistivity (ohm cm)	6×10^{13}	8×10^{14}

(c) Fatty Diamines. Fatty diamines may react with epoxy resins in the same manner as do other aliphatic amines. A typical fatty diamine has the following structure:

$$R_{soya}\,NH(CH_2)_3NH_2$$

These materials are used in conjunction with epoxy resins required for low-temperature work. Unfortunately, they produce resins with poor thermal and chemical resistance.

(d) Polyesters. These act as nonreactive plasticizers similar to those used for vinyl resins and take no part in the curing reaction. The effect is

not permanent and the method is not widely used.

However, a variety of polyols such as supplied by Union Carbide are very useful for formulating in the room temperature curing of cycloaliphatic epoxides. Polycaprolactones, which were described in the previous chapter, can be used and provide the ultimate in heat distortion and toughness for hydrated aluminum field systems. A typical product is as follows:

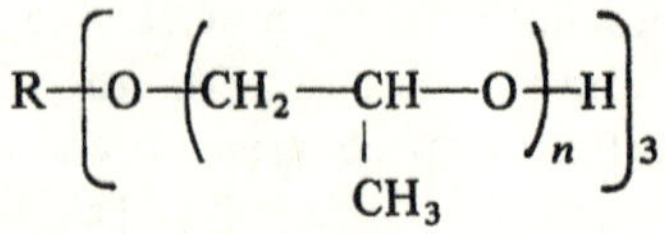

"Niax" polyol LHT 240

These materials are regarded as flexibilizers which are permanent and provide improved toughness while still maintaining a high level of electrical characteristics and heat distortion temperature.

FILLERS

Fillers are normally added to epoxy resins to reduce cost, dilute exotherm during curing, reduce shrinkage, and lower the coefficient of thermal expansion. A wide range of fillers may be used and because the uncured resin is inert the possibility of a reaction between filler and resin is remote. The following are fillers that have been suitable by experimentation for epoxy resins: mica (200 mesh), silica, or quartz (120 mesh), marble flour, alumina, iron oxide, slate powder, coarse silica sand, and aerosil. The effect of using different types and amounts of fillers on the thermal properties of epoxy resins are listed in Tables 6.5 and 6.6.

Table 6.5

COMPARISON OF THERMAL CONDUCTIVITY OF FILLED "EPIKOTE" 828* CURED WITH "EPIKURE" Z* AND STEEL AND ALUMINUM

Filler	*Thermal conductivity* (Btu/h/ft²/ft/°F)
None	0.13
Alumina (325 parts)	0.82
Silica (150 parts)	0.44
Steel	26.0
Aluminum	116.0

Table 6.6

COMPARISON OF THE THERMAL COEFFICIENT OF EXPANSION OF EPIKOTE 828*/MPD CASTINGS USING VARIOUS AMOUNTS AND TYPES OF FILLERS WITH STEEL AND ALUMINUM

Filler		*Linear coefficient of expansion* $\times 10^{-5}$ *(per °C)*
None		4.8
Silica (250 parts per hundred of resin)		2.61
Silica	(30 p.h.r.)	4.5
Calcium carbonate	(200 p.h.r.)	2.3
Calcium carbonate	(35 p.h.r.)	4.6
Alumina	(150 p.h.r.)	3.0
Alumina	(50 p.h.r.)	4.1
Steel		*ca* 1.0
Aluminum		1.84
Brass		1.88
Copper		1.68

(MPD–meta-phenylenediamine)

*Registered trademark of Shell Chemical Company Limited.

RELEASE AGENTS

These are normally required where the epoxy resins are being used for laminating applications. Silicone release agents, polyethylene film, cellophane, and various waxes have all proved suitable.

PROPERTIES

Epoxy resins possess many outstanding characteristics including extreme toughness, low shrinkage on cure due to the absence of volatiles, and excellent thermal, chemical, solvent, and water resistance. They do, however, exhibit fairly low impact strength in the unmodified form and for certain applications, such as thin films, they have to be modified with flexibilizers. These resins also exhibit a high degree of adhesion when bonded to a wide variety of surfaces such as metal and wood. This property is correlated with their polar nature and their ability to 'wet' the surfaces of various substrates.

Other important properties include outstanding electrical insulating qualities maintained over a comparatively wide range of temperatures (Table 6.7).

Table 6.7

TYPICAL PROPERTIES OF EPOXY RESINS

Test	*Molding resin (unfilled)*	*Casting resin (unfilled)*
Specific gravity at 25°C	1.7–1.9	1.1–1.3
Tensile strength (lb/in.^{2} × 10^3)	8–10	10–15
Elongation (%)	15	2.5
Compressive strength (lb/in.^{2} × 10^3)	12–20	14–18
Flexural strength (lb/in.^{2} × 10^3)	15–18	15–20
Impact strength (Izod notched ft lb/in. notch)	1	0.5–2.0
Coefficient of thermal expansion × 10^{-5}	5	4–6
Heat distortion temp. (°C)	140–150	120–250
Volume resistivity (ohm cm)	1×10^{14}	1×10^{17}
Dielectric strength (V/mil)	400	400
Dielectric constant (60 c/s)	3–5	3.5–5
Power factor (60 c/s)	0.35	0.002
Water absorption (24 h immersion %)	0.05–0.1	0.1
Chemical resistance		
Effect weak acids	Resistant	Resistant
Strong acids	Attacked	Attacked
Weak alkalis	Resistant	Resistant
Strong alkalis	Resistant	Resistant
Organic solvents	Resistant	Resistant

These properties are also typical of epoxy films but they can be further upgraded by reacting with other resins (e.g. fatty acid polyamides). Molding powders can be modified by filling with glass fiber, etc., and the finished moldings have high dimensional stability, low water-absorption, and good resistance to tracking. They also exhibit good heat resistance and moldings are claimed to be capable of withstanding 200°C without breakdown. Properties of a typical epoxide molding powder are illustrated in Table 6.8.

Table 6.8

PROPERTIES OF A TYPICAL EPOXIDE MOLDING COMPOSITION
(B.S.771 Test Methods where applicable)

Property	*Units*	*Value*
Specific gravity	–	1.8–2.1
Flexural strength	10^3 lb in.$^{-2}$	13–19
Mold shrinkage	in./in.	<0.002
After-shrinkage 48 h at 105°C	%	negligible
Water absorption .	mg	5–10
Dielectric constant 800 c/s 10^6 c/s	–	4.5–5.5
Power factor 800 c/s 10^6 c/s	–	0.01–0.02
Volume resistivity	ohm cm	10^{14}–10^{15}

APPLICATIONS AND PROCESSING TECHNIQUES

Adhesives

On curing, epoxy resins possess excellent bonding characteristics, and the reaction takes place without the evolution of volatiles and, often, even without heat or pressure. They may be used to bond various metals, rubbers, wood and glass with satisfactory results although with certain plastics such as PVC and polyethylene the bond is usually poor.

The relatively high price of epoxy resins has limited their widespread use for bonding processes in the wood and allied industries. Urea, melamine, and phenol-formaldehyde are normally used for plywood and the many other furniture applications, although for certain structural applications where wood laminates of high strengths are required the epoxies are generally to be preferred. Metal/metal bonding is very successful with epoxy resins which are now used extensively in the aircraft industry where, in addition to good bond requirements, good thermal resistance is also required.

Some of the epoxy resins are used not only for their good adhesive properties alone but also combined with toughness especially for floor-surfacing compounds in the construction of new floors and in resurfacing old ones. These compounds may be laid on concrete, metal, or wood substrates, and by suitable formulation a range of colors, surface finishes and thicknesses can be made to suit various industrial and domestic requirements. For this type of application they are normally three-component systems consisting of resin, curing agent, and aggregate.

Epoxy resins are also used as adhesives in the bonding of microscopic hollow spheres of phenol-formaldehyde resin, and known as *syntactic foams*. These foams are used extensively as core sandwiches and for replacing rotten wood in old buildings. They are supplied in putty form and may be trowelled into cavities where, eventually, they set to a tough hard mass.

It has been claimed that the cycloaliphatic epoxies show distinct advantages over glycidyl ether systems particularly in their ability to yield high-strength bonds on poorly-cleaned or even oily-metal surfaces.

Surface Coatings

Coatings based on epoxy resins possess good flexibility, adhesion, toughness, and chemical resistance. Their many applications include primers and paints for vehicles, chemical tank linings, and the inside coatings of beer cans. Epoxy coating resins have been blended with certain compatible phenol and urea-formaldehyde resins to give outstanding toughness as well as resistance to chemicals and solvents, and are used for coating the inside and outside of drums and tanks. They are also used in containers of beverages, meat and fish cans, and aerosols. Oil-modified epoxy resins are used quite widely for collapsible tube coatings because of their ability to withstand the container being flexed and their resistance to liquid contents which might, by chemical action, corrode some substances.

Considerable work has been carried out to develop water-thinnable systems. They can be used as primers for aircraft and automobile applications. The maintenance coating area is also important and this is served by the epoxy-coal tar systems which can be cured by polyamines.

The use of epoxies is growing steadily in the field of powder coating and this is discussed in greater detail in the appropriate section later in the book.

Laminates

Epoxy resins are used in the manufacture of both low and high pressure laminates. The processing techniques are not discussed fully here as the methods used are virtually interchangeable with those used for polyesters described in the Chapter 5.

As laminating resins they are used with a variety of reinforcing agents such as glassfiber, asbestos, and certain synthetic fibers. As noted earlier, they cure without water of condensation, which makes them ideally suitable for this application. Epoxies also exhibit outstanding adhesion to glass, and on curing the laminates exhibit excellent mechanical, thermal, and electrical insulating properties. Such laminates also possess excellent chemical resistance.

The use of epoxy laminates has been somewhat restricted in all but some of the highly specialized industrial fields due to the relative high price of the base resin in comparison with some of the more common resins. Very few uses have been found in the domestic consumer field, although they are used in industry where their superior mechanical strength and good heat and chemical resistance are required. Typical applications include structural components for aircraft and missiles, chemical resistant tanks, and laminated tubes and pipes for conveying oil.

CASTING AND TOOLING

Casting techniques are now widely used with epoxy resins in the electronics industry for potting, encapsulating transformers, valves, capacitors and many other electronic components. There are variations of technique, the following being the methods most commonly used (Fig. 6.1).

(*a*) Casting (*b*) Potting (*c*) Encapsulating (*d*) Sealing

Fig. 6.1

(a) Casting. This process is normally carried out by simple gravity techniques. The casting resin must be in the fluid, uncured state and be of sufficiently low viscosity to be conveniently poured into molds of wood, metal, plaster of paris, or certain plastics and elastomers (Fig. 6.1*a*).

(b) Potting. This is a modified form of casting in which the component to be potted may be considered as a nonremovable mold (Fig. 6.1*b*).

(c) Encapsulating and Embedding. In this method a protective shield is cast completely round the component without any impregnation being involved. Encapsulation and embedment are terms used interchangeably to describe virtually the same process (Fig. 6.1*c*).

(d) Sealing and Impregnating. A component may be sealed by placing it in a container and then completely filling the voids with a casting resin.

This method is becoming very popular with the electrical and aircraft industries (Fig. 6.1*d*).

Impregnating involves forcing the liquid resin into the interstices of the component. It may be a completely independent operation or can be used in conjunction with any of the other techniques mentioned above.

Before a resin is considered suitable for casting and/or tooling techniques it must possess certain of the following attributes.

Casting

- Good dimensional stability
- Low volumetric shrinkage on cure
- Shock resistant
- Tough and in certain applications flexible
- Good electrical, chemical, and heat resistant properties
- Good water resistance
- Must not corrode metallic inserts or parts of component
- Low coefficient of thermal expansion on curing
- Sufficiently low viscosity to be pourable

Tooling

- High dimensional stability
- Low shrinkage on curing
- High compressive strength
- Good abrasion resistance
- Thermally stable
- Good adhesion to metal inserts
- Low coefficient of thermal expansion
- Must not exhibit cold flow.

Tooling

Epoxy cast tools are used mainly for making dies, jigs, foundry patterns. plastic processing molds, and Keller models. In many of these applications epoxy resins possess advantages over more conventional metals: reduction in cost and time of manufacture, reduction in weight, ease of manufacture, and modification or repair.

The idea of making epoxy tools with plastics was first considered some fifteen years ago but certain deficiencies in the materials made the idea unattractive and it was not until the epoxy resins became commercially available that interest was once more roused.

MISCELLANEOUS APPLICATIONS

Their use as a molding material for the compression and transfer techniques was, to some extent, retarded by their high adhesive properties.

Certain developments have taken place which allow incorporations of a built-in release agent and these molding materials are now suitable for conventional molding techniques. They are loaded with mineral fillers, synthetic fibers, or glassfiber. Once more, the relatively high price restricts the field applications.

Epoxy resins may be foamed by incorporating a blowing agent, such as the azo compounds which liberate nitrogen on heating. The bulk density of these foams is controlled by the amount of blowing agent; 2 percent gives a useful balance of properties. Epoxy foams are suitable for use in encapsulation of electronic compounds where strength/weight ratio is critical. They have a fairly low coefficient of thermal conductivity and thus become a useful insulating medium.

SUGGEST FURTHER READING

Epoxide Resins. (1970) W. G. Potter (London: The Plastics and Rubber Institute.

The Use of Epoxy Resins. (1975) W. G. Potter (London: Butterworth).

7.

SILICONES

Silicones are high molecular weight semiorganic polymers built up of long chains of alternate silicon and oxygen atoms. Their history dates back to the nineteenth century, when Dumas predicted the possibility of the existence of organo-silicon compounds.

The first systematic work was carried out by Professor F. S. Kipping and co-workers at University College, Nottingham, during the first half of the twentieth century. He began his work with the preparation of asymmetric compounds of tetra-substituted silicon, and the work developed into a very full survey of the chemistry of silicon. Kipping's main interest was in the chemistry of this element and he described many of the compounds produced in the course of his studies as 'uninviting' and 'glue-like.' It is obvious from his published work that he did not foresee any technical usefulness for silicone polymers and he therefore made no attempt to develop them.

However, just prior to the outbreak of the Second World War certain industrial organizations began to show commercial interest in silicone polymers. A considerable amount of new development work was then undertaken in the U.S.A. particularly in the laboratories of the General Electric Company, the Corning Glass Works and the Dow Chemical Company. The last two companies were so convinced that there was a future for silicone polymers that they decided to join forces and form a new company, Dow Corning Corporation, to manufacture and market silicones. It was from the laboratories of this company that the first commercial silicone was produced.

NOMENCLATURE

Kipping first used the word 'silicone' to denote the organo-silicon analogues of ketones

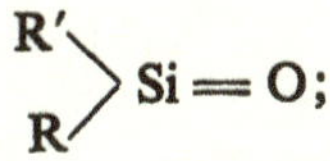

he quickly realized that such materials were polymeric, having the structure

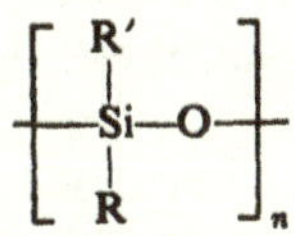

The simplest member of this family is silane

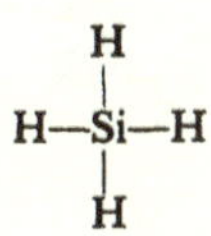

The hydrogen atoms of this material may be replaced by halogen, alkyl, or aryl groups giving compounds such as

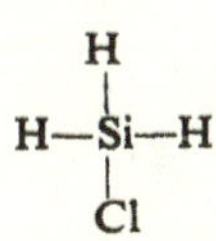

Monochlorsilane

or

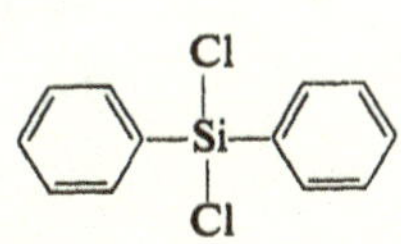

Diphenyldichlorsilane

The basic structure of the building blocks from which silicone polymers are produced is

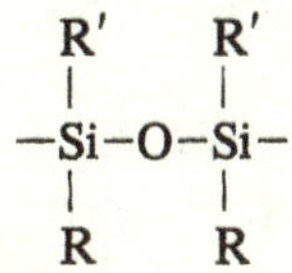

This grouping is designated as the siloxane linkage and is the characteristic structure to be found in all silicone polymers. A compound consisting of a large number of such groupings is referred to as a polysiloxane.

The pendant R groups which are attached to the silicon atoms usually are simple hydrocarbon radicals such as methyl, phenyl, or vinyl. In production many other functional groups have been introduced into the polymer structure including alkyl, cyanoethyl, fluoroalkyl, and nitrile. The introduction of such functional groups has been shown to significantly modify properties as solvent resistance, high and low temperature resistance, solubility, and compatibility.

This particular branch of chemistry does not only involve linear polymer structures but also branched and cyclic ones. Probably the most common cyclic material in this group is octamethyltetrapolysiloxane and unlike most of the other silicone polymers, these cyclic materials have discreet boiling and freezing points and can therefore be separated by distillation processes.

Chain stopping with materials containing such groups as $(CH_3)_3$–Si–O, –CH–CH_2 and –OH. The first group has been found suitable in the manufacture of silicone fluids and the latter two for use where cross-linking is required.

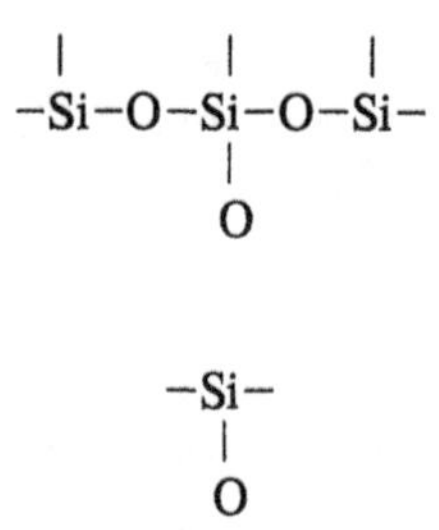

Branched structure

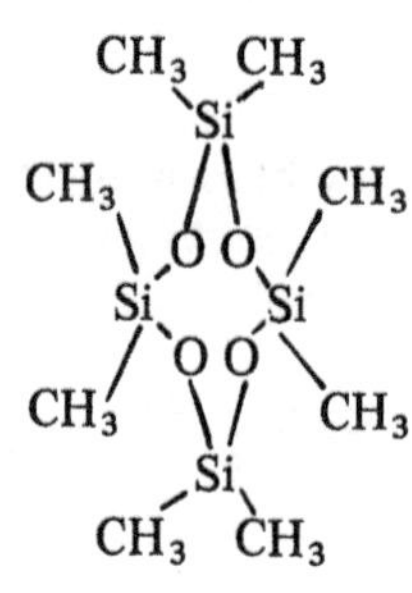

Cyclic structure

RAW MATERIALS

The starting materials for the manufacture of silicones are the substituted chlorsilanes, of general formula $R_nSiCl_{(4-n)}$. There are four methods normally used for preparing chlorsilanes suitable for commercial purposes, namely the Grignard Process, Direct Process, Olefin Addition Process, and Sodium Condensation Process. Of these, only the first two methods are of commercial importance.

1. Grignard Process

This process was widely used by Kipping and was first applied in 1906. The Grignard reagent is prepared by reacting magnesium metal suspended in anhydrous ether with a halide of the organic radical it is required to introduce, e.g., methyl chloride. This is then reacted with silicon tetrachloride to give a mixture of chlorsilanes.

$$\begin{array}{ll} Mg + CH_3Cl & \rightarrow CH_3MgCl \\ CH_3MgCl + SiCl_4 & \rightarrow SiCH_3Cl_3 + MgCl_2 \\ CH_3MgCl + SiCH_3Cl_3 & \rightarrow Si(CH_3)_2Cl_2 + MgCl_2 \\ CH_3MgCl + Si(CH_3)_2Cl_2 & \rightarrow Si(CH_3)_3Cl + MgCl_2 \end{array}$$

A high yield is obtained by this process although the resultant products must be fractionally distilled to obtain single substances in a high state of purity.

2. Direct Process

In this process the element silicon is converted into chlorsilanes, containing both an organic radical and chlorine. An organic halide such as methyl chloride is passed through a mixture of powdered silicon using a copper catalyst and heated to 250-280°C. A free radical mechanism is used to explain the reaction that occurs, but will not be discussed here. The copper first reacts with the methyl chloride to give copper methyl and cuprous chloride. The overall reaction may be written:

$$\underset{\text{Methyl chloride}}{2CH_3Cl} + \underset{\text{Silicon}}{Si} \xrightarrow[\text{catalyst}]{\text{Copper}} \underset{\text{Dimethyl dichlorsilane}}{Si(CH_3)_2Cl_2}$$
$$250\text{–}280°C$$

The phenyl chlorsilanes may also be manufactured in a similar manner by replacing the methyl chloride with chlorobenzene. There are, however, some differences in the manufacturing process and it is imperative to remember that excess chlorobenzene is passed through the reaction vessel to sweep out the high-boiling phenyl chlorsilanes. The unreacted chlorobenzene may then be recovered by fractionation.

The reaction takes place as follows:

$$2C_6H_5Cl + Si \rightarrow Si(C_6H_5)_2Cl_2$$

Chlorobenzene   Silicon   Diphenyl dichlorsilane

3. Olefin Addition

In this process trichlorsilane is reacted under pressure with an unsaturated organic material. It can be used to prepare a wide variety of organosilicon compounds but cannot be used to prepare methyl chlorsilanes.

4. Sodium Condensation Process

This process depends on the reaction between silicon tetrachloride and an organic chloride in the presence of sodium:

$$SiCl_4 + 4RCl + 8Na \rightarrow SiR_4 + 8NaCl$$

It is a tricky process to carry out and there is difficulty in disposing of the unreacted sodium. These drawbacks have to some extent limited its use on a commercial scale.

POLYMERIZATION AND STRUCTURE

(a) Silicone Fluids

Silicone fluids are prepared from dimethyl dichlorsilane with a small amount of trimethyl chlorsilane used as a 'chain stopper.' The first stage in the production of these fluids is the hydrolysis of the chlorsilane:

$$n\,Cl-\overset{\displaystyle CH_3}{\underset{\displaystyle CH_3}{\overset{|}{\underset{|}{Si}}}}-Cl + 2n\,H_2O \rightarrow n\left[HO-\overset{\displaystyle CH_3}{\underset{\displaystyle CH_3}{\overset{|}{\underset{|}{Si}}}}-OH\right]$$

The hydroxyl groups in the silane diol formed are unstable and they therefore react with each other to form linear polymers

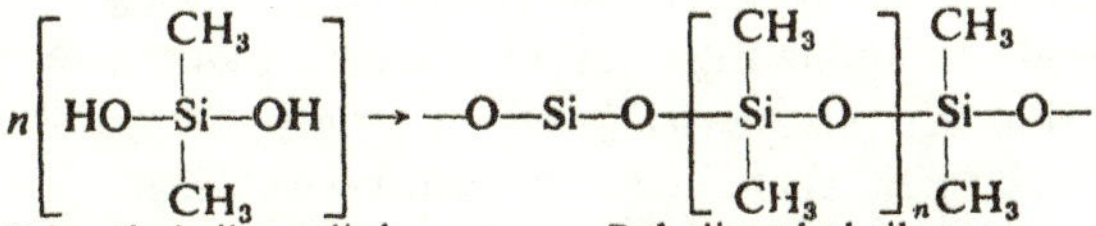

Dimethyl silane diol Polydimethyl siloxane

The molecular weight of the fluids can be varied over a wide range and may be predetermined by careful selection of chlorsilanes.

A relatively new group of silicone polymers is designated silicone polycarbinols and combine the typical properties associated with the silicones plus the alcohol reactivity. Unlike the fluids described previously, these silicone polycarbinols reacted further with a variety of materials. Typical polymers which can be modified with these new silicone polymers are urethanes, polyesters, methylol melamines, acrylics, and SBR.

The silicone polycarbinols take either of two basic structures: (1) an ABA type, such as

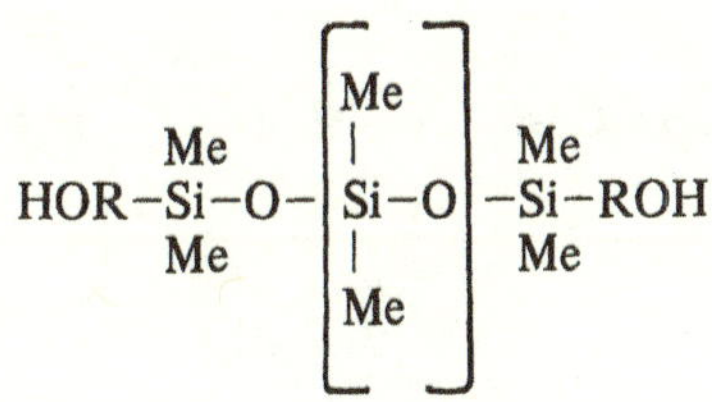

or (2) a graft copolymer, such as

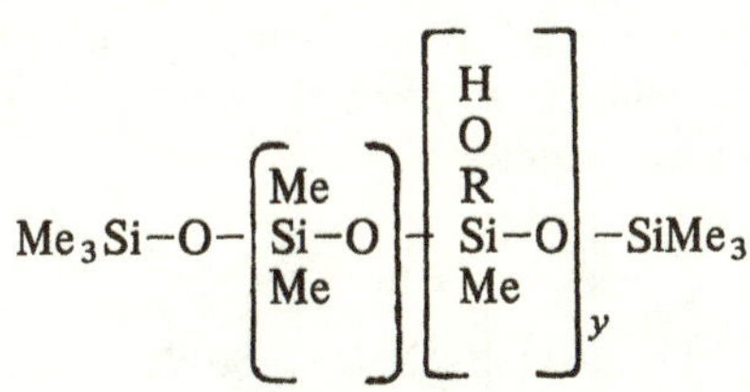

R is an organic spacer and the alcohol moiety is attached to the silicone polymer via a nonhydrolyzable Si–C bond. Dow Corning Corp. offers four materials of which Q2-8026, 1248 and Q4-3557 fluids are graft copolymers and Q4-3667 fluid is an ABA block type copolymer.

The reaction may be continued, and final curing may be effected by heating with a catalyst such as the octoates or naphthenates.

These resins can be copolymerized with certain organic resins such as the epoxides. It was seen in Chapter 6 that the epoxy resins are renowned for their outstanding strength, and by combining this characteristic with the excellent thermal stability and electrical insulating properties of silicones an excellent laminating resin is produced. Silicone resins now available are suitable for polymerizing with a variety of organic resins such as the epoxy, polyester, phenolic and polyol types. These silicone resins are low molecular weight siloxanols with a hydroxyl content of between 3-5 percent. The hydroxyl groups act as weak acids and for this reason the intermediate can be viewed as a highly functional polybasic acid. Copolymerization takes place as follows:

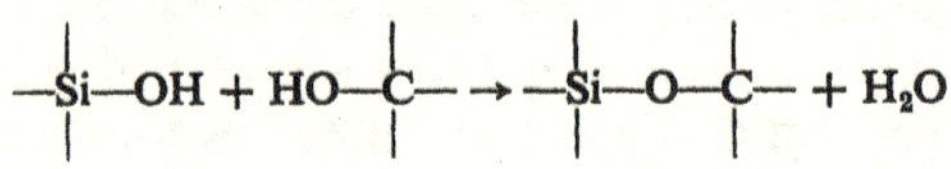

This reaction is to some extent retarded by a side reaction

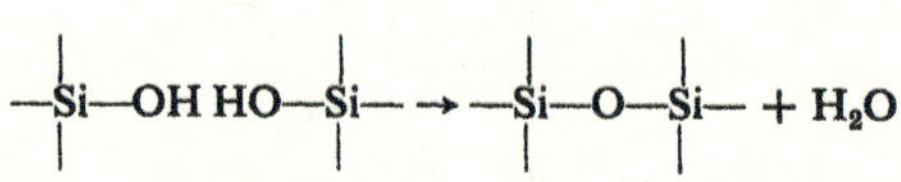

(b) Silicone Elastomers

Silicone elastomers are based upon linear polymers of very high molecular weight. The first silicone elastomers were made from polydimethyl siloxane, of the following structure:

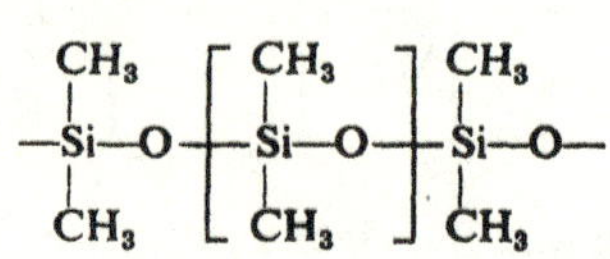

Many variations have since been developed for these elastomers including the introduction of phenyl and vinyl groups to improve certain of the physical properties of the elastomer, viz. low temperature brittle point and compression set respectively.

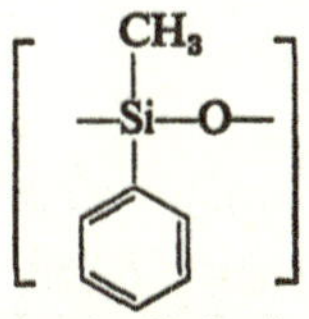

Phenyl substitution (improves low temperature brittle point)

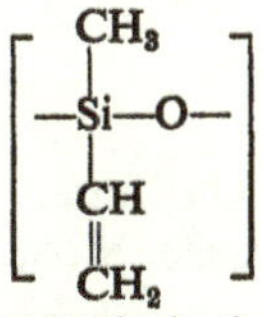

Vinyl substitution (improves compression set)

The introduction of fluorine and nitrile groups into the silicone elastomer chain structure has produced elastomers with outstanding heat stability combined with excellent petrol and oil resistance.

Fluorine substitution

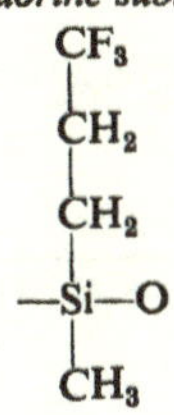

(i)
Trifluoropropylsiloxane

Nitrile substitution

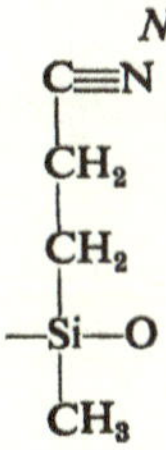

(ii)
Cyanoethyl

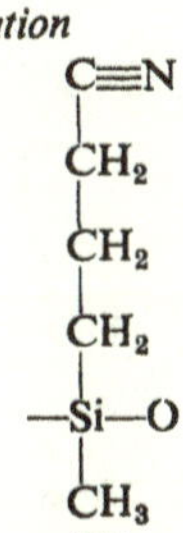

(iii)
Cyanopropyl

Another variation can be obtained by introducing hydroxyl groups into the polymer chain. This produces elastomers which are suitable for curing at room temperature. The curing reaction is normally carried out by means of an organo metallic salt such as dibutyl tin dilaurate or stannous octoate. These elastomers are generally referred to as either cold-curing rubbers or room-temperature vulcanizing rubbers (RTV).

In general, these rubbers are based on low molecular weight silicone polymers with reactive end groups. As with the heat curing types mentioned previously, specific groups such as fluorine, nitrile, and phenyl groups can be built into the polymer structure to impart desired characteristics. Generally the curing systems for these RTVrubbers can be classified into two main groups and the first group which is a condensation type can be further subdivided into sections "with and without moisture."

1. Condensation

(a) Without moisture. With this type of reaction the reactive end group is usually a silanol (≡Si–OH). As would be expected with the Kienle theory of functionality the curing agent must have a number of at least three or greater.

$$\equiv\text{Si}-\text{OH} + \text{OH}-\text{Si}\equiv \xrightarrow[\text{base}]{\text{Organic}} \equiv\text{Si}-\text{O}-\text{Si}\equiv + \text{H}_2\text{O}$$

Alkoxy containing cross-linking agents such as ethyl *o*-silicate requires a tin soap as catalyst.

$$\equiv\text{Si}-\text{OH} + \text{CH}_3\text{CH}_2\text{O}-\text{Si}\equiv \xrightarrow[\text{Soap}]{\text{Tin}} \equiv\text{Si}-\text{O}-\text{Si}\equiv + \text{CH}_3\text{CH}_2\text{OH}$$

Catalysts can be omitted if polyfunctional aminoxy silicone compounds are used.

$$\equiv\text{Si}-\text{OH} + \text{R}_2\text{NO}-\text{Si}\equiv \longrightarrow \equiv\text{Si}-\text{O}-\text{Si}\equiv + \text{R}_2\text{NOH}$$

Industrially these types of cold curing silicone elastomers are normally sold as two-pack systems which are mixed just prior to use. The curing takes place without the presence of moisture and large thicknesses can be cured.

(b) With moisture. These particular types are based on a silanol terminated polymer to which a polyfunctional silicone-containing curing agent is added. Variations include using a polymer which is end-stopped with the curing agent and in general a condensation catalyst is used.

This curing takes place only in the presence of atmospheric moisture so the reaction takes place initially at the surface and the diffusion of the moisture further into the matrix ensures that complete cure takes place throughout the rubber.

A typical reaction, with X representing the reactive groups of the curing agent which terminates the polymer chain, follows:

$$\equiv\text{Si}-\text{X}\ \text{H}_2\text{O} \longrightarrow \equiv\text{Si}-\text{OH}\ \text{HX}$$

$$\equiv Si-OH\ X-Si\equiv \longrightarrow \equiv Si-O-Si\equiv HX$$

These polyfunctional, silicone-containing, curing agents can be based on alkoxy, amino, and ketoximo groups.

2. Addition Cure

This type of reaction takes place without the formation of any volatile or requirements of moisture. The curing system involves the metal ion-catalyzed addition of a polyfunctional, silicone hydride, cross-linking agent to a dimethyl-vinyl siloxy terminated silane polymer.

$$\equiv Si-CH=CH_2 + H-S\equiv \longrightarrow \equiv Si-CH_2CH_2-Si\equiv$$

In general they are two part systems but some use has been made of a one component system using an initiator.

PROPERTIES

Certain properties are common to all silicones, of which the most important are their exceptional stability to heat, cold, and oxidation.

1. Fluids

The influence of temperature on the viscosity of methyl silicone fluids is relatively small. They also exhibit unusually high resistance to mechanical shearing, particularly in hydraulic systems. They are also highly water-repellant and water condenses as discrete droplets on surfaces previously treated by the fluids.

Dimethyl silicone fluids are colorless, odorless, nontoxic and physiologically inert. They are resistant to oxidation, have excellent resistance to a wide variety of chemicals, and are available in a wide range of viscosities. Some physical properties are given in Table 7.1.

The typical properties of the silicone polycarbinols are given in Table 7.2 but as such it is probably more useful to examine the properties obtained when reacting these materials with other more conventional polymers. The following are the properties which can be improved for these

Table 7.1

PHYSICAL PROPERTIES OF SOME SILICONE FLUIDS

Fluid	*Viscosity at 25°C cS*	*Permittivity at 25°C (100 c/s)*	*Freezing point (°C)*		*Boiling point (°C)*	*Flash point min. (°F)*	*Thermal conductivity cal cm. sec cm² °C at 50° C*	*Specific gravity at 25°C/ 25°C*	*Refractive index at 25°C*
MS 200/3	3	2.52	−65		70–100	215	0.00027	0.90	0.39
			Pour point						
MS 200/10	10	2.65	-65		>200	325	0.00032	0.94	1.40
MS 200/20	20	2.68	-60		>200	520	0.00034	0.96	1.40
				°C	*Volatility after 48 h at percent*				
MS 200/500	500	2.75	-50	200	<2	600	0.00038	0.97	1.403
MS 200/12 500	12,500	2.82	*Solidification temp. –46*	200	<2	600	0.00038	0.97	1.403
			Freezing point						
MS 510/50	50	2.7	-70	250	10	525	–	0.99	1.42
MS 510/550	550	2.8	-70	250	6	525	–	0.99	1.43
MS 550	100–150	2.9	-50	250	15	575	0.00035	1.06-1.08	1.49–1.50

organic polymers by the reaction with the silicone types: release, abrasion resistance, water repellancy, water vapor transmission, low and high temperature flexibility and electrical insulative properties. Specifically, in such applications as textile treatment with resins, as described in Chapter 4, the reaction with silicone polycarbinols and methylol derivatives leads to enhanced properties such as softer hand and water repellancy.

Table 7.2

TYPICAL PROPERTIES OF SILICONE POLYCARBINOLS

	Dow Corning brand fluids			
	1248	Q4-3667	Q2-8026	Q4-3557
Structure	Graft	ABA	Graft	Graft
Alcohol functionality	Secondary	Primary	Primary	Secondary
Weight percent OH	0.9–1.1	1.4–1.6	2.0	4.5
Average no. OH groups per molecule	3	2	3	2
Viscosity	160 cs.	320 cs.	465 cs.	40 cs.
Flash	250°F	285°F	400°F	>200°F
Refractive index	1.4110	1.44	1.45	1.43
Pour point	−130°F	65°F	60°F	−80°F
Color	Clear	Clear straw	Clear straw	Clear straw
Average hydroxyl eq. wt.	2000	1200	800	400

2. Resins

The physical properties of silicone resins are dependent to a large extent on the Me.Si ratio, and it has been found that a ratio between 1:3 and 1:5 gives the most useful products. Resins prepared with such ratios exhibit excellent stability at high temperatures and are also resistant to moisture. The excellent thermal and electrical resistance exhibited by silicone resins has made possible the upgrading of the suitability of certain electrical

Table 7.3

TYPICAL PROPERTIES OF SOME SILICONE RESINS

	Coating resins		*Impregnating varnish*		*Bonding resins*	
Property of cured resin	*MS 935*	*MS 994*	*MS 997*	*MS 1360*	*MS 840*	*MS 2103*
Electric strength, film thickness 10 mil (V/mil)	>1000	>1500	>1000	>1500	>1000	>1000
Power factor at 10^2 c/s	0.003	0.001	0.006	0.005	0.003	0.002
Power factor at 10^6 c/s	0.004	0.001	0.007	0.013	–	0.002
Permittivity at 10^2 c/s	2.9	2.8	3.1	3.6	2.9	2.9
Permittivity at 10^6 c/s	2.7	2.8	3.0	3.5	–	2.9
Craze life (min, h at 250°C)	800	3000	1000	–	–	–
Flex life (min, h at 250°C)	100	1000	250	–	–	–
Property of resin solution						
Solids content (percent)	50	50	50	50	80	60
Solvent	Xylene	Xylene	Xylene	Cresylic acid	Acetone	Toluene
Viscosity at 25°C cS	80–150	80–125	100–200	4300–8600	350–1750	60–125
Processing	Require curing at 200–250°C			Cures at 50°C above vinyl formal wire enamels	Will rapidly air dry tack-free to form a soft film. Require curing at 250°C to form hard nonthermoplastic bond.	

equipment that has been impregnated with these resins. These resins possess good chemical resistance especially to dilute mineral acids and alkalis. Their resistance to many organic solvents such as toluene and xylene is not good. The properties of some silicone resins are given in Table 7.3.

3. Elastomers

Silicone elastomers may now be produced with tensile strengths in the region of 1500 lb/in.^{2}. They remain serviceable and stable over a wide temperature range from as low as −70°C to as high as 250°C. They also possess excellent electrical resistance and insulating properties. Typical properties of silicone elastomers are listed in Table 7.4.

Table 7.4

TYPICAL PROPERTIES OF SILICONE RUBBER

Temperature stability (°C)	−100 to + 250
Specific gravity at 25°C	1.0–2.0
Hardness (B.S. deg.)	40–85
Tensile strength (lb/in.^{2})	700–1500
Ultimate elongation (%)	100–400
Compression set (% at 150°C)	10–70
Brittle point (°C)	−110 to −60
Tear strength (lb)	4–5
Electric strength (V/mil)	250–600
Permittivity at Mc/s	3.0–4.0
Power factor (1 Mc/s)	0.001–0.007
Coefficient of cubic expansion/°C	≈ 0.0007
Thermal conductivity (cal cm/cm s)	$\approx 5 \times 10^{-4}$
Specific heat (cal/°C/g)	0.275 to 0.325

Silicone elastomers are generally resistant to a wide variety of chemicals and, in general, will resist the attack of oxidizing agents, weak acids, weak bases, salts and vegetable oils. They swell in the presence of strong acids and bases, chlorinated solvents, and reducing agents but they are not attacked chemically and normally return to the original shape on removal from the source of the chemical. Silicone elastomers are also unaffected by u/v light, ozone and many oxidizing agents. These elastomers do not harden up and become brittle over a wide range of temperature as do the conventional natural and synthetic organic elastomers. This can be seen in Fig. 7.1, where the temperature aging of silicone rubber is compared with chloroprene and natural rubber.

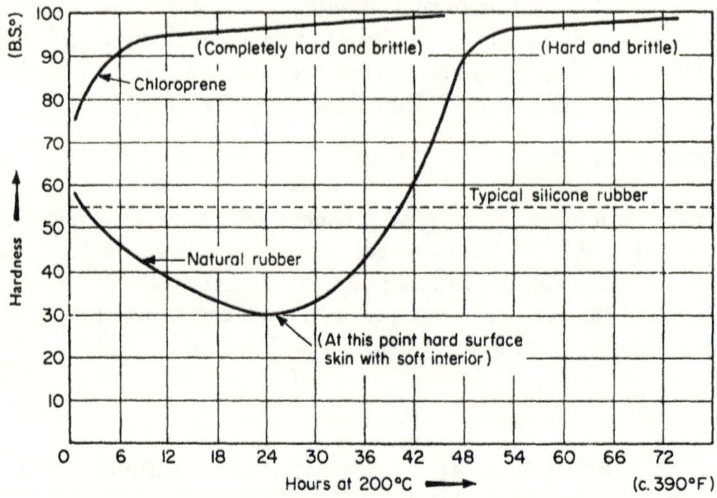

Fig. 7.1

COMPOUNDING

(a) Fluids. Silicone fluids are normally used without fillers or additives; therefore, very little compounding is necessary. However, in fluids for the production of greases, inert fillers such as silica and carbon black as well as metallic soaps, e.g., lithium octoate, are added.

(b) Resins. Relatively little compounding is required with silicone resins and they are generally supplied either solventless or in solutions of xylene and toluene. The resins may be used in conjunction with glass fibers and the processing details vary little from those described for polyesters in Chapter 5.

(c) Elastomers. Unfilled silicone gums have negligible mechanical strength and must therefore be reinforced by the addition of a filler. The fillers are classified as reinforcing or semireinforcing and nonreinforcing. The reinforcing fillers commonly used are precipitated and fume silicas. There are numerous semireinforcing fillers of which the following are frequently used and include diatomaceous earth, quartz, calcium carbonate, and precipitated calcium silicate.

Special additives are frequently used in conjunction with silicone elastomers to provide certain enhanced properties including improved heat stability and reduction in compression set. Heat stability is normally improved by the addition of ferric oxide, ferric octoate, or barium zirconate. In the early days of the development of these elastomers the compression set properties were improved by adding various quinones, but these chemicals have now been generally rendered obsolete. The new technique is to introduce a few molar percent of vinyl groups and thus tailor the polymer chain as mentioned previously.

Silicone elastomers may be readily pigmented by the addition of heat-stable inorganic oxides such as iron, nickel, chromium, titanium, or cadmium.

PROCESSING TECHNIQUES FOR SILICONE ELASTOMERS

A silicone elastomer stock is compounded and processed on the same type of equipment as used for organic rubbers. It can be mixed on a two-roll mill, in a Banbury or in a dough mixer. The curing agent should always be added on a mill to avoid premature setup. Unvulcanized silicone elastomer is softer and weaker than organic rubber, and a scraper blade attachment on the mill is recommended for removing the silicone compound. After compounding, the next step is to cure the rubber. The first stage of curing is carried out in a press, hot air chamber or steam vulcanizer, the process being referred to as set-cure. The second stage is carried out in an air-circulating oven and is especially necessary in order to develop the optimum physical properties required for many applications and for maximum resistance to compression set.

Fully compounded silicone stocks may be molded, extruded, laminated, calendered, or applied from solvent dispersion. Conventional rubber processing machinery is used and should include a two-roll mill equipped with scraper blades, a steam or electrically heated molding press operating at temperatures between 100–150°C with provision for water cooling. A steam vulcanizer capable of operating at pressures up to 100 $lb/in.^2$ should also be available. In addition, a forced air-draught circulating oven is essential to obtain an adequate final cure for optimum physical properties.

APPLICATIONS

1. Fluids

Because silicone fluids have small viscosity change with temperature and nonsticking properties, they are now being used where such properties are of prime importance. The silicone fluids have certain advantages over hydrocarbon oils and are therefore being used more and more in aircraft hydraulic systems. Antisticking characteristics have led to at least two large uses, namely in polishes and as mold release agents in plastics and rubber molding.

Silicone oils may be used in the form of an aqueous emulsion or solvent solution for mold release applications. They offer a unique combination of properties, viz. inertness, heat resistance, low surface tension combined with good lubricating power. These materials have now replaced many of the more conventional mold release materials such as soap and talc. A typical formulation for the preparation of a mold release agent would be as follows:

Dimethyl silicone fluid . . .	35 parts
Emulsifier	2 parts
Water	63 parts

The above ingredients are mixed together and then diluted to a concentration of around 1 percent.

The use of silicone fluids in car and furniture polishes has grown in popularity over the past few years. Normally about 2-4 percent is added to either wax or waxless formulation and in the case of the wax polish the silicone fluid serves to lubricate the microcrystalline wax. The following are suitable formulations for a wax and a waxless polish.

		Waxless (%)	*Wax (%)*
A	MS 200*	4	3
	White spirit	19	94
	Wax	—	3
	Kerosene	2	
	Oleic acid	2.5	
	Morpholine	0.75	
	28% NH_4OH	15.75	
	Water	14	
B	Silica	40	'B' is stirred
	Water	Suspension	into 'A'

*Registered trademark of Midland Silicones Limited.

Silicone fluids are now widely used in the paint industry as antiflotation and antisilking agents. The addition of a small amount of a low viscosity dimethyl silicone fluid (1-50,000 ppm) assists in the grinding and dispersion of the pigment as well as inhibiting flotation and flow lines. It has also been established that a small addition of a silicone fluid proves effective in reducing 'foaming' in certain materials susceptible to this phenomena. They are also used in the production of polyurethane foams to control certain characteristics such as cell size. Silicone antifoams are now widely used in food manufacture and sewage disposal.

One of the earliest uses of silicone fluids was as a water-repellant treatment for a wide variety of surfaces. They are now being used extensively for treating textiles, paper, and leather. In the textile treatment the silicones not only confer excellent resistance to water but also to staining.

As stated previously, the silicone polycarbinols are also being used for treating textiles but in conjunction with other resins such as hexamethylol melamine. The demand for improving the basic performance characteristics of urethanes is increasing as the market acceptance of the urethane family grows. The suppliers of the silicone polycarbinols (Dow Corning) recommend that these polymers afford a simple unique method for upgrading the physical properties of urethanes to meet the ever increasing new demands. Benefits claimed are improved scuff resistance, wear, water repellancy, stain resistance, weatherability, low temperature flex, hydrolytic stability, and rate of water vapor transmission. Other claims are that the blocking and rustle of fabric coatings are reduced.

2. *Resins*

The film-forming properties of silicone resins are not as good as those of the more conventional organic polymers, but they do possess greater heat stability. Paints based on silicone resins and aluminum pigments can effectively withstand direct heating to around 700°C. Their overall cost may be reduced by blending them with organic coating materials such as glyptals, urea, and phenol-formaldehyde as well as the epoxy resins. The aluminum-based silicone paints are used for coating certain parts of jet engines which are subjected to high temperatures.

The silicone resins have found extensive use in high-temperature insulation for electrical equipment. Silicones are classified according to Class H insulation which has a top working temperature of 180°C. They are therefore used for high-temperature resistant coatings for electrical equipment. The main advantages of Class H insulation is to permit an increased output per unit volume of meter. Such a property is very important where space and weight saving are required in traction motors, aircraft and submarines.

Other interesting applications for silicone resins include the waterproofing of masonry and utilization as a release agent. Silicone resins containing solvent may be sprayed on brickwork, and on the subsequent evaporation of the solvent the bricks are left with a coating of silicone resin and are thus water repellant. As a mold release agent they offer an advantage over the fluids in that their effect is semipermanent. The baking industry has used silicone resins for coating bread tins and they have found that they are able to obtain many hundreds of satisfactory releases before the treatment wears off.

Silicone resins have also been used in preparing molding powders suitable for the compression and transfer molding techniques. These resins are expensive, and although the overall price may be reduced by addition of fillers such as asbestos powder or chopped glass wool, their use is still somewhat limited.

Another use for these resins is in the preparation of laminates with glass fiber, asbestos, and mica as the reinforcing material. Such laminates may be manufactured by press molding at both high and low pressures, by vacuum bag molding, or by filament winding. The resin is normally applied from a solvent solution (usually of the high-boiling aromatic type such as xylene or toluene) and is precured on the reinforcing material, although solventless silicone laminating resins are also available. The main field of application is in the aircraft industry, where heat resistance is required, or in the electrical industry where the insulating properties are of prime importance. The fact that the laminates have relatively poor strength characteristics has been overcome by the copolymerization with epoxy resins, etc., as mentioned earlier, and this has opened a wider field of applications.

3. *Elastomers*

Undoubtedly the growth and popularity of the use of silicone elastomers can to a large extent be attributed to the requirements of the aircraft and electrical industries. It is often the unique combination of properties of silicone elastomers that influences branches of certain industries to utilize these high-price elastomers in preference to the conventional organic types.

The main uses of silicone elastomers in the aircraft industry are for door seals, hot air ducts, flexible sleeving, gaskets, and vibration dampers. The fluorinated silicone elastomers are now widely used for such applications where their outstanding resistance to chemicals and fuels may be used effectively.

A relatively big market for these elastomers is for the insulation sleeving used on power and control cables in ships. There are many advantages in using silicone rubber as an insulator, namely resistance to oxidation, ozone and a variety of chemicals. One particular advantage that is gained for use with cables in ships is that in the advent of a fire on board, the silicone would burn, leaving a silica ash which would have sufficient insulating value, possibly, to allow the ship to return to base.

The comparatively high price of silicone elastomers has somewhat restricted their use in domestic and industrial appliances. However, in certain instances where specialized properties are required these rubbers have been used. Typical appliances include use as oven-door seals, cable insulation, belting, gaskets and caulking pastes for sealing joints (e.g., joints to be found in a domestic steam iron).

More and more uses are being found for silicone elastomers in the medical field and here the physiological inertness of silicones has been fully exploited. Applications already include tubing, vial stoppers, gaskets, teats, and heart valves. An important use is for tubing on blood transfusion equipment and it has been found in service that such tubing is unaffected by 30 heat sterilizations involving exposure to steam at 20 $lb/in.^2$ pressure for thirty minutes each time. Natural rubber tubing is rendered unserviceable by comparatively few sterilizations. Other advantages for using silicone elastomers in such applications include the fact that this material has no effect on the quality of the blood and possesses a surface to which the blood does not adhere as readily as it does to natural rubber.

An interesting advance has occurred in the medical field, made possible only by the availability of silicone elastomers. This can be seen in Fig. 7.2, which is a diagram of a baby's head with an implanted silicone automatic brain valve, the most promising cure yet developed for hydrocephalus (water on the brain). The silicone elastomer valves have certain advantages over the conventional plastic brain valve. They can be sterilized, and their complex mechanism does not go out of adjustment. They consist of two-side slit directional nipples accurate to 0.0003 in., inside stainless-steel cases instead of steel springs or sapphire ball-checks in an acrylic housing. Connected with a silicone rubber housing, the nipples open at the slightest input pressure, but provide a double seal against fatal backflow of blood. Thus, the system can drain into the jugular vein instead of the stomach, greatly easing installation. The double valve gives the assembly a pump action which can be checked by hand for accuracy and functioning.

A property of equal significance is that of the incompatibility of silicone elastomers to organic bodies over a wide range of temperatures. This

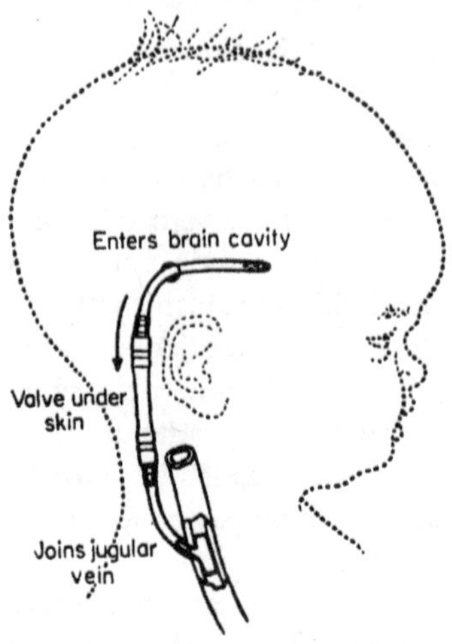

Fig. 7.2

property has been utilized in the fabrication of silicone rubber-covered rollers for use by the plastics industry as guide and pressure rollers during the calendering and embossing of plastic films and coated fabrics.

The cold-curing silicone elastomers are also being used extensively in a variety of applications including sealing, caulking, potting, encapsulating, and impregnating. They are excellent materials for encapsulating delicate and electronic components especially where severe humidity conditions have to be met. Because they do not require a heat cure they are used for encapsulating heat-sensitive components or where temperature precludes high-temperature curing. Electrical equipment such as resistors, capacitors, and transformers may also be encapsulated quite easily.

A relatively new material has been introduced to the electronic potting and encapsulation field, called silicone dielectric gel. This material is supplied as a two-component system, containing no filler, being a water-white viscous liquid. It exhibits unusual properties in that it does not exhibit cold-flow, is self-sealing and remains in place even after curing, although it still remains soft and flexible. This has the advantage that in the event of an electrical breakdown occurring with the potted component it is possible to explore into the potting medium and thereby locate the point of weakness. This material is also an excellent damping medium and has found application in dash-pots.

SUGGESTED FURTHER READING

Silicone Polycarbinols. (1975) Dow Corning Corp. RTV Silicones (GEC Technical Bulletin).

8.

POLYURETHANES

The term 'polyurethane' refers to a group of polymers which have found widespread application over the past quarter of a century. In general, they are produced as a result of the reaction between an aromatic diisocyanate and a compound rich in hydroxyl groups. The initial developmental work was carried out on polyurethanes produced from hydroxyl-enriched polyesters but over the past few years considerable interest has been given to the use of polyoxyalkylene glycols. These latter materials are commonly referred to as polyether polyols.

The foundation for the now well-known polyurethane field was the original isocyanate synthesis carried out by Wurtz in 1849

$$R_2SO_4 + 2KNCO \rightarrow 2RNCO + K_2SO_4$$

The preparation and chemical reactions of isocyanates have therefore been known for over a hundred years but it was not until the middle 1930's that interest in their industrial importance grew to any proportion. Much of the original developmental work was carried out in Germany in an attempt to produce polymers competitive with those being produced by Du Pont at that time, viz. polyamides. In 1937, Dr. Otto Bayer experimented with the addition products of diisocyanates. The object of this work was to produce synthetic fibers not covered by the Du Pont patents. His work quickly bore fruit when it was shortly announced that he had discovered the diisocyanate poly addition process. A German patent No. 728 981 covering all aspects of this work was obtained and today the number of patents on these polymers runs into many hundreds. Thus it is to Bayer and his coworkers that most credit must go for the development of these interesting materials.

Polyurethanes exist in various polymeric forms, of which the most important are the foams, elastomers, and coatings. They are also used as adhesives, molding compounds and fibers, although such applications are discussed in less detail in this chapter.

A relatively new class of urethane polymers are those which are designated as ionomers. They are block polymers which are linear, containing ionic centers at wide intervals and may be regarded as hetero polymers having a pronounced segment structure. The main field of application for this group of products appears to be in aqueous dispersions for coating of textiles, paper, and plastic film.

RAW MATERIALS

The production of polyester resins has been described in Chapter 5, the type in general use for the production of polyurethanes being the adipic acid and ethylene or propylene glycol. For polyether polyurethanes used in the manufacture of flexible foam the polyether is normally obtained by the alkali-catalyzed polymerization of propylene oxide, ethylene oxide, or tetrahydrofuran.

These reactions can be initiated by the addition of small amounts of hydroxy bearing materials, e.g.,

$$HOCH_2CHOH + nCH_2{-}CH_2\,(\text{-O-}) \longrightarrow HO[(CH_2)_2O]_nCH_2CH_2OH$$

Preparation of Diisocyanates

The diisocyanates used in the production of polyurethanes are normally aromatic materials, the most common being tolylene diisocyanate, a mixture of 2:4 and 2:6 isomers (often referred to as TDI). There are many ways of making isocyanates but the following has found most commercial use and consists essentially of reacting an amine or amine salt with phosgene in an inert medium such as *o*-dichloro benzene.

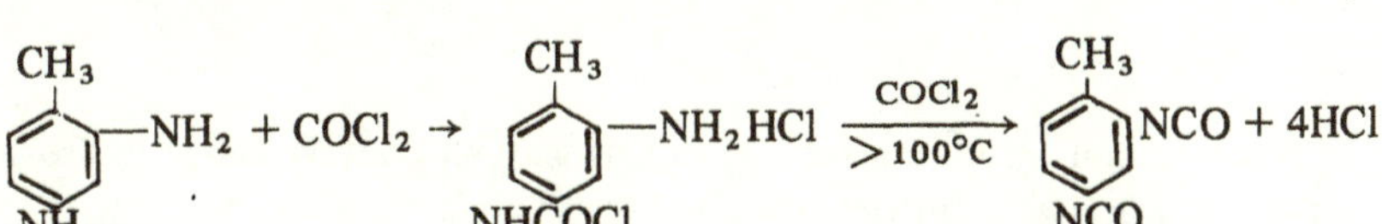

m-Tolylene diamine | Phosgene carbamyl chloride | Tolylene 2,4 diisocyanate

This diisocyanate is generally preferred although others such as diphenyl methane–4,4 diisocyanate, (MDI), *m*-phenylene diisocyanate, 1,6-hexamethylene diisocyanate are used as well as many others.

The MDI is made by the condensation of aniline with formaldehyde in the presence of hydrochloric acid followed by phosgenation.

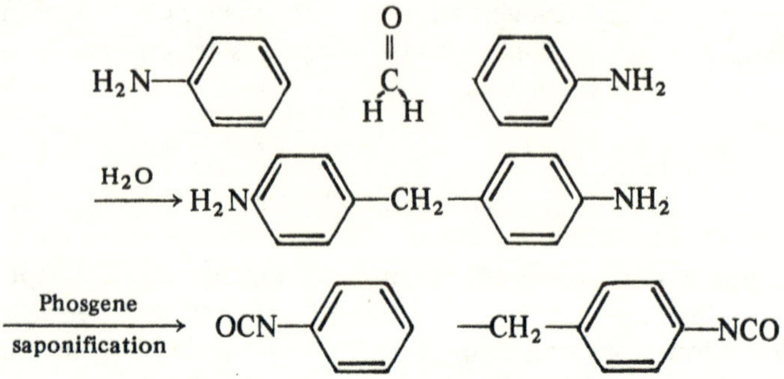

MDI

General Reaction of Isocyanates

Isocyanates can react with all compounds containing reactive hydrogen atoms, and the basic reaction may be represented as

$$\text{—N=C=O} + \text{H} \rightarrow \text{—NHCO—}$$
Active hydrogen

In such reactions the hydrogen normally becomes attached to the nitrogen atom and the remainder of the compound links to the carbon atom of the isocyanate group thus

$$\text{R—N=C=O} + \text{H—R} \rightarrow \text{RNHCOR}$$

The reaction products formed in this way all contain active hydrogen atoms which enable further reactions to take place. The reactivity of the urethane hydrogen is comparatively low, and this has certain advantages as will be seen later in the preparation of prepolymers.

Typical Reactions

(i) Isocyanate and Water. The reaction between an isocyanate and water leads to the formation of an unstable substituted carbamic acid which breaks down very quickly to give an amine and carbon dioxide.

$$\mathrm{R{-}N{=}C{=}O + H_2O \rightarrow RNHCOOH \rightarrow RNH_2 + CO_2}$$

Isocyanate	Water	Substituted carbamic acid	Amine	Carbon dioxide

(ii) Isocyanate + Hydroxy compound. This reaction leads to the formation of a urethane

$$\mathrm{R{-}N{=}C{=}O + R'OH \rightarrow RNHCOOR'}$$

Isocyanate	Hydroxy compound	Urethane (carbamic acid ester)

(iii) Isocyanate + Amine. This reaction which would, of course, follow on from the reaction with water leads to the formation of a substituted urea.

$$\mathrm{R{-}N{=}C{=}O + R'NH_2 \rightarrow RNHCONHR'}$$

Isocyanate	Amine	Diisocyanate urea

(iv) This could be an extension of the reaction with a hydroxy compound and leads to the formation of an allophanate

$$\mathrm{R{-}N{=}C{=}O + R'NHCOOR'' \rightarrow RNHCONR'COOR''}$$

(v) Isocyanate + Di-substituted Urea. This reaction could possibly follow from the reaction with an amine to form a substituted biuret.

$$\mathrm{R{-}N{=}C{=}O + R'NHCONHR'' \rightarrow RNHCONR'CONHR''}$$

Isocyanate	Disubstituted urea	Substituted biuret

(vi) Isocyanate + Acid. This reaction could lead to the formation of an amide, as follows:

$$\mathrm{R{-}N{=}C{=}O + RCOOH \rightarrow {-}NHCO{-}}$$

Amide

It is interesting to note that reactions (i), (iii), (iv) and (v) all take place during the production of a polyurethane foam, and the production of the allophanate can be prevented by strict control of the pH. Reaction (vi) is indicative of what can take place should acid impurities be present.

(vii) Isocyanate + Oxazolidine. This reaction takes place by the oxazolidine ring opening thereby providing active hydrogen sites to react with the isocyanate to yield urethanes. The chemistry is more fully described under section 5 on Surface Coatings.

POLYMERIZATION AND STRUCTURE

We have seen in Chapter 1 that if one of the reactants taking part in a polymerization mechanism has a functionality greater than two ($f = 2$) and the other reactant has a functionality of two ($f = 2$) then cross-linking occurs. This assumes that one of these coreactants is not present in large excess. Polymers can therefore form from a 2, 3 system which are thermoset, infusible, and highly insoluble materials and by continuing the interaction a highly cross-linked structure results. This forms a basis for preparing many of the polyurethane coatings which can be cross-linked by tri- or polyfunctional coreactants.

In the general reactions of isocyanates described earlier only monomeric materials have been represented, but it follows that plurality of reactive centers can lead to polymerization; therefore, if several mols of a polymeric hydroxy compound and a diisocyanate are used the resultant product is not a urethane but a polyurethane. This is represented as follows by showing only functional groups.

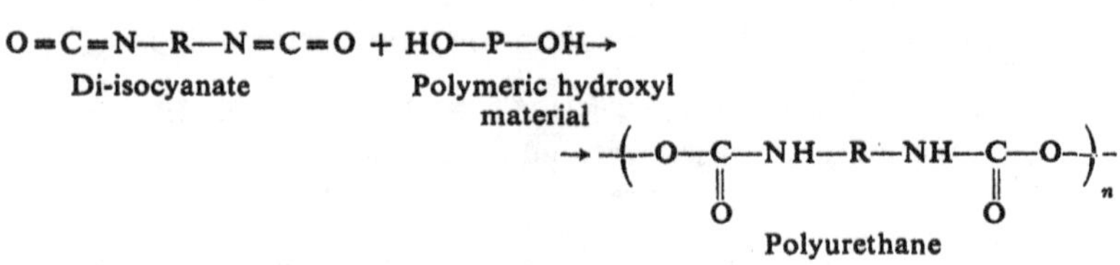

The ionomers are produced by reacting diisocyanates with ionizable nitrogen compounds. The reaction with mono- or bifunctional quaternary agents yield linear or cross-linked elastomers.

MANUFACTURE

Foams

There are two main types of polyurethane foam, viz, the rigid and the flexible. At present there are two techniques used to produce foamed polyurethanes. One method consists of forming a prepolymer (mixing a polyhydroxylic compound with a polyisocyanate) and then reacting it at a later stage with water, catalyst, and other additives to produce the final foam product. The catalyst is normally a tertiary amine and, in addition, an emulsifying agent (e.g., octyl-phenol/ethylene oxide condensate) as well as a structure regulator (e.g., silicone oil) are added. This process is sometimes referred to as the two-stage process. This is in contrast to the other method which is a one-stage process and is commonly termed 'the one-shot technique.' Essentially this latter process consists of mixing together a polyhydroxylic compound, a diisocyanate and an aqueous activator mixture. Water is generally necessary to produce the foaming which is effected by the liberation of carbon dioxide although foaming can also be produced by the addition of certain fluorocarbons. The water is mixed with the activator system which, in addition to the catalyst, contains small quantities of a silicone oil (stabilizer/structure regulator) and organometallic compounds to give quick gel strength. The resultant mixture may then be discharged in the liquid state into the mold, where the foaming takes place. It must be pointed out that the density of the foam is controlled largely by the ratio of the diisocyanate and water.

Rubbers

Much of the initial developmental work was carried out on reacting polyesters containing trifunctional alcohol groups with isocyanates. Unfortunately, elastomers with poor tear resistance were obtained. Similar results were also achieved from reacting triisocyanates with linear polyesters. Further work was subsequently carried out using a short-chain polyester with terminal hydroxyl groups and a bulky diisocyanate, the resultant product being of high enough mechanical strength to warrant commercial interest.

It is interesting to look at the steps involved in the building up of a polyurethane elastomer.

(*a*) An excess of a polyhydroxy alcohol (e.g., glycols, glycerol, etc.) is condensed with a dibasic acid (e.g., adipic acid) to obtain a short-chain linear polymer with a degree of polymerization around eight to ten, having terminal hydroxyl groups. If we let 'P' represent the bulk of the polyester molecule then we have:

$$\underset{\text{Ethylene glycol}}{HO{-}(CH_2)_2{-}OH} + \underset{\text{Adipic acid}}{HOOC(CH_2)_4COOH} \rightarrow \underset{\text{Polyester}}{HO{-}P{-}OH}$$

(*b*) An excess of a suitable bulky diisocyanate when added to this polyester results in chain-lengthening, through the urethane linkages so formed, giving long complex chains having terminal isocyanate groups.

(*c*) The terminal isocyanate groups so formed are reacted with a polyhydroxy alcohol (glycol) and this results in further chain-lengthening, and active hydrogen atoms are formed which are capable of being cross-linked.

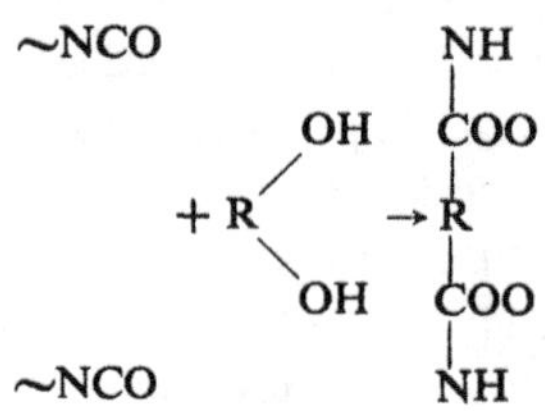

Urethane link

If water is used instead of the alcohol, carbon dioxide is formed, as we have seen previously from the foamed products, and it is therefore essential to remove all traces of water in the preparation of polyurethane elastomers.

In the early 1960's, the textile coating industry used one component urethanes. Technically these can be considered copolymers in which every molecule of long soft segments (polyester, polyether) is combined with short hard segments (urethane and urea groups). The hard segments form bridges between the molecular chains, resulting in an elastomer structure like vulcanized rubber. The reaction product of hexamethylene diisocyanate and 1,4 butane diol shows strong intermolecular bonds similar to the superpolyamides described in Chapter 12. Even though they are still relatively low in molecular weight, they have great difficulty in being converted to solutions and are only soluble in such solvents as cresol and formic acid.

(*d*) The cross-linking mechanism which takes place takes one of two forms; it can either be at the urethane or at the urea bridges which are formed. This latter method is effected by the addition of surplus diisocyanate. In actual practice it is usual to modify and catalyze the reaction by the addition of diamines or amino alcohols such as ethanolamine.

There are two different methods used for preparing polyurethane elastomers. One method consists of producing a prepolymer containing sufficient isocyanate to effect complete cure. These are termed cast rubbers. They have limited self-life, a defect which has been overcome in the second method by carrying out chain extension of the polyester or polyether with a diisocyanate. The proportions are calculated to give no free isocyanate end-groups and the final curing is effected by the addition of more isocyanate or by the conventional rubber-curing agents such as sulfur or certain organic peroxides.

Molding Compounds

Polyurethane molding compounds have certain elastomeric properties and could possibly have been included in the last section. They were originally based on the reaction between castor oil and a diisocyanate, used either in a one-shot or prepolymer technique. It is now obvious that in the first method the complete reaction between the diisocyanate and the other reactants are carried out by the end-user, whereas with the prepolymer system the diisocyanate and castor oil are partly reacted before being delivered to the customer. A typical formulation found suitable is:

Anhydrous castor oil	1.66 parts
Tolylene diisocyanate	3.5–4.0 parts
2-Ethyl hexane diol	1 part
Ethyl diethanolamine	0.2 part

The anhydrous castor oil and isocyanate are simply mixed together and degassed in vacuo, the resultant mixture is then poured into a female mold and then either allowed to cure by the action of a catalyst or by heating to somewhere in the region of 100°C.

These products based on castor oil have been replaced to a large extent by polymers obtained as the reaction product of tetramethylene glycol and hexamethylene diisocyanate. It has now become possible to injection mold and extrude these polymers with the resultant finished molding being harder than the castor oil types.

Surface Coatings

The production of surface coatings based on polyurethanes generally takes either of the following:

(i) Urethane Oils. This type of isocyanate coatings is one of the earliest reported uses of such materials in the surface-coating field. The diisocyanates combine with unsaturated drying oils and then dry in the same manner as conventional drying oils, that is by air oxidation. The oils are prepared by converting a conventional drying oil into a mixture of mono- and diglycerides and then reacting the free hydroxyl groups with a diisocyanate. If we take, for example, the conventional material linseed oil diglyceride which we can represent by 'L' then the reaction proceeds as follows:

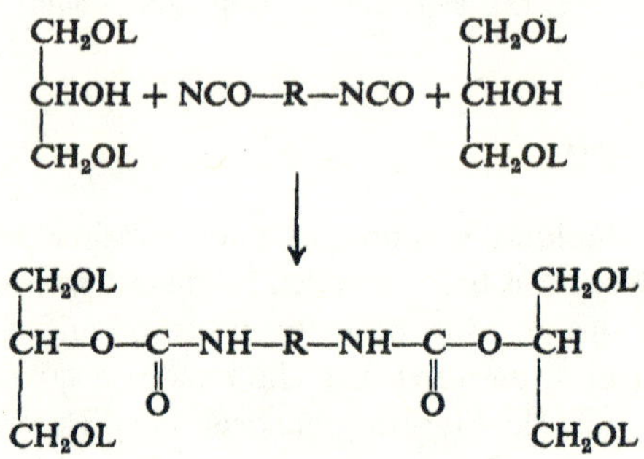

These oils so formed may be used as clear varnishes or pigmented and used as paints. As mentioned previously, the curing takes place by oxidation of the unsaturated centers in the drying oil residues, and autoxidation catalysts, usually cobalt salts (e.g., cobalt naphthenate) are used to promote oxidation and subsequent polymerization.

(ii) Moisture-cured Polyurethanes. These consist of adducts of hydroxy containing resins with a large excess of polyisocyanate. They are prepared by refluxing a polyol with a polyisocyanate or, alternatively, the polyol and polyisocyanate may be simply mixed and allowed to react at room temperature over a period of days. The adduct so formed is then dissolved in water-free solvents to form a lacquer. (N.B. Water must be absent at this stage otherwise premature curing takes place.) On applying this lacquer to a surface the moisture present in the atmosphere reacts with the isocyanate groups present in the resin, and subsequent reactions lead to gelation and the formation of a cured paint film.

(iii) Two-pot Systems. These are systems in which the high molecular weight polyester polyether is reacted and cured with a polyisocyanate at room temperature. When the two components are mixed the subsequent

coating will, of course, have a limited shelf-life. It is important to point out that at least a portion of the hydroxyl-containing material must contain three or more hydroxyl groups so that cross-linking results, as previously indicated. The degree of cross-linking can therefore, of course, be controlled by the amount of trihydric material present which, in turn, provides a means of 'tailoring' the final properties of the coating.

The chief characteristics of such coating systems are excellent gloss, hardness, good solvent and fuel resistance, and good chemical and water resistance. They may also be used in combination with two-pack epoxy primers and, as such, make ideal coatings for reinforced plastics.

A later development is the addition of a polyester with a molecular weight of 2000 to an adduct of three molecules of TDI with one mol of a trihydroxy compound. The resultant product is stable under storage and can be conveniently cross-linked for coatings as shown by a typical polyfunctional TDI based isocyanate. Other coatings for substrates such as textiles utilize amino-modified and TDI-preextended glycol adipates from solution in ethyl acetate or methyl ethyl ketone cross-linking with an addition of TDI-based NCO-containing isocyanurates. A typical commercial product is Desmodur TH (Bayer) which is claimed to be an adduct from 2,4 toluene diisocyanate (3.0 mols) and 3- methylol pentylene glycol- 2,4 (1 mol).

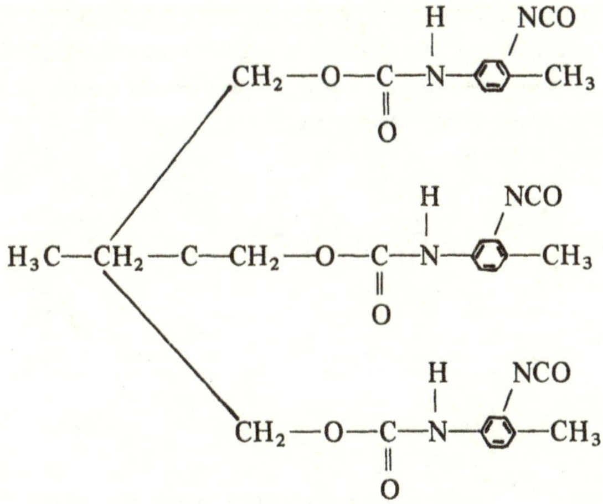

(iv) Blocked Isocyanate Coatings. We have seen previously that an isocyanate reacts with a monohydric alcohol or phenol to form a urethane which is not reactive with hydroxyl or other functional groups at room temperature. This is not so when heat is applied, when the urethane de-

composes to make the isocyanate group available for reaction. A blocked isocyanate prepared from phenol when mixed with a polyester or polyol gives a medium which has unlimited shelf-life. On heating to around 150°C isocyanate groups are set free and films are formed which exhibit excellent hardness and chemical resistance. The cleaved blocking agent diffuses into the surrounding surface of the coating or volatilizes into the atmosphere.

The formation and thermal cleavage of blocked isocyanates is shown as follows:

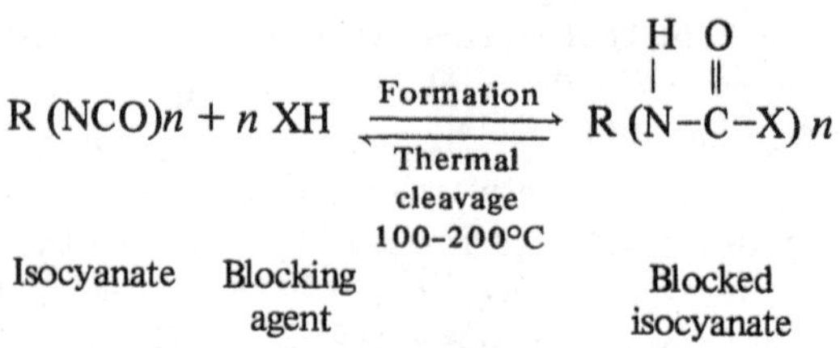

A few typical blocking agents are ethyl malonate, novolaks (PF resins), dimethyl amino methyl phenol, caprolactam, pyrocatechol, phenol, and ketoximes.

Commercial materials include Hylene MP from Du Pont which is a phenol blocked MDI, Mondur S from Mobay, a phenol blocked TDI adduct, and Isonate 123P from Upjohn which is a caprolactam blocked PAPI. A relatively new group of products are the aminimides from Ashland Chemical Company and it does not have the hazardous phenol as a by-product. Their chemistry is as follows:

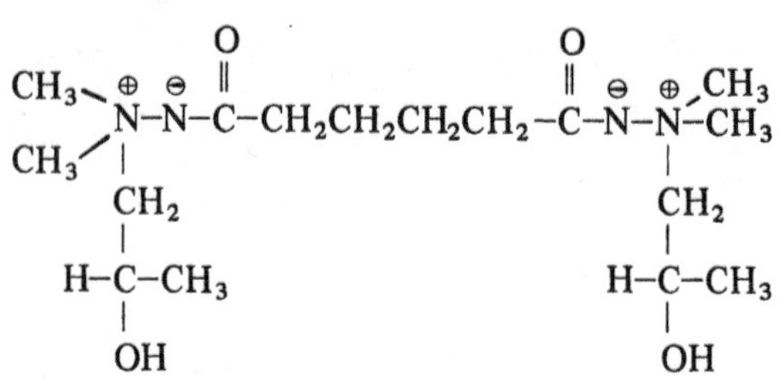

These materials are advantageous, apart from not having phenol as the by-product, because they can be used in conjunction with aqueous dispersions. The above material is sold commercially under the number AL-X-300.

The following are the structures for two blocked isocyanates in common use.

(i) Hylene MP (Methylene-bis-phenyl-(4-phenyl carbamate).

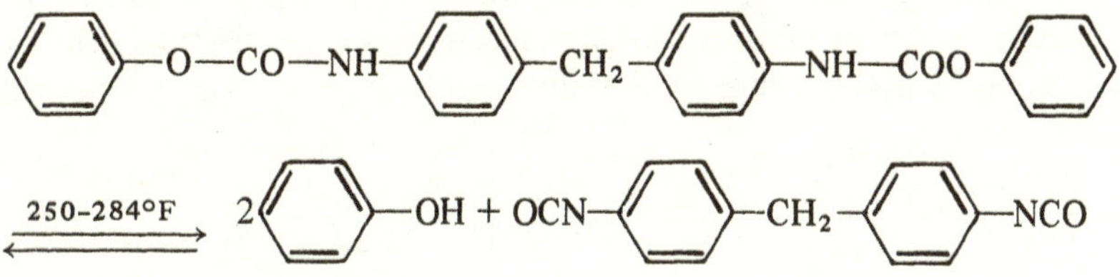

Phenol MDI

(ii) Isonate 123P (ε-caprolactam-blocked PAPI).

PAPI–polymethylene polyphenyl isocyanate.

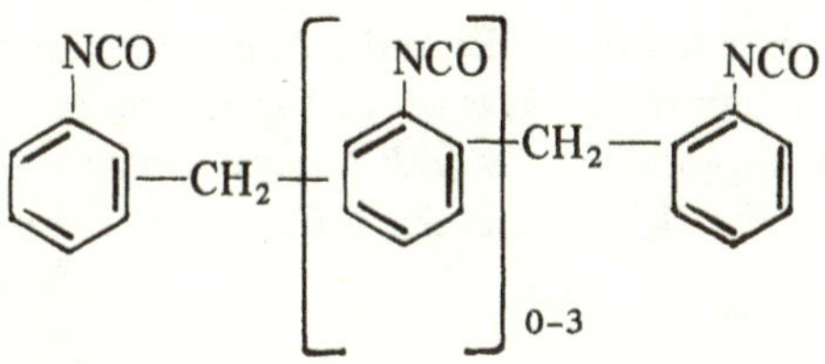

Structure of blocked PAPI.

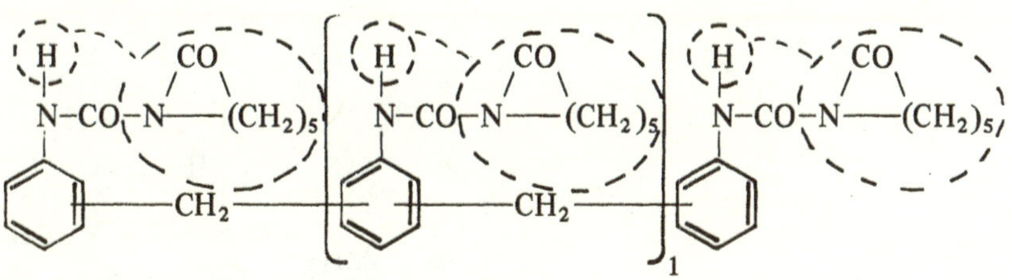

Acrylic Urethanes

These polymers may also be termed oxazolidine-acrylics.

Many polymers are now commercially available which react with isocyanates to form useful coatings. A new class of materials is claimed to have unique properties and offer fast cure without the addition of a catalyst. Such polymers possess a unique functionality as a site for cross-linking with isocyanate polymers. This is stated in a patent by W. D. Emmons, U.S. Patent 3,743,626 (1973) assigned to Rohm and Haas Company.

This grouping is known as oxazolidine and reacts with isocyanates when activated by moisture. In the absence of moisture the equilibrium of the oxazolidine reaction is heavily in favor of the blocked reactive site so that minute amounts of moisture do not cause premature activation. The following reaction shows this equilibrium:

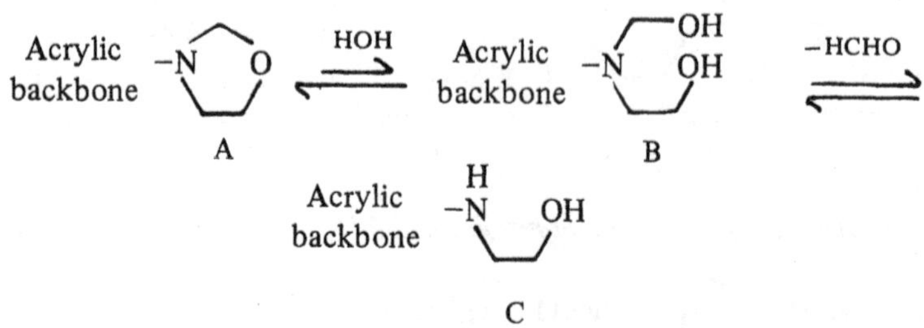

The ratio of components A, B, and C is approximately 99:0.5:0.5 at equilibrium. When an isocyanate is added to the system, it has been found to react rapidly and preferentially with the amine-active hydrogen, pulling the equilibrium to the right and activating more of the oxazolidine functionality (A).

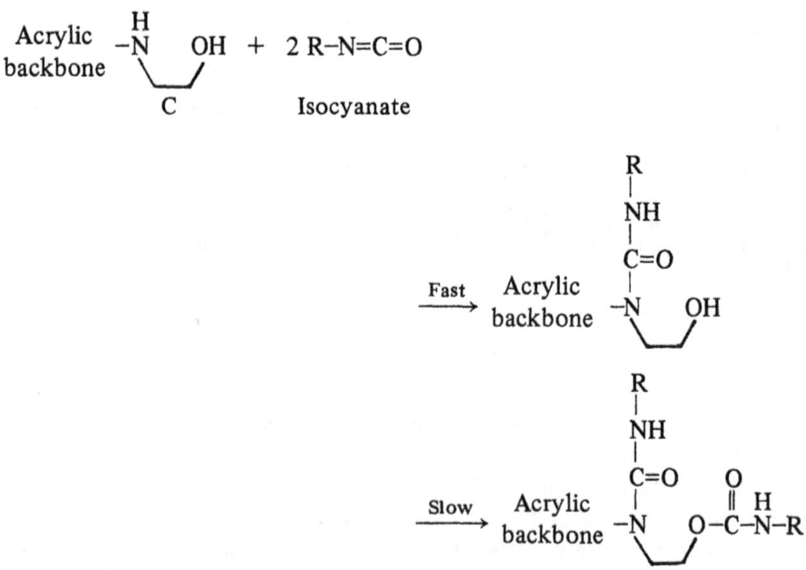

Urea-Urethane intermediate

D

This reaction has been found to be much faster than a conventional hydroxyl-isocyanate reaction and proceeds without the need for a catalyst at a very fast rate. The reaction of intermediate D with more intermediate C leads to cross-linking.

Water is a necessary component of the oxazolidine reaction with isocyanates. A sufficient amount for a room temperature cure is normally available in the atmosphere. With heat curing systems, it is advisable to inject steam into the oven, otherwise rapid cross-linking does not take place.

Adhesives

Polyurethane adhesives can be prepared by dissolving an hydroxylic compound such as based on adipic acid and a triol, in 80 percent ethyl acetate and then mixing three parts of the resultant product with one part of an isocyanate (e.g., tolylene diisocyanate). The adhesive is then first spread on the substrate being bonded and the solvent is allowed to evaporate. Final curing is carried out either at room temperature or accelerated by heat.

Fibers

It is probably true to say that polyurethane fibers have yet to reach any major degree of commercial importance. Some use has been found, however, as monofilaments which are obtained from the reaction of hexamethylene diisocyanate with tetra methylene glycol or butane diol. Polyurethane fibers have a softening point around 170°C (cf. polyamides 280°C) and a melting point as low as 183°C which has severely restricted their use for many textile applications. But they do exhibit certain outstanding characteristics which are of interest, including good abrasion, oxidation and acid resistance as well as elasticity. They may be used either alone or in admixture with other synthetic fibers.

The early preparation of polyurethane fibers involved the reaction of a diisocyanate with a diamine to give a polyurea. They were infusible and strongly hydrophilic and therefore were of limited interest for fiber application. In later developments, when the diamine was replaced with a diol of the same general molecular weight, a polymer was obtained which had useful properties. A typical reaction would be between 1,4 butane diol and hexamethylene diisocyanate. This gave the following structure:

$$-[(CH_2)_4OOCNH(CH_2)_6NHCOO(CH_2)_4OOCNH(CH_2)_6NHCOO]-$$

PROCESSING

Foams

As stated earlier, rigid and flexible polyurethane foams can be prepared by accurately mixing an hydroxylic rich material with an isocyanate, catalyst, water, and other additives and finally dispersing them into specific shaped areas. The actual mixing procedure can be carried out either by hand or with equipment which can be operated intermittently or continuously. A typical machine might feed the resin at the rate of 4 liters/minute; the diisocyanate at 2 liters/minute at the rate of 9000 injections/minute; and the activator solutions at 0.4 liters/minute at the rate of 3000 injections/minute. The material is then thoroughly mixed in the machine and is then dispersed into the mold where it foams (*see* Plate 9). If the molds are for making slabs of material, which is often the case, they move past the mixing head continuously while the head transverses the width of the mold. After foaming is complete, the slabs may be allowed to either set and self-cure or they are placed in curing ovens. The slabs are then cut according to requirements. One great advantage of these materials is that they can be foamed *in situ* which enables them to be used to fill awkward voids and cavities.

Rubbers

The liquid-cast rubbers are formed in a two-stage process which does not permit the incorporation of fillers. Processing techniques are similar to those used in the casting of epoxy and polyester resins.

Certain polyurethane rubbers now available are suitable for processing on conventional rubber machinery using conventional techniques (*see* Chapter 20). They are cured by means of a high boiling point liquid isocyanate and they differ from the conventional natural and synthetic rubbers in that they do not require accelerators in processing. Unlike the cast rubbers, these materials may be compounded using fillers such as carbon black and silica.

Reaction Injection Molding (RIM)

This process has been developed in an attempt to combine the best features of conventional fast injection molding with the liquid casting techniques to produce large elastomeric polyurethane moldings. It has been discussed previously that a polyurethane is produced as a result of the reaction of a polyester or polyether (polyol) with a polyisocyanate. Cellular products are produced using a polyol of molecular weight in the region of 3000 and an isocyanate such as TDI or the more reactive MDI, with a catalyst and blowing agent. For the RIM process, precise amounts of these reactive components are metered to a mixing head where the liquid streams are homogenized and injected under relatively low pressures into a closed mold. Immediately the materials begin to react and foam slightly giving a micro cellular structure. Reaction times of two to three seconds are used and mold residence times of one minute can also be achieved. As an example, a 10-pound component could be molded with a shot to shot cycle time of under two minutes.

This technique has high economic advantages in its ability to mold large parts of high surface quality at production rates competitive with injection molding, while offering lower tooling and equipment costs. Other advantages include the ability to mold thin-walled articles and the requirement for only a low energy consumption in molding.

There has been considerable work carried out in developing new chemicals for this process and recently new materials have been introduced with a high flexural modulus of up to 300,000 psi. These materials have become contenders for sheet metal replacement, (fenders and body panels) on automobiles. The original RIM materials were only suitable for automotive facia panels and self-skinning low density foams.

The ability of this new process to make large parts quickly, in high automotive production, suggests many other possible applications for production of both low and high modulus parts. It is claimed that the ultimate goal for production cycle time is two minutes but even at the four minutes which is achievable at present puts the cost about equal to injection molded parts. One big attraction of this technique is that it requires less expensive equipment than injection molding techniques, which require high torque extruders and very high tonnage clamps to process high viscosity melts and mold large parts. The molds can also be fabricated from less expensive materials such as aluminum-field epoxy toolings rather than steel.

PROPERTIES

Foams

Many of the foams produced today use polyethers for those of the flexible kind and polyesters for the rigid ones. The following is a range of typical properties exhibited by polyurethane foams, although it must be emphasized that such properties have a bearing on their selection for particular end uses.

Foams can be produced with densities ranging from as low as 1.5 lb/ft^3 to as high as 40 lb/ft^3, depending on the choice of the base raw materials. For the cushioning field, however, flexible open-cell foams of density about 2 lb/ft^3 are to be preferred. The load capacity of these foams is some four times as high as that of natural rubber latex foam of equivalent density (Fig. 8.1). Tensile strengths varying from 18 to greater than 300

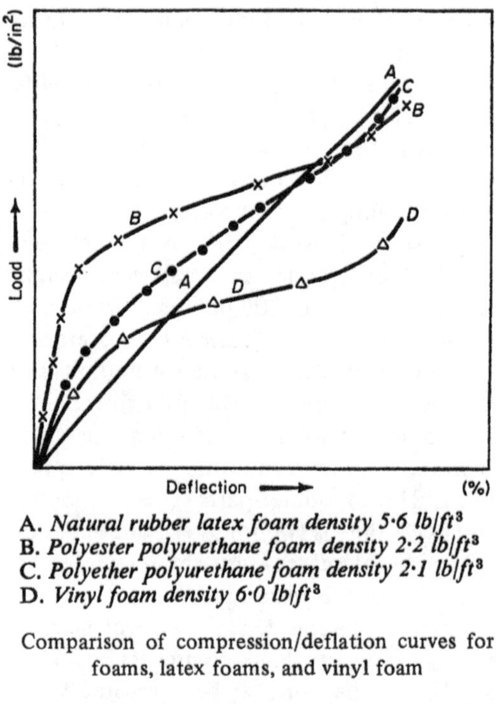

A. *Natural rubber latex foam density 5·6 lb/ft³*
B. *Polyester polyurethane foam density 2·2 lb/ft³*
C. *Polyether polyurethane foam density 2·1 lb/ft³*
D. *Vinyl foam density 6·0 lb/ft³*

Fig. 8.1 Comparison of compression/deflation curves for polyurethane foams, latex foams, and vinyl foam

lb/in.^{2} covering the range flexible to rigid, can be achieved. 25 lb/in.^{2} is a common figure for the flexible foams prepared from polyether polyols as against 18-20 lb/in.^{2} for polyester-based foams. Foams with an elongation of 200 percent are possible, and foams with an elongation as high as 300 percent have been formulated for the cushioning field. Flexible polyurethane foams are claimed to have considerably higher tear strength than natural rubber foams. A property that is of prime importance in the cushioning field is that of compression set. Figures as low as 10 percent are possible with polyurethanes (Fig. 8.2).

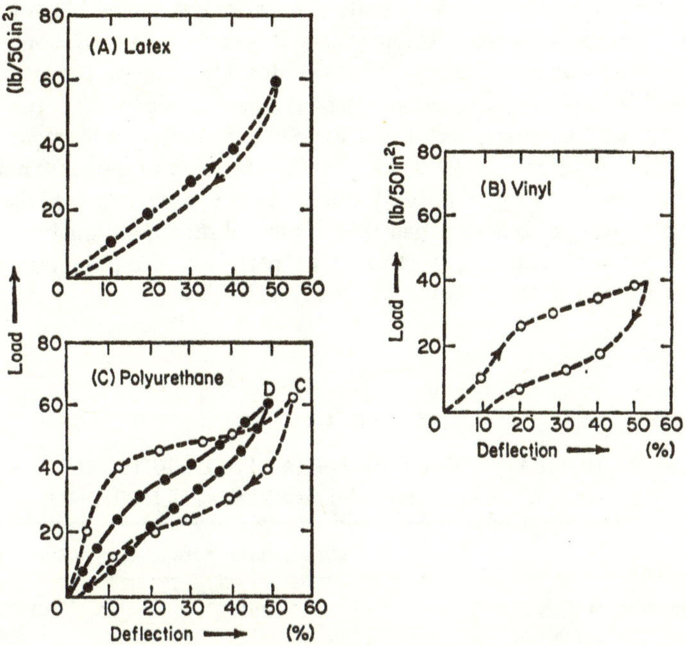

A. *Natural rubber latex foam density 6·9 lb/ft*^{3}
B. *Vinyl foam density 6·0 lb/ft*^{3}
C. *Polyester polyurethane foam density 2·7 lb/ft*^{3}
D. *Polyether polyurethane foam density 2·7 lb/ft*^{3}

Fig. 8.2 Comparison of hysteresis curves for polyurethane foams, latex foams, and vinyl foams

These foams have good resistance to rot, vermin, and many organic solvents. They burn only in the presence of a flame and can be made self-extinguishing. The polyurethane foams have good thermal and acoustic insulating properties and also good radar, radio, and X-ray transmission properties.

Rubbers

Polyurethane rubbers, if correctly compounded and cured, exhibit excellent physical characteristics, including high tensile and tear strengths. The resistance to mechanical abrasion is high and they also have excellent shock-absorption properties. Their main disadvantage is the relatively poor resistance to hydrolytic agents (e.g., acids), steam and high temperatures. They do, however, exhibit excellent resistance to oxidation, uv light and they do not burn readily. These elastomers also have the outstanding characteristic of extreme resistance to aromatic hydrocarbons (oils, etc.). They are not, however, very resistant to chlorinated hydrocarbons such as methylene chloride and will show a volume increase of over 300 percent if immersed in this solvent for only a few days. The resistance of these elastomers to ketones is better than that of natural rubber although inferior to the polychloroprenes. The physical properties of the polyurethanes are compared with those of natural rubber in Table 8.1.

Table 8.1

COMPARISON OF PHYSICAL PROPERTIES OF POLYURETHANE AND NATURAL RUBBER

Test	*Polyurethane rubber*	*Natural rubber*
Tensile strength (lb/in.^{2})	5000	2500
Elongation at break (%)	650	700
Shore hardness	65	40
Tear strength (lb/in.^{2})	2200	1500
Resilience at R.T. (%)	65	70
Oil and chemical resistance	Good	Fair

MOLDINGS

Moldings obtained from polyurethane resins will absorb a considerable amount of shock and exhibit little shrinkage on curing. The typical values obtained on measurement of the physical properties of certain polyurethane molding materials are given in Table 8.2.

Reaction Injection Molding

Both the conventional and high modulus types have essentially a solid skin with a cellular core. Typical specific gravity is 1.0 (about 62 lb/ft^3) although developmental work is being carried out to produce foamed and nonfoamed products with densities down to 0.6. Micro cellular RIM materials in general have density reductions of 5-10 percent and this may be

Table 8.2

PHYSICAL PROPERTIES OF POLYURETHANE MOLDING MATERIALS

S.G. at 25°C	1.1-1.5
Shrinkage on cure (%)	2–3
Hardness shore A	50–80
Tensile strength (lb/in.^{2})	800-2000
Elongation at break (%)	200-500
Dynamic resilience	8
Working temperature range (°C)	–80–+125
Water absorption (%)	0.1
Volume resistivity cohm (m)	1×10^{12}
Dielectric strength (V/mil)	350–450
Dielectric constant	3–4
Chemical Resistance	
Concentrated acids	Resistant
Concentrated alkalis	Resistant
Oils and fuels	Resistant
Steam	Resistant
Solvents	Resistant
Chlorinated hydrocarbons	Attacked

Table 8.3

PROPERTIES COMPARISON OF ELASTOMERIC AND HIGH-MODULUS RIM URETHANES

	Elastomer (fascia-type)	*High modulus (RIM 125*)*
Density (lb/ft^3)	62	62
Shore D hardness	50–54	70
Tensile strength (psi)	2500–3000	4100–4500
Elongation (%)	200–300	100–120
Die C tear strength (pli)	400–500	750–850
Flexural modulus, 10^3 psi		
–20°F	60–70	220–260
75°F	19–24	120–125
158°F	8–11	37–45
Modulus multiple (–20 to 158°F)	6.8	5.8
Sag (4" OH, 1 h @ 250°F)–in.	0.3–0.5	0.3–0.6
Dart impact break (°F)	–50	–
Izod impact (ft-lb/in.)	–	5.5

*Union Carbide Corp. urethane system

reduced even further by the addition of nitrogen or air. The hardness is generally between 65–75 Shore D. The properties can be modified by glass reinforcement as with conventional thermoplastics and the preferred reinforcement for these new cellular products is milled glass fiber. Expected improvements include higher stiffness, heat-deflection temperature, and tensile strength with lower mold shrinkage. The properties of elastomeric and high modulus RIM urethanes are given in Table 8.3 and the effect of adding glass fiber is illustrated in Table 8.4.

Table 8.4

HOW GLASS FIBERS AFFECT PROPERTIES OF SF-4*

Glass level in polymer (wt %)	*0%*	*10%*	*20%*
Overall density (lb/ft^3)	66.0	72.7	76.2
Tensile strength (psi)	4300	4350	4000
M_{20}	3500	3800	
M_{50}	4100		
Elongation (%)	75	30	20
Set %	31	9	3
Shore D hardness	71	72	76
Flex mod. @ 73°F (psi)	100,000	130,000	218,000
Die C tear (pli)	360	350	280
Sag (4" oh, 1 h @ 250°F)–in.	.20	.19	.09
Linear coefficient of expansion (in./in./°F)	7.11×10^{-5}	4.75×10^{-5}	2.48×10^{-5}

*Mobay Chemical Corp. experimental high-modulus urethane system

SURFACE COATINGS

These materials exhibit excellent adhesion to many surfaces (including leather and rubber), high gloss, and abrasion resistance. The weather resistance of polyurethane coatings is also good and they are therefore finding many uses in outdoor finishes. They have good electrical insulating properties, low permeability to gases, and good chemical resistance. They are regarded as superior to alkyds in many respects, although they are somewhat more expensive.

APPLICATIONS

Foams

The major part of the polyurethane flexible foams goes into cushioning materials, although certain of the polyester-type foams are not used due to their high hysteresis properties (i.e. poor recovery after compression—*see* Fig. 8.2). They are now firmly established in furniture cushioning, where both slab and molded shapes are used for seating, chair backs, and arms. They are also used extensively in automobiles for car upholstery, crash pads, instrument panel trim, and for arm rests, which can also be used in conjunction with a 'vinyl' skin. Carpet underlay also utilizes these foams and the packaging industry is a big potential market. A recent development has been the lamination of the foam to textiles, giving what are commonly known as 'foam-backs.' These have become important in the clothing industry where such materials are used for interlinings or as thermal insulation against the outer cloth surface.

Most of the applications found for rigid polyurethanes have utilized their combination of good thermal insulation properties, structural strength, buoyancy characteristics, and shock resistance combined with light weight. They have, therefore, been used as a lightweight reinforcement in conjunction with a skin of polyester/glass laminate in the aircraft industry. Such applications include radome and antenna housings. Sandwich construction of rigid polyurethanes with metals and certain plastics have a high potential in the building industry. Rigid foams, blown by using fluorocarbons have been used both in industrial and domestic refrigerators as insulating materials.

Rubbers

One of the largest potential markets available to polyurethane elastomers is the rubber-tire market. The consumption of both natural and synthetic rubber for tires in Great Britain is over 150,000 tons, and this presents a potentially large market. Tires have already been produced from a polyurethane in U.S.A. but have been more of a research project than a commercial proposition for cars although now they are used on the wheels of fork-trucks. Their potentialities are obvious and the high abrasion resistance exhibited by such elastomers is of course of prime importance for use as a tire material. Unfortunately, this is not sufficient to offset the high cost of this material in comparison with conventional rubbers. Other disadvantages which must be overcome for this application include the low resistance to heat build-up of this elastomer.

The shoe industry has utilized the high mechanical properties of these elastomers and it has been reported that rubber heels fabricated from a polyurethane give a life of some eight times that of conventional rubber heels. Printing rollers are also produced from the cast polyurethane elastomers and give excellent performance under severe working conditions. It has also been suggested that such rollers give nearly three times the life of the conventional 'composition' rollers.

Miscellaneous applications include mallet heads, shock absorption pads, oil-seals, diaphragms, and a host of aircraft applications.

Surface Coatings

The four main fields of application for polyurethane coatings are as clear finishes, industrial finishes, marine finishes, and floor finishes. They can be used for coating wires, rubber, steel, reinforced plastics, and wood. They are used for coating rubber in order to improve ozone resistance and also impart chemical resistance to substances which hitherto were prone to such attack. One use of particular interest is that of polyurethane coatings to achieve a patent leather finish.

While they are not specifically surface coatings, the aqueous dispersions based on polyurethane ionomers are included under this section. As a result of interactions between chains, the properties of these materials are similar to coatings based on cross-linked elastomers. These ionomers, when placed in polar organic solvents, spontaneously form aqueous dispersions on the addition of water. The physical properties of the dried films are not satisfactory and have not gained acceptance in the coating industry. Now however, they are being used for tie coats in textiles and films.

Adhesives

Urethane polymers can be used to bond metal to metal, rubber to fabric (tire cords), porcelain to porcelain, and phenol-formaldehyde to a wide variety of other plastic substrates. Considerable use is now being made of these polymers for laminating textiles to textiles and textiles to foam. The resultant bond is both washable and dry cleanable. The process for bonding normally consists of reacting a two-component system of polymer and isocyanate in solvents such as ethyl acetate or methylene chloride. The addition and reaction of the isocyanate *in situ* leads to chemical bonding rather than the simple mechanical types that result in a high level of adhesion being obtained.

Miscellaneous

Polyurethane fibers have yet to fulfill the early promise and their uses fall far short of the polyamides. Uses recommended for these urethane fibers include filter cloths, ropes, and cable insulation. Over recent years they have found use in hosiery and underwear and there is an increasing application for them as the elastic component of ladies' swim-wear and foundation garments. One such product which has made an impact on the textile field is Spandex fiber. This is a generic name for a fiber-forming substance composed of at least 85 percent of a rubbery polyurethane.

Another use which is yet to fulfill the earlier promise is that in poromeric coatings for shoe uppers. These are thermoplastic polyurethanes which have been dissolved in a solvent such as dimethyl formamide and then coagulated to give a porous structure which enables the coating to breathe like leather with the added claim of improved abrasion and scuff resistance. Corfam was one of the earlier types of these materials introduced by Du Pont in the U.S.A. but was withdrawn later for a variety of reasons and now an English product named Porvair is being used extensively as well as in the U.S.A. by Inmont Corporation which has an arrangement with the English company.

Thermoplastic polyurethanes are used in a variety of coating applications for apparel because of the appearance and "hand." Such coatings of these polymers are claimed to provide better breathability than PVC and are more dry cleanable.

The polyurethane molding materials mentioned earlier in the chapter based on castor oil were largely used for potting and encapsulation. The latter types which are harder and have good resistance to abrasion with low coefficients of friction between them and other materials are finding uses for pinions, bushings, and high precision parts in electrical and mechanical systems. They are advantageous where humid conditions exist and are used in washing machines.

SUGGESTED FURTHER READING

Advances in Urethane Science and Technology. (1976) Edited by Professors Kurt C. Frisch and S. L. Reegan (USA: Technomic Publishing Co.).

Urethanes in Elastomers and Coatings. (1973) Journal of Elastomers and Plastics (USA: Technomic Publishing Co.)

Urethanes in Coated Fabrics. (1973) Journal of Coated Fabrics (USA: Technomic Publishing Co.).

The Development and Use of Polyurethane Products. (1971) E. N. Doyle (New York: McGraw-Hill).

Section B Thermoplastics

9.

POLYOLEFINS

The class of materials formed by the polymerization of certain unsaturated hydrocarbons, known as 'olefins,' is an extremely important one. The main members are low- and high-density polyethylenes and polypropylene together with some copolymers. They are conveniently studied under one chapter heading since, on the whole, they form a family of polymers with a particular range of properties. Technically, polyisobutylene and polyisoprene are also polyolefins, but their properties render them more suitable for study in the chapter on synthetic rubbers (Chapter 21). Common usage has, in fact, made the term 'polyolefins' synonymous with the polymers and copolymers of ethylene and propylene, and this meaning is normally implied wherever the term polyolefins' occurs in the text.

Two distinct types of polyethylene are known and these are dealt with separately. Although there is now available a broad spectrum of polyethylenes, due to copolymerization with small amounts of other olefins such as butene-1, or to the blending of the two types of polyethylene, the respecctive methods of manufacture deserve separate mention. The influence of molecular structure on physical and chemical properties is also well illustrateed by the two polymers.

Since one of the basic differences in the two polyethylenes is that of density they are often referred to as low-density and high-density polyethylene. Another system of nomenclature is derived from the fact that the manufacturing conditions are very different. In particular, one in made under conditions of high pressure (15-30,000 lb/in.^{2}), while the other uses pressures which are normally below 500 lb/in.^{2} Unfortunately, the high-pressure process produces the low-density polymer and vice-versa, and this has often led to confusion in the past.

Future reference in the text to polyethylene uses the terminology 'low-density' and 'high-density', to describe the two main types of polymer.

POLYETHYLENE

History of Development. Low molecular weight polymers of ethylene were prepared as early as 1879, while a condensation process using diazomethane was described in 1900.

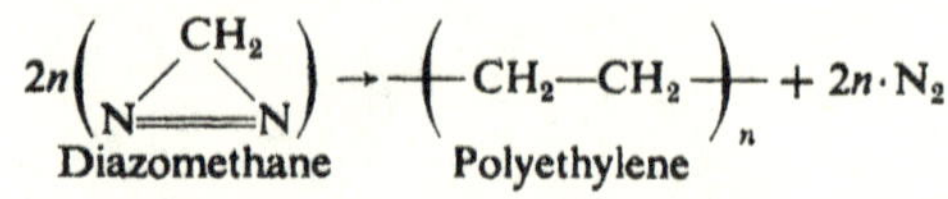

In 1913, Ipatieff polymerized ethylene using metallic halides as catalysts:

$$\underset{\text{Ethylene}}{n\cdot CH_2{=}CH_2} \xrightarrow[\text{metallic halides}]{400^\circ C \ 70 \text{ atm.}} \underset{\text{Polyethylene}}{\left(CH_2{-}CH_2 \right)_n}$$

Again, only low molecular weight products were obtained.

The first high molecular weigh product was obtained by Imperial Chemical Industries Ltd., in 1933, during a pure research program devoted to the effects of very high pressures on chemical reactions. Pressures of the order of thousands of atmospheres were used and special equipment had to be developed to withstand these pressures. During the course of experiments on the reaction between ethylene and benzaldehyde, a white powder was discovered in the reaction vessel and this was later found to be a polymer of ethylene.

Repetition of the experiment using ethylene alone led to the explosive decomposition of the ethylene and damage to the apparatus. Two years later, work was again initiated on the project using improved apparatus which had been developed in the meantime. It was here that chance intervened, as so often happens in even the best planned research. A fall in pressure occurred during the reaction and fresh supplies of ethylene were added to restore the pressure to its original value. When the experiment was completed, about 8 grams of a white powdery solid (polyethylene) was obtained. The success of this experiment was eventually traced to the fact that the replacement ethylene had contained, by chance, the right amount of oxygen to catalyze the formation of the polyethylene. The real element of chance in this whole affair was the occurrence of the pressure drop, which was greater than could be accounted for by polymerization and which must, therefore, have been due to a leak at one of the joints in the apparatus.

It was not until 1935 that sufficient material was obtained for extensive evaluation of the polymer but it was soon realized that the material had a great many properties of interest to the electrical industry, particularly as an insulating material for underwater cables. The first full-scale manufacturing unit came into use in September 1939 and its output was used almost exclusively for radar equipment, thereby making a vital contribution to the war effort.

After the war the search for new uses for polyethylene led to the development of extruded film (for packaging, agriculture and buildings), bottles, toys, household goods, cold-water piping, and many other uses.

The next major step came in the early 1950's when Professor Ziegler, in the course of work on organo-metallic compounds, discovered catalysts that made possible the polymerization of ethylene at near atmospheric pressures. At about the same time, two other low-pressure methods were developed in the U.S.A. One of these was associated with the name of Phillips Petroleum Co. and the other with Standard Oil of Indiana. Apart from the natural interest attaching to a completely novel method of producing polyethylene, the discovery was of immense importance because the resultant product was found to differ in several important respects from the conventional polymer. As already mentioned, one difference was in density, and this has become the commonest means of distinguishing between the two polymers.

In recent years (1960 onwards) there has been a great deal of process development designed to prodouce low density polyethylene either by *low pressure* gas phase polymerization or by liquid phase processes similar to those used for producing high density polyethylene. Although these newer processes did indeed produce low density polyethylene, there are still differences between the conventional low density polyethylene and the newer polymers. These latter are known as linear low density polyethylenes (or LLDPE) and are similar in structure to high density polyethylene but with slightly longer and more numerous side chains. LLDPE is a copolymer, the main comonomers being butene, hexene and octene.

PREPARATION OF MONOMER

Ethylene has the formula $CH_2 = CH_2$. The double bond has a functionality of two so that the ethylene forms predominantly linear molecules. It is a gas, boiling at −104 °C at atmospheric pressure, and can be prepared by several methods, some of which are given below:

(a) In the laboratory, ethylene can be prepared by the dehydration of alcohol.

$$\underset{\text{Ethyl alcohol}}{C_2H_5OH} \xrightarrow[\text{acid.}]{\text{Conc. sulfuric}} \underset{\text{Ethylene}}{CH_2{=}CH_2} + H_2O$$

Recovery of the sulfuric acid is difficult and phosphoric acid is sometimes used instead. Even better is the use of toluene p-sulfonic acid which, with alcohol vapor and steam, is passed over aluminum oxide at a temperature of 350 °C. The acid can be recovered by crystallization.

(b) Industrially, ethylene can be obtained by the cracking of natural gases. For example

$$\underset{\text{Propane}}{CH_3{\cdot}CH_2{\cdot}CH_3} \longrightarrow \underset{\text{Ethylene}}{CH_2{=}CH_2} + \underset{\text{Methane}}{CH_4}$$

(c) Ethylene is also obtained by the thermal cracking of light oil fractions from oil refineries.

The latter process is the one in common use in the U.K. In all cases very careful purification of the ethylene is necessary since purities of at least 99.8 percent are required for polymerization. Oxygen is removed by passage over a reduced copper catalyst or by treatment under pressure with molten sodium at 150 °C. Acetylene (which is a cross-linking agent) is removed by catalytic hydrogenation.

POLYMERIZATION

As already mentioned, there are two fundamentally different processes for the production of polyethylene and examples of each of these are described below.

High-pressure Process

The polymerization of ethylene can occur under a wide range of pressures and temperatures. Most commercial high-pressure processes, however, utilize pressures between 1000 and 3000 atmospheres and temperatures between 100 and 300 °C. Temperatures above 300 °C are not normally used as the polymer tends to degrade. Various catalysts have been used, oxygen being the first. The concentration of oxygen is critical and optimum yields are obtained with about 0.05–0.06 percent oxygen. Other catalysts used are organic peroxides (such as benzoyl peroxide and ditertiary butyl peroxide), metal alkyls, and azo compounds. The reaction is highly exothermic and one of the early difficulties is this process was the removal of excess heat from the contents of the reaction vessels.

The highly purified ethylene is first led over a reduced copper catalyst to remove oxygen. The precise amountn of oxygen necessary to act as an initiator is then added and the gases are compressed in multistage compressors. They are then pumped into the reaction vessel by compressors of special design. Two main types of reaction vessel are used, the autoclave and tubular. In both cases provision must be made for the close control of catalyst concentration, temperature, and pressure. Unconverted ethylene is separated from the molten polymer and is recycled. The polymer is extruded as a continuous band or ribbon, solidified by cooling and cut into granules. Where further homogenization is needed as, for instance, in film and certain other grades, the product may undergo further processing in refiners, Banburys or screw extruders. A simplified flow-sheet of the high pressure process is given in Fig. 9.1. The vigorous

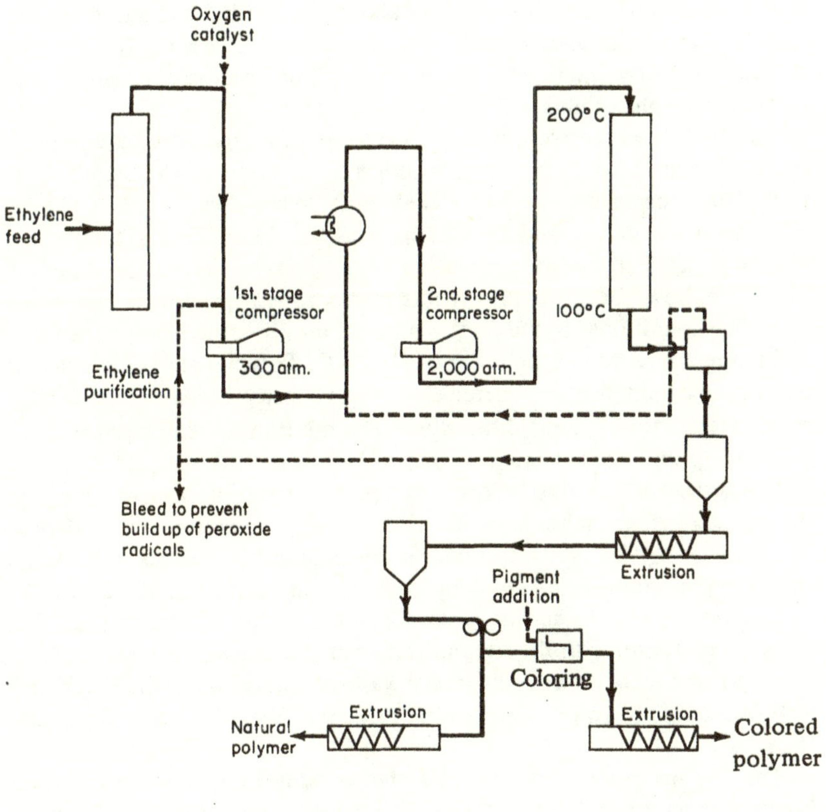

Fig. 9.1 Simplified flow scheme for the high-pressure polymerization of polyethylene

nature of the reaction at high pressures leads to a great deal of chain branching and this has as important bearing on the properties of the product.

Low-pressure Processes

The essential difference between the low-pressure and high-pressure polymerization of ethylene lies in the catalyst used. The catalysts discovered by Professor Ziegler to give polymerization at near atmospheric pressures stemmed from his work on organo-metallic compounds, particularly the organo-aluminum compounds. The first catalyst system to give satisfactory results was a combination of aluminum triethyl and a titanium derivative such as the tetrachloride. The active catalyst is actually a reaction product of the two materials since neither will polymerize ethylene to high molecular weight products if used alone.

The Phillips process utilizes such catalysts as partly reduced chromium oxide supported on steam activated silica/alumina, while the Standard Oil of Indiana process makes use of nickel oxide on charcoal.

In both the Ziegler and the Phillips process, the catalyst system is suspended in a liquid hydrocarbon through which the ethylene gas is passed. There are, however, some differences in subsequent procedure and in the operating conditions. In the Ziegler process, pressures are near atmospheric and temperatures are of the order of 50-75°C. Polymer granules settle out and the resultant slurry is stirred until the viscosity becomes high enough to interfere with efficient dispersion. The mixture is then passed through working-up and solvent recovery stages. In general, the sequence is catalyst deactivation, catalyst decomposition and removal, solvent recovery, drying, extrusion, and granulation (Fig. 9.2). Complete removal of catalyst residues is extremely important as the electrical properties of the polymer may be seriously affected by their presence. One method used for Ziegler polymers is to add dry hydrochloric acid gas which forms alcohol soluble complexes with the titanium (the titanium being the more difficult metal to remove). After preliminary washing with alcohol, the polymer is well washed with water. The resultant slurry is filtered or centrifuged and then dried.

The Phillips process operates at higher pressures (400-500 lb/in.^{2}) and higher temperatures (100-175°C). The liquid medium is usually cyclohexane and at the temperatures used, the polyethylene dissolves, and is removed as an approximately 10 percent solution in cyclohexane. The cata-

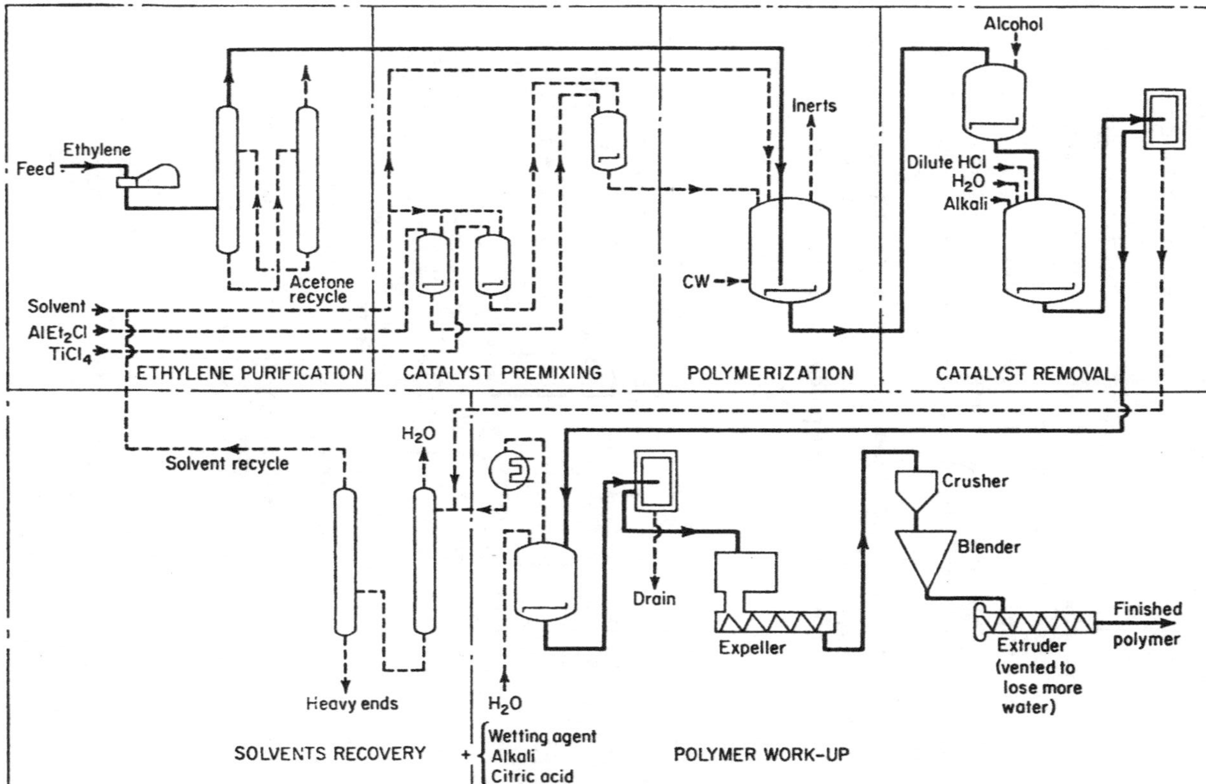

Fig. 9.2 Simplified flow scheme for polyethylene manufacure by Ziegler process

lyst is removed quite simply by hot centrifuging. The polymer is obtained by cooling, to effect precipitation, filtration, and drying (Fig. 9.3). Alternatively, the solvent is removed by water and steam stripping and the polymer air-dried.

More recently (during the late 1960's), Union Carbide Corporation has introduced a gas phase process for the manufacture of high density polyethylene based on a catalyst of their own development.

Ethylene, a small quantity of hydrogen, catalyst, and comonomer (if used) are fed continuously to a gas phase reactor where a polymerization takes place at about 285 lb/in.^{2} pressure and at a temperature within the range 85-100°C.

The reactor is discharged to a product receiver where unreacted ethylene is recovered and recycled to the reactor. The polymer product is transferred from the product receiver to a product purge tank and then through

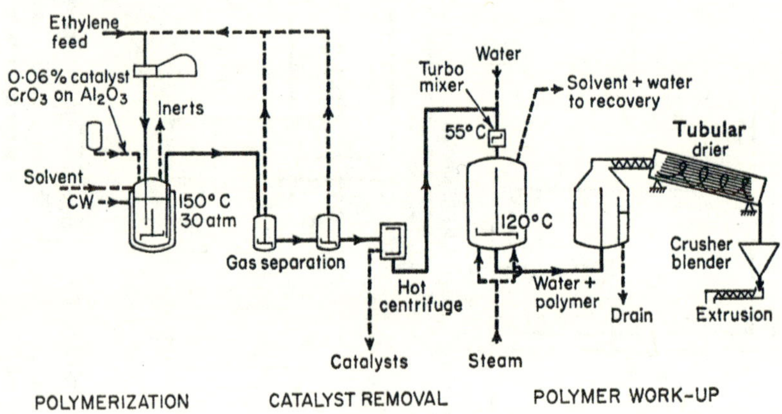

Fig. 9.3 Simplified flow scheme for polyethylene manufacture by Phillips process

a gas lock valve to a conveying system which takes it to the storage silos. From the silos the product is withdrawn for mixing, extrusion, granulation, etc.

A simplified process block diagram is shown in Fig. 9.4. Because gas

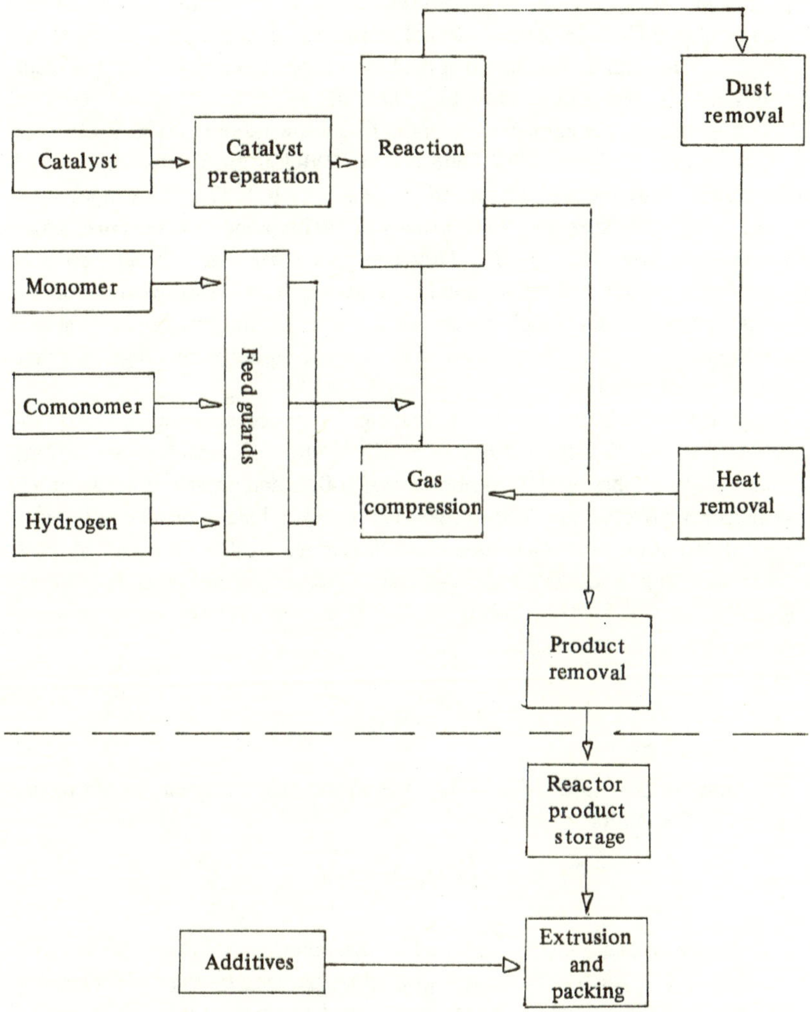

Fig. 9.4 High density polyethylene gas phase process (Union Carbide Corp.)

phase polymerization requires no solvent there is no need for the separation of the solvent from the polymer, or for solvent recovery. The catalyst efficiency is also said to be sufficiently high as to render residual catalyst removal unnecessary. Thus there is no necessity to wash and dry the polymerization product and so there is no recovery of washing solvent and no effluent.

A typical gas phase process for producing linear low density polyethylene is that of Union Carbide Corporation. The process is based on a fluidized bed of polyethylene. Ethylene is fed in at the base of the fluidized bed and the resultant polymer is taken out continuously, keeping a constant fluidized bed level within the reactor. The temperature is relatively low (about 100 °C while pressures are markedly lower (689-2068 kN/m^2-100-300 lb/in^2) than in the high pressure gas phase process developed originally by ICI. The efficiency is lower than in high pressure gas phase processes, being about 2% polymerization per pass, compared with 15-30%. However, plant size can be greatly reduced by eliminating much of the equipment necessary in high pressure operation and the capital costs are also much lower. A somewhat similar process is the stirred bed system. A Ziegler type catalyst is used and a higher percentage polymermization is achieved.

There are basically two types of liquid phase process, namely, slurry and solution. One slurry process is that of Phillips Petroleum which uses a chromium catalyst and operates at a temperature of 100 °C and pressures between 689 and 4826 kN/m2 (100 and 700 lb/in^2). The polyethylene is produced in particle form. Solution processes have been developed by Du Pont and by Dow Chemical, using a Ziegler type catalyst. Operating conditions are around 180-250 °C and 2757-4137 kN/m^2 (400-600 lb/in^2). Additives can be incorporated directly in these processes.

STRUCTURE

The simplest structure for polyethylene would be a completely unbranched chain of —CH_2 units, as shown:

$$\left(CH_2-CH_2 \right)_n$$

This structure is closely appraoched by the polymers produced by the low-pressure process. Such a structure allows close packing by the molecules, and hence a higher density and greater rigidity than is possible with the branched molecules produced by the high-pressure process. The latter process produces molecules which contain long and short branch-chains which effectively interfere with a close approach by the molecules. Examples of branching in a low-density polyethylene molecule are shown below.

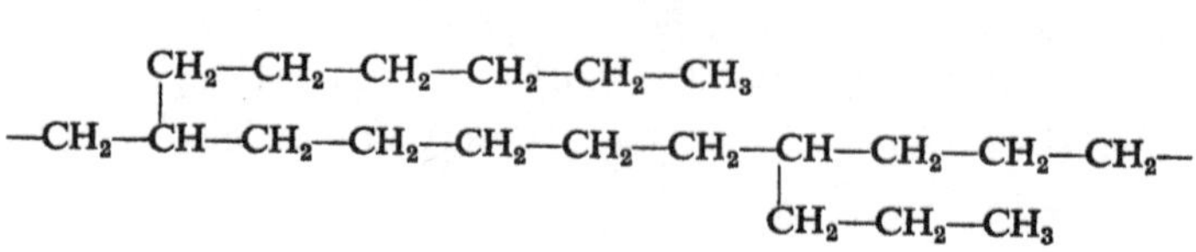

It will be noticed that each of the branch-chains contains a terminal methyl group ($—CH_3$). A convenient way of measuring branching is by the number of methyl groups per 100 carbon atoms and this can be done by infrared spectroscopy. Some short branches occur in the high-density polymers, particularly in those having densities around 0.95 g/cm³ but the molecules in polyethylene 0.96 g/cm³ are very largely linear. Mention must also be made here of the production of copolymers of ethylene with about one percent butene-1. This is a method of introducing controlled amounts of branching into the essentially linear polymers normally produced. The density of these copolymers is usually about 0.95 g/cm³.

In addition to density, one other important property that is affected by the structure of the molecule is crystallinity. Because of their length, polymer molecules become entangled with each other. This prevents complete crystallization and there are thus disordered areas between the crystallites. These crystallites are areas where the linear molecules are closely packed in an orderly arrangement. One theory states that a single molecule can pass through both ordered and disorderd states and can, therefore, be part of more than one crystallite (see Fig. 9.5). More recent work seems to point to molecules folding

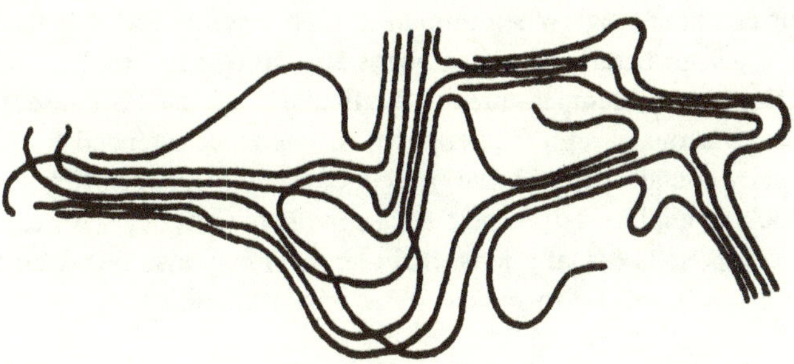

Fig. 9.5

on themselves in a parallel arrangement to form lamellae. Crystallization spreads as other molecules align themselves into position and also start to fold. Irregularities in molecular structure cause this crystalline growth to occur in more than one direction with the consequent formation of larger structures, known as spherulites. In either case, the close approach of the molecular chains in the crystalline areas leads to the formation of much stronger intermolecular forces in them than in the amorphous ones. The presence of branching tends to inhibit crystallization and low density polyethylene usually has a crystallinity of about 55–77 percent whereas that of high density polyethylene varies from 75–95 percent.

COMPOUNDING

For certain applications, it may be necessary to incorporate pigments, carbon black, antioxidants, slip additives, etc., with the polymer and further processing is therefore carried out. Such compounding is done on a variety of equipment the choice of which is governed by the scale of operations and the quality of end-product desired. In some cases the components of the compound are thoroughly premixed in a tumble or ribbon blender before being fed to the compounding equipment where the polyethylene is melted and then subjected to high shear stresses.

For larger-scale operations the most popular equipment is the Banbury mixer, but compounding may also be carried out in an extruder, provided that most of the heating is supplied by the mechanical work done on the polyethylene in the extruder. If too much external heat is supplied then extrusion is facilitated and can occur before complete mixing has been achieved.

Carbon black is incorporated in cases where resistance to weathering is required, and about two percent is usually sufficient. It should be well dispersed and the particle size should preferably be under 25 m The use of polyisobutylene or butyl rubber blends to improve stress-cracking resistance is another example of compounding. When preparing such a compound it is usual first to prepare a master batch containing about 50 : 50 of butyl rubber and polyethylene. Higher final temperatures in the Banbury are used because the melt remains very viscous at high temperatures. Compounds intended for film extrusion usually contain slip additive (to reduce friction between film surfaces) and antiblocking additives (to promote ease in opening of bags, for example). Antistatic compounds (usually long-chain amines) may also be added to film and bottle blowing grades to reduce static dust attraction.

PROPERTIES

The properties of polyethylene depend not only on chemical constitution and molecular weight but on degree of crystallinity (and, hence, density). In most cases, therefore, there is a range of properties corresponding to the range of densities now available. The molecular weight distribution can also affect polymer properties. In particular, polyethylenes with a narrow molecular weight distribution have better resistance to low-temperature brittleness and environmental stress cracking.

General

For any particular density of polyethylene there is a range of polymers of different molecular weights. These materials vary, particularly in processability and in mechanical properties and it is, therefore, essential that the converter should have some measure of the molecular weight of the particular polyethylene in which he is interested. A simple method which depends on measurement of the melt viscosity of the polymer was evolved by I.C.I. The apparatus is basically an extrusion plastometer with a standard orifice. The weight, in grams, of material extruded in ten minutes under a constant dead-weight load, at a temperature of 190 °C, is called the 'Melt Flow Index.' A low melt flow index corresponds to a high-melt viscosity. Since melt viscosity is related to the molecular weight of the polymer, a low MFI corresponds to a high molecular weight. A strict relationship applies only to polymers of similar density. For polyethylene of a higher density, a particular melt flow index will correspond to a slighly smaller molecular weight. More details of the measurement of melt flow index are give in Chapter 31.

The relationship between melt viscosity and melt flow index is logarithmic. Thus, for a melt flow index of 0.2 the melt viscosity of a 0.92 density polyethylene would be 300,000 poises, while for a melt flow index of two, it would be 30,000 poises. Similarly, a melt flow index of 20 corresponds to a melt viscosity of 3000 poises. The melt flow index is effectively used as a method of grading polyethylenes and, together with density, is a good guide (though not a complete one) to the type of polyethylene under consideration.

Physical Properties

Low-density polyethylene is a tough, slightly translucent, waxy solid. Increasing the density has the effect of raising the softening point, increasing the rigidity and surface hardness, and decreasing the impact strength. The higher density materials are also less waxy in appearance and feel. One of the most significant differences between low- and high-density polyethylene is that of softening point since one is below the boiling point of water while the other is above. Articles manufactured from the high-density material can thus be steam-sterilized, thereby opening up many applications previously closed to polyethylene.

The increased rigidity of high-density polyethylene means that equivalent rigidity can be attained with lower wall thickness and consequent savings in raw material costs. In many cases now, materials of densities intermediate between those of low- and high-density polyethylene are used. These are produced either by modified high-pressure processes (giving polymers with densities up to about 0.935g/cm^3) or by blending low- and high-density materials.

The range of physical properties available with polyethylene is shown in the Table 9.1.

Chemical Properties

In general, polyethylenes are resistant to most chemicals except oxidizing acids, free halogens, and certain ketones. At room temperature polyethylene is insoluble in all solvents. At elevated temperatures the solubility of polyethylene in hydrocarbons and chlorinated hydrocarbons increases quite sharply. The solubility in any particular solvent is dependent on the density of the polymer, the high-density polyethylene having the lower solubility. Although insoluble at room temperature, polyethylenes slowly absorb hydrocarbons and halogenated hydrocarbons with subsequent swelling. Here again, the effect is less for the high-density materials. Table 9.2 shows the resistance of low- and high-density polyethylene to certain common chemicals.

Table 9.1

PHYSICAL PROPERTIES OF POLYETHYLENE

Property	*Test method*	*Units*	*Value*	
			Low density	*High density*
Specific gravity			0.915–0.925	0.945–0.965
Yield stress	B.S.2782: 301.E.	lb/in.^{2}	1250–2000	3000–4600
Yield elongation	B.S.2782. 301.E.	%	16–20	11–16
Ultimate stress	B.S.2782: 301.E.	lb/in.^{2}	1000–2300	3100–5500
Ultimate elongation	B.S.2782: 301.E.	%	200–600	50–600
Impact strength (Izod)	B.S.2782: Pt. 3.	ft lb/in. notch	Does not break	2–12
Low-temperature brittleness point		°C	<-70	<-70
Hardness (Shore)	A.S.T.M. D.955-51	D Scale	41–46	60–70
Softening point (Vicat)		°C	85–87	127
Power factor 10^6 c/s	B.S.2782: Pt. 2.		<0.0002	0.0003–0.0005
Dielectric constant 10^6 c/s	B.S.2782: Pt. 2.		2.25–2.35	2.25–2.35
Volume resistivity	B.S.2782: Pt. 2.	ohm/cm	$>10^{18}$	$>10^{16}$

Table 9.2

EFFECT OF VARIOUS CHEMICALS ON POLYETHYLENE (AT NORMAL TEMPERATURE)

Chemical	*% Weight increase (30 days immersion)*		*Remarks*	
	Low density	*High density*	*Low density*	*High density*
Acetic acid (glacial)	0.85	0.75	Some attack	Some attack
Acetone	1.25	0.65	Some loss of strength	Some loss of strength
Alcohol (ethyl)	0.20	0.15	Some loss of strength	Some loss of strength
Ammonium hydroxide	Negligible	Negligible	Unaffected	Unaffected
Carbon tetrachloride	36.0	14.5	Swelling and loss of strength	Swelling and loss of strength
Ethyl acetate	3.1	2.3	Slight attack	Slight attack
Motor oil	3.4	0.7	Slight attack	Satisfactory
Nitric acid (conc.)	0.35	0.3	Some attack (yellowing)	Some attack (yellowing)
Petroleum	13	6	Swelling and loss of strength	Slight swelling and loss of strength
Sulfuric acid (conc.)	Negligible	Negligible	Some attack	Some attack

Owing to the paraffinic nature of the polyethylene molecule, it is extremely resistant to water, the water absorption at room temperature being so low as to be negligible. In this respect, there is very little difference between the low- and high-density polymers. The water vapor permeability is also low but in this case the high-density material is a better barrier than is the low-density, by a factor of about seven times. Water-vapor permeability is closely affected by crystallinity and this is primarily a matter of molecular structure, although it can be modified by extrusion conditions. The higher the crystallinity, the lower the permeability although the relation is not linear. Permeabilities of low- and high-density polyethylene to water vapor and certain gases are shown in Table 9.3.

An important phenomenon which must be taken into account when considering the chemical resistance of polyethylene under conditions of stress is that of 'environmental stress cracking.' When polyethylene is stressed multiaxially, in contact with certain polar liquids or vapors, cracking may occur. The substances

Table 9.3

PERMEABILITY OF POLYETHYLENES CM^3 (N.T.P.)/CM^2/S/CM HG/MIL $\times 10^8$ AT 25°C

Gas or vapor	*Permeability*	
	Low density	*High density*
Oxygen	15	3
Carbon dioxide	55	13
Nitrogen	5	1
Water vapor	420	60

which cause this 'stress cracking,' may not be solvents for, or even be more than slightly absorbed by, the polyethylene and if the polyethylene sample is unstressed no cracking occurs in contact with these very same substances. The probability of stress-cracking occurring decreases with decreasing melt flow index (increasing molecular weight), while the addition of 5–10 percent of polyisobutylene also confers protection. The influence of density is more difficult to assess. Figures may be encountered which appear to prove that high-density polyethylenes are more liable to stress crack, but these are largely due to the fact that the tests were carried out under constant deformation conditions. The greater rigidity of the high-density polymer means that the test sample is under conditions of higher stress concentrationss than a similarly deformed low-density sample. (A brief account of a test for stress cracking resistance is given in Chapter 31.) Table 9.4 gives a classification of certain materials as stress cracking agents for polyethylene.

Table 9.4

CLASSIFICATION OF MATERIALS AS ENVIRONMENTAL STRESS CRACKING AGENTS FOR POLYETHYLENE

Active	*Inactive*
Metallic soaps	Water
Sulfonated and sulfated alcohols	Sugars
Polyglycol esters	Inorganic salts (acid and neutral)
Liquid hydrocarbons	Polyhydric alcohols
Silicone fluids	Rosin
Organic esters	

The resistance of low-density polyethylene to oils and greases is rather poor but, again, improvement is noted with increases in molecular weight and density.

Polyethylene is susceptible to oxidation, particularly in the presence of uv light. Prolonged exposure to heat also accelerates oxidation with consequent deterioration of electrical and mechanical properties and appearance. The addition of antioxidants (such as substituted phenols, aromatic amines, etc.) overcomes oxidation due to high temperatures but photo-oxidation degradation is best combated by the incorporation of about 2 percent of carbon black which screens out the uv light. For outdoor uses in tropical countries, further protection is obtained by the additon of 0.1 to 0.2 percent of an antioxidant. Reports have been published of polyethylene, protected in this way, being serviceable after six or more years in the tropics.

Linear Low Density

The properties of LLDPE will vary according to the particular process employed in the same way that the properties of LDPE or HDPE vary. However, a major feature of LLDPE is that its molecular weight distribution is more narrow than that of LDPE. In general, the advantages of LLDPE over LDPE are improved chemical resistance, improved performance at both low and high temperatures, higher surface gloss, higher strength at a given density, and a greater resistance to environmental stress cracking. In film form, LLDPE shows improved puncture resistance and tear strength. Typical figures for puncture resistance at a thickness of around 75 um (0.003 in) are 0.4 ft lb/in^2 (834 J/m^2) for LDPE and 0.9 ft lb/in^2 (1877 J/m^2) for LLDPE. Similarly, the tear strengths are about 350 gf (0.77 lb.f) for LPDE and 1050 gf (2.31 lb.f) for LLDPE (mean of machine and cross directions). At a denisty of 0.92 g/cm^3 the melting points of LDPE and LLDPE are 100 °C and 125 °C respectively.

The differences in molecular structure affect the rheology of the two materials. LLDPE is more viscous at extrusion shear rates and requires greater power to extrude. In addition, a wider die gap is required when extruding 100% LLDPE in order to prevent melt fracture. For best results some minor modifications have to be made to the screw and die gap of extruders designed for ordinary LDPE. Equipment has been designed to handle LLDPE, while another approach is the use of additives that allow the extrusion of LLDPE without the occurrence of melt fracture.

APPLICATIONS

Polyethylene is an extremely versatile polymer and has a very wide range of applications. Among the fabrication techniques which can be used for polyethylene are extrusion (into pipe, sheet, cable covering, or film), injection molding, blow molding, extrusion coating, vacuum forming, sintering, and flame spraying, and it is the ease of processing, coupled with its toughness and good chemical and electrical properties which have led to its position as one of the foremost plastics in the world today.

Linear low density polyethylene has also been used for the manufacture of heavy duty sacks where its greater elongation and ultimate tensile strength make it particularly suitable. Its higher melting point also opens up the possibility of hot fill applications, one suggested use being for cement sacks.

Film

The largest single outlet for polyethylene is film and most of this is consumed by the packaging industry. One very large use for low density polyethylene film is the manufacture of heavy duty sacks. The first products to be packed in these sacks were compound fertilizers which were previously sold in multiwall paper sacks with an inner ply of 0.001 inch-thick, low-density polyethylene. The main advantage claimed for the all plastic sack was that the filled sack could be stored in the open air, thereby freeing valuable storage space for the farmer. In addition, the fact that the plastic sacks of fertilizer could be kept in the fields where they would later be required, would save labor in distributing sacks from a central storage point. Although this market now consumes many thousands of tons of polymer per annum, many difficulties were encounterd in the early days of development. One of the first was that of handling. When the open mouth sack was filled and sealed it assumed a pillow-like shape that was difficult to stack. The stack instability was often wrongly attributed to a lower coefficient of friction for low density polyethylene but in fact polyethylene on polyethylene has a higher coefficient than paper on paper and it was soon recognized that the main trouble was the small area of contact between the pillow-shaped plastics as opposed to the large area between block-shaped paper sacks. A solution was found to be the provision of a few microperferations along the top of the bag. These allowed the air inside the sack to escape and the contents of the bag were consolidated under the influence of impacts received during handling. The holes were small enough to exclude moisture except during heavy rainfall. This meant that the sacks could not be stored in completely unprotected conditions but a light tarpaulin or sheet of polyethylene film over the stack was found to be sufficient protection. This was still better than for paper sacks that lose strength under conditions of high humidity.

There are also many situations where the available filling equipment demands valved sacks so that these had to be developed for polyethylene sacks in order to compete fully with paper. Today, polyethylene valved sacks are commercially available, with a block-bottom construction so that neither filling nor stacking is a problem. Low-density polyethylene sacks are also used for the packaging of peat, plastic granules, and even coal.

Another large outlet for low-density polyethylene film is shrink wrapping. Shrink wrapping is based on the fact that a plastic film that has been stretched under heat during its manufacture reverts to its original dimensions when reheated. The technique was first applied to the packaging of oven-ready turkeys for deep freeze storage using vinylidene chloride/vinyl chloride copolymer as the film and hot water as the shrinkage medium. Nowadays, low density polyethylene is the most popular film for shrink wrapping and the heating is usually carried out by means of hot air tunnels. The applications for shrink wrapping have also greatly increased in numbers. One popular field of use is that of transit packaging, where a shrink wrapped package of, say, one or two dozen tins or jars, collated on one or two trays, is used instead of a fiberboard case. The mechanical protection given by the shrink wrap is not as great as that given by a well-designed fiberboard case but is usually completely adequate. In addition, a shrink wrapped package can be seen to be fragile and is usually handled more carefully as a result. Advantages to the retailer are:

(1) There is less packaging material to be disposed of, after use.
(2) The package takes up progressively less space on the warehouse shelf as the goods are sold, whereas a fiberboard case takes up the same amount of room whether it contains one can or twenty.

Low-density polyethylene film is also used for the shrink wrapping of whole pallet loads, giving stable loads with good weather-proofing. Miscellaneous applications include the wrapping of awkwardly shaped objects such as cucumbers, rolls of wallpaper and central heating radiators. Low density polyethylene is used for these applications because it is cheap, tough, and waterproof. Where a clearer or more sparkling film is required, then polypropylene or PVC are the more likely choices.

Some of the low density polyethylene's shrink wrapping markets have been taken over by stretch wrapping. The principle of stretch wrapping is that the film is stretched around the article to be wrapped and is then heat sealed. The residual tension of the film then gives a tight contour wrap. Advantages of stretch wrapping compared with shrink wrapping include energy savings (because no shrink tunnel is used). This also reduces both capital and running costs. The use of stretch wrapping also allows the use of standard film widths for a range of different product collations which enables cost savings to be made in the ordering of film. Pallet wrapping is one market where there has been appreciable movement from shrink wrapping to stretch wrapping. One technical advantage here is in the overwrapping of pallet loads of packs that have already been individually shrink wrapped. If such a load is subsequently shrink wrapped there is the danger of welding occuring between the pallet wrap and the film covering the individual packs. The use of stretch wrapping as the outer wrap solves this problem completely.

Linear low density polyethylene is also making inroads into the stretch wrap market because of cost/performance advantages. One disadvantage is that LLDPE has considerably less inherent cling than other stretch films like PVC and EVA. One way of overcoming this is by the incorporation of cling additives. Another is to eliminate the need for cling by such devices as closure by mechanical crimping. LLDPE is also being used in stretch wrapping in the form of a laminate with other plastic films. One example is a triple laminate with conventional LDPE as the outer layers and an inner core of LLDPE. The advantages claimed here are a greater tear and puncture resistance and a higher tensile strength and elongation, compared with LDPE. The overall thickness can, therefore, be reduced and the stretchability is greater. The film weight per pallet load can, therefore, be reduced with a consequent cost saving. Other films have been developed that are based on blends of LLDPE with other polymers such as EVA. It should be noted that in shrink wrapping (as opposed to stretch wrapping) the lower hot melt strength of LDPE makes it the more suitable material.

One of the earliest uses for low density polyethylene film was the packaging of fresh fruit and vegetables. In the general trend toward the packaging of foodstuffs, fresh produce had tended to lag behind but the advent of a cheap, tough film made the packaging of fresh produce an economic and technical feasibility. The good water-vapor barrier characteristics of polyethylene made it eminently suitable for preventng the drying out of green vegetables while still enabling the housewife to clearly see the contents of the package. However, although polyethylene is comparatively permeable to oxygen and carbon dioxide, it was found that the permeability was insufficient to maintain free interchange of the carbon dioxide developed inside the bag with atmospheric oxygen outside the bag. In the absence of such interchange, necessary for the respiration of the produce, off-odors were developed although the produce might still look green and fresh. Ventilation of the polyethylene bags by means of a few small holes, about ¼ inch in diameter, was found to eliminate the trouble without causing drying out of the produce.

Ventilation is also employed in bags used for the packaging of potatoes, although for a different reason. The potatoes are washed before packaging and are usually still damp when packed. Since polyethylene is an efficient water-vapor barrier, condensation of moisture occurs on the inner surface of the film unless ventilation is provided and such condensation would look unsightly and reduce the sales appeal of the package. Excess moisture in the bag could also lead to rotting of the produce.

The toughness of low density polyethylene at low temperatures and the fact that it can be formed into bags on a form/fill/seal basis has led to extensive uses in the packaging of frozen foods. The first outlet was the packaging of frozen peas. Frozen peas were originally packed in cartons, the carton being filled and then frozen between plates. The peas were received by the housewife as a solid block. In the case of the form/fill/seal bags, the peas are frozen in a blast of air and remain loose enough to be filled into the polyethylene bag formed from flat film round the filling mandrel. The use of polyethylene film has spread to frozen soft fruits, brussels sprouts, and other produce able to be handled by this type of filling operation.

The general inertness of low density polyethylene film has led to widespread use in many areas of food packaging including made up bags for use in the home for the storage of food in the refrigerator or deep-freeze.

Another market is for the packaging of sliced bread. Initial progress was slow because the traditional wrapping material—waxed paper—was cheap and difficulties were experiences in handling a limp, totally fusible film as opposed to the stiffer waxed paper in which only the wax melts during heat sealing. Low-density polyethylene's success came when equipment was developed for packaging sliced loaves into ready-made-up bags. These are closed with wire ties or plactic clips thereby overcoming the problem of heat sealing. Easy reclosure is another advantage from the housewife's point of view.

The use of high density polyethylene film in packaging was restricted for many years to areas where its higher softening point was necessary, i.e., in boil-in-the-bag food packaging. The development of ultra-thin high density polyethylene film (down to about seven microns) opened up many new markets. Such film, in the range of seven to ten microns, is similar to tissue paper in feel and appearance and also has the property of grease-proofness. It is produced using blow-up ratios of from 4 : 1 to 7 : 1 and draw ratios of around 10 : 1. Among the advantages that it has over paper are waterproofness, low gas and water vapor permeability, and good aroma retention. It is heat sealable, can be printed and is readily self-colored. It can be made into bags either by heat sealing or by the use of adhesives. High density polyethylene tissue has been used as a wrapping material for a number of food-stuffs, including meat, wet fish, meat pies, and dried fruit. There are also a number of nonfood packaging uses, one good example being the wrapping of cut flowers. The flowers are less prone to wilting through loss of moisture and the wrap itself does not wilt or drip water even after a long journey. It is also used for the wrapping of delicate instruments to keep them dust-free and to prevent corrosion and staining which can occur with paper, unless expensive chloride and sulfur-free grades are used.

The slightly thicker film (20–30 microns) is also used as a wrapping material. Its barrier properties are even better by virtue of its increased thickness and it is particularly suitable for the packaging of frozen meat. In Greece, it has been attractively printed and sold as a gift wrap while in Belgium it is sold in packs of twelve sheets for household wrapping uses. More generally, it is replacing paper bags for retail use. Finally, even thicker film (around 100 microns) is beginning to find use as a free-standing supermarket carrier bag.

Nonpackaging uses for polyethylene film include applications in building, agriculture, and horticulture. In building, for instance, low density polyethylene film is used as a cover for wet concrete to promote more even curing, as a damp-proof course and for the protection of buildings and workers so that work can continue under cover during bad weather.

Agriculture and horticultural uses are extremely varied and include silage covers, linings for temporary reservoirs, seed-bed covers, and greenhouses. The latter is an interesting application since it makes use of the fact that the light transmission characteristics of polyethylene are similar to those of glass in the visible region while having an absorption band in the ultraviolet. It is necessary to enclose an air space between to layers of polyethylene if heat losses at night are to be eliminated, since polyethylene is more transparent than glass to infrared radiation; therefore, the ground loses heat through the film by radiation. Black low density polyethylene film makes an excellent mulch for plants such as strawberries. The ground is covered and the seedlings planted out in prepared holes in the film. The black film absorbs heat and encourages earlier growth of the seedlings but inhibits weed growth in surrounding soil by excluding the light. Rain water percolates through the holes in the film but general evaporation is retarded. Low density polyethylene film has also been used as an impermeable cover for soil that has to be fumigated with poisonous fumigants such as methyl bromide.

Film Tapes and Fibers

Film tapes are widely used in the manufacture of woven sacks. Polypropylene is the most popular polymer but high density polyethylene is often used and a range of sacks is marketed in the U.K. based on high-density polyethylene coated with low-density polyethylene. One application is for the packaging of expensive grass seeds for export. The woven polyethylene sacks are 50 percent more expensive than the film sacks previously used but elimination of burst sacks more than compensates for this in product savings.

Coatings

The use of polyethylene for the coating of various substrates is notable for the many new applications available for such traditional materials as paper, fiberboard, aluminum foil, and regenerated cellulose film. Practically all coating with polyethylene is done by extrusion coating techniques (*see* Chapter 26—Extrusion) and low-density resins are used almost exclusively. Apart from the higher temperatures which have to be employed with high-density polyethylene, which present no insuperable difficulties, it is not always possible to make economic use of high-density polyethylene's better barrier properties. This is especially the case with paper where there is a danger of individual fibers penetrating very thin coatings and thus vitiating the moisture vapor resistance by 'wicking.' This can be obviated to some extent by using paper with glazed finishes. It is sometimes necessary to have a minimum thickness of coating in order to give adequate heat seal strength thereby not utilizing high-density polyethylene's superior barrier properties to the fullest. This is not to say that high-density polyethylene has no future as a coating material but only to suggest that its main uses are where low-density polyethylene cannot do the job except at heavy coating weights.

In the U.S.A. one of the largest outlets for polyethylene-coated board is the packaging of milk. In a land where doorstep delivery of milk has largely given way to collection from the local supermarket, cartons very soon replaced the glass bottle. At first, wax-coated board was used but the advent of polyethylene-coated board (which solved the problem of wax flaking off and floating in the milk) led to a changeover to the latter. Some polyethylene-coated cartons are used for milk in the U.K. but the different pattern of distribution favors returnable glass bottles. In the U.S.A. and Canada there has since been a further change towards high-density polyethylene bottles together with some form/fill/seal low-density ones.

Polyethylene-coated regenerated cellulose film is used in many cases where a vacuum-pack is required, for example, with bacon, cheese, and coffee, especially if transparency is also desired. In this type of coated material, the gas barrier properties of the cellulose film are added to the flexibility, water-vapor resistance, and heat-sealing properties of polyethylene. In addition, the cellulose film can be reverse printed on its inner surface. The printing is thus sandwiched between the cellulose and polyethylene, giving it protection against scuffing.

Injection Molding

In the injection-molding field both low- and high-density polyethylenes have found extensive use in household articles such as buckets, bowls, laundry baskets, baby baths, and many other items including such large moldings as dustbins. The laboratory and hospital also benefit from polyethylene's toughness and chemical inertness and a whole range of equipment has been molded, including beakers, funnels, measuring cylinders, jugs, and petrie dishes.

Polyethylene has also made extensive inroads into the toy market because of its relative unbreakability and its possibilities for color. The flexibility of the lower-density polymers also makes them excellent materials for manufacturing construction kits where parts fit into each other by means of the 'snap-in' principle.

In the packaging field, both low- and high-density polyethylene are used for closures, spouts, and dispensers. Low-density polyethylene is used for snap-on caps or plug fittings where its resiliency ensures a good seal, while high-density polyethylene is used for screw caps where rigidity is needed.

Linear low density polyethylene has made a significant penetration into markets such as lids and housewares where superior properties such as stiffness, heat distortion resistance, and stress crack resistance can be coupled with higher melt indexes to produce faster cycle times.

Blow Molding

Blow-molded articles provide a very large market for polyethylene. Bottles, jerricans, drums, and containers of all shapes and sizes are manufactured by the process and the polyethylene bottle has almost become the symbol of the plastic age.

The high-density polymer is favored when a high softening point, better barrier properties, or greater rigidity are required. Its better barrier properties are utilized in the bleach bottle (where a low permeability to chlorine is required) and for the manufacture of jerricans for paraffin (kerosene). Other applications for high-density polyethylene bottles include household and car polishes, lubricating oils, and ethylene glycol antifreezes. In addition to packaging applications, the blow-moldling technique has also opened up many other outlets for low- and high-density polyethylene. Toys, in particular, show cost savings when blow-molded in one piece as against injection molding of component parts with subsequent assembly. Dolls, too, are often blow molded in polyethylene instead of being slush-molded in PVC. Other miscellaneous applications for blow-molded polyethylene include coat hangers, rolling pins, mooring buoys, and pipe fittings.

Pipe

Both low- and high-density polymers have been used for the extrusion of pipe. In agriculture, low-density polyethylene pipe is often used for water supply and for irrigation where its flexibility makes it easy to install and handle. Low-density polyethylene pipe has also been used for domestic cold-water services, one big advantage being that it is not damaged should the water freeze. Polyethylene pipe is also used in chemical plants, the high-density material being preferred because of its rigidity and superior chemical resistance.

Monofilament

High-density polyethylene can be extruded and cold drawn to give monofilament suitable for many applications. In particular, it has been used for filter cloths, woven material for deck-chairs, ropes, tennis-nets, fishing-nets and, in thicker strands, pot-scourers. Polyethylene ropes have the advantage of being water-resistant, and hence they do not become waterlogged and, in fact, float on water since they have a density less than unity.

A remarkable extenstion of the monofilament extrusion technique is extruded plastic netting. One method of producing this is to use a circular die in which the inner core and outer portion of the die can be counter rotated while the molten polymer is extruded through notches cut in the outer surface of the inner core and the inside of the outer die component. Polyethylene was the first polymer used and the resultant netting is widely used for the packaging of oranges, lemons, apples, nuts, onions, etc., as well as for decorative overwrapping of cosmetics for multipacks and as protective sleeving.

Electrical Applications

The first major use of polyethylene was as an electrical insulator, replacing gutta-percha in the manufacture of submarine cables. In addition to its good electrical properties it possesses the virtues of water impermeability and chemical inertness and is thus an excellent material to replace paper insulation in telephone cables. The use of polyethylene in radar and other high-frequency equipment has already been mentioned and is part of history. The experience thus gained has enabled polyethylene to find many other uses in the electrical industry, a familiar example being that of transatlantic and Commonwealth-linking submarine telephone cables.

Miscellaneous Applications

Polyethylene powder can be used in fluidized bed coating and gives coatings of good gloss, chemical resistance, and abrasion resistance. Polyethylene powder is also used for the dispersion coating of irregularly shaped objects by means of spraying with subsequent heating.

Polyethylene may also be used for the powder coating of cloths or carpet backs. The powder is spread on the surface by a doctor blade, then passed through an oven to melt the polyethylene. On leaving the oven, the coated material is nipped between two rollers, that in contact with the molten polyethylene being water-cooled and polished.

An increasingly large outlet for polyethylene is in the sinter molding of large containers, tanks, etc. In the U.K. low-density polyethylene is the most popular, while on the Continent high-density polyethylene is more often used. Among the objects produced by this method are small boats, milk churns, and tote-bins.

ETHYLENE COPOLYMERS

It was briefly noted under 'Structure,' that controlled amounts of chain-branching could be introduced into the substantially linear 0.96 density polymers by means of copolymerization with small amounts of butene-1. One property which is improved by the introduction of a certain amount of branching is that of creep resistance. With a linear molecule such as high-density polyethylene there is only a small resistance to creep under static loads but the introduction of even a small percentage of branching has an appreciable effect by interfering with the free movement of the molecules relative to each other. The introduction of branching also improves the resistance to stress cracking. Other copolymers are dealt with in the following sections.

Polyallomers

Polyallomers are high crystalline block copolymers of ethylene and propylene and were introduced in 1962 by Eastman Chemical Products. These copolymers exhibit a degree of crystallinity usually found only in homopolymers yet they differ in many physical properties from polyethylene, polypropylene, and blends of the two. They are also different from copolymers of ethylene and propylene produced by other processes. These block copolymers are, in fact, examples of allomerism which is defined as a variation in chemical composition without change in crystalline form. The term polyallomer was coined, therefore, to identify this new class of copolymers and distinguish it from the previously known homopolymer and copolymer variations.

The highly crystalline nature of polyallomers is confirmed by their insolubility in hexane and heptane, while infrared spectra indicate that they are block and not random copolymers. Propylene/ethylene polyallomer exhibits many of the properties of both high-density polyethylene and crystalline polypropylene but has a specific gravity lower than both (0.896).

The surface hardness of polyallomer is slightly less than that of polypropylene but its abrasion resistance is greater. Polyallomer is superior to high density polyethylene in flow characteristics, moldability, softening point, and hardness. Like polypropylene it gives excellent molded-in hinges and has an even better resistance to flex fatigue. Polyallomer also offers better color and lower susceptibility to blush marks than polypropylene. Polyallomer tends to crystallize slowly, even after solidification from the melt. From a practical standpoint this means that tough parts can be produced from polyallomer without the necessity for rapid cooling that is essential for manufacturing tough parts from either of the two homopolymers.

Applications include fishing tackle boxes, bowling ball bags, and typewriter cases which utilize its integral hinge property, its toughness and its good moldability. Polyallomer is also used for luggage shells, loose-leaf binders, etc., made from pre-embossed sheet.

Ethylene/Vinyl Acetate Copolymers (EVA)

The proportion of vinyl acetate in EVA copolymers can vary, theoretically, from 1-99 percent but commercially available products normally contain less than 50 percent. Of these, the ones with 21-50 percent vinyl acetate are used as additives for waxes and adhesives, to give increased strength, improved barrier properties, and better processing characteristics.

EVA copolymers containing seven to eight percent of vinyl acetate may be considered as modified low-density polyethylene whereas those with around 15-28 percent of vinyl acetate behave more like flexible PVC. One advantage that EVA has over flexible PVC is that its flexibility is inherent and does not depend on the presence of plasticizers, which could be leached out.

In general, the properties of EVA compare with those of low-density polyethylene as follows:

(1) It has a higher impact strength.
(2) It has a greater elasticity.
(3) It has a higher permeability to water vapor and gases.

(4) It has better low temperature properties.
(5) It has a greater resistance to environmental stress cracking.
(6) It has a higher filler retention.
(7) It has a lower heat seal temperature.
(8) It has a greater flexural life.

Properties to be noted in the case of films are the tackiness of high vinyl acetate content copolymers and a greater coefficient of friction. This means that EVA films have a greater tendency to blocking and lower slip characteristics than low density polyethylene films.

Applications include snap-on caps, the lining of bag-in-box containers and stretch wrapping. In the U.S.A. a copolymer with a fairly high vinyl acetate content has been chill roll cast to give a flexible, high-clarity film which competes with flexible PVC as a meat wrap.

Ionomers

The term 'ionomers' was coined to describe a family of polymers in which there are ionic forces between the polymer chains, as well as the usual covalent bonds between the atoms making up each chain. Basically they are ethylene/methacrylic acid copolymers with some of the acid groups present as a metal salt. The metal ions provide the cationic part of the link while carboxyl groups, located along the polymer chain, constitute the anionic portion. Sodium, potassium, magnesium, and zinc are all suitable metal ions.

A typical ionomer has been found to contain 2.8 percent sodium which is equivalent to 17 sodium ions per 1000 carbon atoms. There were also 9 COO and 5 COOH groups per 1000 carbon atoms.

The ionic interchain links strengthen and stiffen the polymer but, unlike the covalent cross-links present in thermosets, they do not destroy its melt processability. Ionomers do have greater melt strength than polyethylene itself, however, and they are, therefore, very suitable for extrusion coating, skin packaging, and vacuum forming. The ionic interchain forces also have an appreciable effect on crystalline morphology. Low-density polyethylene itself is a semicrystalline polymer and can only be given good clarity by rapid quenching and a consequent inhibition of spherulite growth. The ionic forces in ionomers eliminate all traces of visible spherulites thus giving excellent clarity.

Ionomers have a greater resistance to oils and greases than low-density polyethylene at room temperature although the differences are not so marked at elevated temperatures. Chemically they are resistant to weak and strong alkalis but they are slowly attacked by acids. Environmental stress crack resistance is good and is claimed to be higher than that of low-density polyethylenes of equivalent melt flow index. Other chemical properties of ionomers are similar to those of low-density polyethylene, being resistant to alcohols, ketones, and esters but swelling slightly in hydrocarbon solvents. The use of carbon black as an antioxidant is also necessary for applications involving outdoor exposure.

Gas permeability of ionomers is similar to that of low-density polyethylene but water vapor permeability is somewhat higher as is water absorption. The polar nature of ionomers means that printing is easier than on low-density polyethylene but some pretreatment is still necessary.

Electrical properties are similar to those of low-density polyethylene but the polar nature again has an effect and the volume resistivity is lower (10^{17} ohm cm as against 10^{17}-10^{19} ohm cm for low-density polyethylene). The dielectric strength is higher than that of low-density polyethylene, however, while the dielectric constant is the same.

Transparent vials with integrally hinged snap caps have been molded from ionomer but the majority of uses lie in the realms of films and coatings. It has been used in skin packaging, for example, because of its high resistance to puncturing and its good adhesion to porous board surfaces. Its high melt strength means that it can be drawn down to very thin coatings (down to 12 microns) without pinholing in extrusion coatings. Finally, its grease resistance has led to its use in form/fill/seal pouches for a range of convenience foods with a high added fat content.

Irradiated Polyethylene

Work by Charlesby and others has shown that many long-chain polymers undergo cross-linking when subjected to high energy radiation. Polyethylene has received particular attention in this respect and several improvements in the properties of this polymer have been noted. When polyethylene is irradiated, hydrogen (and smaller amounts of methane, ethane, and propane) is liberated and the polymer becomes increasingly insoluble due to cross-linking of the molecules by C—C bonds. Increasing irradiation leads to yellowing and, eventually, the polyethylene becomes a dark ruby red in color. At first the polyethylene becomes more flexible but later it turns harder and harder, eventually becoming quite brittle.

In the U.S.A. irradiated polyethylene film has been produced by passing ordinary low-density polyethylene film continuously through a high-energy electron field. This converts it to an infusible film which, with appropriate stabilizers, withstands prolonged aging at 150°C or short-term exposure to 230°C. Pipe can also be irradiated with similar results. It is obvious that this idea of converting a thermoplastic polymer to a thermosetting one, after forming, opens up many new possibilities.

Modified Polyethylene

It is possible to modify polyethylene chemically by a number of means including oxidation, halogenation, and chlorsulfonation. Of these the most important, commercially, is chlorsulfonation.

Oxidation. Controlled oxidation of polyethylene can be carried out by heating the molten polyethylene at 160-200°C for some hours on a mixing mill in contact with air or by heating the polyethylene with an oxidizing agent, such as nitric acid, in the liquid phase. The first process yields carboxylic acid derivatives with molecular weights of the same order as the unmodified polymer. The higher molecular weight products are partially cross-linked. The second process yields dicarboxylic acid derivatives, the molecular weights of which vary between about 250-1000. A further process of oxidation is by treating a solution of polyethylene in carbon tetrachloride at 70-78°C with ozone and oxygen. A mixture of products is obtained but the chief ones are partially esterified hydroxydicarboxylic acids. Molecular weights vary between 100-2000. Polyethylenes which contain carboxyl groups can be easily dispersed in water, giving stable emulsions, and are of commercial interest on this account.

Halogenation. The reactions of chlorine, bromine, and fluorine have all been investigated but chlorination has received the most attention. Many processes have been described, including solution and suspension chlorination with photo or chemical catalysis. A wide range of chlorinated polyethylenes has been produced, from soft, pliable, rubber-like materials to rigid, transparent plastics similar in some respects to unplasticized polyvinyl chloride. The patent literature contains references to the use of chlorinated polyethylenes for electrical insulation, sheet, and flooring.

Chlorinated polyethylenes are also used as impact modifiers for rigid PVC. Resins with a chlorine content of around 36 percent offer the best balance between compatibility with PVC and the greatest favorable effect on impact strength.

Chlorsulfonation. The chlorination of polyethylene in the presence of sulfur dioxide effects the introduction of chlorine and sulfonyl chloride groups to the molecule and produces a vulcanizable rubber with good chemical and abrasion resistance.

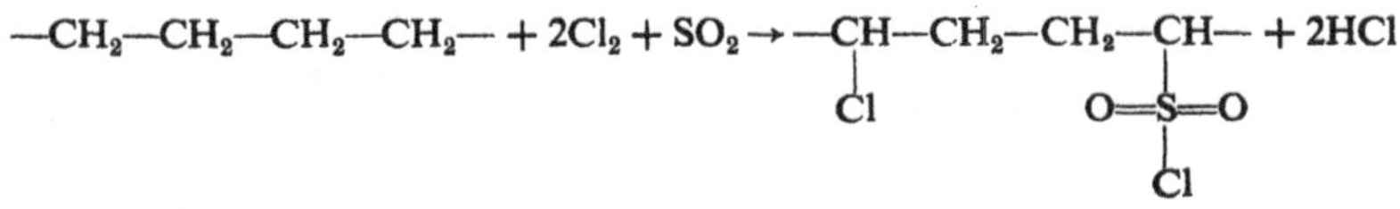

The chlorine confers oil resistance on the polymer while the highly reactive sulfonyl chloride groups enable easy cross-linking (or vulcanization) by zinc and other metal oxides. The presence of water is necessary for cross-linking and the reaction may be visualized as shown below.

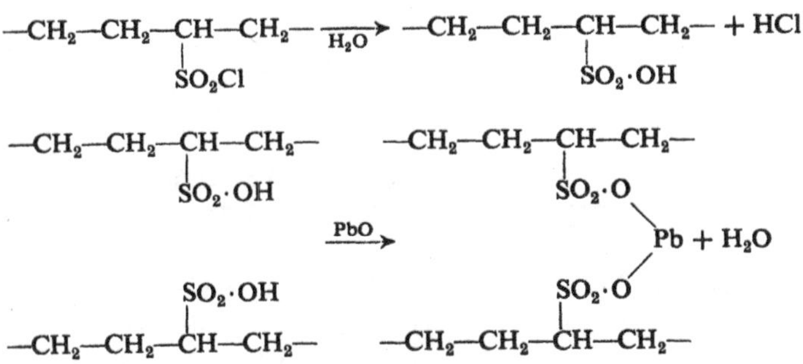

Chlorsulfonated polyethylene has excellent resistance to abrasion coupled with resistance to a wide range of chemicals. It also possesses good weatherability, even in colored grades. This latter property (and the abrasion resistance) has been utilized in the manufacture of white sidewall tires. It is extremely resistant to oils, ozone, and strong oxidizing chemicals such as chlorine bleaches.

Among the applications of chlorsulfonated polyethylene are ventilator strips, spark-plug covers, jacketing of ignition wire, industrial hose for corrosive chemicals such as bleaches and sulfuric acid, and the lining of chemical tanks. The following table of properties (Table 9.5) refers to Hypalon, marketed by E. I. Du Pont de Nemours & Co. Inc.

Table 9.5

PROPERTIES OF HYPALON,
CHLORSULFONATED POLYETHYLENE

Hardness range	40–95 (Shore A)
Temperature range (in use)	−62°–182°C
Tensile strength	over 3000 lb/in.^{2}
Adhesion to fabrics	Good
Adhesion to metal	Excellent
Chemical resistance	Good to excellent
Compression set resistance	Fair
Ozone resistance	Outstanding
Flame resistance	Good
Sunlight and weather resistance	Outstanding
Abrasion resistance	Excellent
Oil resistance	Good to excellent

VERY HIGH MOLECULAR WEIGHT, HIGH DENSITY POLYETHYLENE

During the past few years, very high molecular weight grades of high density polyethylene have been developed. Such grades are also known as ultra-high molecular weight polyethylene, or UHMWPE. Increased molecular weight gives products with high abrasion resistance, low coefficient of friction, high impact strength, and self-lubrication properties. However, as molecular weight increases, processability becomes correspondingly more difficult. Molecular weights range from one to five million. The resin is supplied as a powder and is compression molded into sheets and plates or ram-extruded into bars, profiles, rods, and tubes. In Europe, rods, pipes, and profiles are being produced on twin-screw extruders, using special lubricants and extrusion aids or blends of UHMWPE with lower molecular weight grades.

Applications include starwheels, guide rails, and bushings on packaging machinery, and chute and bin liners, wiper blades, and skirt boards in materials handling equipment. The excellent impact and abrasion resistance also makes it suitable for wear and bearing surfaces in prosthetic devices for hip, knee, and finger repair.

POLYPROPYLENE

History of Development. Early attempts to polymerize propylene using high-pressure processes gave only oily liquids or rubbery solids of no commercial value. Meanwhile, Professor Natta, in Italy, had been investigating the use of Ziegler-type catalysts. He found, for instance, that if in the Ziegler catalyst system, $Al(C_2H_5)_3 + TiCl_4$, the $TiCl_4$ was replaced by $TiCl_3$, a stereospecific catalyst was formed which yielded crystalline high-molecular weight polymers of propylene. A stereospecific catalyst is one which controls the position of each monomer unit as it is added to the polymer chain, thus allowing the formation of a polymer of regular structure from asymmetric monomer units such as propylene. The announcement of Professor Natta's stereospecific polymerization of propylene was greeted as a major breakthrough in polymer technology since this was the first time that polymers with a regular spatial structure had been produced synthetically.

Preparation of Monomer

Propylene, $CH_3 \cdot CH{=}CH_2$, is a gas, boiling at –48°C at atmospheric pressure. It is obtained commercially from refinery gases by distillation from propane/propylene mixtures. Propylene required for polymerization has to be very pure. Water must be absent since this destroys the catalyst, while methyl acetylene, $CH_3C{\equiv}CH$, should also be removed as it acts as a cross-linking agent.

Polymerization

Basically, the polymerization of propylene is similar to the Ziegler process for the preparation of high-density polyethylene. A typical process is described below.

Propylene at a pressure of about 100 atm is led into a reaction vessel in which is a well-stirred dispersion of catalyst in a liquid hydrocarbon. The temperature is kept low enough to ensure precipitation of the polymer as it is formed. Stirring is continued until the polymer content is between 35 and 40 percent, when the slurry is pumped to a flash drum where the unreacted propylene is removed and recycled. The slurry is then centrifuged to remove the liquid hydrocarbon. The liquid hydrocarbon and the temperature conditions are chosen so that most of any 'atactic' polymer formed remains in solution and is removed in the liquid phase. (The definition of 'atactic' is given below.)

Catalyst removal from the polymer is carried out by extracting with a weak solution of hydrochloric acid gas in methyl alcohol. After removal of the extractant, the polymer is washed with water to remove the acid, steam distilled to remove solvent traces, dried, extruded, and pelletized.

More recently (1977) it has been announced that a new catalyst has been developed jointly by Montedison in Italy and Mitsui Petrochemical Industries in Japan. The new catalyst is said to give a polymer needing no catalyst removal and with a high isotactic index, thus eliminating extraction of the amorphous 'atactic' material almost entirely. The new system means simpler and less expensive plants while plant running costs are also reduced.

Structure

The basic structural unit of polypropylene is

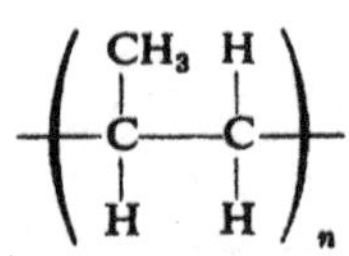

and is thus similar to that of polyethylene with a methyl group substituted for one of the hydrogens. The polymerization of propylene without the use of stereospecific catalysts produced haphazard linkages of monomer units to give a rubbery polymer with what is known as an 'atactic' structure, shown diagrammatically in Fig. 9.6. Such chains could obviously not

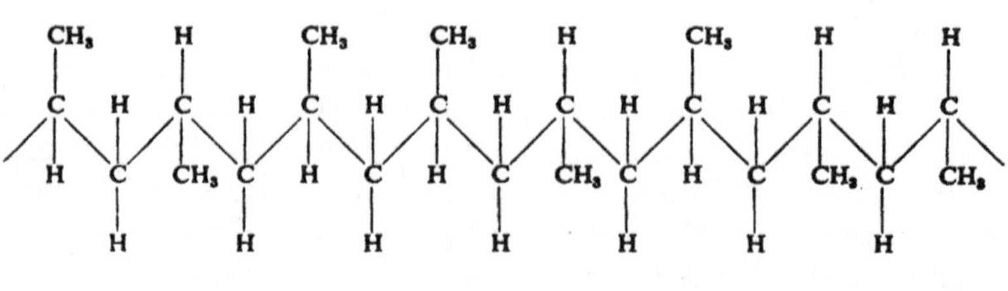

Fig. 9.6 'Atactic' polypropylene

group together to form a crystallizable structure. The name 'isotactic' was given to the regular structure of the polypropylene produced by stereospecific catalysts because it was thought that the methyl groups were always along one side of the polymer chain (*see* Fig. 9.7).

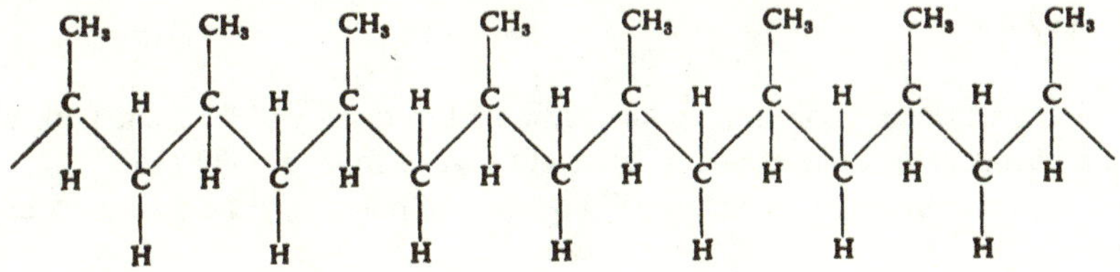

Fig. 9.7 'Isotactic' polypropylene

Later work has shown the existence of a form in which the methyl groups occur on alternate sides of the carbon chain ('syndiotactic') while it is also possible to obtain what is known as a 'stereoblock' polymer, in which short-chain lengths of isotactic configuration occur alternatively. (Fig. 9.8).

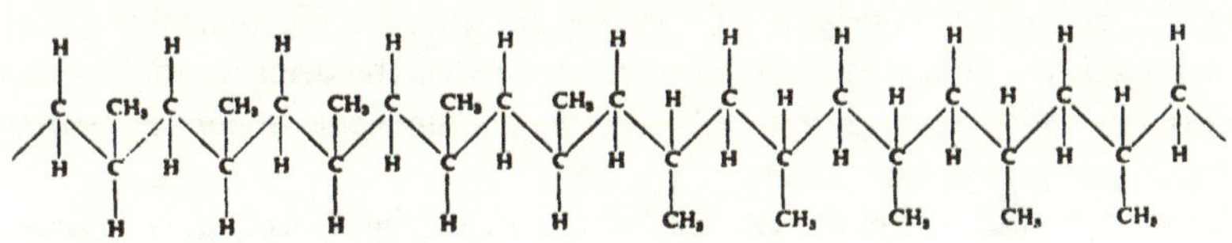

Fig. 9.8 'Stereoblock' polymer

In general, a stereoblock polymer with infrequent inversion in a long-chain possesses properties only slightly different from a high molecular weight isotactic polymer.

The regular structure of isotactic polypropylene allows close approach of the polymer molecules, and crystallization can then occur in a similar manner to that of polyethylene. The presence of the methyl group does restrict movement of the polymer molecules to some extent, however, and crystallization rarely exceeds 65–70 percent even for material having an isotacticity of 95–97 percent. The degree of isotacticity for a particular polymer can be measured by extraction with *n*-hexane, or a similar solvent. The *n*-hexane dissolves the atactic material which can then be recovered and weighed. Any stereoblock polymer remains with the isotactic material.

Copolymers of propylene with other olefins are also produced for particular applications. These can be regular or random. Random copolymers result when a small amount of comonomer (usually ethylene) is polymerized at random intervals along the polypropylene chain.

Properties

Polypropylene is colorless, odorless, has a low density (0.90 g/cm^3), excellent chemical resistance, and a relatively high softening point (150° c).

Physical Properties. The physical properties of polypropylene depend mainly on:

(a) Percentage of isotactic material present.
(b) Molecular weight.
(c) Crystallinity.

Most commercially available polypropylenes have approximately the same isotacticity while the crystallinity of a sample of given isotacticity depends on its thermal history. The main differences between commercial grades lie, therefore, in their molecular weight. As with polyethylene, this is commonly denoted by the melt flow index. In the case of polypropylenethe determination is carried out at 230°C instead of 190°C in order to allow for the higher melting point of polypropylene.

General physical properties are similar to those of high-density polyethylene. Properties which are significantly better are hardness (high-density polyethylene, 65–70; polypropylene, 75—Shore D), and softening point (high-density polyethylene, 122–130°C; polypropylene, 140–150°C), while it has a slightly lower mold shrinkage and gives moldings with a higher gloss.

In thin sections, polypropylene has a very high resistance to repeated flexing. It is thus possible to mold articles with built-in hinges. Integrally molded hinges have been produced which have withstood over a million flexes without failure and even at temperatures down to −70°C, correctly designed and molded hinges survive over 50,000 flexes while at −30°C the figure would be over 200,000.

The impact strength of polypropylene is lower than that of high-density polyethylene although still adequate for most purposes. The impact strength is affected by temperature and there is some reduction in that of polypropylene at temperatules below about 10°C. This reduction in impact strength is much less in the case of rapidly quenched film because of the very fine crystalline structure obtained by such treatment. The impact strength at low temperature can be improved by the addition of small quantities of butyl rubber or polyisobutylene although this entails some sacrifice in rigidity and in resistance to high temperatures. Another method of achieving higher impact strength at low temperatures is by copolymerization of polypropylene with other olefins.

The presence of tlle bulky methyl groups along the polypropylene chain leads to an increased resistance to creep over that displayed by low or high-density polyethylenes. This is particularly noticeable at temperatures above 100 °C and under high dead-weight conditions.

Polypropylene has excellent electrical properties, due to its nonpolar nature, and is similar in this respect to the polyethylenes. The electrical properties are unaffected by isotacticity or molecular weight but may, of course, be influenced by the presence of catalyst residues. The particular advantage of polypropylene for electrical uses lies in the fact that in addition to its excellent electrical properties it has superior thermal and mechanical properties.

Polypropylene is subject to oxidative degradation at the elevated temperatures necessary for processing but the addition of small amounts of an antioxidant is able to stabilize it during processing and at elevated service temperatures. For photo-oxidative degradation, the incorporation of about 2 percent of finely divided carbon black is satisfactory. Table 9.6 shows the physical properties of a typical polypropylene.

Table 9.6

TYPICAL PROPERTIES OF POLYPROPYLENE

Property	*Test method*	*Units*	*Value*
Specific gravity	–		0.90–0.91
Tensile strength	BS.2782, Pt.3	lb/in.^{2}	4000–5200
Elongation at break	BS.2782, Pt.3	%	400–900
Impact strength (Izod) injection molded bar	BS.2782, Pt.3	ft lb/in. notch	2.5–4.0
Hardness (Shore)	ASTM.D.1484–57.T	D scale	71–75
Softening point (Vicat)	BS.2782, Pt.1	°C	150
Power factor (10^6 c/s)	ASTM.D.150–54.T		2.3×10^{-4}
Dielectric constant (10^6 c/s)	ASTM.D.150–54.T		2.0–2.15
Volume resistivity	ASTM.D.257–54.T	ohm/cm	$0.7\text{–}1.0 \times 10^{16}$

Chemical Properties. The chemical resistance of polypropylene is similar to that of polyethylene although here again it possesses the additional advantage of being usable at higher operating temperatures. It has better resistance to oils and greases than has polyethylene and is not subject to environmental stress cracking. The permeability to oxygen, carbon dioxide , and water vapor lies somewhere between those of low-density and high-density polyethylene, but it should be noted that biaxial orientation of polypropylene film reduces its permeability.

APPLICATIONS

Polypropylene's interesting combination of properties has led to its use in a wide variety of products. In the field of injection molding it has been used for domestic holloware, chair shells, crates and tote boxes, domestic appliance parts such as washing machine tops, and a wide range of automotive applications. The latter include distributor caps, radiator fans, gloveboxes, and accelerator pedals utilizing integral hinges. Luggage and a wide range of containers have also been molded with integral hinges.

Talc-filled grades of polypropylene have also found a wide range of uses. In injection molding, talc contents of about 40 percent give high stiffness moldings for use in industrial and automotive fields. A talc-filled sheet grade has also been developed which reduces heat sag during thermoforming, giving faster cycles with reduced mold shrinkage. Because rigidity is also improved it allows the use of thinner sheet. Talc-filled polypropylene sheet also has attractions as a material from which to make cartons. It can be folded to give a good dead fold and can be handled by high speed carton making machinery. Its advantages over fiberboard include the fact that it is sterilizable, crush-resistant, and stands up to deep freeze conditions and to heating in microwave ovens. As it is waterproof it competes with plastics coated boards but it has the additional advantage that the edges of the material are also completely non-absorbent to liquids. Cartons made from this material can be fitted with tear tapes and can be printed by gravure, litho, and flexographic processes.

Polypropylene is emerging as a competitor in the thin-wall injection molding market to polystyrene. The price gap between the two plastics has narrowed since the large increases in crude oil prices because polypropylene is less vulnerable to increases in feed-stock prices. In addition, polypropylene has performance advantages such as a much higher softening point. Another packaging application is screw-on bottle closures where polypropylene's resiliency allows it to be used for linerless closures (with molded-in sealing features-Fig. 9.9) or closures with snap-in decorative inserts.

A rather newer and extremely interesting use for polypropylene is an extruded 'corrugated board.' It is not really a competitor with normal corrugated fiberboard but rather with wooden cases and in the development of new packaging systems needing a reusable, high-duty medium with easy 'knock-down'

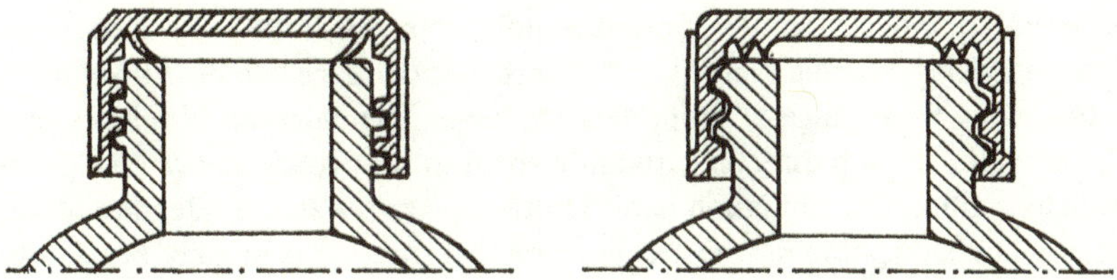

Fig. 9.9 Examples of Polypropylene Linerless Closures

properties. It also offers scope in horticulture for the packaging of wet produce straight from the fields.

Polypropylene can be slit-die extruded on to a chilled roller to give film of good clarity, gloss, and strength. Its permeability in this form to gases and water vapor is a little higher than that of high density polyethylene but biaxial orientation by stretching under heat improves its barrier characteristics and its impact strength at low temperature.

Cast polypropylene film has good transparency and gloss, but increasing thickness makes it more and more difficult to effect rapid cooling and thus prevent the growth of larger spherulites which cause haze. There is, therefore, a consequent loss of clarity with increasing thickness.

The tensile strength of cast polypropylene is about twice that of cast LDPE, but the tear strength is only about half. The elongation at break is high for cast polypropylene film and it can be cold drawn. One limitation of cast polypropylene film is its low impact strength at low temperatures (below about O °C).

The permeability of cast polypropylene film is slightly higher than that of HDPE, but much lower than that of LDPE. The chemical resistance of polypropylene is good and, in particular, it is more resistant to oil and greases than the polyethylenes.

Cast polypropylene films (and blown non-oriented films) find their main outlets in the packaging of textiles, such as hosiery, shirts, sheets, and blankets. Their use in these applications depends largely on their superior clarity vis-a-vis LDPE and their excellent sealing characteristics on side-weld overwrapping equipment. The use of non-oriented polypropylene film for medical packaging is also growing, particularly in the field of medical disposables where its relatively high temperature resistance allows use in autoclaves where temperatures of 121–130 °C are common.

These outlets are relatively small, however, and the large volume outlets require the superior properties of oriented polypropylene (OPP). Because of the wide variations possible in the degree of orientation in the two directions there is an equally wide range of properties available. The balanced film has a tensile strength, in each direction, roughly equal to four times that of cast polypropylene film. Tear initiation is difficult, but tear resistance after initiation, is low. The elongation at break is also low because it has already been fully drawn. Biaxial orientation improves the gloss slightly and greatly reduces haze so that the oriented film has a greatly improved clarity over that of the cast film.

Biaxial orientation also improves the barrier properties of polypropylene film and its low temperature impact strength. Although both moisture vapor and gas barrier properties are improved by biaxial orientation, the gas barrier properties are still not as good as those of regenerated cellulose. However, coated grades of oriented polypropylene are available which are very much better in this respect. Coated grades were originally developed to improve heat sealability, because shrinkage tends to occur when the highly stretched film is heated. The provision of a lower melting point polymer as a coating is one way of solving this problem. One coating which has been developed is acrylic based and does not add anything to the barrier properties of OPP film. However, it does provide a relatively wide sealing range of 45 °C and it adds sparkle and good machine slip characteristics. Its hot-tack seal strength is good enough for it to be widely used for the vertical form/fill/seal packaging of light weight products such as snack foods. Heat seal coatings that are also barriers to water vapor and gases are based on PVDC. Here, again, there are various categories. The first has a fairly wide heat seal temperature range (33–45 °C) with improved oxygen barrier properties. Both one or two sides coated films are available. The one side coated films are used as laminating substrates for use with other OPP films and cellulose films, to package snack foods and confectionery. The two sides coated films are used in unsupported form for bags and carton overwraps. The second category has superior gas barrier properties but a narrower sealing range (28 °C). Because of this such films are nsed as laminating substrates with other materials for the controlled or modified atmosphere packaging of coffee, processed meats and natural cheese. The third category of films have acrylic coatings on one side and a PVDC coating on the other. Such films provide a degree of gas and aroma barrier together with lap sealability.

Coextrusion is the other route to heat sealability. Most coextruded OPP films consist of two or three layers. The major component is the core ply which usually consists of virgin homopolymer and some regrind. One or both sides of the core may be covered with a PP-rich copolymer, or an ionomer, or a vinyl acetate-modified PE layer. PP-rich copolymers of propylene and ethylene are used for their good hot-tack seal strengths and excellent finished seal strengths. Ionomer is used because of its wide sealing range and grease resistance, while vinyl acetate-modified PEs provide good anchoring properties in coating, metallizing and laminating operations.

Opp films have been used for shrink wrapping where extra sales appeal is required. They are more expensive than low-density polyethylene films and their use is only justified where greater clarity and sparkle are considered essential. One such field is in the shrink wrapping of collations of cosmetic and toiletry items.

Polypropylene is also used for the overwrapping of single cartons in order to confer protection against water vapor pick-up. One of the advantages of OPP film in such applications (compared with regenerated cellulose film) is that the water vapor barrier properties do not depend only on the coating. With moisture-proof grades of regenerated cellulose film the barrier properties are often reduced because of damage to the coating at the edges and corners of the carton.

Highly transparent film

Highly transparent blown polypropylene films have been developed as direct competitors to regenerated cellulose film in areas of retail packaging where prime need is consumer appeal, i.e. good surface gloss, high clarity, crackle, etc. The strength of such films is not so high as that of normal orientated polypropylene, but this is seen as an advantage in some retail packaging where complaints are often made that plastics packs are too strong and difficult to open.

A tube is extruded downwards from a rotating spreader die and water cooled after a short (300–500 mm — 12–20 in) passage through air. The tube passes through a calibrator just below the water surface. The blow-up ratio is very low (less than 1:2). The preferred polymers are those with a higher than usual atactic and heptane soluble content.

Pearlized (opaque) polypropylene

Another PP film development is pearlized OPP. One variant is a gassed, expanded foam, oriented film which offers opacity and increased yield because of its lower density. In the U.K. it is widely used for wrapping confectionery and biscuits, and is replacing glassine and foil tissues for these applications. Opaque OPP films can also be formed by other methods. In one of these, homopolymer PP resin is mixed with a small amount of particulate matter and a thick sheet is extruded and then oriented by the tenter frame process. The polypropylene pulls away from each particle, thus creating an air-filled void.

After heat stabilization the OPP film is similar to a micropore foamed film. In another process, the material is produced as a filled film without voids, the opacity being a direct result of the amount of particulate material added.

In the film with air-filled voids the whiteness and opacity are created partly by the encapsulated particles but, to a greater degree, the opacity is also caused by light rays being reflected by the polypropylene cell walls and by the air within each cell. The refractive index of the air is less than that of the polypropylene and it is this difference that creates the opacity.

Polypropylene's high softening point and chemical inertness has promoted its use for ducting, tank linings, and piping for chemical plant, especially where these items may be used at elevated temperatures. Extruded sheet can be thermoformed, using suitable equipment. In general, doublesided heating is necessary and the heaters should be of high capacity. Where more complicated structures are required they can be fabricated from sheet by hot gas-welding techniques.

Rather more novel markets are likely to develop in thin wall moldings via the solid phase forming techniques, such as stretch forming and solid phase pressure forming, described in Chapter 30. Cycle times are lower than for thermoforming and the clarity is greatly improved. Because the polypropylene is not heated to very high temperatures in these processes the containers so formed are free from odor and taint.

The blow-molding technique is also applicable to polypropylene and bottles with freedom from stress cracking and with good resilience (for squeeze action) have been obtained. Tonnages are unlikely to be large because processing speeds are lower than for other polymers. The use of special copolymer grades and specially developed production techniques can give bottles with good clarity and general performance but they will probably remain more expensive than PVC and the polyethylenes.

Polypropylene is used in large quantities in the fields of film tapes and fibers (*see* Chapter 30). Woven polypropylene sacks, for example, are used for the packaging of horticultural produce, grain, and animal feeding stuffs. Ventilation is important in the case of grain because condensation inside the sack would lead to fermentation of the grain. A woven sack is also necessary because of the practice of sampling the grain, with a probe, through the wall of the sack. Additional applications in tropical countries include the packaging of coffee and sugar, fish, cotton, and groundnuts.

Because of the trends towards bulk delivery in countries such as the U.S.A. and the U.K., the market for sacks may remain static but in the developing countries demand is increasing because of new and expensive agricultural products needing efficient protection.

Tufted carpet backing is another large market for polypropylene woven film tapes and a high degree of penetration has been achieved in the American and European tufted carpet industries. This is because woven polypropylene fabrics are much lighter, are rot proof and moisture proof and give a higher pile than can be achieved with jute backings. A similar backing is used for needle-punched 'indoor-outdoor' carpeting where the carpet pile is itself made from polypropylene. For this application the polypropylene tapes are fibrillated, then chopped into staple lengths.

Film fibers have been used as a replacement for sisal in agricultural baler twine, packaging twine, and ropes. Polypropylene baler twine is rot-proof and waterproof and (when ultra-violet stabilizers are added) has a longer life than sisal twine. Sisal has a slightly greater breaking load than polypropylene but the latter has a much higher strength to weight ratio.

Polypropylene can be extruded and 'cold drawn' to give monofilaments of low density, high tenacity, and good abrasion resistance. Ropes made from polypropylene monofilament have the advantage of floating on water, even after prolonged immersion, and they are, of course, rot proof. Polypropylene in fiber form has also found outlets in the field of textiles.

POLY (BUTENE-1)

Butene-1 (or butylene) is the next step up from propylene in the olefin family and is polymerized with the aid of Ziegler-type catalysts to give a stereoregular polymer with the basic structure shown below:

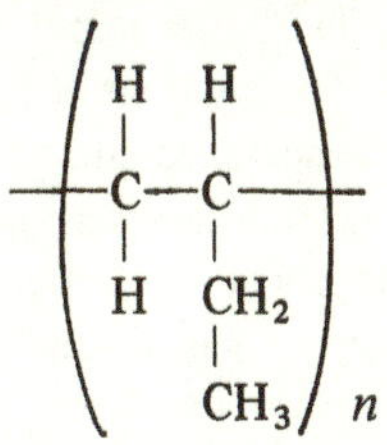

Poly (butene- 1) crystallizes from the melt as the tetragonal crystalline Form II and this changes over a period of three to six days to the more stable hexagonal Form I. The change is accompanied by an increase in density, hardness, rigidity, and strength. Crystallinity also increases from around 30 percent to about 50 percent.

Poly (butene-1) possesses the flexibility of low-density polyethylene coupled with very low creep properties more characteristic of the engineering thermoplastics. Its tensile yield strength is about 2300 lb/in.^{2} At stress levels of around 1150 lb/in.^{2}, (50 percent tensile yield stress) the 1000-hour creep figures are only about one percent at 23 °C. At 75 percent yield stress the 1000-hour creep figure is still only about 6 percent. In some cases poly (butene-1) could be more useful than a higher strength material with poor creep properties. This is particularly true at elevated temperatures where the creep properties of many common thermoplastics fall off rapidly. For example, long term stress tests at 82 °C have shown poly (butene-1) pipe to have approximately the same strength at that temperature as low-density polyethylene has at room temperature. The resistance to creep allied witll a high resistance to environmental stress cracking makes it a very suitable material for pipes.

Poly (butene-1) can be made into film by blow-extrusion or by chill roll casting. The resultant film is tough, with good tear resistance, puncture resistance and impact strength. Its retention of strength at elevated temperatures makes it suitable for hot-fill packaging applications. It is a good water vapor barrier but is not a particularly good barrier to gases. It has been used in the U.S.A. for milk pouches.

POLY (METHYL PENTENE) (TPX)

Methyl pentene polymers are based on 4-methyl pentene-1 as the principal monomer. This is copolymerized with other monomers, using Ziegler type catalysts at atmospheric pressure. Properties distinguishing this polymer from the other polyolefins are transparency (light transmission 90 percent), high crystalline melting point (240 °C), and a low specific gravity (0.83). It also retains useful properties up to about 205 °C. Chemical properties are broadly similar to those of the other polyolefins. Like them it is resistant to acids and alkalis as well as to many solvents. It is softened and swells in contact with hydrocarbons and is subject to environmental stress cracking (like the polyethylenes but unlike polypropylene). Electrical properties are good, in line with other polyolefins, but its permeability to gases and water vapor is much higher.

Polymethylpentene has a wide range of uses, from laboratory and medical ware (syringes, holloware, and connector pieces), lighting diffusers, lenses and reflectors, and packaging (closures and cook-in food container).

SUGGESTED FURTHER READING

Polyolefins, Structure and Properties. (1966) H.V. Boenig (Elsevier: Amsterdam)

Vinyl and Allied Polymers, Vol. 1 — Aliphatic Polyolefins and Polydienes; (1972) G.A.R. Matthews (The Plastics and Rubber Institute)

Fluoro-olefin Polymers (1968) P.D.Ritchie (London: The Plastics and Rubber Institute)

10.

VINYLS

The vinyl family of plastics are made by the addition polymerization of certain substituted ethylenes. The substitution is of one hydrogen atom only, and may be by a single atom or by a group of atoms such as the acetate group, in the case of vinyl acetate. The double bond remains unchanged. In the wider sense, the vinyl polymers include such materials as polystyrene (based on styrene $-CH_2-CH.C_6H_5$) but in general the vinyl polymer series can be divided into three main groups:

1. Polyvinyl chloride (homopolymer).
2. Vinyl chloride copolymers, e.g., with vinyl acetate, vinylidene chloride, ethylene, and propylene as the comonomers.
3. Polyvinylidene chloride, polyvinyl acetate, polyvinyl alcohol, polyvinyl acetals, and polyvinyl ethers.

Of these polyvinyl chloride, vinyl chloride/vinyl acetate and vinyl chloride/vinylidene chloride copolymers are the most important and are described in more detail than any of the others.

The existence of the monomer vinyl chloride was reported as early as 1838 when it was presumably first synthesized by a French chemist, Regnault, from ethylene dichloride. The ability of this monomer to polymerize was not known at that time although Regnault did report that a white substance was formed when the monomer was sealed in a tube and exposed to sunlight. This effect of sunlight on vinyl chloride was later investigated by Baumann, in 1872, who observed that a tough white substance formed as a result. Very little significant advance was then made in this field until 1912 when the Russian chemist, Ostromislensky, reported that sunlight catalyzed the polymerization of vinyl halides to yield resinous products. About the same period a patent was granted in Germany for the manufacture of vinyl chloride by reacting hydrogen chloride gas with acetylene at 180°C in the presence of metal chlorides as the catalysts. This period is regarded by many as being the birth of polyvinyl chloride although its beginnings as a technically useful product stemmed from the patent disclosures of Semon of the B. F. Goodrich Company in 1933 and Reid of Carbide and Carbon Chemicals in 1934.

Considerable developmental work was carried out on the preparation of the chemistry for vinyl derivatives and the decade preceding World War II brought about a technological breakthrough. This was achieved in Germany by the I. G. Farben Company, who developed a process for the handling of high pressure acetylene. The work was directed by Dr. J. W. Reppe and his work was recognized as leading to the development of many new vinyl polymers. Many of the major vinyl polymers can be classified as originating from acetylene, including poly vinyl chloride, poly vinyl acetate, poly vinyl acetals, and the poly vinyl ethers. The latter materials were developed in the U.S.A. in 1948, although they have not reached the major tonnages consumed as some of the other vinyl polymers.

The main disadvantages of these polymers prior to the 1930's were their insolubility in common solvents, but when it was later discovered that it was possible to dissolve them in high boiling solvents such as tritolyl phosphate (then known as tricresyl phosphate) many new applications were opened up. By the start of WW II, the significance of plasticizing PVC was fully realized, as well as the need for considerably more developmental work if the polymer was to become commercially important. It was during the war that the real importance of this polymer became obvious when, due to the acute shortage of rubber, many companies began to look for a substitute. They found that plasticized PVC was suitable for many of the applications that hitherto had been restricted to rubber, including cable insulation and sheathing. After the war many countries returned to using rubbers but many still continued to use PVC and today this polymer is one of the leading thermoplastics.

RAW MATERIALS

Vinyl Chloride

There are basically two sources of vinyl chloride monomer depending on whether acetylene or natural gas is readily available. The PVC produced in Europe has been largely based on acetylene obtained from calcium carbide, and in the U.S.A. from ethylene. At the present time, however, the ethylene route is becoming increasingly important due to the widespread growth of the petrochemical industry. The basic ethylene is, of course, obtained by cracking oil, and the two main routes being represented as follows:

(i) *Direct Addition.* The first method is the direct addition of hydrogen chloride to acetylene using a catalyst such as mercuric chloride.

$$CH{\equiv}CH + HCl \longrightarrow CH_2{=}CHCl$$

(ii) *Chlorination of Ethylene.* This method involves the chlorination of ethylene which is then followed by the abstraction of hydrogen chloride as follows.

$$CH_2{=}CH_2 + Cl_2 \rightarrow \underset{\text{Dichloroethane}}{CH_2Cl{-}CH_2Cl} \xrightarrow{HCl} CH_2{=}CHCl + HCl$$

The dichloroethane is first purified and dried, then it is led through the tubes of a pyrolysis furnace where it is cracked to give hydrogen chloride and vinyl chloride. The reaction mixture at this stage also contains some unconverted dichloroethane, which is therefore led to an absorber where the vinyl chloride and dichloroethane are separated from the hydrogen chloride and the vinyl chloride is separated from the dichloroethane by distillation.

Vinylidene Chloride. This monomer may be prepared by chlorinating ethylene to give 1,1,2 trichloroethane which is, in turn, pyrolyzed to dichloroethylene. The required monomer 1,1 dichloroethylene ($CH_2{=}CCl_2$ boiling point 31.7°C) is separated from the *cis* and *trans* 1,2 isomers.

$$CH_2{=}CH_2 + Cl_2 \rightarrow \underset{+\,HCl}{CH_2Cl{-}CHCl_2} \xrightarrow{CaO} CH_2{=}CCl_2 + CaCl_2 + H_2O$$

Vinyl Acetate. This monomer is normally prepared commercially by the liquid or gaseous state reaction of acetic acid with acetylene. Mercuric sulfate is used as a catalyst in conjunction with hot dilute sulfuric acid, which hydrates acetylene to form acetaldehyde. The intermediate in this reaction is the unstable vinyl alcohol which readily isomerizes to the aldehyde.

$$CH_3COOH + CH{\equiv}CH \rightarrow CH_2{=}CHOCOCH_3$$

Vinyl Alcohol

This monomer does not exist as such and is a tautomer of acetaldehyde with the equilibrium in favor of the acetaldehyde. Therefore polyvinyl alcohol is prepared by the hydrolysis of polyvinyl acetate. Polymers prepared by such techniques are high molecular weight resins containing varying percentages of hydroxyl and acetate groups.

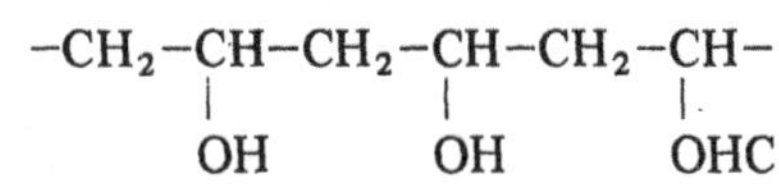

Processes used involve alcoholysis in methanol or ethanol using an acid (e.g., sulfuric acid) or alkaline (e.g., sodium methoxide) catalyst.

Poly vinyl alcohols are usually classified as partially hydrolyzed (15-30 percent poly vinyl acetate groups) and completely hydrolyzed (0-5 percent acetate groups).

Vinyl Acetals

Acetals are formed by the well-known classical reaction between aldehydes and alcohols. The addition reaction of one molecule of an alcohol with one molecule of an aldehyde produces what can be regarded as a 'half acetal.' These are rarely isolated because of their inherent instability. They are therefore reacted with another molecule of alcohol to form a stable acetal.

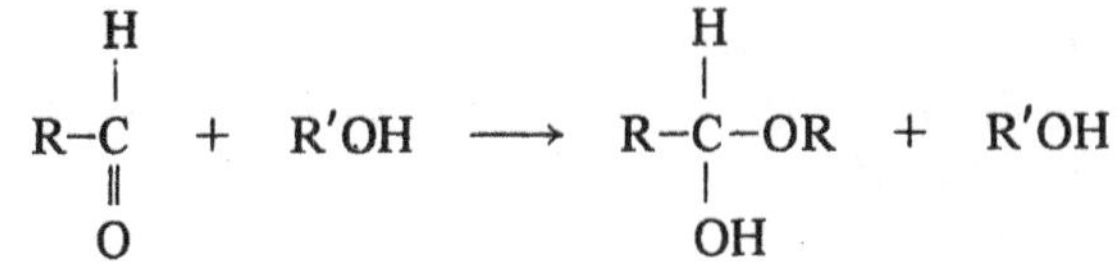

Aldehyde Alcohol Half acetal Alcohol

$$\longrightarrow \underset{\text{Acetal}}{\text{R}-\overset{\text{H}}{\overset{|}{\text{C}}}(\text{OR}')_2} + \text{H}_2\text{O}$$

The overall reaction of one molecule of aldehyde and two molecules of alcohol can be illustrated as follows:

$$\underset{\text{Aldehyde}}{\text{RCHO}} + \underset{\text{Alcohol}}{2\text{R}'\text{OH}} \longrightarrow \underset{\text{Acetal}}{\text{R}-\text{CH}-(\text{OR}')_2} + \text{H}_2\text{O}$$

Vinyl Ethers

Two methods have been used for the commercial synthesis of vinyl ethers.

(1) The pyrolysis of the corresponding acetal.

$$\text{CH}_3-\text{CH(OR)}_2 \longrightarrow \text{CH}_2{=}\text{CHOR} + \text{ROH}$$

(2) Direct vinylation of the alcohol with acetylene.

$$\text{HC}{\equiv}\text{CH} + \text{ROH} \xrightarrow{\text{KOH}} \text{CH}{=}\text{CHOR}$$

This latter process which can be regarded as a vinylation process probably offers what must be regarded as the easiest and most economical approach.

Typical reaction conditions are

Temperature: 130–170°C
Catalyst: 1–3 percent potassium hydroxide (the potassium alcoholate is thus formed *in situ*)
Acetylation Pressure: 100–150 psi (reduced with nitrogen)
Yield: 90-95 percent

Polymerization. The polymerization of vinyl chloride monomer is carried out by a free radical process, the free radicals being provided by the decomposition of an initiator.

$$R{-}R' \rightarrow \qquad R + R'$$

$$R + CH_2Cl{=}CH_2 \rightarrow R{-}CHCl{-}CH_2 \xrightarrow{VC} R{-}CHCl{-}CH_2{-}CHCl{-}CH_2$$

The chain continues to grow until it combines with another free radical such as a chain stopper, e.g., carbon tetrachloride. This process is referred to as termination.

There are two main methods of polymerizing vinyl chloride; by an emulsion or a suspension technique. Other methods are also used but only on a very small scale. Only brief details are given.

Another process of vinyl addition which is of interest is that of ionic catalysis. This can produce spatially regular structures in polymer chains. In this technique the vinyl bond is activated either as a carbonium ion, i.e., by cationic Friedel Crafts acid catalyst or as a carbonium ion, i.e., ionic basic catalysts. Both Professors Ziegler and Natta found that certain heterogenous catalyst systems, e.g., titanium tetrachloride plus aluminum triethyl are particularly useful for the production of stereo regular polymers.

Suspension Polymerization. This method is undoubtedly the predominant process for the preparation of PVC in the world today. It consists essentially of suspending the monomer in water to form droplets, this state being maintained by a suspension agent such as gelatin, ethyl, or methyl cellulose and polyvinyl alcohol. The suspending medium of water helps in controlling the heat formed during the polymerization reaction. A wetting agent such as sulfonated castor oil is also used and the free radical polymerization is initiated by an oil soluble/water insoluble type catalyst such as benzoyl or lauroyl peroxide. Strong agitation is necessary to keep the reactants in suspension, and polymerization occurs in each individual monomer droplet. The speed of stirring is a very important factor and controls the size of the particles; the greater the speed of stirring the smaller the particle size and vice versa. The polymerization is carried out for a predetermined time, normally between 12 and 24 hours, and on completion of the reaction the polymer is in the form of a slurry. The slurry is then passed to a stripping tank to remove any unreacted vinyl chloride, filtered, and finally dried. Normal methods of drying are to pass the material through a heated continuous rotary drier and then finally flash-dried.

Emulsion Polymerization. As with the suspension process this technique uses water as a suspending medium to control the heat produced during the polymerization reaction. The actual mechanism is different and produces polymers with completely different properties. The monomer is first emulsified with a fatty acid or a cationic soap and the initiation is carried out with inorganic materials such as potassium and ammonium persulfate or hydrogen peroxide. In contrast to the previous suspension methods these initiators are soluble in the water phase. The vinyl chloride diffuses through the water layer to the emulsifier micelles, thereby acting as the nuclei for polymerization. This process also requires the use of protective colloids such as carboxy methyl cellulose. On completion of the polymerization the emulsion is either spray-dried or coagulated by adding an electrolyte, then filtered and finally flash dried. Various washing processes are necessary for the removal of the soap if polymers of high purity are to be obtained.

Bulk Polymerization. This is not a very satisfactory method of polymerization and produces polymers of low molecular weight. A commercial method, however, involves the liquifying of the monomer, then dissolving the initiator into the monomer and finally polymerizing. The polymer so formed is insoluble in the monomer and is therefore precipitated as it is formed, to give low molecular weight polymers. The final stage is therefore relatively simple and consists mainly of separating the polymer from the monomer phase. The polymers produced by this method have excellent purity with good clarity although they have poor mechanical strength. This, coupled with the problem of heat dissipation makes the method unattractive.

Solution Polymerization. Modifications have been made to the bulk polymerization technique, one method being to add solvents, such as acetone, during the polymerization. This gives the finished polymer a higher molecular weight. The process requires the use of chain stoppers which somewhat limits the choice of the solvent, and since the process is a hazardous one it is seldom used.

Vinyl Acetate

The polymerization of this monomer also takes place by the free-radical addition reaction described previously. The free radicals are generated by the decomposition of organic peroxides such as benzoyl peroxide or inorganic salts such as potassium or ammonium persulfate. Conventional methods of suspension, emulsion, bulk, and solution are used. This monomer also copolymerizes with vinyl chloride, vinylidene chloride, alkyl maleates, and fumarates. There are many other comonomers too numerous to name. A relatively new development has been in the copolymerization of vinyl acetate with ethylene. These are either emulsions or solid polymers and are described in Chapter 9.

Vinyl Acetals

Polyvinyl acetals are prepared from aldehydes and polyvinyl alcohols. It has been stated that there are two basic types of polyvinyl alcohols, (15-30 percent and 0-5 percent acetate groups) and both types in various molecular weights are used in producing commercial polyvinyl acetals.

The conditions for the acetal reaction and the concentration of the particular aldehyde and polyvinyl alcohol used are closely controlled to yield polymers containing predetermined proportions of hydroxyl acetate and acetal groups. The end product may be shown by the following schematic structure.

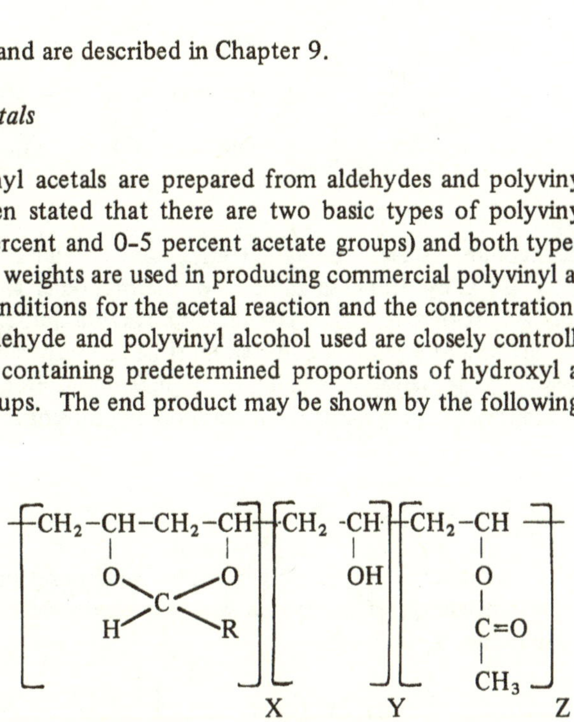

PV Acetal PV Alcohol PV Acetate

The X, Y, and Z segments are distributed randomly along the molecule backbone. The polyvinyl acetal is a cyclic acetal because the hydroxyl groups are attached to the same chain. It may also be cross-linked by heating or with a trace of mineral acid. It seems in general, that cross-linking takes place by transacetalization but it may involve more complex reactions such as that between the acetate or hydroxyl groups on adjacent chains.

Vinyl Ethers

Polymers can be produced from this monomer by free radical processes, however, only very low molecular weight products are obtained. Acid-type catalysts have been found to yield low to high molecular weight polymers, the low temperatures tending to favor higher molecular weights. Catalysts which have been reported as being suitable are BF_3 (hydrate or etherate) $SnCl_4$, $FeCl_3$, $AlCl_3$, SO_2, and iodine.

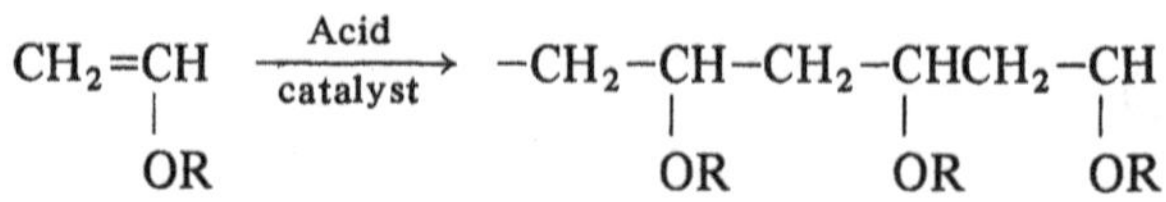

Low and medium viscosity polyvinyl ethers are obtained by a bulk technique at 0°-10°C by dropwise addition of the monomer and catalyst to a brine cooled, stirred reactor. The reaction is rapid and exothermic and therefore the rate of monomer addition must be carefully controlled.

High viscosity grades have been prepared of polyvinyl isobutyl ether at much lower temperatures, e.g., –50°C. A low bonding solvent such as propane can be used to control the exotherm.

Propylene-Vinyl Chloride

Compositions containing two to ten percent by weight of propylene are made by copolymerization of propylene with vinyl chloride. The structure of this copolymer is

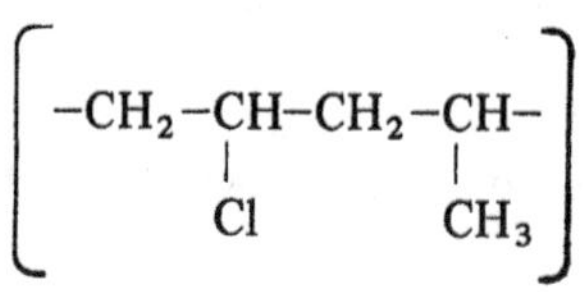

The introduction of stable hydrocarbon structure into the vinyl chloride chain results in an increased resistance to thermal degradation and reduction in dehydrochlorination.

MANUFACTURE OF VINYL CHLORIDE/VINYL ACETATE AND VINYLIDENE CHLORIDE COPOLYMERS

These copolymers are usually manufactured with a vinyl acetate content between 3-30 percent, using solution techniques. The correct choice of solvent for the monomer and catalyst allows the copolymer to precipitate as soon as the required molecular weight is reached. Polymerization is normally carried out between 30-40°C with vigorous stirring for about 48 hours or more. For paint applications these copolymers may be further modified by the addition of maleic anhydride and/or by partial hydrolysis of some of the acetate groups. It has been established that both these additional processes improve adhesion and reactivity and at the same time also give an increase in water resistance. The copolymerization with vinyl acetate or other such similar comonomer serves to reduce the fusion temperature and can be regarded as internal plasticizing.

Vinylidene chloride polymerizes more readily than vinyl chloride, which makes it somewhat more difficult to obtain homogenous polymers. Strict control is therefore essential during polymerization, by the addition of further vinylidene chloride and–or the removal of vinyl chloride.

CLASSIFICATION OF PVC POLYMERS

Various methods are used in classifying thermoplastics and with those containing a high proportion of crystallinity such as polyethylene; it has been relatively simple to carry out such a classification by molten viscosity measurements (MFI). This is, of course, related to the molecular weight and density of the polyolefin. PVC and its copolymers are almost amorphous, therefore density does not have the same significance, hence molecular weight measurements have been used to classify the polymers. The viscosity of PVC at high temperatures is considerably higher than that of the polyolefins and therefore measurement is somewhat more difficult. Other alternative methods of measurement have been investigated and those based on solvent solution techniques have been found the most reliable. Such results have also been found to give a good indication of the molecular weight and of ultimate end polymer properties, but unfortunately processing properties tend to be unreliable.

The various methods of determining molecular weights on a solution viscosity method have been discussed in Chapter 1. Viscosity measurements are used to classify PVC by what is termed the 'K' value, which is obtained directly from viscosity measurements of a solution of the polymer in a solvent such as cyclohexanone or ethylene dichloride (concentration normally 1 g/100 ml of solvent). The K value can then be calculated from the equation developed by Fikentscher (Ref: *Cellulose Chem.* 13.60. (1932)) as follows:

$$\frac{\log \eta_r}{c} = \frac{75k}{1 + 1.5kc} + k \qquad (1)$$

where η_r is the relative viscosity and c the concentration (g/100 ml).

From equation (1) we obtain

$$K = 1000k \tag{2}$$

The actual K value is not dependent on concentration but on the solvent, and the measurement in cyclohexanone is some 4 units higher than if measured in ethylene dichloride.

STRUCTURE OF PVC

The structure of the vinyl chloride polymers is different from that of the polyolefins in that it is substantially amorphous and does not readily crystallize, a fact which is confirmed by X-rays. The basic structure is as follows:

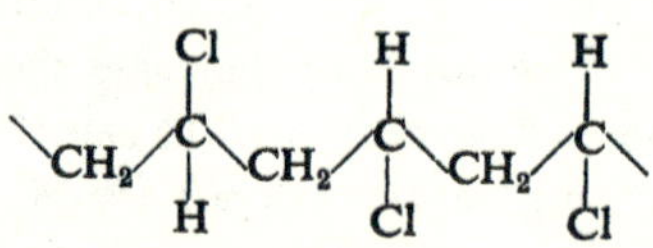

It can be seen that the bulky chlorine atoms are on alternate sides of the chains and the units are repeated in a head-to-tail arrangement. There are strong cohesive forces present in PVC due largely to strong dipole moments caused by the chlorine atoms. Such polarity causes incompatibility with nonpolar hydrocarbons and thus PVC is resistant to petrol and mineral oils although it has the effect of making the polymer soluble in strong polar solvents (e.g. ketones and chlorinated hydrocarbons).

It is also interesting to consider the structure of PVC when heated to 80°C and above. Heating not only produces a structural change but also a series of color changes, from light yellow, through dark yellow, orange and, ultimately, to reddish brown and even black. The release of HCl during this reaction suggests that degradation has resulted, yielding the following possible polyene structure:

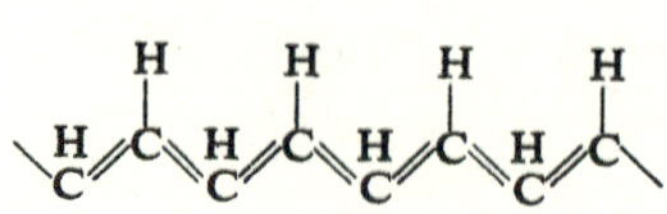

The presence of this unsaturated structure has been confirmed by absorption studies and is indicative of color formation.

PROCESSING

In view of the scope of this book and the wider variations of processing for PVC, only this homopolymer and its related copolymers are discussed in any detail.

In general, most thermoplastics are delivered in a form suitable for processing by the many techniques described later in the book. This is not so with PVC which is in a form which usually requires compounding by the processor. Compounding is necessary because of the relatively low thermal stability and high melt viscosity of this material. By correct choice of compounding additives it is found that processing is made easier and, in some instances, the cost of the compound can be reduced. In general, the compounding techniques can be classified according to whether the polymer is plasticized or not. Polyvinyl compounds normally contain the base polymer together with stabilizer, lubricant, filler, and pigments and in the plasticized type, plasticizer is, of course, added.

METHODS OF COMPOUNDING

The fact that it is necessary to add such additives makes it imperative to ensure that these materials are mixed thoroughly. The initial mixing is done in a simple mixer which exerts a low friction action on the mass. The most homogenous compound is obtained if the dry materials, e.g., base polymer, stabilizer, lubricant are mixed first and the liquid ingredients added as required, e.g., plasticizers. Extenders may be added when all the dry materials are fully dispersed, and at this stage the compound is referred to as a premix and may also be prepared on machinery which exerts a high friction action on the mass. The next stage is to flux the PVC either in a two-roll mill or by an internal mixer technique (e.g., Banbury). This process serves to gel and homogenize the compound which is then either charged on to a mill where it is sheeted, cooled and chipped ready for feeding into conventional fabricating equipment, or fed directly to a mill and transferred as a hot strip to the calender.

The premix may also be fed directly into a continuous mixer/extruder, and such machines are suitable where long runs of specific compounds are required. The extruders used for this purpose are single or double screw although the extruding action is usually modified to improve mixing efficiency. Temperatures required for these methods depend to a large extent on the compound formulations. However, for most purposes a Banbury temperature of 130–150°C and a mill temperature of 125–150°C suffice.

In contrast to these compounding methods, which depend upon dispersion and gelation of the various compounding ingredients, there is another method based on mixing the compounding ingredients to yield a free-flowing powder.

The process consists of blending the dry ingredients, i.e. polymer, stabilizer, and lubricant in the cold for a few minutes, then adding the liquid components by spraying them on the agitated dry blend in a jacketed, steam-heated mixer. Heating is applied until the temperature reaches 100°C, at which it is maintained for a few minutes, then the mass is cooled to 50°C, discharged from the mix, and finally sieved to remove any agglomerated particles that may have occurred. This final compound is then referred to as a dry-blend powder.

The use of compounding extruders is now generally preferred for compounding rigid vinyl formulations. The absence of plasticizer also leads to the absence of difficulties in the feeding of the extruders. The process of compounding PVC is regarded as an art in itself and many tailor-made formulations can complicate the choice of a particular compound for a specific application. Table 10.1 gives formulations that can be used to give a suggested end-product; two main factors to be borne in mind when making a choice are desired properties and cost. It must be noted than when processing the copolymers, lower temperatures are generally used.

CALENDERING

Of all the processing methods associated with PVC, calendering is probably the most widely used. Products obtained from calendering include unsupported film and sheet, coated fabric and paper, and flooring. This process is described in detail elsewhere in the book: it consists basically of passing the material between two or more pairs of rotating rolls, each pair of rolls forming a nip through which the material is drawn or extruded.

The calendering process may be considered as a combined forming and finishing operation, the physical properties of the finished product depending on the processing temperature, which can vary considerably, depending upon the nature of the base polymer, and may range from 130–

Table 10.1

TYPICAL PVC FORMULATIONS

End-use	*Designed properties*	*Polymer (parts)*	*Plasticizer (parts)*	*Stabilizer (parts)*	*Lubricant (parts)*	*Filler (parts)*	*Pigment (parts)*
Wire insulation	Flame resistant, flexible, insulator with good aging properties	100 (suspension) electrical grade	40 dioctylphthalate 20 tritolyphosphate	6 lead carbonate or tribasic sulfate	1 calcium stearate	15 china clay	1–2 as required
Floor tile	Good abrasion resistance, low creep, low cost	Copolymer v.c./v.a. 100	20 dialphanol phthalate 5 polymeric plasticizer	2 lead carbonate 6 lead salicylate	1 lead stearate	100 asbestos 250 calcium carbonate also 10 chlorinated paraffin wax	10 titanium dioxide
Calendering	Rapid gelling	100 (emulsion)	45 dialphanol phthalate	1 organo tin stabilizer	1 lead stearate	–	1 as required
Rigid transparent sheet	Easy processing polymer	100	–	2 dibutyl tin mercaptide	1 calcium stearate	–	–
Gramophone records		Copolymer 100 v.c./v.a.	–	–	1½ lead stearate	–	1½ carbon black
Extruded hose	Stable, fairly stiff mix	100	45 dialphanol phthalate	7 lead carbonate	1 calcium stearate	30 china clay, 20 chlorinated paraffin wax	
Mine belting	(Flame resistant–antimony compounds added to improve this characteristic)	100 (suspension)	35 tritolyphosphate 20 dioctylphthalate 5 dioctyl sebacate	9 barium ricinoleate, 1 cadmium laurate, 1 calcium stearate		10 calcium carbonate 5 antimony oxide	5 carbon black
Nontoxic applications		100	40 dioctyl phthalate	2 diphenyl urea	2 calcium stearate		

(N.B. These formulations are given only as a guide and it does not follow that they are used on a commercial basis.)

200°C. A final press polishing at higher temperatures than these would improve the clarity of the sheet and the gloss, and increase the tensile strength.

The calender generally used for PVC is invariably a four-roll machine, the rollers, or bowl as they are more often called, arranged in the form of an inverted L or having a Z-type arrangement. Calenders used for producing floor tiles, etc., normally consist of two rolls in an I shape, used either singly or in series. The actual thickness of the end-product is largely governed by the gap between the bowls, which is referred to as the nip.

Plasticized film formulations are normally based on PVC grades of medium to high K values. For thicker plasticized sheeting and even for rigid sheet it is normal to use lower K values or even vinyl chloride/vinyl acetate copolymers. Another point to note is that with rigid sheet, higher calendering temperatures are required than used for plasticized polymers. Mass temperatures above 200°C have been reported. The thickness of the sheet obtained by calendering is from 4-24 mil (0.1 to 0.6 mm). Sheet with thickness of over 40 mil (1 mm) is obtained by laminating the calendered sheet. By the extrusion method, either flat sheet is extruded directly from the extruder or a cylinder is extruded and then cut and opened to a flat sheet. A schematic diagram of the production of rigid PVC is given in Fig. 10.1.

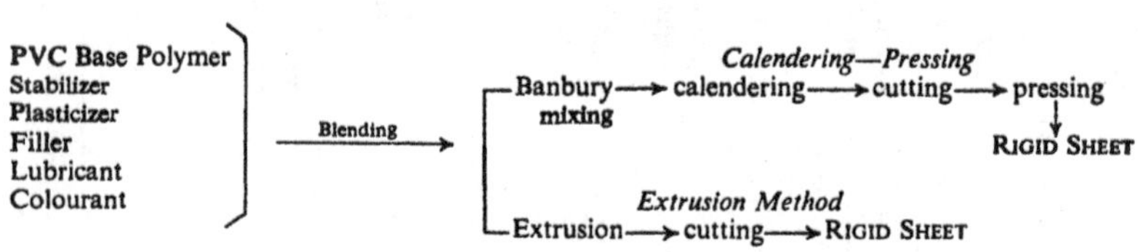

Fig. 10.1 Production of rigid PVC sheet

For coating fabrics it is first necessary to preheat the fabric immediately before it enters the nip to ensure that it is approximately the same temperature as the stock of the calender. The actual coating process can be carried out by one of two methods; the fabric can either be fed through the final nip of a four-roll calender or post-calendered by passing the calendered film and fabric through a further pair of rolls for subsequent lamination. A precaution that must be observed with paper coating is that of drying the paper stock 24 hours before coating.

For the production of PVC flooring, the following sequence of processes is used: internal mixing, milling on a two-roll mill, calendering, and finally pressing to give a product which can be sold either in continuous form or as individual tiles.

EXTRUSION

This process is second in importance to calendering and consists of forcing the uniformly heated and softened compound through a die and then cooling it. Both rigid and flexible sections can be extruded to give the following applications: pipe, tube (hose), rod, and profile extrusion; cable and wire covering; film and bottle blowing; sheet and filament.

Suitably plasticized PVC compounds can be processed readily on single- or double-screw extruders using either the granulate or dry-blend materials discussed earlier. The actual extrudate is normally cooled by passing through a water trough, when it is finally wound on to rollers. Plasticized extrusion is used to produce garden hose, certain profiles and cable covering.

PVC film and sheet are usually manufactured by the calendering process, therefore the film-blowing technique is used only when a seamless bag is required for such applications as packaging. As yet the process is not often used for processing the homopolymer but it is used extensively for producing film from polyvinyl chloride/vinylidene chloride copolymers. Film in tubular form is obtained from the extruder, quenched below 20°C to an amorphous condition, and then biaxially stretched by air inflation at 20-25°C. During this orientation process crystallinity is induced and as a result the tensile strength and solvent resistance of the film is increased. Monofilament can also be produced by similar extrusion techniques. The extrusion of rigid PVC is somewhat more difficult than with plasticized PVC and polyethylene. This is due largely to the high viscosities obtained at normal extrusion speeds, leading to difficulties in homogenization. Decomposition tends to take place and it therefore is essential to use a powerful stabilizing system. Despite such difficulties unplasticized polymer is extruded to form rigid pipes suitable for use in electrical conduit, chemical plant, and for the transportation or disposal of water, gas, and sewage. Formulations suitable for extrusion of these pipes are as follows:

1. PVC polymer	100 parts
Dibasic lead stearate	2 parts
Dibasic lead phosphite	½ part

Stearic acid	½–1 part
2. PVC polymer	100 parts
Tribasic lead sulfate	5 parts
Lead stearate	2 parts
Stearic acid	½–1 part

It is generally recommended that the exact amount of lubricant, i.e., stearic acid in these formulations should be determined experimentally for the particular equipment to be used.

There are two general methods of processing open to the manufacturers of rigid PVC pipe: direct extrusion from a powder premix, or a two-stage process whereby partial precompounding and pregelling precedes extrusion. The above formulations can be used as powder premix for the production of pipe for noncritical applications such as electrical conduit, in a twin or single-screw extruder with a high length/diameter ratio, preferably 20 : 1. However, in order to obtain higher throughput of pipes with optimum physical properties, the two-stage process involving partial precompounding and pregelling is preferred by many processors. It must be pointed out, however, that it is now also possible to use powder premixes to produce pipes. The extrusion of rigid vinyl sheet is complicated still further by difficulties in die design, which, however, have been largely overcome by using fishtail dies. For best results it has been found necessary to use screws of the long compression type, i.e., about two-thirds of the total effective length. This is in contrast to the polyolefins which require screws of the short compression type.

The blow-extrusion of rigid PVC to form bottles and similar containers is becoming increasingly important. With the plasticized compounds very little difficulty is experienced and nearly all compounds can be blown into containers of one form or another. Higher temperatures are, of course, essential with rigid PVC and articles can be blown by either extrusion/blowing or injection blowing techniques; another technique which has certain advantages is the preformed parison.

INJECTION MOLDING

Plasticized compounds can be injection molded without any difficulty on conventional ram-type machines. Temperatures in a range of 150–170°C are normally employed and, provided the compound is adequately stabilized, good quality moldings can be obtained. Difficulties are experienced with rigid PVC due to the base polymer softening over this very wide temperature range, and it does not obtain a very flowable state until heated to a temperature somewhere in the region of 190°C. Even at this temperature it is stable only for a short time and decomposes rapidly if heated to too high a temperature for too long a period. These conditions must be avoided if production runs are to be successful. Injection machines providing pressures around 20,000 lb/in.^{2} are normally quite adequate for most thermoplastics but due to the high melt viscosity of the rigid PVC, machines capable of providing pressures of 30,000 lb/in.^{2} are necessary. Suspension polymers with a lower K value are usually preferred for injection molding because they can be processed at slightly lower temperatures, whereas emulsion polymers stabilized with conventional materials degrade far too early. By using injection machines with a screw preplasticizer instead of the ram, it is possible to obtain more uniform frictional heat. Lower injection pressures can then be used with polymers of normal K value. High plasticized PVC compounds suitable for footwear can be processed by extrusion-injection molding techniques in which injection pressure is developed by rotation of a screw rather than by a reciprocating action. This permits low injection pressures and the use of light alloy molds which are cheap in comparison to conventional molds.

COMPRESSION MOLDING

This process, although more commonly associated with thermosetting polymers can be used also for PVC. The process, of course, suffers from the fact that the mold requires cooling to a temperature that allows the molding to be rigid enough to remove. The technique, nevertheless, is of great importance in the production of gramophone records from vinyl chloride/vinyl acetate copolymers. It is also used to a small extent in the production of floor tiles but the extrusion process is generally preferred.

PROPERTIES

The base polymer is a white granular material with a density of 1.4 containing approximately 56 percent chlorine. The range of physical properties for the homopolymers and copolymers are given in Table 10.2. Its properties depend on molecular weight, the lower molecular weight polymers being soluble in a wide range of solvents, while the higher molecular

Table 10.2

TYPICAL PROPERTIES OF PVC AND COPOLYMERS

Property	*Flexible* General purpose	*Flexible* Electrical	*Rigid* Normal impact	*Rigid* High impact
Physical				
Specific gravity	1.2–1.6	1.2–1.6	1.4–1.6	1.3–1.4
Tensile strength (lb/in.^{2})	1500–3000	1500–3000	6500	9000
Elongation at break (%)	100–500	100–500	25	35
Compression strength (lb/in.^{2})	1250	1250	10,000	5-9000
Shore hardness	A50–95	A60–95	D75–85	D75–80
Specific heat (cal/cm)	0.3	0.4	0.25	0.25
Coefficient of linear expansion (°C × 10^{-5})	1	1	5	5
Maximum operating temperature (°C)	54–80	80–104	70	65–70
Water absorption (24 h %)	0.25	0.5–1	0.1–0.4	0.05–0.2
Flammability	←——— Self-extinguishing ———→			
Electrical				
Volume resistivity ohm/cm	10^{11}–10^{13}	10^{11}–10^{14}	10^{12}–10^{16}	10^{12}–10^{14}
Power factor (60 c/s)	0.08–0.15	0.08–0.15	0.02	0.03
Chemical				
Effect of weak acids	←——— Resistant ———→			
strong acids	←——— Resistant ———→			
weak alkalis	←——— Resistant ———→			
strong alkalis	←——— Resistant ———→			
solvents	Resistant to alcohols, aliphatic hydrocarbons, and mineral oils, soluble in ketones, esters, and to a certain degree aromatic hydrocarbons and chlorinated hydrocarbons			

weight materials are insoluble. The following indicates the effect of molecular weight on end-uses.

Molecular weight	*End-uses*
9,000	Surface coatings
10,000	Calendering, coatings
12,500	Compression molding
21,000	Cable sheathing
22,000	Textile fibers

The high chlorine content of the base polymer accounts for the polar nature of PVC which makes it resistant to petrol and mineral oils, and, on the other hand, soluble in solvents such as acetone and chlorinated hydrocarbon. PVC is also resistant to most acids and alkalis, and the high chlorine content also confers on it a high degree of flame-resistance.

Polyvinyl acetate

These resins are normally supplied as thermoplastic solids or as aqueous emulsions at solids content around 50 percent. The solid resins are odorless, tasteless, nontoxic, essentially clear, and colorless. The emulsions have a tendency to smell of the characteristic odor of the free unpolymerized monomer. Most grades of these resins have a somewhat broad molecular weight distribution. They do not possess a sharp melting point and soften over a relatively wide temperature range. These resins are neutral and non-corrosive, unaffected by sunlight, ultraviolet, and air. The various grades have good heat stability up to 100°C but at 150°C slight discoloration occurs and at 200-250°C decomposition takes place. It will burn slowly and soften at 50°C. They can be dissolved in acetone, methyl ethyl ketone, ethyl acetate, etc., and may be plasticized with a wide variety of plasticizers including the phthalates.

The chemical properties and reaction of polyvinyl acetate are those typical of stable esters. Some of the properties of polyvinyl acetate are listed in Table 10.3.

Table 10.3

PHYSICAL PROPERTIES OF POLYVINYL ACETATE

Density, at 20°C	1.19 g/ml
Refractive index, at 20°C	1.466
Water absorption	2%
Thermal coefficient of linear expansion	8.6×10^{-5} per °C
Thermal conductivity	38×10^{-5}
Specific heat	0.39 cal/g/°C
Dipole moment	1.85 Debye units
Dielectric strength at 30°C	1000 v/mil

Polyvinyl Alcohol

The properties of aqueous solutions of this polymer vary with the percent hydrolysis and the molecular weight of the intermediate polyvinyl acetate resin. Low molecular weight, low (70-80) percent hydrolyzed grades dissolve rapidly in water at room temperature. High molecular weight, high (95-100) percent hydrolyzed grades are best dissolved by dispersing in cold water and heating to 85-90°C in the stirring.

The arrangement of hydroxyl groups is predominantly 1,3 glycol units and the possibilities for hydrogen bonding in solution and film form leads to a high degree of versatility. It has been seen that one of the most useful reactions of polyvinyl alcohol is the condensation with aldehydes to form acetal resins. They may be unsolubilized by reacting with formaldehyde, furfural and certain amino resins. Plasticization is possible by adding glycerol, glycols, and sorbitol.

Table 10.4 lists some typical properties of polyvinyl alcohol.

Table 10.4

PHYSICAL PROPERTIES OF POLYVINYL ALCOHOL

Physical form	White to light straw granular powder
Specific gravity	1.19-1.27
Bulk density:	30-40 lb/ft^3
Refractive index, n_D, 25°C	1.51
Tensile strength, 50% relative humidity	to 22,000 psi
Elongation, 50% relative humidity	to 300% unplasticized; to 600% plasticized
Thermal coefficient of linear expansion	1×10^{-4}/°C, at 0-50°C plasticized
I.C.I. Sward hardness	27-57% *vs* glass at 50% relative humidity
Heat-sealing temperature	110-150°C
Heat stability	Darkening, slow degradation >100°C. Rapid degradation, decomposition >200°C
Effect of light	Negligible
Burning rate	Slow
Gas impermeability	High
Mold resistance	Good
Organic solvents	Generally insoluble

Polyvinyl Acetal

The properties of polyvinyl acetal resins depend chiefly on the types and amounts of functional groups present and on the resin molecular weight. As a general rule, the substitution of butyral or formal groups for acetate groups results in a more hydrophobic polymer with a high heat distribution polymer. Coupled with this is an increase in the polymers' toughness and adhesion to a wide variety of surfaces. The outstanding specific adhesion of the acetal resins is believed to be a result of their multifunctional constitution, because each molecule presents the choice of three different functional groups to a surface. This increases the probability of adhesion to a wide variety of substrates.

These resins are thermoplastic and soluble in a range of solvents. They may also be cross-linked by heating in the presence of a mineral acid. The mechanism of cross-linking has been discussed previously. The various reactions are shown as follows:

(1) Reaction with Isocyanates

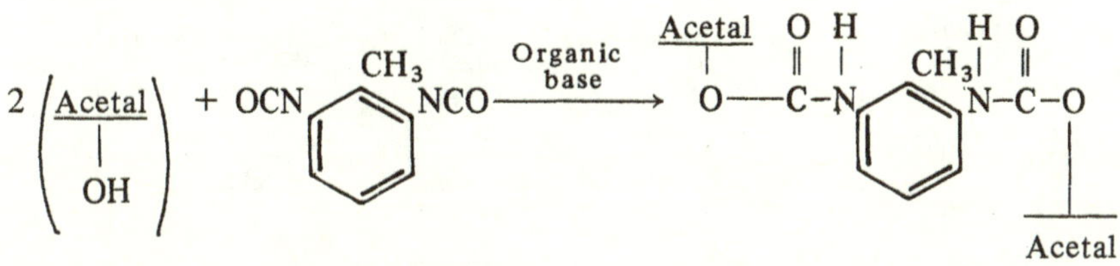

(2) Reaction with Phenolics

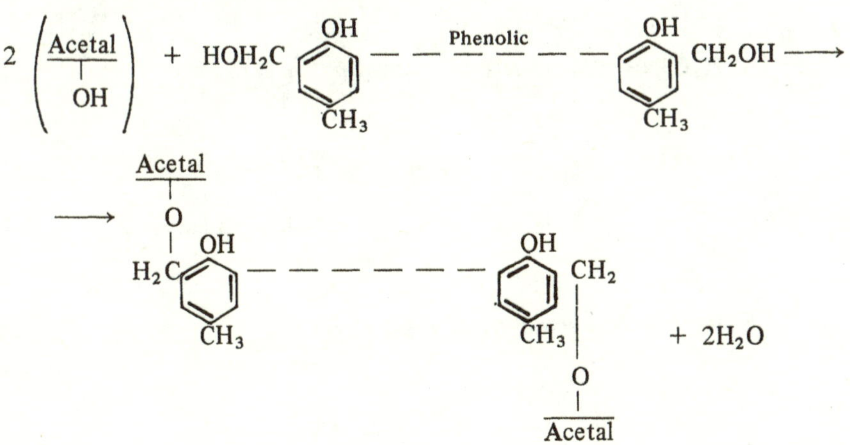

(3) Reaction with Dialdehydes

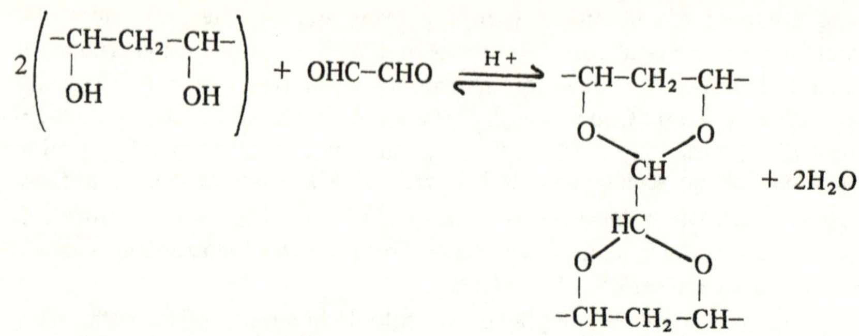

(4) Reaction with Melamines

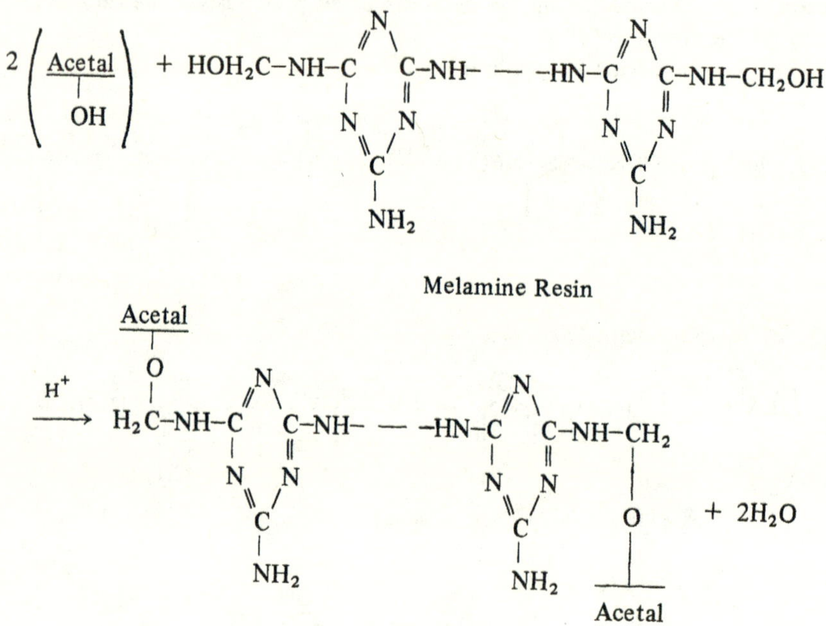

(5) Reaction with Epoxies (Anhydride Cure)

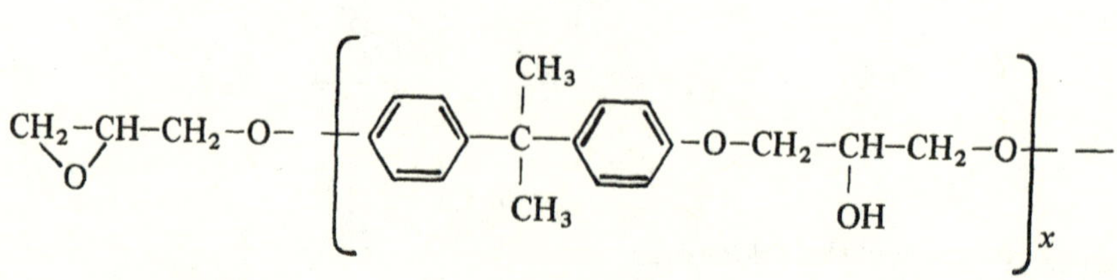

Typical Epoxy Resin

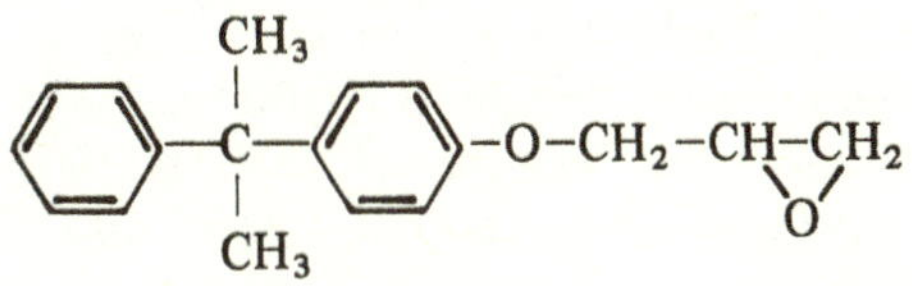

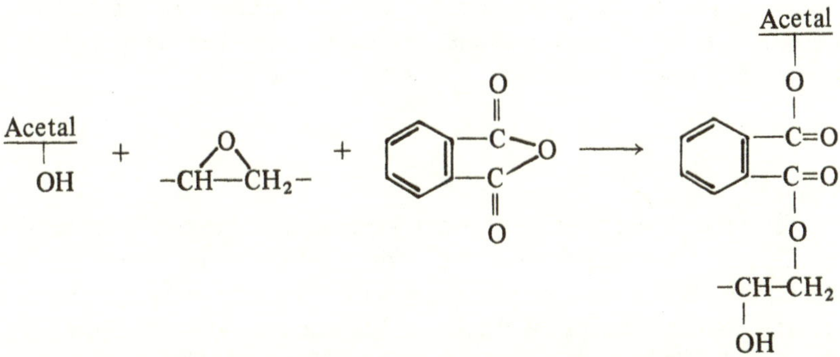

The vinyl acetal coatings are characterized by high resistance to aliphatic hydrocarbons, mineral, animal, and some vegetable oils. They withstand strong alkalis but are attacked by strong acids. By cross-linking, the stability to acids increases and they can withstand heating up to 200°F for prolonged periods.

A common solvent for all of the polyvinyl butyral and formal is a combination of 60 parts toluene and 40 parts ethanol (95 percent) by weight. Typical plasticizers for polyvinyl butyral are dibutyl phthalate, tricresyl phosphate, dibutyl sebacate and rosin derivatives. For polyvinyl formal there are diphenyl and diethyl phthalate, dicyclohexyl phthalate and tricresyl phosphate.

Polyvinyl Ethers

Typical materials in this group are polyvinyl methyl ether (PVM), polyvinyl ethyl ether (PVE), and polyvinyl isobutyl ether (PVI). All these polymers are essentially of the atactic noncrystalline type. The low or medium viscosity grades are sticky, viscous gums. The high molecular weight grades are tough rubber solids. They are compatible with phenolics and rosin derivatives.

Propylene-Vinyl Chloride

Most of the copolymers used are compounded and densities range from

1.3 to 1.4. Molded parts can be used up to temperatures of 60°C with room temperature tensile strengths of 7000 psi, high flexural modules (340,000 psi), and high impact strength (14 ft lb/in.). They also have low water absorption and measured figures report less than 1 percent after 30 days at 24°C.

The compounds have good chemical resistance especially to acids, alkalis, alcohols, aliphatic hydrocarbons, greases, and oils. Exposure to chlorinated solvents has little effect on stress cracking properties.

PASTES

So far we have only dealt with vinyl compounds in the powder or semipowder form. Another form of PVC which is used quite extensively is based on the dispersion of the polymer in a volatile liquid and is referred to as a paste. Its development goes back over 100 years ago when, in Germany, it was found that high molecular weight vinyl resins could be ground in an organic media to give stable dispersions. This is, of course, subject to the dispersing phase having some solvating action on the resin without actually dissolving it.

Vinyl pastes can be described as dispersions of the polymer in nonaqueous liquids which do not dissolve the resin at ordinary temperatures. Basic requirements of the polymer for good paste-forming properties is that it should be derived from small spherical particles (average size, 2μ) with a fairly wide size distribution. It must also be emphasized that, as with other PVC materials, it is essential to compound the paste by adding plasticizer, stabilizer, fillers, etc.

When the paste is heated to between 145–175°C, fusion takes place and this can be regarded as a mutual solubilization of resin and plasticizer. On cooling to 50°C it becomes a tough vinyl material exhibiting all the usual characteristics associated with the normal vinyl molded compounds.

These pastes can be divided into two main categories: plastisols, which are based on 100 percent solid mixtures of polymer, plasticizer, and compounding ingredients; and organosols, which are low viscosity mixtures of paste polymer and compounding ingredients dispersed in a mixture of plasticizer, and an organic diluent having a solids content varying between 60 and 80 percent. The volatile solvents are removed during the fusion process. Other types of paste are used but only to a small extent. These include the plastigel, which is based on a polymer dispersed in a plasticizer but with the addition of a thickening agent. The thickening agent increases the viscosity of the compound and produces a paste with butter to putty-like consistency. Rigisols are another form of paste; they have a reduced plasticizer content and can be used in slush molding.

The viscosity of the paste does not remain constant at varying rates of shear and, therefore, does not behave like a Newtonian fluid. It exhibits either dilatant or pseudoplastic properties, and for such fluids the rate of shear, *dv/dr*, is directly proportional to the shear force.

$$F = \eta \frac{dv}{dr}$$

where F = applied shear force per unit area (dynes/cm^2)

$$\frac{dv}{dr} = \text{shear rate } (s^{-1})$$

$$\eta = \text{viscosity (dyne/s/cm}^2 = \text{poise)}$$

A dilatant liquid exhibits the phenomenon that with increasing shear force the increase of shear rate is less proportional. Therefore, at high shear rates the flow of the liquid becomes relatively more difficult. This is often referred to as shear thickening, and some vinyl pastes are prone to this. However, in general, the majority of these pastes can be formulated to become shear thinning, particularly so if phthalate plasticizer are used. Such pastes are then termed pseudoplastics, their viscosity decreasing as the shear rate is increased. The types of flow illustrating shear-rate dependence, the effect of temperature and storage time on plastisol viscosity, and aging are shown respectively in Figs. 10.2, 10.3, and 10.4.

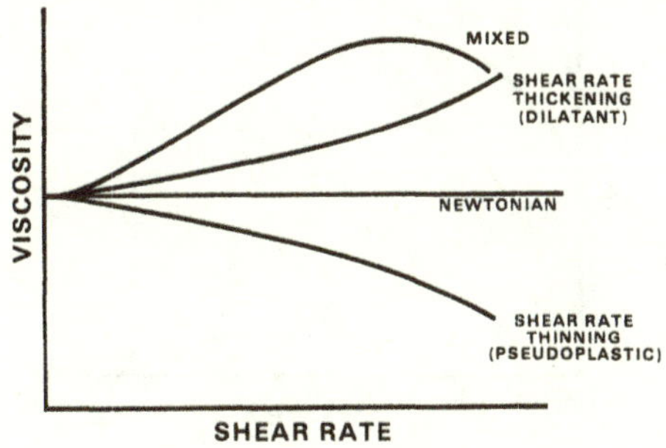

Fig. 10.2 Types of flow illustrating shear-rate dependence

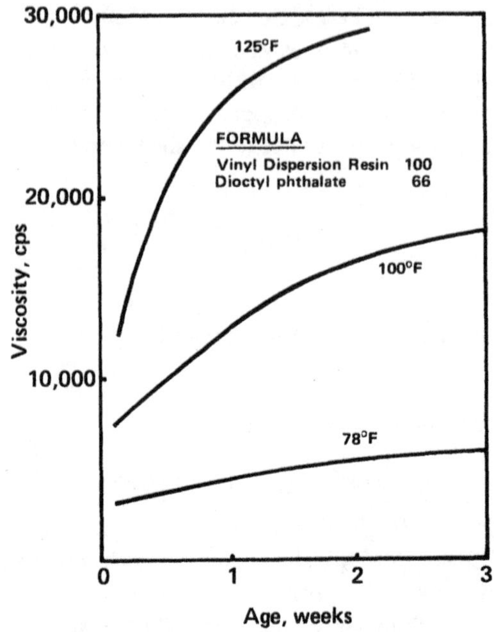

Fig. 10.3 Effect of temperature and storage time on plastisol viscosity

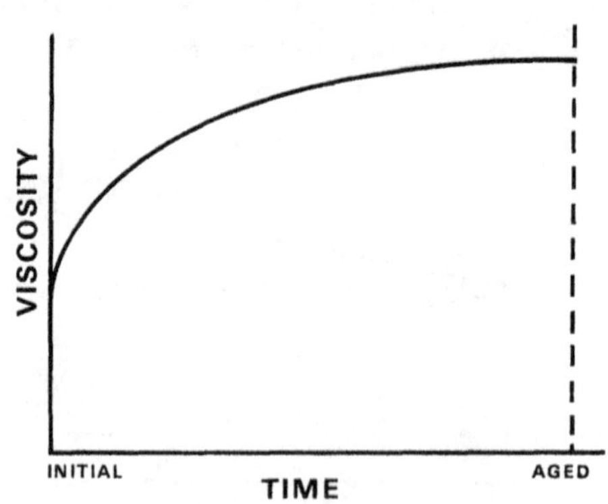

Fig. 10.4 Viscosity aging of vinyl dispersions

Processing

The vinyl pastes are first prepared by thorough mixing in a dough or paddle-type mixer with sufficient plasticizer to form a thick paste. This will allow the maximum shearing effect from the mixing action to be obtained. Other compounding ingredients are then added and the remainder of the plasticizer is finally added when all the other ingredients are well dispersed. Unfortunately, during the mixing process, air bubbles are entrapped in the paste, and these must be removed either by allowing to stand for 5-10 hours or by mixing under vacuum. One very important aspect to bear in mind with such pastes is that of storage. Due to the mutual interaction between the polymer and the plasticizer, the viscosity tends to increase, and the higher the storage temperature the more rapidly the viscosity increases, emphasizing the need for the paste to be stored in a cool place (*see* Fig. 10.3). The methods of processing vinyl paste are relatively simple and easy to carry out; they include spreading, dipping, slush molding and rotational casting, and the production of foams. These are nearly all described in Chapter 28 and no further elaboration is needed here. Table 10.5, however, gives the specific requirements of the paste for each particular processing method, with an indication of the end-use.

The fusion temperature is an important characteristic of vinyl pastes and depends on the molecular weight, amount and type of comonomer, plasticizer level, and polymer-plasticizer interaction parameter. Fusion of pastes is accomplished normally by heat alone and the particle size, shape, and distribution of the vinyl polymer particle affect this characteristic.

APPLICATIONS

Many of the initial applications for PVC were merely direct replacements for cellulose nitrate because the high degree of inflammability of this latter polymer made certain applications hazardous. The introduction of PVC meant that a polymer was available with a high degree of flame resistance and the opportunity was taken quickly to replace these older materials in such applications where flame resistance was required. It is by its high chlorine content that PVC and its copolymers are distinguished, and this is reflected in various important properties shown in applications.

The excellent chemical resistance of PVC enables tubes, rods, and sheets to be used by constructional engineers in designing chemical plant with corrosion problems in mind. Rigid PVC sheet is used extensively for

Table 10.5

PVC PASTE PROCESSES–REQUIREMENTS AND END-USES

Method	*Type*	*Requirement*	*End-use*
1. Spreading Coating	Plastisol and/ or	Medium viscosity No dilatency–thixotropic	Leather cloth Conveyor belting
(Knife and roll coating)	Organosols	Pastes are preferred because they are more mobile at	Coated paper Unsupported film
Dipping		high shear rates Gel temperature 180°C	Wire goods, gloves
2. Casting			
(*a*) Normal	Plastisol	Low-medium shear with good pourability	Shoe soles
(*b*) Slush molding			
Conventional	Plastisol	Low-medium viscosity	Hollow articles, e.g., toys (dolls)
Rotational	Plastisols/ Rigisols	as above but must exhibit special temperature/ viscosity behavior	Playballs
3. Foams	Plastisols	Low-medium viscosity can be chemically or mechanically blown	Flotation, cushioning

the fabrication of covers, ducts, tank linings, and many other items of chemical plant equipment. Because such sheet is dimensionally stable it also lends itself to precision engineering. In the actual fabricating process the sheet is softened by heat, shaped on formers, cooled, and then joined by welding. It can readily be built up into large units which, incidentally, need no painting. Pipe and fittings made from rigid PVC have many industrial uses, including the transportation of chemicals and in effluent disposal systems. It is also attractive for the transportation of water and gas because of its noncorrodability in acid soils and freedom from scale formation.

The weather resistance of PVC is good if the correct polymer/stabilizer system has been chosen. It is useful for garden hoses, likely to be left lying around in the open. Another application that has made use of the good weather resistance of PVC is corrugated roofing. The sheets developed over the past few years are now competitive with acrylic and reinforced polyester sheets.

By far the largest use of PVC at present consists of film and thin flexible sheet, which has already been discussed. Both film and sheeting can be made in all colors or with excellent clarity, they can be embossed or printed to give many different kinds of finish, and an added attraction is their relative ease of heat sealing. Uses of these films are growing quite extensively and already include the packaging of shampoos, collapsible containers, and drum liners. It is likely that PVC sacks for fertilizers and many other such materials will become more and more popular and that PVC will take a large share of the ever-growing packaging market.

Laminates of PVC cloth, made from cotton, nylon, jute, terylene, and other synthetic fibers by spreading, dipping, or calendering techniques, provide a product suitable for use in a wide range of applications. PVC leathercloth is being extensively used for upholstery fabrics and interior decoration in all types of passenger transportation such as buses, coaches, ships, boats, and aircraft. Its use for upholstery, interior trim, car hoods, and covers in automobiles is now firmly established. Other uses for coated fabrics include substitutes for leather, such as document cases and wallets, and for protective and foul-weather clothing. Coatings can also be applied by dipping techniques to form end-products such as industrial gloves and wire goods for industrial and domestic use.

The replacement of rubber coal conveyor belting with nonflammable belting made from PVC has been one of the most important developments in the coal-mining industry in recent years. Unlike natural and certain synthetic rubbers this polymer does not propagate flame, and therefore fire risk in the mine is reduced.

Cable coatings and sheathing form the biggest outlet for PVC after the film and sheet applications. It is because of this polymer's excellent electrical properties, good aging characteristics, and toughness that it is used widely, mainly in domestic and industrial cables and wires. Diverse uses range from bell-flex to large cable sheathings and the fact that PVC can be produced in many contrasting colors is an added advantage in that it aids the identification of complex circuits. Its specific uses include industrial and domestic wiring, telephone exchange wiring, and telephone installations, radio and television sets.

The fact that PVC does not rot or perish and that it is relatively tough makes it a suitable material for soft flexible toys. These may be produced by rotational casting, and include dolls and play balls.

The use of PVC for domestic applications such as furnishing and interior decorations has grown quite considerably; it is by no means unusual to find PVC floor tiles in practically any new building. PVC curtaining is available in an unlimited range of patterns and colors, and because it does not shrink in steamy conditions it can be used in the bathroom and in the kitchen. Other domestic uses include resilient carpet backing, wall coverings, gay shelf and table covers in a wide range of attractive styles and designs. The low-flammability characteristics of PVC are particularly attractive to architects when specifying components to be used extensively in buildings; PVC floor-covering and wall tiles are now used extensively in buildings, shops, and factories in addition to stair treads and covers for handrails.

As a foam material PVC has relatively few end-uses due to its somewhat high cost and its dependence of flexibility on temperature changes. Uses include cushioning material in cars and furniture upholstery, carpet underlays, mattresses, and shop mountings. Foamed PVC is now raising considerable interest as a buoyancy and insulation material and the market is expected to grow quite considerably in the next few years. These foams have excellent heat-sealing qualities and are therefore used where they have an advantage over rubber latex and polyurethane, viz. abrasion and chemical resistance coupled with fire resistance.

Various developments have taken place in the past few years in the technology of PVC, particularly in injection and injection/extrusion molding. Moldings as widely different as vacuum-cleaner components and raincoat buttons can now be produced at high speeds, while other applications include electrical components such as flexible terminal blocks and hand grips for cycles and tools. Molded footwear, formerly produced by casting techniques, can now generally be made by low-pressure injection/extrusion techniques. This allows wider possibilities in design, and inner linings can be applied during the actual molding operation. PVC compounds can also be extrusion blow-molded to produce containers which have the advantage of excellent clarity, good rigidity, high resistance to a wide variety of chemicals and also a low permeability to gases. In addition, printing on these articles can be accomplished without the necessity for pre-treating the surface. Typical applications include packaging of inorganic chemicals and solutions, petrol, motor oil, paraffin, and white spirit based materials, lotions, perfumes, and food flavoring and essential oils.

Vinyl chloride/vinyl acetate copolymers

Copolymerization with vinyl acetate adds a bulky unit to the main chain and this tends to keep the individual chains further apart. In some ways, therefore, the copolymer is similar in its behavior to a plasticized PVC but there is no possible loss of plasticization in service. If extra flexibility is required, however, additional plasticization can be carried out in a normal way. One use for the plasticized copolymer is for sheet, which is then thermoformed into refrigerator trays.

One very large market for vinyl chloride/vinyl acetate copolymers is in the field of sound reproduction. These materials have the required combination of molding properties when hot, together with good acoustical and mechanical properties when cold. The precision molding necessary can be seen from the fact that the records contain hundreds of grooves per inch.

Poly(vinyl acetate)

As mentioned earlier, vinyl acetate can be polymerized by itself as well as being copolymerized with vinyl chloride. The resultant polymer is not suitable as a molding powder but is extremely important as an adhesive and as the basic resin in many emulsion paints.

Poly(vinyl acetate) adhesives are stable suspensions of the resin particles in water. The suspensions also usually contain water soluble protective colloids such as poly(vinyl alcohol) or 2-hydroxyethyl cellulose ether, and may also contain plasticizers, fillers, solvents, anti-foaming agents, and preservatives.

Poly(vinyl acetate) adhesives are used in a wide range of packaging applications such as laminating cellulose film to paper, spiral tube winding, case sealing, and overlap sealing for high gloss can labels. They are so widely used because they can be produced with a wide range of specific properties, and have fast setting speeds. They are also low in odor, taste, color, and toxicity, and have excellent aging properties. Non-packaging uses for these adhesives include wood, leather, tiles, and many porous substrates.

Poly(vinyl acetate) is also used in surface coatings for painting wood, plaster, and many similar surfaces. A limited use has been found in the textile industry for finishing nonwovens.

Vinylidene chloride copolymers

Vinylidene chloride is copolymerized with various other monomers to form resins with extremely high gas and water vapor barrier properties. Such materials are often referred to as poly(vinylidene chloride) or PVDC but in fact the homopolymer is not used commercially. The most common comonomer is vinyl chloride, other comonomers being methyl acrylate, methyl methacrylate, acrylonitrile, and methacrylonitrile.

These copolymers are important as fibers, films and coatings. As fibers they are used in car upholstery because they do not absorb moisture and so do not stain and are easy to clean. Other uses are in carpets, furniture upholstery and industrial filter cloths, although wider use has been restricted by the growth of polyamide, polyester and polyolefin fibers.

Both casting and blow extrusion methods are used for the production of VDC copolymer films, although blow extrusion is preferred for the production of oriented film. The barrier and strength properties of VDC copolymer films depend on a number of factors including the amount and type of comonomer used, the degree of molecular orientation, and the direction of orientation in the finished film. A high VDC content gives a high gas and water vapor barrier while a comparatively high degree of orientation gives improved tensile strength.

In unsupported film form VDC copolymers are used for household wraps, drum liners, and food bags. Multilayer films, which are usually coextrusions with polyolefins, are used in the packaging of cheese, meat and other gas or moisture-sensitive foods.

VDC polymers are used as barrier layers in thermoformed containers, the sheet being produced either by coextrusion or by laminating with semi-rigid polystyrene or polyolefin substrates. They can also be used as the barrier layer in multi-layer bottles (see Chapter 28).

Finally, VDC copolymers are used to coat a wide variety of substrates, such as paper, regenerated cellulose film, and polypropylene film. Such coatings confer barrier and heat seal properties onto the particular substrate.

Poly(vinyl alcohol)

Vinyl alcohol itself is unstable and poly(vinyl alcohol) is made by the hydrolysis of poly(vinyl acetate). It is an unusual polymer in one other respect, being soluble in water. It is often used, therefore, in the manufacture of film sachets used to give controlled dosage in water. The sachet, plus contents, are simply added to the required amount of water, the sachet dissolves and the contents are released. This method of dosage is particularly valuable where the contents are toxic or where there are other reasons for not touching them.

Poly(vinyl alcohol) is also used in adhesives, textile sizing, emulsifiers and protective colloids. Miscellaneous applications include cosmetics, pharmaceuticals, photosensitive coatings, and release coatings.

Ethylene/Vinyl Alcohol Copolymers (EVOH)

EVOH copolymers are highly crystalline with properties that are very much dependent on the relative concentrations of the comonomers. Some idea of the range of properties is given in the Table (10.6). Their most outstanding characteristic is their ability to provide a barrier to gases such as oxygen, nitrogen, and carbon dioxide. The use of EVOH copolymers as an oxygen barrier is particularly important in enhancing flavor and quality retention. The gas barrier properties are also valuable in gas-fill packaging techniques.

The extremely high gas and solvent barrier properties of EVOH are reduced under humid conditions. However, because the barrier properties are so high, it is feasible to use the material in very low gauges and it is coextruded with other polymers, such as HDPE and PP on the outside of the multi-layer structure, thus giving protection to the EVOH and reducing the cost.

EVOH resins have high mechanical strength, surface hardness, and elasticity. They are also highly resistant to abrasion and have excellent weatherability. EVOH films have a very high gloss and low haze so that their clarity is excellent. Resistance to oils and solvents is very high and this makes EVOH resins an excellent choice for the packaging of oily foods, edible and mineral oils and organic solvents.

EVOH based films are produced in a number of ways, including emulsion coating, lamination and coextrusion, either tubular or flat. Biaxially oriented multilayer films and coextrusion barrier coatings are also being developed. Structures in use include PP/tie layer/EVOH/tie layer/recycled scrap/PP and nylon/EVOH/nylon/ionomer.

Table 10.6. RANGE OF EVOH RESINS

Property	*Range*
Melt index	0.7–20
Density (g/cm^3)	1.13–1.21
Ethylene content (mol. %)	29–48
Melting point (°C)	158–189
Oxygen permeability (cm^3/100 in^2/d/atm. – 1 mil film)	0.010–0.091 (PVDC = 0.150)

The first and most important use of polyvinyl butyral is as the interlayer for automotive safety glass. In this application the excellent adhesion, impact strength and light stability of this polymer has been put to full use. Increasing applications include structural adhesives, textile coatings, surface coatings (wood and metal), washed primers, and lately as an additive in the compounding of polymers for powder coatings. The most important use of polyvinyl formal is as wire enamel to provide electrical insulation on magnetic wire.

The main use for the polyvinyl ethers are as adhesives and surface coatings. The adhesives are essentially sensitive, hot melts, and remoistenable types.

SUGGESTED FURTHER READING

Vinyl and Allied Polymers Volume 2. Vinyl Chloride and Vinyl Acetate Polymers. (1972) G. A. R. Mathews (The Plastics and Rubber Institute).

Encyclopaedia of PVC Vol. 1 (1976) *Vol.2* (1977) *and Vol. 3* (1977) Ed. L. I. Nass (New York: Marcel Dekker).

Plastisols and Organosols. (1972) M. A. Sarvetnik (New York: Van Nostrand).

Polyvinyl Chloride Technology. (1963) L. P. Whittington (Society of Plastics Engineers).

11.

POLYSTYRENE AND COPOLYMERS

Styrene monomer is the simplest aromatic compound containing an unsaturated side-chain and was first isolated by Neuman, at the end of the eighteenth century. Work continued on the extraction of styrene from natural resins such as storax and dragon's-blood (an exotic name for the bright red gum which exudes from a kind of palm fruit!) but in 1866, Berthelot reported the synthesis of styrene by the simple process of passing benzene and ethylene through a red-hot tube. In 1839, a German apothecary, named Simon, observed that styrene solidified on storage or when heated. The solid product was at first believed to be an oxidation product but the true nature of the reaction (i.e., polymerization) was recognized in 1845 by Blyth and Hoffman.

However, styrene and polystyrene were to remain chemical curiosities for eighty years until 1925, when commercial production of styrene monomer began in Germany and the U.S.A. By 1939, both styrene and polystyrene were well-established products although production rates of the styrene monomer were relatively low. With the outbreak of hostilities in 1939 came a sudden increase in demand for styrene monomer in order to supply the enormous quantities needed for the production of styrene/butadiene rubbers. After the war the styrene/butadiene synthetic rubber program slowed up, thus releasing much of the styrene monomer capacity for the manufacture of polystyrene. Since that time the production of polystyrene has risen steadily and it is now one of the major thermoplastics.

PREPARATION OF MONOMER

Styrene, otherwise known as vinyl benzene, has the formula

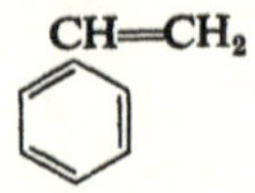

Pure styrene is a liquid, boiling at 145°C and polymerizing slowly at room temperature. It can be kept for long periods if a small amount of an inhibitor, such as hydroquinone, is added. Prior to polymerization the inhibitor is removed by washing or distillation.

Styrene can be prepared in many different ways, the most important (historically or commercially) being given below.

(i) By the slow distillation of cinnamic acid.

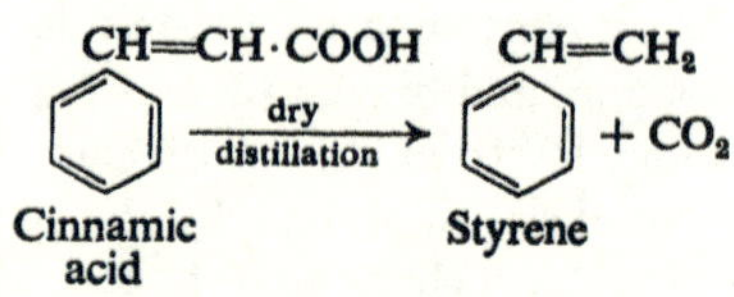

This is a laboratory method, used when very pure styrene is required.

(ii) By the dehydration of phenyl ethyl alcohol.

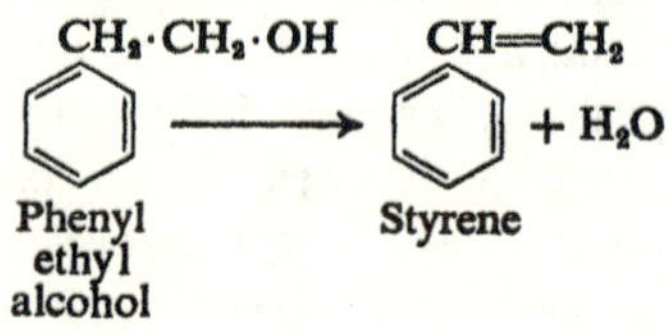

(iii) By the action of chlorine on ethyl benzene and subsequent heating of the α-chloro-ethyl benzene so formed. Although high yields of styrene are obtained by this method it has never been commercially developed, due to the impossibility of completely removing chlorinated hydrocarbons.

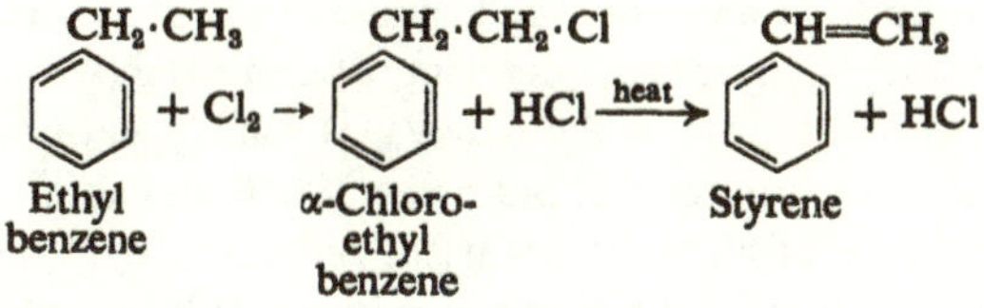

(iv) By the passage of benzene vapor and ethylene through a red-hot tube.

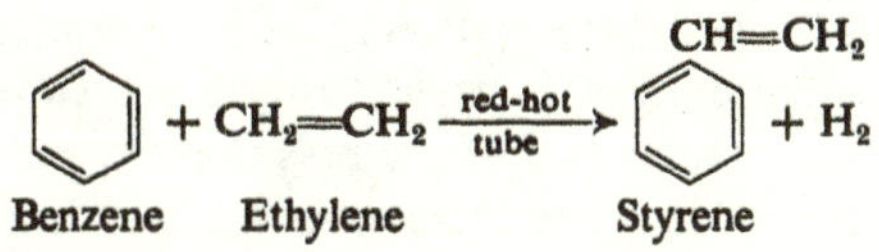

This method is, of course, the classical one used by Berthelot.

(v) By the dehydrogenation of ethyl benzene.

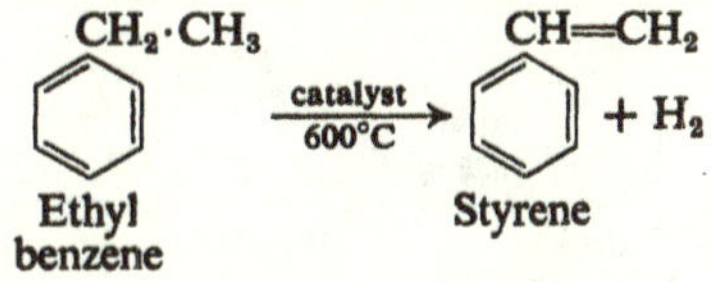

This is the usual process for the industrial preparation of styrene. The ethyl benzene itself is obtained by the direct alkylation of benzene in either the liquid or the gaseous phase.

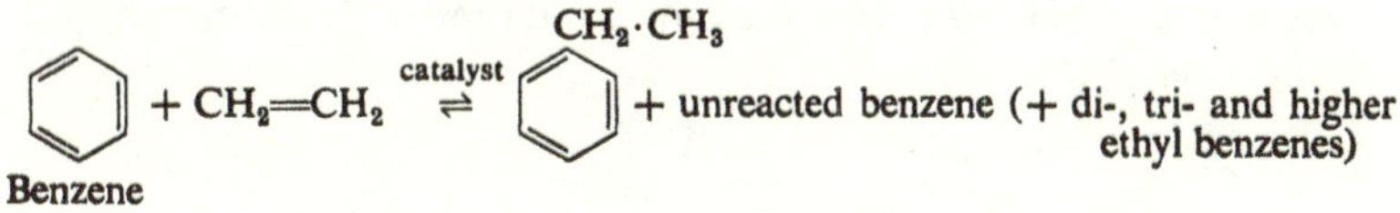

The liquid phase reaction is carried out at 85°C and 5 lb/in.^{2} pressure in glass-lined towers. The catalyst is aluminum chloride and, in practice, small quantities of di-, tri- and higher ethyl benzenes are added to the benzene in order to diminish the possibility of their formation during the reaction. The benzene used must be dried before being fed to the reactor as the presence of water increases the consumption of catalyst.

In the gaseous phase process the catalyst is usually phosphoric acid or silica alumina, the operating conditions being 300°C and about 600 lb/in.^{2}.

Whichever method of alkylation is used, the next step is the separation of pure ethyl benzene from the reaction mixture, which mixture is fed to a distillation unit consisting of three columns where benzene, monoethyl benzene, and polyethyl benzenes are removed from the tops of the first, second, and third columns respectively. The benzene is recycled to the benzene feedstock line in the alkylation plant while the monoethyl benzene is passed on to the dehydrogenation plant for conversion to styrene.

The dehydrogenation of ethyl benzene to styrene is carried out catalytically at temperatures around 600°C. Low pressures favor the conversion as the reaction is accompanied by an increase in volume. In practice the partial pressure of the reactants is reduced by the addition of steam. This has the further advantage of supplying the heat for the reaction, thus avoiding the possibility of degradation of the ethyl benzene, as might happen in a direct fired furnace. Mixed oxide catalysts are normally used, examples being zinc oxide/aluminum oxide/calcium oxide, and iron oxide promoted with potassium carbonate and chromium oxide (Shell 105).

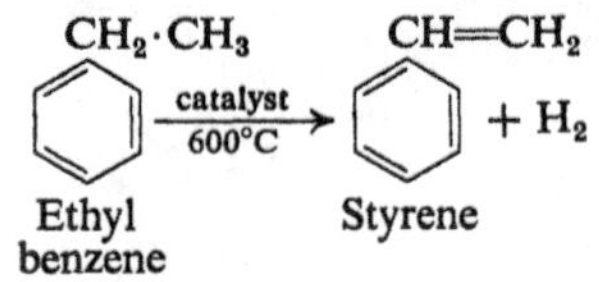

The separation of styrene from the reaction mixture is complicated by the fact that the boiling points of styrene and ethyl benzene are very close. Another problem is the fact that the polymerization of styrene monomer is easily catalyzed by quite small quantities of various impurities. The recovery of styrene is now carried out by vacuum distillation with addition of polymerization inhibitors.

POLYMERIZATION

1. *Unmodified Polystyrene*

The polymerization of styrene can be carried out by all four methods.

(a) Bulk Process. Either batch or continuous processes are possible but in batch processes, the removal of the heat of reaction and control of the mass temperature are both difficult. Average molecular weights tend to be low and the spread of molecular weight, large. The removal of styrene monomer, which is necessary to produce grades with high, heat distortion temperatures, is also difficult.

Continuous processes are based on batchwise prepolymerization in stainless steel or aluminum kettles at 85°C. At about 35 percent conversion, the resultant viscous solution of polystyrene in styrene monomer is passed down a tower with a temperature gradient from 100 to 200°C. The molten polymer is taken off at the bottom of the tower, extruded, and granulated. Heat is usually the only catalyst employed, and the absence of catalyst residues gives a resultant product with good electrical properties.

(b) Solvent Process. The styrene monomer is dissolved in an inert solvent such as benzene or cyclohexanone. Stirring of the high viscosity solutions so obtained is difficult and problems of heat transfer still exist, though to a lesser degree. Removal of solvent traces is difficult and expensive and, in practice, the resultant polymers are often used in solution form for applications such as coatings and impregnating compositions.

(c) Emulsion Process. The monomer is mixed with water, a water-soluble catalyst, and an emulsifying agent such as five percent sodium oleate. Buffers are also added, as the process is very dependent on the pH value. Heat transfer problems are negligible since the water phase assists in heat dispersion. After precipitation, the product is washed to remove emulsifying agents, but in practice some contamination always remains. This method gives the highest molecular weight product but the clarity and color are somewhat poorer than those obtained by bulk polymerization. Because of this they are often employed in the manufacture of the high-impact materials.

(d) Suspension Process. The styrene monomer is suspended in water with a styrene-soluble catalyst (such as benzoyl peroxide) and a stabilizer. When the mixture is stirred, the styrene breaks into small droplets each containing some catalyst so that the process is akin to the 'bulk' one without the disadvantage of difficult mass temperature control. The product is obtained in the form of small granules which are easily washed and dried.

In all the above processes the presence of divinyl benzene must be avoided. This compound has a functionality of four and thus forms a thermosetting resin by cross-linking. The existence of about one percent of divinyl benzene in the styrene monomer produces a hard 'glass,' impervious to solvents, while even as little as 0.003 percent has a noticeable effect on the properties of the resultant polystyrene and causes 'crazing.'

In its right place, however, divinyl benzene has many uses. In small

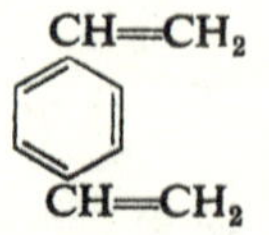

quantities (about 0.5 percent) it gives better-processing butadiene/styrene rubbers, while cross-linked bead polymers of styrene and divinyl benzene are extensively used for the production of ion-exchange resins. The resultant polymers are reacted with sulfonating agents giving solid, insoluble aromatic polysulfonic acids. Divinyl benzene has also been copolymerized with unsaturated polyesters to give products of value in the fields of coatings and adhesives.

2. *Toughened Polystyrene*

In its unmodified state, polystyrene is a crystal-clear, hard, rigid plastic. It has many excellent properties but for some applications its impact strength may not always be adequate. There has been much research into this problem and many improved polymers have been produced. One of the more successful approaches has been the incorporation of rubbery materials into the polystyrene to give a toughened polymer. Two basic methods are available for the preparation of toughened polystyrenes. These are by mechanical mixing of the rubber and polystyrene and by polymerization of a rubber-in-styrene solution.

(*a*) The mixing of the rubber and polystyrene can be achieved by a variety of methods, chief among which is mechanical mixing of dry polymers on a two-roll mill, in an internal mixer, or in a suitable extruder.

Solutions of the rubber and the polystyrene can also be blended, the solvents then being removed by vacuum evaporation. This method is a costly one because an expensive plant is needed for the vacuum evaporation and subsequent recovery of the solvents.

Another possible method is to blend the rubber and resin latices and to follow this by coagulation and drying.

(*b*) Interpolymerization of rubber dissolved in styrene can be achieved by bulk, solution, or suspension processes. Bulk processes may be batch or continuous, the latter being carried out in pipe or tower reactors.

Suspension polymerization is carried out by dissolving the rubber and catalyst in the styrene followed by dispersion by stirring in water with a suitable suspension agent such as polyvinyl alcohol. Polymerization is then completed by heating the suspension.

One of the requirements for successfully improving the impact strength of polystyrene is that there should be good adhesion between the rubber and the polystyrene. This is helped by the presence of similar molecular types in both of the components, which can be achieved by utilizing a butadiene/styrene synthetic rubber. In practice one containing about 25 percent styrene is found to be satisfactory. There should also be good dispersion of the rubber and its particle size should be small, since it is basically incompatible with the polystyrene. The amount of rubber which should be incorporated has to be a compromise between the increasing impact strength obtained, and the reduction in moldability, softening point, and surface finish which occur concurrently. Between 5 and 20 percent is normally used.

More recently the trend is to use butadiene rubbers as reinforcing agents. One of the advantages offered by butadiene rubbers is better retention of impact strength at low temperatures. This is due to the lower glass transition temperature of butadiene rubbers, i.e., –80 to –100°C against approximately –30 to –40°C for butadiene/styrene rubbers.

3. Expanded Polystyrene

Polystyrene is available in an expandable form which can be processed by steam molding to give light-weight articles with many outlets, particularly in the fields of packaging and thermal insulation. A patent on the expansion of polystyrene was taken out in the U.K. as long ago as 1943 but expanded polystyrene's commercial debut was not until 1955 when it was introduced in Germany by BASF. Initially it was used for thermal insulation but its cushioning properties, attractive appearance, and light weight soon led to its adoption as a packaging material.

There are two basic methods for the production of expandable beads, namely, by polymerizing and gassing the beads in one step, or by polymerizing the styrene, then impregnating the polystyrene beads at elevated temperature and pressure. The gassing agent in both cases is pentane.

An American process for the manufacture of expanded polystyrene slabs utilizing methyl chloride is also known. In this process polystyrene beads are gassed in an extruder while the slabs are being produced. Capital equipment costs are high, as are transport costs for the finished product.

Molding of expandable polystyrene beads is carried out in three stages: preexpansion, maturing, and molding. Preexpansion is achieved by heating the expandable beads in steam. This has the dual effect of increasing the pressure of the blowing agent within the beads and of softening the polystyrene. The beads thus expand, the degree of expansion (and hence the final density) being controlled by the temperature and the duration of heating. Preexpansion can be carried out as a batch operation or continuously.

The preexpanded bead consists of noncommunicating cells of 80–150 microns diameter with about nine million cells to the cubic inch. The bulk density of the original expandable polystyrene beads is about 42 lb/ft^3 (673 g/ℓ) and expanded articles can be produced with bulk densities ranging from just below this figure down to 12 oz/ft^3 (12 g/ℓ) or even lower. However, at bulk densities of 1 lb/ft^3 (16 g/ℓ) and below, the cell walls are only one to two microns thick and have almost reached the elastic limit of the polystyrene. If articles of very low density are produced by molding the requisite weight of expandable beads in a mold of the required volume, poor fusion and excessive shrinkage normally occur. This is due to the beads expanding almost to their limit with the available blowing agent and filling the mold but without retaining sufficient expansion power to fill the voids between the expanded beads and thereby form a completely homogeneous fused mass.

After preexpansion the beads are allowed to cool, when the residual blowing agent condenses inside the bead, thus causing a partial vacuum. At this stage the beads are extremely weak and if pressed between finger and thumb they collapse. During the maturing period air permeates through the multicellular structure of each bead until equilibrium with the atmosphere is achieved. The beads are then resilient and do not collapse when squeezed between finger and thumb. In this condition the beads are ready for molding since expansion of the air within the bead during the molding process can contribute to the pressure generated by the residual blowing agent and so ensure good fusion.

For a given density there is a minimum and maximum maturing time. The minimum time is determined by the rate of air diffusion into the beads while the maximum time depends on the rate at which the residual blowing agent permeates out to the atmosphere. If excessive loss of blowing agent occurs, the expansion characteristics during molding are impaired. Maturing times normally range from 9 to 24 hours.

When maturing is complete, the preexpanded beads are transferred to the mold, which is then closed and heated by injecting steam. The residual blowing agent and the air which entered the beads during maturing both expand and the polystyrene is again softened. Since the beads are confined in a closed mold, the expansion in volume of the blowing agent and air within the beads causes them to distort and fill the voids between the beads. Individual beads merge into the mass, giving a coherent, noncommunicating microcellular structure (*see* Fig. 11.1). The mold is then cooled, opened, and the molding removed.

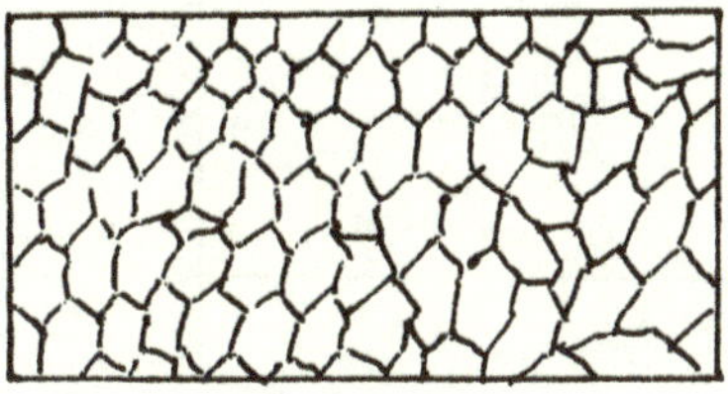

Fig. 11.1

It will be seen from the above that it is important to achieve the correct bulk density during the preexpansion process. The only change that occurs during the molding process is the expansion of the beads into the air spaces present before the mold is heated. The weight and volume of the molding produced is thus the same as that of the preexpanded beads used to fill the mold.

Preexpansion can be carried out in equipment as simple as an oil drum fitted with a wire gauze or muslin support for the beads. Steam is introduced in the bottom of the drum and passes up through the beads. Any condensate is allowed to drain through holes in the bottom of the drum. A more suitable apparatus for batch preexpansion is shown in Fig. 11.2. It consists essentially of a stirred autoclave into which the beads are introduced via a pipe. Steam is led in through the false bottom. The preexpanded beads are extracted from the base of the autoclave by means of a pneumatic conveyor. The preexpansion process may also be made continuous. One method utilizes a free space steam heater which consists basically of a stirred vertical cylinder with horizontal take-off points at various heights up the cylinder.

The expandable beads are introduced into the base of the cylinder by either a steam injector or a screw feed. As the beads expand they are swept upwards and out of the preexpander. The required density is normally obtained by controlling the feed rate of expandable beads to the base of the equipment, allowing air to bleed into the steam supply to re-

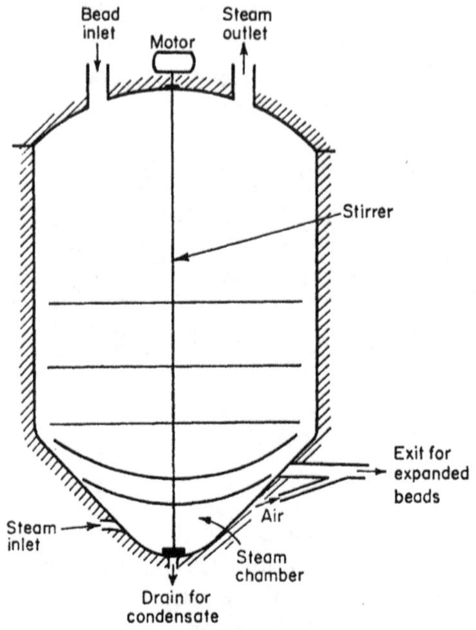

Fig. 11.2 Autoclave batch pre-expander

duce temperature, or varying the height at which the preexpanded beads are removed.

The molding of blocks from which sheets can be cut for thermal insulation is usually carried out by filling the beads into a mold consisting of an open-ended steel frame, closable at each end by perforated steam chests which are then clamped into position. After steaming and cooling, the ends are removed and the fused block extracted.

Molding of shaped articles such as packages or molded carton inserts is carried out by one or other of two basic methods. In the autoclave method, the mold is filled with preexpanded beads, closed, and placed in the autoclave. Fusion of the beads is achieved by steaming for a given period at 10–20 lb/in.^{2} steam pressure. A convenient method of mold filling is to use an air injector. With air line pressures of about 100 lb/in.^{2} the mold is completely filled with beads. In addition, a certain amount of precompression is achieved which ensures that in the molding operation maximum use is made of the expansion power of the beads to fuse completely into a homogeneous mass.

Automatic molding machines are also available, working on similar lines to a conventional injection molding machine. The preexpanded beads are injected into the mold by means of an air injector and then fused by steam. The mold is in two parts and steaming starts while the mold is still closing. This preheating avoids condensation forming during molding. The beads are then injected and when the mold is filled, steam at 15-20 lb/in.^{2} pressure is introduced into the male section and at 30-40 lb/in.^{2} pressure into the female section. The steam from the male portion of the mold displaces the air from the voids between the beads and the mold is then completely closed. Steaming is then continued until fusion of the beads is complete. At the end of the steaming cycle, the steam is quickly vented to atmosphere and cooling is carried out by circulating cold water through both parts of the mold. The mold is then opened and the article ejected automatically. A simple representation of a horizontal automatic molding press is shown in Fig. 11.3.

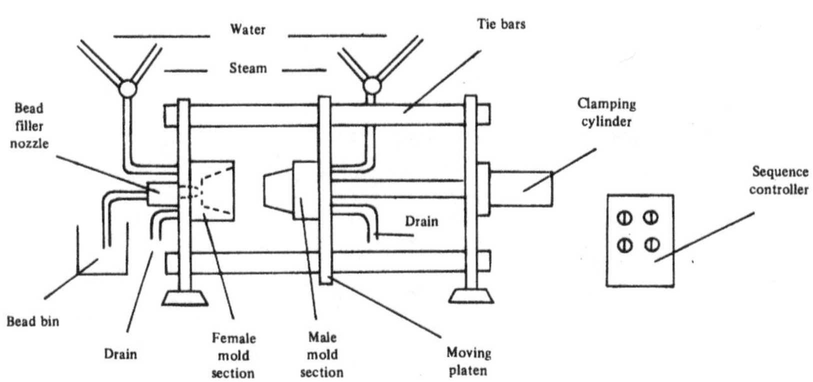

Fig. 11.3 Horizontal automatic molding press

Although steam heating is still the most widely used heating medium for expandable polystyrene molding machines, other processes have been developed. In the U.S.A. one process employs dielectric heating instead of superheated steam. One of the main advantages is that it allows the use of reinforced polyester molds because of the lower operating temperatures. Such molds are much cheaper than equivalent metal molds. Another advantage is that faster production rates are possible due to reduced heating and cooling cycles.

Expanded polystyrene is also available in the form of extruded sheet. Techniques for producing it were developed in America between 1955 and 1960. The first patent applications were dated 1958 and were in the name of Koppers Co., Inc. Extruded expanded polystyrene sheet first became commercially available in the U.S.A. in 1959.

In appearance the extruded sheet has an attractive sheen with a feel very similar to that of paper. In fact the original expectations of this material were that it would replace paper for many applications, including writing papers, wrapping papers, and labels. Later it was believed that there were better opportunities in the fields of thermoformed sheet to provide trays, boxes, and box inserts, and as a laminate with paper (*see* under 'Applications').

The extrusion of expanded polystyrene sheet can be carried out in two main ways. One utilizes ordinary polystyrene, gassed in the extruder under pressure, while the other utilizes pregassed beads. Further details of the extrusion of expanded polystyrene sheet are given in the chapter on 'Extrusion.'

CHEMICAL STRUCTURE

Polystyrene has been shown to have the following formula:

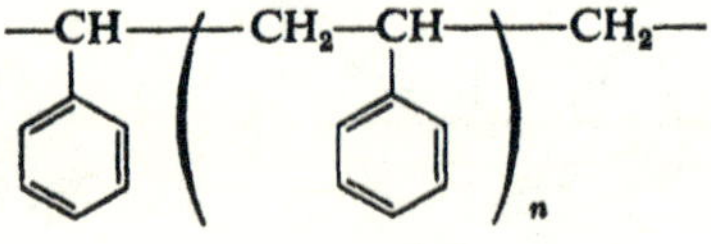

When prepared by the methods already described polystyrene is completely amorphous. This is because the bulky nature of the benzene rings in the above structure does not permit the close approach of the polymer chains; a random arrangement of the chains therefore results. This should be compared with the high crystallinity obtained in an almost completely linear polymer such as high density polyethylene (Chapter 9).

However, it should also be noted that Professor Natta has prepared a crystalline polystyrene by the use of stereospecific catalysts (cf. polypropylene–Chapter 9). Such catalysts cause the benzene groups to assume a regular arrangement along the basic carbon chain and the various molecules can then take up an orderly arrangement to produce a crystalline structure. The actual configuration of each molecule is a helical distribution of the benzene rings along the carbon chain. Among the interesting properties of this 'isotactic' polystyrene is a much higher crystalline melting point of about 230°C (cf. the properties of the normal amorphous polystyrenes later in this chapter).

PROPERTIES

Polystyrene is a colorless, transparent thermoplastic with a distinctly metallic ring when dropped onto a hard surface. It is hard, has a fairly high tensile strength (6-10,000 lb/in.^{2}) and a high refractive index (1.59). It softens at about 90-95°C, while at 140°C it is a mobile liquid, excellent for injection molding. Intrinsically it is a rather brittle material and, as mentioned earlier in this chapter, toughened (or high-impact) grades have been developed for the more critical applications.

It has outstanding electrical properties including a very low power factor, high dielectric strength and high volume resistivity. Chemically, it is resistant to strong acids and bases and is insoluble in aliphatic hydrocarbons and the lower alcohols. It is, however, soluble in esters, aromatic hydrocarbons, higher alcohols, and chlorinated hydrocarbons.

A great number of grades of polystyrene have been manufactured for various specific purposes. Thus we have quick processing, general-purpose and heat-resistant grades, as well as the toughened grades already mentioned. Leaving the question of toughened grades for the moment, the other grades are obtained from polymers of differing molecular weight distribution and average molecular weight, together with varying amounts of lubricant. Among the lubricants in general use are such compounds as esters, organic acids, and mineral oils. In addition to improving the flow of the polystyrene, these lubricants also reduce the brittleness, lower the softening point, and increase the setting-up time. The presence of low molecular weight polystyrenes also has the effect of improving the flow, but in this case the brittleness is increased. Another similarity between low molecular weight polystyrenes and internal lubricants is in their reducing effect on softening point. However, the magnitude of the effect is not so great in the low molecular weight polystyrenes.

It is in the balancing of these various effects to produce the grades required by the market that the skill of the formulator is evidenced.

Quick processing grades, for example, can be produced either by speeding up the setting time or by improving the flow. Fast-setting resins usually contain very little lubricant and are often of a slightly lower average molecular weight than a general-purpose grade. This lower molecular weight does confer slightly improved flow characteristics as an additional aid to quick processing. The easy flow grades, on the other hand, contain fairly large amounts of lubricant and a large proportion of low molecular weight polystyrenes. The latter, of course, increases the brittleness of the finished molding but this is counteracted by the amount of internal lubricant added.

Heat-resistant grades are of high average molecular weight, contain no internal lubricants, and have a minimum of residual styrene monomer or low molecular weight products. The flow is, of course, reduced, and slightly more difficulty is experienced in molding. In order to reduce any stresses which may have occurred in the molded articles, annealing is often carried out by placing them in warm water immediately on removal from the mold.

The increase in impact strength of the toughened polystyrenes is achieved at the expense of a loss in clarity, and crystal-clear grades cannot be obtained. Apart from the increase in impact strength, other physical changes occur. These include increases in elongation at break and reductions in tensile strength, surface hardness, and softening point. The increased flexibility of toughened polystyrenes allows the use of metal inserts in moldings since the flexibility compensates for the differences in the coefficients of expansion between the metal and the plastic. The increased flexibility also facilitates the machining of toughened polystyrene. As far as the chemical properties of the toughened polystyrenes are concerned they can be taken, for all practical purposes, as similar to those already given for the basic polystyrenes.

The properties of typical basic and toughened polystyrenes are summarized in Table 11.1.

APPLICATIONS

Polystyrene can be injection molded, extruded into sheet and film, or blown into bottles. Its excellent electrical properties (such as very low power factor and high dielectric strength) ensure its extensive use in the electrical and radio industries.

The reduced level of low molecular weight fractions and the absence of internal lubricants in heat-resistant grades leads to their having better elec-

Table 11.1

TYPICAL PROPERTIES OF POLYSTYRENE

Property	*Method of test*	*Units*	*Unmodified polystyrenes*			*Toughened polystyrenes*	
			General purpose	Quick processing	Heat resistant	High impact	Extra high impact
Specific gravity	ASTM. D.792–50	–	1.05–1.07	1.05–1.07	1.05–1.07	1.05–1.07	1.04–1.06
Tensile strength	ASTM. D.638–56 T	lb/in.^{2}	6000–7500	6000–7500	7000–8000	4500–5500	2500–3500
Elongation at break	ASTM. D.638–56 T	%	1–2.5	1–2.5	1.5–2.0	12.5–22.5	25–45
Impact strength							
Izod, notched 1/8 in. inj.-molded bar	ASTM. D.256–56	ft lb/in. of notch	0.3–0.6	0.3–0.6	0.3–0.6	1.0–2.0	1.5–2.5
Izod, notched ½ in. comp.-molded bar	B.S. 1493–1958	ft lb/in. of notch	0.2–0.35	0.2–0.35	0.2–0.35	0.6–1.0	1.0–2.0
Hardness	ASTM. D.785–51	Rockwell, M. Scale.	65–80	65–80	75–80	45–60	25–35
Machinability	–	–	Good	Good	Good	Excellent	Excellent
Specific heat	–	cal/°C/g	0.32	0.32	0.32	0.32	0.37
Softening point	B.S. 1493–1958	°C	85–90	82–84	100–103	85–90	78–83

Table 11.1 *(cont'd.)*

Property	*Method of test*	*Units*	*Unmodified polystyrenes* General purpose	Quick processing	Heat resistant	*Toughened polystyrenes* High impact	Extra high impact
Refractive index	ASTM. D.542–50	N_d^{25}	1.59	1.59	1.59	–	–
Light transmission	ASTM. D.1003–52	%	88–90	88–90	88–90	–	–
Power factor at 1 Mc/s	B.S. 1493-1958	–	0.0001–0.0010	0.0001–0.0010	0.0001–0.0005	0.0002–0.0010	0.0001–0.0003
Dielectric constant at 1 Mc/s	B.S. 1493-1958	–	2.45–2.55	2.45–2.55	2.50–2.70	2.45–2.65	2.50–2.90
Volume resistivity	ASTM. D.257–57 T	ohm/cm	$>10^{13}$	$>10^{13}$	$>10^{13}$	$>10^{13}$	$>10^{13}$
Water absorption	ASTM. D.570–54 T	%	<0.05	<0.05	<0.05	<0.25	<0.3
Resistance to–							
Alkalis			Excellent	Excellent	Excellent	Excellent	Excellent
Acids			← Attacked by strong oxidizing acids →				
Water			Excellent	Excellent	Excellent	Excellent	Excellent
Oxygen			Excellent	Excellent	Excellent	Good	Good
Solvents			← Soluble in Esters, Aromatic Hydrocarbons, Higher Alcohols, and Chlorinated Hydrocarbons →				

trical properties than a general-purpose grade. This combination of improved heat resistance and good electrical properties is often very useful, particularly for radio cabinets, which may be subjected to relatively high temperatures. Cabinets molded from general-purpose grades have been found to sag after exposure to temperatures of 77°C for two hours whereas heat-resistant grades can withstand temperatures of 88°C for as long as 20 hours. Coil formers, frequency transformers for radios, and electrical insulating film for capacitors are among the many other uses for heat-resistant grades in the electrical industry. Frequency transformers for radios were originally made from general-purpose grades of polystyrene but were found to distort in service at temperatures around 65°C. Heat-resistant grades, on the other hand, can be made to withstand temperatures up to 95°C.

Heat-resistant grades are also of importance for packaging uses such as preserve jars, where the product is hot-filled. For this, and similar applications, an improved impact strength is also a useful asset and high molecular weight, heat resistant grades would be used. Other uses for such grades include drinking glasses and tumblers.

Where heat resistance is not important but where superior resistance to vibration, slight flexing, or shock is required, high molecular weight grades are used. Examples of the use of such are: for molding clothes hangers (for light-weight articles such as drip-dry shirts), clothes pegs, thin-wall packages, and hooks. High molecular weight grades are also used for such applications as brushes, where the moldings are subjected to drilling and tufting operations since these grades have better machining properties than other grades of polystyrene.

Polystyrene has long been a popular plastic for packaging applications and many thousands of tons are used annually for the packaging market. Many foodstuffs, for instance, including jam, marmalade, honey, and butter are packed in individual pots injection molded from basic polystyrene. Other, larger, tubs and pots are injection molded from toughened polystyrene and are used for a wide variety of foodstuffs such as cream, yogurt, glace cherries, chopped peel, honey, cottage cheese, etc. The use of toughened polystyrene gives the extra impact strength required by these applications and superflow grades have been developed to give the thin walls required for competition, cost-wise, with vacuum formed tubs. Many other outlets for polystyrene have been found in the packaging of cosmetics. Powder compacts are a well-known example, the excellent color range of polystyrene, together with its high gloss, making it particularly attractive for articles which rely so heavily on 'sales appeal.' Other examples are injection molded pots for face and other creams. When blown by methods such as the injection blowing process which reduces residual strain, general purpose polystyrene gives bottles of good impact strength. So far such bottles have been used largely for talcum powder because of their attractive appearance when colored. This is usually enhanced by the use of gold blocked printing. In the closely allied fields of pharmaceuticals and toiletries, polystyrene is also widely used. For example, injection molded tubes are extensively used for the packaging of tablets although the poor water vapor barrier properties of the material make it unsuitable as a pack for hygroscopic tablets or for export to tropical areas. The advantages include the ease of printing on polystyrene compared with glass and the absence of the risk of glass fragments due to accidental breakage. Jars, made by a modified injection blow molding process are used for the packaging of petroleum jelly. The freedom from glass splinters is very important here because the product is widely used on babies' skins. Toughened polystyrene has been used for injection molding containers for ointments but care must be taken as certain ointments, such as those containing methyl salicylate (oil of Wintergreen), cause chemical attack of the polystyrene.

Polystyrene has been extensively used for bottle closures, especially on bottles made from low or high density polyethylene where the hardness of the polystyrene gives good 'bite' on the neck of the bottle. Care must be exercised in material choice because certain chemicals, such as ketones and hydrocarbons attack polystyrene. Polypropylene or urea formaldehyde might then be chosen. On glass bottles, where there is a danger of cracking of the caps due to excess closing torque after filling, a toughened grade of polystyrene is often used.

A wide range of housewares is also manufactured in polystyrene although in this field there has been much successful competition from the polyolefins (particularly polyethylene). The clarity of polystyrene has led to its use for food canisters where visibility of the contents is required, while the gloss and excellent color range of the pigmented material gives it the edge for those foodstuffs where transparency would be undesirable. Other uses include light fittings, kitchen scale pans, clock casings, tape recorder decks and spools, vacuum flask outer shells, fruit bowls, and toilet cisterns. Polystyrene can also be extruded into filaments which, when oriented by drawing, are ideal for use in brushes owing to their high strength and flexibility. In certain other houseware uses, such as measuring jugs, there is a trend towards the use of materials with better chemical resistance, such as the styrene-acrylonitrile copolymers.

Very many toys were formerly made from polystyrene but a number of applications are now being replaced by the polyolefins because of the latter's much greater toughness. In toys for younger children there is also the risk of injury by sharp edges should the polystyrene be broken.

One area of the toy industry where polystyrene is still extensively used is in the field of model kits, for aircraft, vintage cars, and a host of other items. The intricate moldings required call for the excellent injection molding performance of polystyrene. Where extra toughness is required, rubber modified grades are used. Another reason for using polystyrene rather than say, one of the polyolefins, is that polystyrene can easily be painted. One excellent example of a toy that makes full use of the properties of polystyrene is the 'Spirograph.' This consists of a series of intermeshing toothed rings and strips, with holes into which a pen or pencil can be fitted. By using different components and moving them inside the master gear, the most complicated designs can be built up, the results being parti cularly beautiful when different colored inks are used. All the components of this set are molded in polystyrene because this particular plastic is hard enough to stand up to the frictional wear of long use and is transparent so that one can tell how the design is progressing at all stages.

Many photographic items are also molded from polystyrene including tweezers (rust-free and easier to keep clean than metal), exposure meter cases and transparency viewers (usually in toughened grades). Other photographic items, such as trays for developing are made, not by injection molding but by the vacuum forming of the polystyrene sheet. This method is also useful for the manufacture of many display items. Toughened polystyrene grades are particularly well-suited for use in sheet form, a very large outlet being for refrigerator liners. Deep draws can be achieved, the resultant article being rigid enough for use in even large-volume refrigerators. The wide range of colors available is an additional selling point for this application. It should perhaps be noted here that with the advent of very large injection molding machines a number of liners were manufactured by injection molding. Mold costs are very high and large runs are necessary for economic molding. Such large runs are rarely attainable in a market containing so many different models and sizes.

Thin-gage toughened polystyrene sheet is extensively used for packaging applications. Very deep draws are possible with toughened polystyrene sheet, an excellent example being the Philip Morris cigarette pack. Deep-drawn containers for fruit squashes are also among the many outlets for thin-gage sheet, the container being of rectangular cross section and 'stepped' for greater rigidity. Perhaps the largest outlet for deep-drawn sheet is that of vending cups, the latest development being a double-walled container designed to allow hot liquids to be held comfortably without burning the hand. Vacuum formed tubs have also been used for cream and yogurt but the development of the superflow grades of toughened polystyrene mentioned earlier has made it possible to injection mold containers with walls thin enough to successfully compete in these markets. A large market has also been developed for thermoformed retail packs for six eggs. Toughened polystyrene sheet is not itself a cushioning material but the design of the pack is such that the eggs are protected from direct impact. In addition, the sheet is translucent so that the eggs are visible without the need to open the pack. Housewives can, therefore, see whether the pack contains white or brown eggs, or whether any eggs are broken. More shallow draws are used for preformed trays for apples, peaches, biscuits, etc., the thickness of the sheet being about five thousandths of an inch.

Basic polystyrene sheet is too brittle for packaging uses although it has been put into use for electrical applications such as capacitors. It can also be obtained in an oriented form. After extrusion the polystyrene is stretched simultaneously in the machine in transverse directions and then quenched. The degree of orientation retained in the finished sheet is governed by the quenching conditions, the rate of stretching, and the degree of elongation. The biaxial orientation produces a superior product, having crystal clarity and satisfactory impact strength. It is a poor barrier to moisture vapor although this is sometimes turned to good account, as in the packaging of fresh fruit and vegetables where moisture loss is required in order to prevent condensation and 'fogging' of the sheet. In thicker sections it is often thermoformed into cups, cake covers, and tubs of excellent clarity and is also used for blister packs.

Expanded Polystyrene Applications

Expanded polystyrene has found widespread use as a thermal insulant. Cold stores, refrigerated transport, chill rooms and freezer display cabinets are just a few of the many outlets for this material. Other fields where the thermal insulation of expanded polystyrene has been utilized to good effect are the insulation of factory roofs, farm buildings, broiler houses, and other buildings where fuel costs are an expensive item. For domestic use, glass fiber or mineral wool are somewhat cheaper than expanded polystyrene in the insulation of lofts of existing dwellings. In new dwellings, however, expanded polystyrene can be used in place of plasterboard which is normally nailed to the underside of the ceiling joists and subsequently plastered to form the ceiling. The expanded polystyrene serves the same purpose as the plasterboard but has the additional advantage of thermal insulation.

The insulation value is usually expressed as the K value, the units being $Btu/ft^2/h/°F/in.$ thickness. The value of expanded polystyrene as an insulating material can be seen by comparing its K value (0.24) with that of some other commonly used material such as cork (0.28-0.32). From this it can be seen that six inches of expanded polystyrene is equivalent to between seven and eight inches of cork.

Because of its noncommunicating cell structure, expanded polystyrene can be used as a buoyancy material. One cubic foot of expanded polystyrene supports a weight of about 50 pounds in water. Except for use in buoyancy lockers, where it needs no additional protection, expanded polystyrene should be covered with plywood, metal, or a resin/glass cloth laminate.

Expanded polystyrene is also finding many applications in the fields of packaging and display because it can be molded to close tolerances. This enables efficient contour packs to be produced with maximum protection against vibration and impact. The material can be self-colored, painted, or decorated with flock, thus opening up many possibilities where packaging must be combined with display. Its clean and nonabrasive surface eliminates the need for overwrapping, while its neutral pH is of great importance in the packaging of metal articles.

In spite of the widely different requirements of industrial packaging and that used for consumer products, expanded polystyrene has been successfully used for both types. In the choice of packaging for cameras, typewriters, hair dryers, transistor radios, etc., appearance and customer acceptance are all-important. Cost is also important but may not always be paramount. On all these counts expanded polystyrene offers advantages. In addition to its intrinsic cushioning properties, the fact that expanded polystyrene can be molded into intricate shapes means that even awkwardly shaped objects can be given a close tolerance contour pack thus distributing impact loading over the whole surface of the object so packaged. At the same time the sparkling appearance of the material offers possibilities for display. Consumer acceptance is usually good because the contour molding makes it easy to remove the contents and in many instances offers a convenient storage space for them.

The question of cost is an interesting one. On a pure replacement basis, piece for piece, other materials such as corrugated board may well be cheaper although if labor costs are taken into account the picture is usually different. For example, one American company manufacturing portable television sets found that as many people were employed in folding and inserting the protective corrugated fiberboard cushions as were engaged in product assembly. Replacement of the corrugated cushioning by molded expanded polystyrene inserts gave reductions in material plus labor costs of 30 percent. In addition, freight charges were reduced by 13 cents for every 100 pounds of original weight. Total savings amounted to 26,000 dollars on an annual output of 200,000 sets while the protective performance of the package was unchanged. Another example was the case of a domestic appliance where no fewer than seven corrugated fiberboard fitments were replaced by two identically shaped moldings in expanded polystyrene at a cost saving of 10 percent. There were also weight savings of about 4 pounds per package and a 25 percent reduction in labor. Savings may be even greater where several items are packaged together, as with a camera with accessories such as a lens hood, filters, and a flash gun where a large number of cushioning pieces would be necessary. In such an example it would be possible to dispense with the fiberboard outer container and inserts altogether and use two expanded polystyrene contour portions held together by adhesive tape.

Industrial products requiring cushioning may range from small electronic components or glassware to a complete dentist's drill. For fragile, complex assemblies the lower densities of expanded polystyrene are more suitable since they give the most efficient cushioning or, in other words, the lowest peak deceleration values. For more robust articles, or glassware, where it is usually sufficient for surface stresses to be reduced by the use of contour packaging, the most important consideration is for the cushioning to survive the journey without breakage. Breakage of the cushioning material might not affect its ability to reduce peak deceleration values but might well allow the contents to move about so that subsequent impacts are not evenly distributed over its surface. For large but relatively light articles a complete, overall molding might be uneconomic or even unnecessary and two-part moldings, fitting over the opposite ends of the product, may often be used to advantage.

A more recent development is the use of expanded polystyrene trays or boxes for horticultural produce and for wet fish. The principal advantage of expanded polystyrene compared with wood, for the packaging of wet fish, is its thermal insulation. In one test, fish packed with crushed ice were sent on a journey of 400 miles in both wooden and expanded polystyrene boxes. When the wooden boxes were opened at the market there was no ice left, the fish were slimy and the temperature within the boxes averaged 7°C. In the expanded polystyrene boxes the fish were fresh and a little ice still remained. The temperature was about 3°C.

In the packaging of horticultural produce using nonreturnable boxes, the use of expanded polystyrene, by reason of its high yield, is gaining popularity. In Europe, expanded polystyrene boxes are being used for grapes, cherries, and tomatoes. By increasing the density of the material to about 2½ lb/ft^3 larger boxes can be made which stand up to more than one trip. Such large boxes are already being used for carrying fish in Denmark and Sweden.

Expanded polystyrene extruded sheet has a closed cell structure, each cell being between 200 and 300 microns in diameter. Densities down to about 3 lb/ft^3 are obtainable after post-expansion by heating the sheet with infrared radiation. The properties which have led to its wide range of applications are an attractive finish, ability to be self-colored, ease of thermoforming, and good cushioning. In addition it has excellent grease and fat resistance, high resistance to water and to bacterial and fungal growths. One of the earliest large scale uses for expanded polystyrene sheet was for interleaves used for the packaging of apples exported from South Africa to the United Kingdom. Apart from the attractive appearance of expanded polystyrene interleaves, the cushioning properties give greater protection to the apples. This was proved during early trials when the fruit arriving in the market was found to obtain a better price than similar grades of apples packed with pulp interleaves. The answer appears to lie in the greater compressibility of the expanded polystyrene sheet which leads to larger areas of contact between the fruit and the interleaf. Other large markets which have developed are supermarket trays for meat and fresh produce. For produce such as apples and tomatoes the cushioning properties of expanded polystyrene sheet reduces bruising, especially when the individual items are prevented from moving in the trays by shrink-wrapping (usually with PVC film). The use of expanded polystyrene trays for meat is a little more complicated. The molded pulp trays formerly used were able to absorb any blood present but the closed cell structure of expanded polystyrene prevents this. One early solution was to place an absorbant tissue pad at the bottom of each tray but this increased both material and labor costs. The solution finally adopted was to mold an embossed pattern in the base of the tray where the blood is held by capillary attraction.

Expanded polystyrene sheet also competes with toughened polystyrene sheet in the egg pack market. Its intrinsic cushioning properties means that protection from breakage can be attained more simply but it does not give the visibility of the contents achieved by toughened polystyrene.

However, the silky surface sheen and the pastel colors available do give added sales appeal in a competitive market.

Sheet can also be formed by slicing veneers from a block of expanded polystyrene and in this form it has found use as a heat-insulating lining for walls, prior to the application of conventional wallpapers.

An important difference should here be noted between the extruded sheet and that produced by the veneer process. The latter, of course, has the same density as that of the block from which it was cut and can thus be obtained in densities down to 1 lb/ft^3. It is not usually possible to produce extruded sheet in densities below 3 lb/ft^3, even with post-expansion and the more usual range is from 6–10 lb/ft^3. An additional difference is in the surface finish. The extruded sheet has a satinized finish whereas the veneers are naturally rougher because of the rupture of the individual cells. This rupture of the surface cells reduces the insulation properties so that extruded sheet has taken a large share of this market in spite of its higher density.

Styrene Copolymers

One important material in this class is the copolymer of styrene with butadiene but because of its properties it is best considered in the chapter on synthetic rubbers (Chapter 21). Other than this, the most important copolymers are styrene/acrylonitrile and styrene/acrylonitrile/butadiene (ABS). Copolymers of styrene and α-methyl styrene

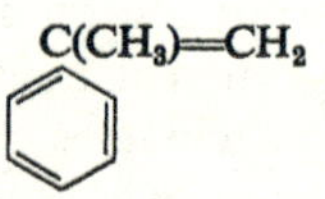

have also been produced. These have an increased temperature resistance, the other physical properties being similar to those of polystyrene.

PREPARATION OF MONOMERS

Acrylonitrile

Acrylonitrile, or vinyl cyanide has the formula $CH_2{=}CH{\cdot}CN$. The liquid, boiling at 79°C may be prepared by the following methods:

(*a*) By dehydration of ethylene cyanohydrin, formed from ethylene oxide and hydrogen cyanide.

$$\underset{\text{Ethylene oxide}}{\overset{CH_2-CH_2}{\diagdown_{O}\diagup}} + \underset{\text{Hydrogen cyanide}}{HCN} \rightarrow \underset{\text{Ethylene cyanohydrin}}{CH_2\cdot OH\cdot CH_2\cdot CN} \xrightarrow{-H_2O} \underset{\text{Acrylonitrile}}{CH_2{=}CH\cdot CN}$$

(*b*) By addition of acetylene and hydrogen cyanide in the presence of a catalyst.

$$\underset{\text{Acetylene}}{CH{\equiv}CH} + \underset{\text{Hydrogen cyanide}}{HCN} \xrightarrow[400\text{–}500^{\circ}C]{CuCl_2} \underset{\text{Acrylonitrile}}{CH_2{=}CH\cdot CN}$$

(*c*) By the ammoxidation of propylene using a bismuth phosphomolybdate catalyst at 290–540°C and 2–3 atm.

$$C_3H_6 + NH_3 + \tfrac{3}{2}O_2 \rightarrow CH_2{=}CH\cdot CN + 3H_2O$$

Butadiene

1 : 3 Butadiene has the formula $CH_2{=}CH{-}CH{=}CH_2$.

It can be made in a variety of ways but the main commercial processes are based on alcohol or petroleum as raw materials. However, during WWII, the raw material used in Germany was acetylene. For interest's sake, the two German processes are shown schematically below:

(*a*) *Aldol Process* $CH{\equiv}CH + H_2O \rightarrow CH_3 \cdot CHO$
Acetylene

$$2CH_3 \cdot CHO \rightarrow CH_3 \cdot CH(OH) \cdot CH_2 \cdot CHO$$
$$CH_3 \cdot CH(OH) \cdot CH_2 \cdot CHO + H_2 \rightarrow CH_3 \cdot CH(OH) \cdot CH_2 \cdot CH_2 \cdot OH$$
$$CH_3 \cdot CH(OH) \cdot CH_2 \cdot CH_2OH \xrightarrow{-2H_2O} \underset{\text{Butadiene}}{CH_2{=}CH{-}CH{=}CH_2}$$

(*b*) *Reppe Process*

$$\underset{\text{Acetylene}}{CH{\equiv}CH} + 2H \cdot CHO \rightarrow HO \cdot CH_2 \cdot C{\equiv}C \cdot CH_2 \cdot OH$$
$$HO \cdot CH_2 \cdot C{\equiv}C \cdot CH_2 \cdot OH + 2H_2 \rightarrow HO \cdot CH_2 \cdot CH_2 \cdot CH_2 \cdot CH_2 \cdot OH$$
$$HO \cdot CH_2 \cdot CH_2 \cdot CH_2 \cdot CH_2 \cdot OH \xrightarrow{-2H_2O} \underset{\text{Butadiene}}{CH_2{=}CH{-}CH{=}CH_2}$$

(c) Butadiene from Alcohol. A commercial process based on alcohol that was used in America during the WWII. Alcohol was dehydrogenated to acetaldehyde, using a chromic oxide catalyst. The acetaldehyde was then mixed with further alcohol and reacted with a catalyst consisting of 2 percent tantalum oxide on silica gel.

(d) Butadiene from Petroleum. Petroleum can be utilized in three different ways:

(i) By the high temperature thermal cracking of petroleum fractions.
(ii) By the dehydrogenation of *n*-butane.
(iii) By the dehydrogenation of *n*-butene.

The choice of process is governed by such factors as the feedstocks available, plant location, capital available, etc., but in general, the thermal cracking process gives a low yield of butadiene and is not economic unless it can form part of an integrated set-up utilizing the propylene and ethylene which are formed as the main products.

Two typical dehydrogenation processes are described, briefly, below:

1. Houdry Process. This process is based on the dehydrogenation of butane. The gas is heated to about 595°C and passed through the catalyst (18–20 percent chromium oxide on aluminum oxide) at an absolute pressure of 100–150 mm of mercury. The resultant mixture of butane, *n*-butene and butadiene contains about 12 percent of butadiene. After quenching, the butadiene is separated from the butane and *n*-butene which are then mixed with fresh butane and recycled to the reactor.

The overall efficiency of this process is between 45-55 percent.

2. *Standard Oil of New Jersey Process.* This process utilizes *n*-butene as the feedstock. The *n*-butene is diluted by steam as it passes over the catalyst, in order to reduce the residence time at high temperature and to supply the necessary heat of reaction. The butene and steam are heated separately to 595° and 705°C respectively and then passed into the reactor. The mixture passes down through a catalyst bed and then through a water quench. After further cooling, hydrogen and carbon dioxide are removed and the C_4 fraction is sent to the butadiene extraction unit. Typical catalysts are Shell 205 (Fe_2O_3, Cr_2O_3, K_2CO_3) and Dow Type B (calcium nickel phosphate).

POLYMERIZATION

(1) Styrene/Acrylonitrile

Styrene/Acrylonitrile copolymers are normally produced by suspension or emulsion polymerization techniques similar to those used for polystyrene.

(2) Acrylonitrile/Butadiene/Styrene

In one method, styrene and acrylonitrile are added to polybutadiene and the mixture stirred and warmed to 50°C. A water-soluble initiator such as potassium persulfate is then added and the mixture polymerized. The emulsion so formed is coagulated or spray-dried to a white powder.

Another method of producing ABS polymers is by blending butadiene/acrylonitrile rubber and styrene/acrylonitrile resin. The blending may be carried out either by mixing the dry ingredients on a two-roll mill or an internal mixer, or by blending the polymer lattices with subsequent coagulation and spray-drying. Such blends are often referred to as 'alloys.'

Styrene/acrylonitrile resins and butadiene/acrylonitrile rubbers are compatible because of the presence of acrylonitrile in each. The rubber thus plasticizes the styrene/acrylonitrile, giving low softening-point mixtures and little improvement in impact strength. Cross-linking the rubber prevents this plasticization, and high softening-point, high impact-strength materials are obtained. The cross-linking is achieved by masticating on a hot mill with a catalyst such as tertiary butyl hydroperoxide.

STRUCTURE

In both types of ABS, it is thought that the rubber exists in the form of discrete particles on the resin. In polymerized ABS (sometimes referred to as Type B) it has been postulated that the polymer is a mixture of polybutadiene, polybutadiene grafted with styrene and acrylonitrile, and styrene/acrylonitrile copolymer. Mixtures of polybutadiene and styrene/acrylonitrile are very weak and the graft polymer acts as a link between them.

PROPERTIES

(1) Styrene/Acrylonitrile

These copolymers have better chemical resistance, surface hardness and mar resistance than normal polystyrene. However, natural styrene/acrylonitrile has a yellowish tinge and a bluing agent is incorporated to overcome this. Resistance to weathering is good and allows the copolymers to be used for outdoor applications.

The properties of styrene/acrylonitrile copolymers vary according to the ratio of the two components. In general, an increase in the acrylonitrile content increases the toughness and chemical resistance but also leads to stiff-flowing, difficult-to-mold materials with an increased tendency to yellowness.

(2) Acrylonitrile/Butadiene/Styrene

Like toughened polystyrene, acrylonitrile/butadiene/styrene (ABS) is translucent. ABS polymers have high impact strength and are extremely tough and hard. Outdoor exposure has little effect on the properties and performance of ABS over extended periods except for some loss of surface gloss. Protection can be obtained by pigmentation or lacquering to reduce the effects of ultra-violet radiation. ABS sheet has excellent melt strength which enables large areas of sheet to be thermoformed without sagging.

Table 11.2 shows the main properties of styrene/acrylonitrile and acrylonitrile/butadiene/styrene. The chemical resistance of ABS is given in Table 11.3.

Table 11.2

TYPICAL PROPERTIES OF STYRENE ACRYLONITRILE AND ABS

Property	*Method of test*	*Units*	*Styrene acrylonitrile*	*ABS type-A (blend or alloy)*	*ABS type-B (interpolymer)*
Specific gravity	ASTM. D.792–50		1.06–1.08	1.05	1.01
Tensile strength	ASTM. D.638–58T	lb/in.^{2}	10,000–11,000	5000	4500
Elongation at break	ASTM. D.638–58T	%	3	20–50	20–40
Impact strength. Izod notched ½ in comp.-molded bar	B.S. 1493–1958	ft lb/in. of notch	0.35	9	5
Hardness	ASTM. D.785–51	Rockwell	M.80	R.100	R.85
Softening point	B.S. 1493–1958	°C	100	100	100

Table 11.3

RESISTANCE TO CHEMICALS OF ABS COPOLYMERS

Chemical	*Effect*
Acids (up to 50% conc.)	No effect
Conc. sulfuric acid	Surface hardening
Other conc. acids	Slight swelling and darkening
Alkalis	No effect
Methyl ethyl ketone	Disintegrates
Alcohols	No effect
Esters	Severe attack
Lubricating oils	Very slight softening
Vegetable oils	No effect
Animal oils	No effect
Detergents and Soaps	No effect

APPLICATIONS

(a) Styrene/Acrylonitrile

These copolymers are used extensively for cups, tumblers, typewriter keys, and refrigerator parts, because of their high tensile strength, good moldability and good chemical resistance. They are also used for certain kitchen equipment, especially where strength and chemical resistance are required. Orange-peel oil, for instance, attacks toughened polystyrene but styrene/acrylonitrile is unaffected. The improved chemical resistance of styrene/acrylonitrile copolymers, notably their resistance to soaps, hair oils, etc., means that they can be used to mold combs and brush backs which are resistant to crazing and cracking.

They are also resistant to light oils and hydrocarbons so that they can be used for oil-filter bowls, vacuum cleaner components, and so on.

(b) Acrylonitrile/Butadiene/Styrene

ABS copolymers have been extruded into pipes and used for transporting corrosive chemicals, natural gas, drinking water, salt water, etc. Their high impact strength, heat and chemical resistance and dimensional stability render them suitable for many light engineering uses, radio and camera cases, safety helmets and fertilizer injectors. They are also used for refrigerator liners where the insulation is polyurethane, foamed *in situ*, since

ABS copolymers are more resistant than toughened polystyrene to the Freon foaming agents which are used, while they have greater impact strength than rigid PVC.

They are highly suitable for component parts of household equipment such as floor-scrubbers and vacuum cleaners which are liable to impact during use, while they have also found extensive use in lightweight luggage. Tote-boxes and crates have also been made in ABS copolymers, their high impact strength being particularly useful. Facia panels of cars are often made from ABS although there is now some competition from polypropylene.

Miscellaneous Styrene Copolymers

The success of styrene/acrylonitrile (SAN) and acrylonitrile/butadiene/styrene (ABS) has led to the development of other copolymers designed to give improved properties in specific areas.

Acrylonitrile/Styrene/Acrylic elastomer (ASA)

The composition of this terpolymer is similar to that of ABS with the acrylic elastomer substituted for butadiene. The acrylic component is present to the extent of about 30 percent. The substitution of the acrylic elastomer for butadiene eliminates a double bond between styrene and butadiene that is excited by uv radiation in sunlight. The resistance of ASA to uv degradation and oxidative embrittlement is, therefore, higher than that of ABS. Applications making use of this improved weatherability include thermoformed signs and boat hulls.

Methacrylate/Butadiene/Styrene (MBS)

The properties of MBS are very similar to those of ABS except for one important advantage, namely, clarity. Its light transmission is 85 percent compared with polycarbonate's 90 percent and acrylic's 92 percent. Uses include stackable tables and various toys where resistance to breakage coupled with transparency is important.

Styrene/Methylmethacrylate (NAS)

This material has a light transmission value of 90 percent and a heat distortion temperature of 100°C. NAS meets FDA specifications for food packaging and has found some applications in that area although its brittleness debars it from some uses. It is less brittle than basic polystyrene and has been used for some housewares (including tumblers) where its heat distortion temperature allows it to be washed in dishwashers.

SUGGESTED FURTHER READING

'Polystyrenes', in Chemical and Process Technology Encyclopedia, pages 914–917 (1974) Edited by D.M. Considine (New York: McGraw-Hill Inc.)

12.

POLYAMIDES

The synthetic thermoplastic polymers characterized as polyamides are better known under the name of nylons. The discovery of the first nylons arose out of the classic researches of W. H. Carothers in the U.S.A. in 1928. He and his coworkers were searching for new synthetic fibers and, in the course of their work, investigated a large number of chemical reactions. These researches culminated in the discovery that the condensation products of diacids with diamines gave long-chain polymers with the desired properties. The first nylon to be developed commercially was the condensation product of hexamethylene diamine, $NH_2(CH_2)_6NH_2$, and adipic acid, $COOH(CH_2)_4COOH$. It was called Type 6,6 (six-six) nylon, the name being derived from the number of carbon atoms in the parent diamine and diacid. A further type to be developed was Type 6,10 (six-ten), this being made from hexamethylene diamine and sebacic acid, $COOH(CH_2)_8$ $COOH$.

Later, particularly in Europe, a different method of making nylon polymers was discovered, namely, by the condensation of certain ω-amino acids. Nylons prepared by this method are described by a single number, which is derived from the number of carbon atoms in the parent amino-acid. Commercial examples of this type are Nylon 6 prepared from ω-amino-caproic acid, Nylon 11, prepared from ω-amino-undecanoic acid, and Nylon 12 prepared from dodecyl lactam (laurilactam).

Although the polyamides were originally developed as fiber-forming polymers, they have since attained a growing importance in other fields, notably, injection molding and extrusion. To a smaller extent, nylons have also been used for the production of bottles by the blow-molding process.

Over recent years considerable use has been made of Nylon 11 and 12 in powder form for coating metals and the copolymers of Nylon 6/66/12 for textiles.

PREPARATION OF RAW MATERIALS

Adipic Acid

As was mentioned above, adipic acid is one of the starting materials for Nylon 6,6 and is derived from phenol by the following sequence of operations. The phenol is first hydrogenated to cyclohexanol, which is then catalytically dehydrogenated to cyclohexanone. The cyclohexanone is then reacted with nitric acid in the presence of a vanadium or copper catalyst, the result being a splitting of the ring and oxidation of both ends to a carboxylic acid grouping.

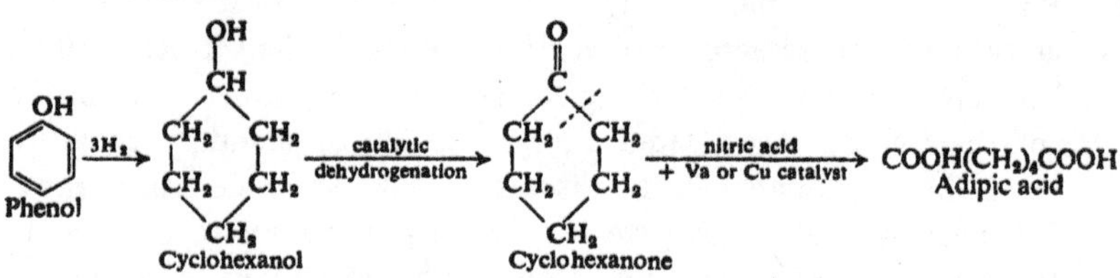

Hexamethylene Diamine

Hexamethylene diamine is the other starting material for Nylon 6,6 and is prepared from adipic acid (*see* above). The sequence of operations is shown by the equations below:

$$\underset{\text{Adipic acid}}{COOH(CH_2)_4COOH} \xrightarrow[350°C]{NH_3} \underset{\text{Adipamide}}{CONH_2(CH_2)_4CONH_2} \xrightarrow[\text{125°C. 600 atm pressure}]{\text{Co/Cu catalyst}} \underset{\text{Hexamethylene diamine}}{NH_2(CH_2)_6NH_2}$$

Sebacic Acid

Sebacic acid has the formula $COOH(CH_2)_8COOH$, and is the dibasic acid from which 6,10 nylon is prepared. It is obtained by the alkaline extraction of castor oil. The details of the preparation are described in Chapter 5.

ω-Amino-Caproic Acid

Since the acid is normally used in the form of its anhydride (ε-caprolactam) it is more convenient to describe the method of manufacturing the latter. The starting point is again phenol, and this is successively reacted to give cyclohexanol and cyclohexanone as in the preparation of adipic acid described earlier. After purification by distillation, the cyclohexanone is reacted with hydroxylamine sulfate and the cyclohexanone oxime, so formed, is then heated with sulfuric acid containing free sulfur trioxide, followed by the addition of ammonia. This causes a molecular rearrangement to give ε-caprolactam.

The relationship between an ω-amino acid and its anhydride is exemplified by ω-amino-caproic acid and ε-caprolactam.

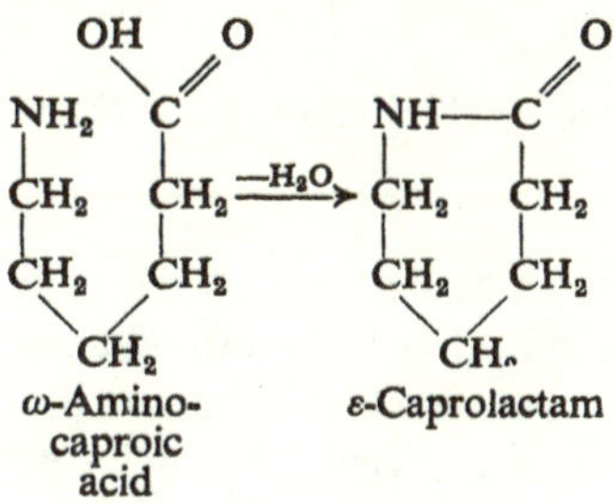

ω-Amino-Undecanoic Acid

This acid, which is the starting point for the manufacture of Nylon 11, is obtained from castor oil by the following procedure.

Castor oil is first treated with methyl alcohol to give methyl ricinoleate. Pyrolysis of this gives methyl undecylenate, which is then hydrolyzed to give the acid. This in turn is reacted with hydrogen bromide to yield the 11-bromo and not the 10-bromo compound. Finally it is reacted with ammonia to give the 11-amino undecanoic acid. This acid is obtained after recrystallization as an aqueous paste which is suitable for charging directly into a continuous polymerization unit. Phosphoric acid has been found as a suitable catalyst for this reaction.

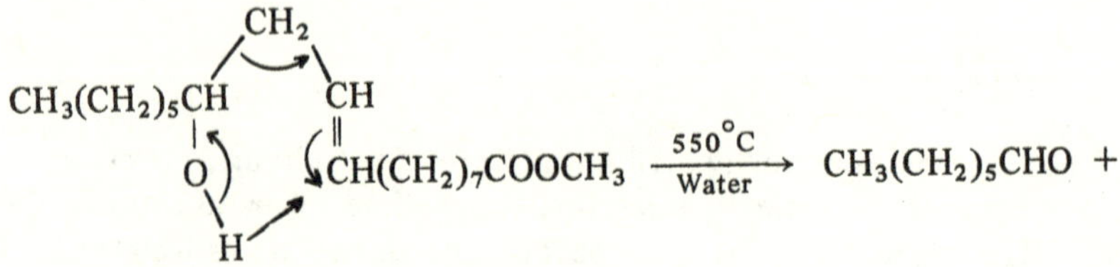

Methyl ricinoleate

$$CH_2{=}CH(CH_2)_8COOCH_3 \xrightarrow{\text{Hydrolysis}} CH_2{=}CH(CH_2)_8COOH$$

Undecylenic acid

$$\xrightarrow{\text{HBr, air}} Br(CH_2)_{10}COOH \xrightarrow[\text{Water}]{NH_3} H_2N(CH_2)_{10}COOH$$

11-Aminoundecanoic acid

MANUFACTURE OF POLYMERS

Nylon 6,6

The reaction of adipic acid and hexamethylene diamine to form Nylon 6,6 is carried out in two stages. Both materials are dissolved in boiling methanol, whereupon a 'salt' is precipitated. This 'nylon salt' is then recrystalized

(i) $\underset{\text{Hexamethylene diamine}}{NH_2(CH_2)_6NH_2} + \underset{\text{Adipic acid}}{COOH(CH_2)_4COOH} \rightarrow$

$$\rightarrow \underset{\text{Nylon salt}}{NH_2(CH_2)_6NH \cdot CO(CH_2)_4COOH} + H_2O$$

(ii) $n[\underset{\text{Nylon salt}}{NH_2(CH_2)_6NH \cdot CO(CH_2)_4COOH}] \xrightarrow{\text{heat}}$

$$\rightarrow \underset{\text{Nylon 6,6 polymer chain}}{\left[NH(CH_2)_6NH \cdot CO(CH_2)_4CO \right]_n} + n \cdot H_2O$$

from water and charged into a glass or stainless-steel lined vessel. The vessel is purged with nitrogen and the contents heated to 280°C. The reaction may be followed by viscosity measurements, by electrical conductivity, or by checking on the ability of the material to 'cold draw.' The reaction can be stopped at any point by the addition of a monofunctional acid or amine. When the reaction is complete, the water formed during the reaction must be removed by refluxing under a vacuum. The reaction product is then pressure-filtered while still hot, and is then in a suitable form for spinning into fibers, for conversion into pellets for injection molding or extrusion.

Dodecyl Lactam (laurilactam)

The polymer Nylon 12 was first produced on a commercial scale in Germany around the middle of 1965. The monomer is prepared by Chemische Werke Huels AG as follows:

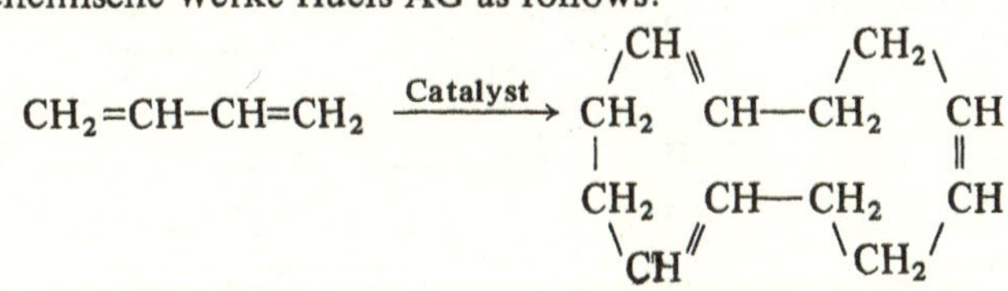

Butadiene Cyclodecatriene

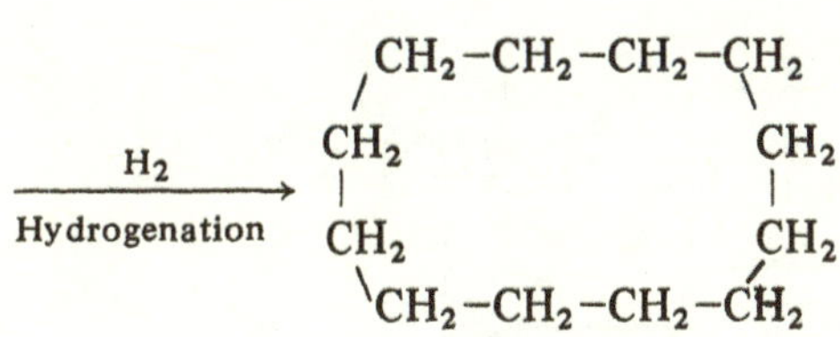

Cyclodecane

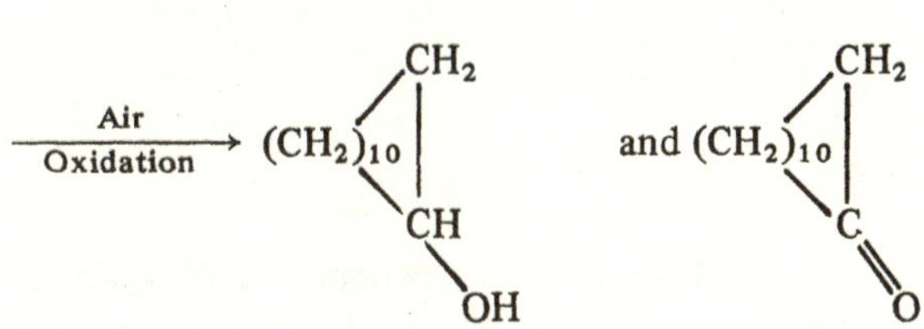

Cyclodecanol Cyclodecanone

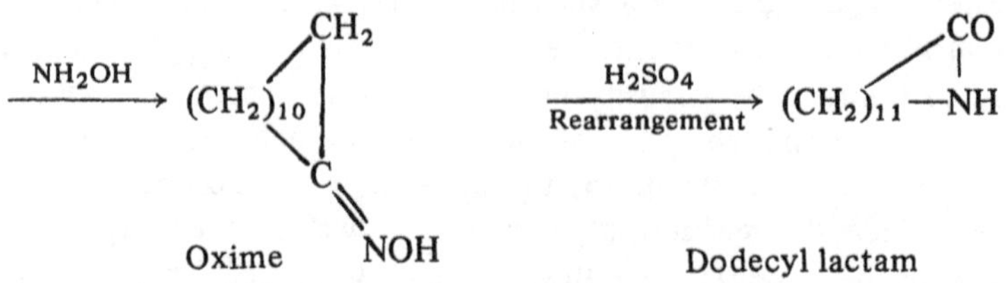

Several other routes to the lactam have been announced including those attributed to Badische Anilin- und Soda-Fabrik (Germany), Aquitaine Organico (France), and Mitsubishi Toray (Japan).

BASF Method

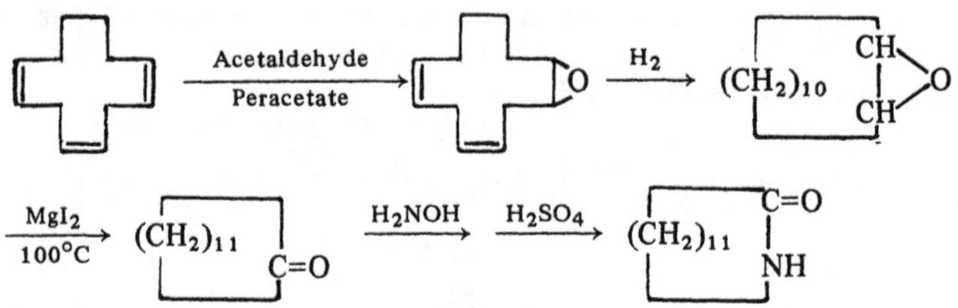

Aquitaine Organico and Toray Method:

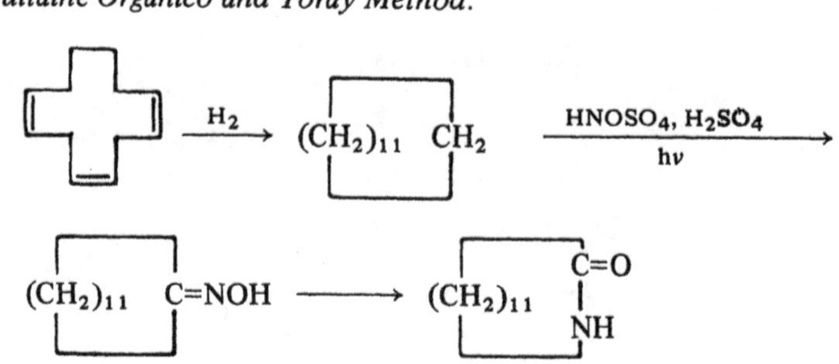

Nylon 6

The condensation of ε-caprolactam to give Nylon 6 can be carried out either batch-wise or continuously. In the batch process, jacketed reaction vessels fitted with spade stirrers can be used. The reaction vessel is charged with the caprolactam, water (as a catalyst), and a stabilizer to limit the molecular weight. The mixture is kept under a blanket of oxygen-free nitrogen gas and heated for 12 hours at a temperature of about 250°C. An equilibrium mixture is eventually obtained which consists of about 90 percent polymer and ten percent low molecular weight products, and this is forced out by nitrogen pressure through an exit cock at the base of the reaction vessel. It is then in the form of a thick monofilament but, on cooling, it is broken up into the more convenient form of small chips about 1/12 in. square. If colored material is desired, the color is introduced at this stage by dry blending, followed by extrusion and chipping. The low molecular weight constituents mentioned earlier are then washed out and the chips dried and packed into air-tight tins.

In the continuous process, a stream of caprolactam, water, and a stabilizer is maintained to the reaction columns which are kept at 250°C. A continuous stream of polymer leaves the column, is cooled, cut into chips, washed, dried, extruded, chipped, and packed. The low molecular weight contents referred to earlier, consist mainly of the monomer which is fed back to the distillation plant employed for the purification of the caprolactam.

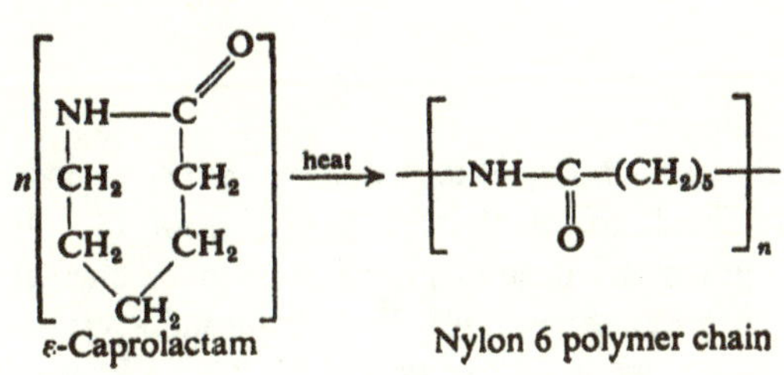

Nylon 11

The polymerization of ω-amino-undecanoic acid to Nylon 11 is a relatively simple process consisting merely of heating the monomer acid to just over 200°C, while distilling off the water formed as a reaction byproduct. Catalysts may be added if desired. Towards the end of the reaction, the pressure is reduced in order that a high molecular weight polymer may be produced. Another essential for the production of a high molecular weight is the use of a pure starting material (French Patent No. 885, 694).

The molten polymer can either be spun directly into fibers or extruded into a ribbon, which is cut into chips and packed into air-tight tins.

Nylon 12

The polymerization of dodecyl lactam (laurilactam) to Nylon 12 is a relatively simple process akin to that of the Nylon 11. However, because of the high degree of stability of the ring the polymerization is much slower. Acid catalysts are used to speed up the reaction but even so temperatures of 300-350°C are required to achieve an acceptable molecular weight. If, for example, laurilactam is heated in the presence of water or small amounts of acid, such as acetic acid, at the usual polymerization temperature of 260°C, 50 percent of the lactam remains unconverted even after a polymerization time of 30 hours. A French Patent Specification No. 1,261,286 states that the polymerization of laurilactam can be carried out between 315 and 320°C in the presence of aliphatic mono- or dicarboxylic acids. This lactam has a low polymerization of heat −1.4 k cal/mol and as stated previously, the ring has a high degree of stability which accounts for the high temperatures required to obtain acceptable degrees of conversion.

Copolymers

These materials are finding extensive use as adhesive coatings for fusible interlinings. Suffice it to say that the manufacture of polyamide copolymers simply requires the desired ingredients to be charged to a reaction vessel and simply polymerized. All the diamines, diacids, amino acids, and lactams mentioned previously can be used in virtually any kind of combination provided of course the ratio of amine to acid grouping is kept within the desired guidelines. Commercially 6/66, 6/610, 6/66/610, and 6/66/12 are the most popular, particularly when used as heat sealable coatings for textiles.

One of the earliest materials was a copolymer of Nylon 6 and 66 in the ratio of 70/30. The softening points of such copolymers is in the region of 170°C which was too high for this particular application. Therefore it was found necessary to dry blend up to equal parts of ortho-para toluene sulfonamide to this copolymer in order to reduce the softening point to a satisfactory level. The following is a list of softening points of a wide range of polyamides. The difference between external plasticization and internal plasticization by copolymerizing with different monomers is illustrated.

Material	*Softening Point °C*
Nylon 6/6	265
6/10	220
6	215
11	185
6/66	170
6/66/610	120
6/66/12	120
6/66 (plasticized with equal amounts of *o-p* toluene sulfonamide)	105

Other satisfactory plasticizers are *n*-ethyl sulfonamide, butyl benzene sulfonamide, urea, and proprietory esters of hydroxy benzoic acid.

The formation of intermolecular hydrogen bonds is controlled by selection of the basic monomers and the respective ratio, which in turn controls the degree of crystallinity. The triangular coordinates of Fig. 12.1 (repre-

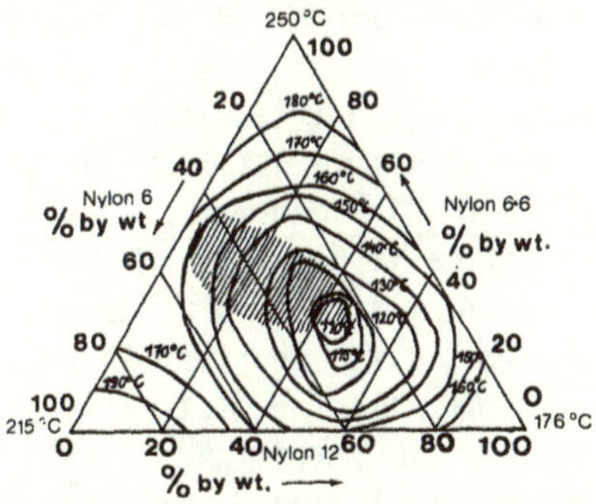

Fig. 12.1 Fusion temperature (°C) triangular coordinates of copolyamide 6,6-6,12. All polymers of the same melting temperature are joined by isotherms. Shaded areas are compositions soluble in 85 : 15 alcohol : water solutions.

senting the fusing temperatures of various ratios of Nylon 6, 66, and 12) show the correlation between the ratio of monomer components and the resulting softening temperatures. The shaded area covers these terpolymers which are soluble in an 85/15 alcohol and water solution. The greater the proportion of Nylon 12, the lower the water absorption but the greater the degree of resistance to chlorinated hydrocarbon solvents such as perchlorethylene and trichlorethylene. The melt viscosities can be lowered by controlled techniques but this leads to a deterioration of other characteristics. The properties of such copolymers are discussed more fully in Chapter 29.

Versamids

Versamids are condensation products of dimerized vegetable oil acids and polyamino compounds such as polyalkylene polyamines. General Mills Inc. in the U.S.A. has produced these type of polyamides for many years under U.S. Patent No. 2,379,413. A cross licensing agreement covering resin research, manufacture, and sales has been in operation between General Mills and Cray Valley Products Ltd., in the U.K. They are marketed under the General Mills reference Versamid, Versalon, and Milvex. A typical reaction for the manufacture of Versamids is as follows:

$$n\text{HOOC–R–COOH} + n\text{H}_2\text{NR}'\text{–NH}_2 \longrightarrow \text{HO(–OC–R–CONHR}'\text{–NH–)}_n\text{H}$$

R is a hydrocarbon group of an indeterminate configuration containing 34 carbon atoms. The polyamine used in the manufacture of Versamid polyamide resins 900, 930, 940, and 950 is ethylene diamine, and R′ is $—CH_2CH_2—$.

The ingredients are blended at low temperatures in an autoclave fitted with an efficient agitator. The temperature is raised until water is being distilled off at such a rate that the temperature does not rise much above 100°C. When distillation reaches the stage where it is so slow as to be negligible (reaction temperature 150-250°C) a vacuum is applied to complete the reaction. An inert gas is then added and the product cooled slightly and drawn from the reactor in the form of a molten mass. The resin may be either liquid or solid, depending upon the degree of reaction.

Liquid Versamids are packaged in metal containers while the solid products may be allowed to harden in fiberboard drums, or may be flaked or chipped.

Milvex type polyamides are based on different feedstock and are higher molecular weight materials resembling the copolyamides in characteristics. They are discussed in a little more detail in Chapter 29.

STRUCTURE

The general structure of the nylons (polyamides) is an alternation of hydrocarbon groups (in this case methylene groups) and imido (—NH · CO—) groups. This is typified by Nylon 6,6 (*see* below).

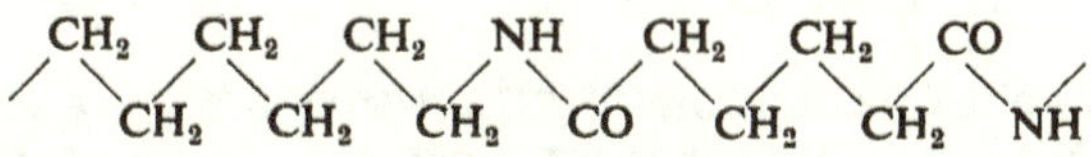

The presence of the polar imido groups leads to a strong attraction between the molecules, due to hydrogen bonding, thus conferring a high melting point on the polymer. The hydrogen bonding also tends to make the chains lie parallel to each other, especially after cold drawing. The consequent increase in crystallinity is shown up practically by X-ray diffraction photographs.

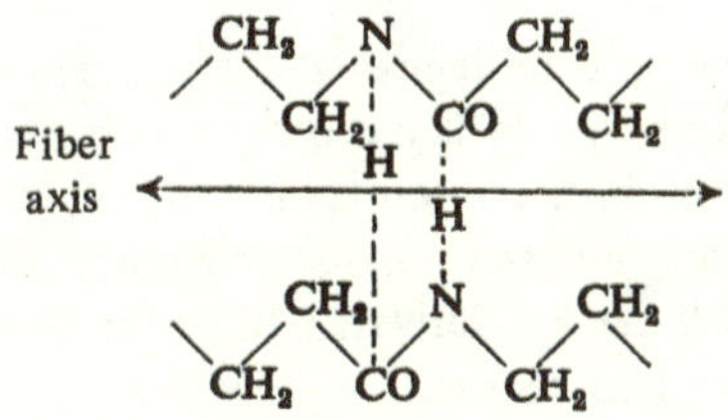

In the representation of two polyamide chains after cold drawing, the hydrogen bond links are shown by dotted lines. A full exposition of hydrogen bonding is beyond the scope of this book but, briefly, it may be considered to be due to the tendency of the hydrogen to resonate between (in this case) the two structures I and II below.

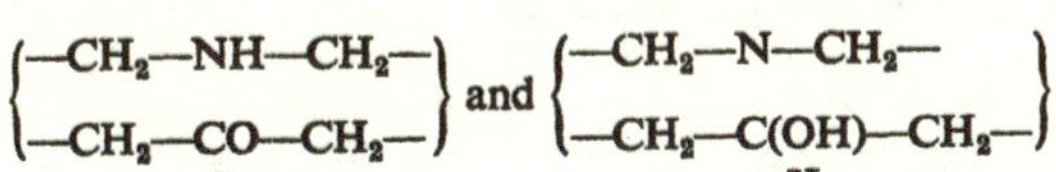

I II

An alternate link is thus formed between the CO and N groups as shown by dotted lines in the preceding diagram.

With this picture in mind it is possible to visualize the way in which the properties of the polyamides can be altered. The longer the hydrocarbon part of the chain, for instance, the less the attraction between the chains and, therefore, the lower the softening point because the hydrogen bonding is, in effect, 'diluted' along the chain. Thus, Nylon 6,10 has a lower softening point than Nylon 6,6.

$$\text{cf. —}(CH_2)_6\cdot NH\cdot CO(CH_2)_4CO\cdot NH\cdot (CH_2)_6\cdot NH\cdot CO\text{— Nylon 6,6 S.Pt. 265°C}$$
$$\text{—}(CH_2)_6\cdot NH\cdot CO(CH_2)_8\cdot CO\cdot NH\cdot (CH_2)_6\cdot NH\cdot CO\text{— Nylon 6,10 S.Pt. 220°C}$$

Longer hydrocarbon groups also lead to a reduction in water absorption and, for this reason, Nylon 6,10 is often preferred for the manufacture of tooth-brush bristles.

Substituting an alkyl group for a hydrogen in the molecule chain lowers the melting point and increases the solubility of the molecule. The explanation for this effect lies in the fact that the bulky alkyl groups prevent close approach of the molecules, thus reducing the hydrogen-bonding between adjacent chains.

The properties of the polyamide may also be altered by substituting the hydrogen in the —NH groups by some other grouping. This can be done either in the polymer or in the intermediate 'nylon salt.' The hydrogen bonding is thus removed and the polymer becomes rubbery, elastic, and amorphous. Thus, the *N*-methyl derivative of Nylon 6,6 is a viscous syrup, setting to a 'glass' at low temperatures.

$$\text{—}(CH_2)_6N\cdot Me\cdot CO(CH_2)_4\cdot CO\cdot N\cdot Me(CH_2)_6\cdot N\cdot Me\cdot CO\text{—}$$

N-methyl Nylon 6,6

Substitution of the amide hydrogen by an alkoxyl group can also be achieved: for instance, the reaction of a polyamide with formaldehyde in the presence of an alcohol results in the substitution of methoxy methyl groups on the amide nitrogen as follows:

$$\begin{matrix}\backslash & & & & \backslash & & \backslash \\ NH & + CH_2O + ROH & \xrightarrow[\text{catalyst}]{\text{Acid}} & & N{-}CH_2OR & \text{or} & N(CH_2O)_2R \\ / & & & & / & & / \\ CO & & & & CO & & CO \\ \backslash & & & & \backslash & & \backslash\end{matrix}$$

Methyl methoxy nylons have been commercially available for some time and about one third of the —NH groups have been substituted. These materials are soluble in lower aliphatic alcohols such as ethanol and absorb up to 21 percent of moisture when immersed in water. On heating this product with 2 percent citric acid at elevated temperatures cross-linking takes place. Typical conditions are 20 minutes at 120°C with the following reaction taking place:

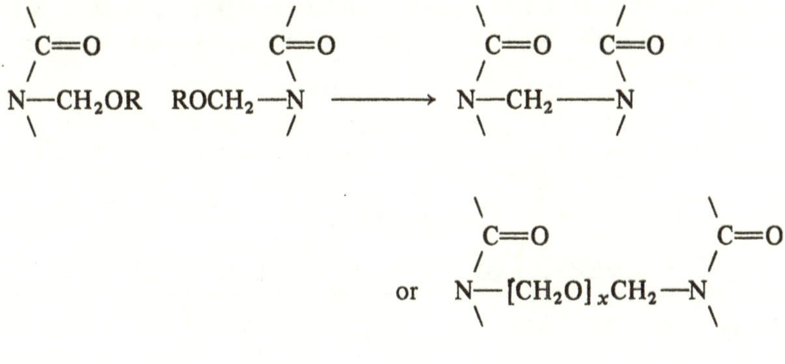

PROPERTIES

Nylons, generally, are produced in the form of hard, horny chips, creamy in color and slightly translucent. As was explained in the section on Structure, the various types of nylon can vary in their properties, but the following generalizations can be made for Nylons 6, 11, 6,6, and 6,10.

Physical Properties

Nylons are tough materials with high tensile strengths and a marked resistance to abrasion. Because of their high crystallinity, they possess sharp melting points which are appreciably higher than most other thermoplastics. However, destruction of the crystallinity by substitution in the —NH group alters the behavior of the polymer completely. In contrast to the majority of thermoplastics, nylons exhibit very little cold flow.

Nylons are slightly hygroscopic and the material must be dried by heating before use. The amount of moisture absorbed is a function of the relative humidity of the atmosphere. The mechanical properties of nylons are reduced somewhat by the absorption of water but the effect is not permanent and the full mechanical properties are restored when the nylon is dried.

The electrical properties are also affected by the ambient relative humidity. For instance, at low humidities nylons have reasonably good electrical insulation properties but these deteriorate as the humidity increases. As with the mechanical properties, the electrical properties are restored to their full value when the nylon is redried. In spite of this disadvantage, nylon moldings find many uses in electrical applications because of the additional advantages of toughness and high temperature resistance. Some typical values for the physical properties of various nylons are given in Table 12.1.

Table 12.1

PROPERTIES OF A RANGE OF POLYAMIDES

Property	*Units*	*Type* 6,6	6,10	6	11
Specific gravity	–	1.14	1.09	1.13	1.05
Tensile strength	$lb/in.^2$	11,000	8000	12,000	8500
Elongation	%	90	100	300	120
Hardness (Rockwell)	R scale	118	111	118	101
Dielectric strength	V/mil	385	470	420	430
Dielectric Constant (at 10^3 cycles)	–	4.0	4.5	4.0	3.5
Volume resistivity (at 23°C and 50% R.H.)	Ohm/cm	4.5×10^{13}	4×10^{14}	4.5×10^{13}	4×10^{14}
Power factor (at 10^3 cycles)	–	0.02	0.04	0.06	0.03
Water absorption	%	1.5	0.4	1.6	0.4
Modulus of elasticity	$lb/in.^2$	400,000	260,000	300,000	180,000
Melting point	°C	265	220	215	185

It may be worth reiterating at this point that some copolymers have also found commercial use and have somewhat different properties. In general, they are more soluble, less stiff and softer than the homopolymers. They also have lower melting points and tensile strengths. This is due to the irregularity set up in the molecules, with consequent interference with the intermolecular forces of hydrogen bonding. The irregular spacing of the amide groups in the molecule also reduces the tendency to crystallize. The effect on polymer melting point brought about by using substituted diamines or dibasic acids is shown in Table 12.2.

Table 12.2

EFFECT OF SUBSTITUTION IN THE PARENT DIACID OR DIAMINE ON THE MELTING POINT OF NYLON 6,6

Dibasic acid	*Diamine*	*m.p.°C*
Adipic acid	Hexamethylene diamine	265
α-Methyl adipic acid	Hexamethylene diamine	166
Adipic acid	3-Methyl hexamethylene diamine	180

Amide groups react with formaldehyde and, in the presence of an alcohol, formation of an alkoxy methyl group occurs on the *N*-atom. The effect of this substitution on the melting point of Nylon 6,6 is shown in Table 12.3.

Table 12.3

EFFECT OF *N*-ALKOXY METHYL SUBSTITUTION ON THE MELTING POINT OF NYLON 6,6

Polymer	*m.p.°C*
Nylon 6,6	265
20% *N*-alkoxy methyl substituted nylon 6,6	190
30% *N*-alkoxy methyl substituted nylon 6,6	135

The Versamids have sharp melting points and the characteristics of these hard polyamides are little affected by temperature until very near their melting points and at room temperature they are nonblocking. They exhibit excellent flexibility which is due to the long fatty chains spaced between the functional groups which are an integral part of the resin chain. Adhesion of these resins to a wide variety of difficult substrates is excellent. The typical properties of a range of different type Versamids is given in Table 12.4.

Table 12.4

PROPERTIES OF A RANGE OF VERSAMID RESINS

Resin	100[(1)]	920	930[(1)]	950[(1)]	1117	1124	1138[(1)]	1139	1140	1160	1200	SR21	1235[(1)]
Ball & Ring Softening Point (°C)	semi-solid	147	110	95	117	124	140	138	140	160	200	160	200
Suggested Application Temperature°C	–	–	150–160	105–115	150–160	170–180	205–215	245–255	245–255	200–210	240–250	215–225	–
Specific Gravity	0.98	0.97	0.97	0.95	0.98	1.00	0.99	0.98	0.99	0.97	0.99	0.97	1.00
Viscosity at 205°C (poise)	–	–	–	–	8	13	62	140	100	43	136	58	1g-6g[(4)]
Viscosity at 150°C (poise)	12	24[(2)]	55	21[(3)]	54	90	–	–	–	–	–	–	–
Elongation at 25°C (%)	–	–	20	10	120	520	560	680	890	370	400	430	400
Tensile Yield at 25°C (psi)	–	–	1800	870	310	260	460	410	340	1900	2100	2000	2800
Tensile Break at 25°C (psi)	–	–	1800	870	180	130	450	730	450	1650	3400	2000	6000
Tensile Shear-Steel (psi)	–	–	750	480	300	300	950	1200	1700	1650	1800	1600	3200
Tensile Button–Steel (psi)	–	–	725	1100	1560	1410	1840	2200	2100	1000	1700	960	–

(1) Can meet the statutory requirements under paragraph 121.2520, Subpart F, of the Federal Food, Drug and Cosmetic Act for adhesives

(2) Viscosity at 160°C (3) Viscosity at 130°C (4) Melt Index Range

Chemical Properties

In general, nylons are inert to inorganic reagents, including liquid ammonia and liquid sulfur dioxide, although they are attacked by oxidizing agents such as hydrogen peroxide and chlorine bleaches. Aqueous solutions of inert inorganic chemicals still exert some effect on the physical properties of molded nylon specimens, however, because of the plasticizing effect of the water itself. Nylons are attacked by concentrated mineral acids at room temperature but dilute acids have little effect except when hot.

Organic acids have little or no effect (especially in low concentration), notable exceptions being 90 percent formic acid and glacial acetic acid, which cause swelling and solution.

Nylons are resistant to caustic alkalis, even at concentrations up to 20 percent, while they are even more stable towards other alkaline solutions such as trisodium phosphate, sodium carbonate, and potassium carbonate.

It is in their resistance to organic solvents that the nylons are of particular interest, however. In general, they are inert to alcohols, ether, acetone, petroleum ether, carbon tetrachloride, benzene, xylene, hydraulic fluid, and aviation oil. They are, however, attacked by phenols, hot formamide, formaldehyde, hot benzyl alcohol, and hot nitrobenzene. It should be realized, however, that certain copolymers or plasticized grades may be less resistant to certain chemicals. A copolymer of 6, 6,6, and 6,10, nylons, for instance, is attacked by alcohols. Nylons are relatively unaffected by prolonged service indoors at room temperature but at elevated temperatures oxidation occurs with consequent yellowing and embrittlement, while strong sunlight also has an adverse effect on their mechanical properties.

The Versamid polyamide resins have excellent resistance to alkalis, aliphatic hydrocarbons, mineral and vegetable oils; good resistance to acids reasonable resistance to esters and ketones; only fair resistance to alcohols due to their alcohol solubility.

APPLICATIONS

Film

Blown nylon film has been used for the packaging of oils, fats, greases, and oil-based products such as putty. It has also been used for food packages such as boil-in-the bag foods (where nylon's high softening point is useful and vacuum-packed cheese, bacon, etc., where low gas permeability is essential. Blown nylon film has also been used for tapes and cable insulation. Biaxial orientation of nylon film improves certain properties, including stiffness, mechanical strength and transparency. The permeability to gases and water vapor is also reduced.

Oriented nylon films can also be vacuum metallized. Such films have excellent flexibility, oxygen barrier properties, anti-static properties and pleasing appearance. They are used in such applications as institutional coffee pouches, metallized balloons and bag-in-box liners.

Coatings

The Versamids, particularly, are used in alcohol or alcohol/hydrocarbon solvents as coatings for paper, aluminum foil, regenerated cellulose, etc. These coatings are used to give heat sealability and to improve the grease and water barrier properties of the substrate. Versamids are particularly useful for coatings because of their flexibility, elongation, and adhesion. Versamids with reactive amine groups can be combined with epoxy resins to give thermosetting materials, and these have found application as drum linings, wire coatings, and coatings for all sorts of industrial articles.

Nylon 11 and more recently Nylon 12 have become widely used in the powder coating of metals. The reduction of these polymers to the powder form is very difficult and industrially it is carried out by cryogenic techniques. These coatings impart to the object all the chemical resistance and other excellent characteristics associated with this family of polymers. Further details of these powdered polymers are given later in the book.

Fibers and Monfilaments

The first commercial application of polyamides was for toothbrush bristles. This was in 1938 and was followed, in 1939, by the appearance of nylon stockings. Other fiber uses quickly followed, including such diverse items as parachute cords, safety belts, and monofilament for rackets.

The success of nylon filaments for toothbrush bristles has been followed by their use for tufting brushes of all sorts from toilet brushes to heavy industrial brushes. For paint and distemper brushes, taper monofilaments are available. The main advantages of nylon for this application lies in its durability, and nylon paint brushes are said to last three to five times as long as those made from natural bristles.

Uses for continuous monofilament are also extremely numerous and include fishing lines, surgical sutures (which can be sterilized by boiling or by treatment with steam under pressure), pot scourers, bead stringing, spectacle manufacture, corrosion-resistant screens or sieves, tire cords and various types of rope. Tow ropes for gliders are a particulalry good example of the use of nylon's special advantages: nylon tow ropes are capable of absorbing the shock during glider pick-up, owing to the fact that they are capable of stretching up to 40 percent of their length.

Textile uses of nylon are well-known. Apart from women's hosiery, where it has displaced silk, nylon is used extensively (either alone or blended with other fibers) in the manufacture of parachute cloth, underwear, night attire, negligees, gloves, corsetry and men's ties, hose and shirts. In addition, waterproof nylon is used for umbrellas, raincoats, lightweight tents, and shower curtains.

Nylon has also found many uses as a staple fiber for making into yarn for swim-suits, coats and sweaters. Nylon 'fur,' in particular, makes an attractive and relatively inexpensive substitute for many natural furs.

The types of nylon particularly useful for fiber forming are 6, 11, and 6,6 for textile fibers (although 6,10 is also used) and 6, 11, 6,6, and 6,10 for filaments and bristles. Types 6,10 and 11 are preferred for the manufacture of bristles, sutures, and fishing lines because of their lower water absorption. Another factor leading to the preferred use of Nylons 6,10, and 11 for brushes is their high rigidity. High molecular weight polymers are necessary for fiber-forming properties, together with a tendency to orientation on cold drawing. The latter property is closely related to the polymer's potentiality for hydrogen bonding, as explained in the section on 'Structure.' For textile fibers it is also important that the polymer should have a relatively high softening point in order that the fabric should not melt when ironed.

Nylon fibers can be coated with various materials to give certain properties not otherwise available. Melamine formaldehyde resins, for instance, confer a crisper finish, and nylon taffetas are produced with degrees of stiffness ranging from very slight to the parchment-like finish required for petticoats.

Adhesives

Polyamide based adhesives may be of the hot-melt or the solution type. The resins most frequently encounterd in adhesive composition are vegetable oil polyamides (Versamids), copolymers (such as the 6, 6,6, 6,10 interpolymers) or modified nylons such as the *N*-alkoxy nylons produced by the reaction of nylon with formaldehyde in the presence of alcohol. It will be noticed that all the resins have low melting points (thus giving ease of application as hot-melt adhesives) and ready solubility in alcohol, alcohol/hydrocarbon mixtures, etc. (for solvent based adhesives).

For hot-melt adhesives, various modifying agents are added. The basic polymers are solid thermoplastics which, when heated at temperatures above their softening points, quickly form low viscosity liquids which can be applied with ease to a variety of substrates. If a second surface is brought into contact with the molten adhesive a simple, yet highly effective bond is rapidly formed as the resin cools. Plasticizers are necessary to give low temperature flexibility while waxy materials are sometimes added to prevent blocking or sealing of the adhesives at room temperatures. Other polymers or resins may also be added in certain instances to improve moisture vapor resistance, initial tack, and adhesion to particular surfaces or to reduce costs without decreasing performance. Solution-type adhesives may also contain modifying agents.

Polyamide-based adhesives have a wide range of application including heat-seal adhesives for bread wrappers, dried soup packets, cigarette packages, and laminating adhesives for many flexible packaging materials such as glassine, aluminum foil, regenerated cellulose, paper, and cellulose acetate. Polyamide adhesives have also been used as side-seam cements in tin cans, as pressure-sensitive adhesives, shoe cements, and bookbinding adhesives.

Polyamide-based thermosetting adhesives have also been prepared from highly branched amino-polyamide resins such as certain of the Versamids. These contain amine end groups so that reaction can take place with epoxy or phenolic resins. In use, a blend of the polyamide resin and epoxy resin is applied to the surfaces to be joined together. These are held in position until cured. Such curing may be at room or elevated temperatures.

Polyamide/phenolic adhesive compositions differ from the epoxy ones inasmuch as pressure is necessary during the bonding operation due to the fact that water is produced as a by-product of the curing reaction. In addition, high curing temperatures are necessary (150–205°C).

Uses for the polyamide thermoset adhesives include wood bonding, bonding of aluminum honeycomb structures for aircraft, bonding copper foil to epoxy/glass or paper-reinforced phenolic laminates, as in printed circuit bases, and as adhesives for steel.

The polyamide copolymers have found widespread use over the last ten years in the heat sealing of textiles. The adhesive bonding of textiles has grown into a very large industry since its early beginning in the U.K. in the middle 1950's. This is covered in detail in Chapter 29.

Printing Inks

When Versamid polyamide resins are combined with suitable pigments and other modifying agents they can be used with great success in flexographic inks. Gloss, adhesion, and pigment-wetting properties are among the factors making polyamide resins suitable for these inks, coupled with high alcohol solubility. The latter property is of importance in flexographic ink formulations because alcohols do not attack the rubber rollers which are normally used in such printing processes. Solvents such as aromatic hydrocarbons and ketones used in some other printing-ink compositions do attack rubber and are not, therefore, suitable.

Polyamide-based inks can be used for the printing of paper, polyethylene, regenerated cellulose, etc., giving good adhesion and resistance to abrasion.

Moldings

High tensile strength and moderately high melting point are usually demanded in molded applications for nylon; thus high molecular weight resins of the 6, 6,6, 6,10, and 11 types are normally used. Applications where nylon's great mechanical strength and toughness and its high melting point are particularly useful are more often found in the field of engineering. A particularly useful property of nylon is its self-lubrication. Thus, it is widely used for gears, bushes and bearings, cams, etc., especially in the food and textile industries where contamination by lubricants is a problem. In addition, nylon mechanical components are quiet-running and are light in weight, the density of nylon being less than half that of even light metals such as aluminum. Examples of engineering uses of nylon are almost too numerous to mention but include speedometer gears, connecting rods in electric shavers, pump impellers, cams and ratchets on calculating machines, timing gears, camera parts, lock-nut inserts, meter wheels, sound projector gears and conveyor chains.

Nylon has also found many outlets in the domestic field, outstanding among which are nylon combs, curtain runners, slide fasteners, soda siphon tops, furniture castors, door catches for refrigerators (and cars), and brush backs.

Nylon's high heat resistance has led to its use in surgical and clinical moldings such as hypodermic syringes, which can be sterilized by boiling water or steam. However, it should be noted that nylon is more sensitive to dry heat than to wet and continuous exposure to dry heat in excess of 120°C could lead to embrittlement.

An interesting use of nylon's strength and abrasion resistance is found in a French-made ski boot where the toepiece cover is molded in Du Pont's Zytel ST.

Another, somewhat unusual application is a range of garden blower and sprayer units. The blowers are used to clean grass clippings, leaves and other debris from outdoor surfaces. The largest model in the range doubles as a sprayer for watering large surfaces or for spraying insecticides, fertilizers and herbicides on lawns and gardens. The housings of these machines are molded from a mineral reinforced nylon 6 copolymer. The blowers are driven by two-stroke gas engines so that the housings must stand up to the operating temperatures and to gas spillage. The housing of the blower/sprayer model must also withstand attack from a variety of chemicals. Good impact strength and surface finish are also factors that contributed to the choice of nylon.

Extrusions (other than film)

Nylon has been used for the extrusion of abrasion-resistant sheathing for communication cables and for heat- and oil-resistant sheathing on cables in aircraft and similar motors.

Tubing and extrusion blow-molded bottles have also been used for a wide range of chemicals and pharmaceuticals because of their chemical inertness. Other useful properties include strength, flexibility, and sterilizability, the latter property being utilized particularly in babies' feeding bottles. Fine-bore nylon tubing has also been found very suitable for blood transfusion equipment and for the draining of wounds.

Finally, an interesting application demonstrating the successful marriage of traditional and new materials, is the manufacture of heavy duty driving belts by laminating leather to oriented nylon strip.

Nylons of ultra-high viscosity are commonly used in blow molding applications because of their excellent impact strength, toughness, heat resistance and chemical resistance. Applications include industrial containers, moped fuel tanks and automotive oil reservoirs.

Nylon has also been used, in the form of a blow molded, seamless liner inside a coventional aerosol can. The idea is that the liner separates the product from the propellant, allowing a much wider range of products to be dispensed, over a wider range of viscosities.

Glass Reinforced Nylon

The useful range of nylon moldings can be increased by the addition of chopped glass fiber to the molding powder. This increases the elastic modulus, tensile strength and heat distortion temperature and reduces thermal expansion and dimensional changes due to water absorption. Creep is also improved (particularly at elevated temperature). Such compounds can produce moldings which are often capable of replacing light metals and, in particular, zinc alloys in die casting. Applications include door handles, rollers for sliding doors, and hinges on domestic equipment. The oil resistance of such moldings has been made use of for the manufacture of oil dip-stick tubes while the strength and rididity are utilized in window fasteners and door fittings.

The strength and rigidity of glass reinforced nylon are also factors in its uses for the wheels of heavy duty wheelchairs. The wheels are 25 inches in diameter with seven spokes and are said to be two and a half times stronger than heavy duty spoked metal wheels. The nylon wheels have a sealed bearing which prevents damage from dirt or grime. Also, because the wheel is molded, the spokes cannot become loose or break, unlike metal spokes which are often replaced on metal wheelchair wheels.

A 33 percent glass reinforced nylon 6,6 has been used for the gearbox and body moldings of a power tool. The properties needed include mechanical and impact strength, fatigue resistance, heat resistance and electrical and heat insulation. The handle molding also utilizes nylon.

Glass reinforced nylon has also opened up applications in the field of sheet stamping. The sheet is supplied to the processor as blanks, in a size and weight appropriate to the required molding. The blanks are heated in a conveyorized infra-red pre-heat oven to slightly above their melt temperature. After rolling, to achieve more effective flow characteristics the heated sheet blank is placed into an open, heated mold. The mold is closed, then opened again after a short dwell time and the formed part is removed. Total cycle times are of the order of 25-60 seconds.

Glass reinforced nylon stampings are likely to replace steel and aluminum for a variety of automotive, appliance and other high strength structural and weight sensitive parts. Applications include an oil pan, and a torque chain cover for cars, a swimming pool filter and a motor housing for an electric lawnmower.

Amorphous Nylons

A relatively new member to the family of polyamides is the amorphous type which has been introduced by Emser Werke AG in Switzerland and now being offered by Union Carbide Corp. in the U.S.A. under the tradename Amidel. Other similar materials are Trogamid T supplied by Dynamit Nobel AG and PACP 9/6 supplied by Phillips Petroleum.

Most polyamides are crystalline and opaque but this particular type is transparent and has been developed for applications where this property is essential. Such a plastic is more economical than glass because it is possible to mold complex parts at low costs and at the same time cut weight and the hazard of breakage.

The exact nature of the chemistry of these new polyamides is obscure because of the scant information offered by the manufacturers but certain claims are made for their properties. Like all other polyamides these materials have a high degree of resistance to most solvents and are better than other clear thermoplastics such as the polycarbonates. In particular, they are resistant to fuel, oil, grease, with the added advantage of having a low humidity absorption. They also have outstanding resistance to stress-cracking.

The claims for the applications for these new materials are numerous and include sight glasses for measuring instruments, caps and covers for fuse and junction boxes in the electrical industries, valves, filters and flow meters. At the moment these polymers are under test by many large consumers of plastics and it is anticipated that the field of application will widen in the not too distant future.

Another recent development is that of the amorphous nylon, Selar PA from Du Pont. This material is said to have a significantly lower water vapor transmission rate than those of semi-crystalline nylons such as nylon 6 or 6,6. A surprising property is its gas permeability, which becomes lower under conditions of high relative humidity. At 100 percent the oxygen barrier properties of Selar PA are said to be comparable to that of an EVOH copolymer. It has been used to produce cast film and both monolayer and coextruded blown films.

Casting

Where large parts are required, the processes of extrusion or injection molding are not suitable and *in situ* polymerization, or monomer casting, is utilized. This process offers the advantage of lower cost and is effective for parts ranging in weight from one to several hundred pounds. Where smaller parts are required they can be machined from rod or sheet.

Lactam-type monomers are used in the casting process, the polymerization being effected with anionic co-catalysts. Lactam flakes are melted under conditions of controlled temperature and atmospheric humidity. The melt is divided into two portions to which are added the catalyst and co-catalyst (or activator) respectively. The solutions are then thoroughly mixed and poured into the molds. Additives, such as colorants, must be thoroughly dried prior to use in order to prevent incomplete polymerization. Cooling of the cast part is allowed to proceed very slowly in order to prevent the development of internal stresses. Packing in insulating material for up to 24 hours, cooling in an oven or annealing in a bath of highly refined mineral oil have all been used, the last-mentioned giving the best results.

Cast nylon has a higher molecular weight than molded or extruded nylon and has a greater crystallinity. It has, therefore, a higher modulus, higher heat deflection temperature, lower moisture absoption, and better dimensional stability.

Applications for nylon castings include gears, slide bearings, and heavy duty rotational bearings where the material's lubricity, strength, and wear resistance are important. Nylon's chemical resistance has been utilized in pulleys for overhead gates in dilute chlorine environments and for chain wear strips in contact with caustic soda. Other factors influencing cast nylon parts, such as gears, are reduced noise levels and reduced use of oils, greases, and other lubricants.

SUGGESTED FURTHER READING

Nylon Plastics. (1973) Ed. Melvin I. Kohan (New York: Wiley).

Synthetic Fibre Materials (1994) Edited by H. Brody (London: Longman).

13.

ACRYLIC POLYMERS

POLYACRYLATES AND POLYMETHACRYLATES

The laboratory history of acrylic monomers began as early as 1843 when acrylic acid was first synthesized.

This was followed in 1865 by the preparation of ethyl methacrylate by Frankland and Duppa, while in 1877 Fittig and Paul noted that it had a tendency to polymerize. By 1900, most of the common acrylates had been prepared in the laboratory and some work had also been done on their polymerization. In 1901, Dr. Rohm, in Germany, started systematic work in the acrylic field and later took an active part in the industrial development of the acrylic ester polymers in that country. Polymethyl acrylate was the first acrylic polymer to be produced industrially (by Rohm and Haas, in 1927). It was marketed as a solution of the polymer in an organic solvent and was used mainly in lacquers and surface coatings. Later, Rowland Hill (of I.C.I.) studied methyl methacrylate and its polymerization in detail, while Crawford (also of I.C.I.) developed an economic method of manufacturing the monomer.

Preparation of Monomers

In order to avoid possible confusion between acrylic and methacrylic esters and acids, the basic acrylic formula should be noted. This is:

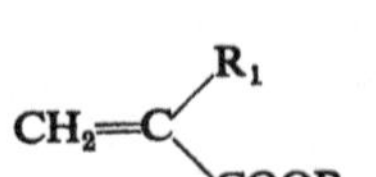

Replacement of R_1 or R_2, or both, by hydrogen, or an aliphatic group such as methyl (CH_3) or ethyl (C_2H_5), yields a wide range of acrylic monomers. Replacement of R_1 and R_2 by hydrogen, for instance, gives acrylic acid, while if R_1 is replaced instead by a methyl group, we have methacrylic acid.

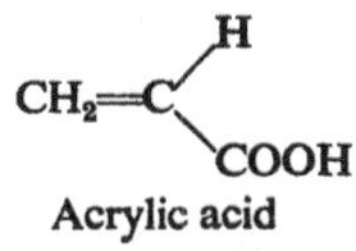
Acrylic acid

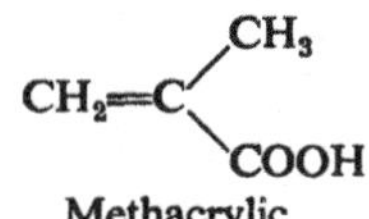
Methacrylic acid

All the monomers dealt with in the next two sections are esters of one or the other of these two acids. The formulas of the methyl esters, for example, are given below:

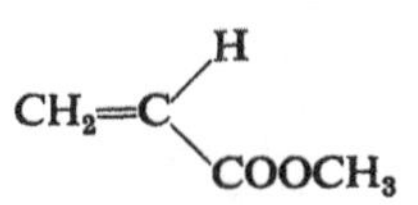
Methyl acrylate

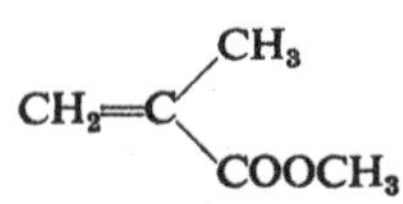
Methyl methacrylate

Preparation of Acrylates

Methyl acrylate is a liquid, boiling at about 80°C, which can be prepared as follows:

(*a*) One commercial synthesis is based on the reaction of ethylene oxide with hydrogen cyanide which yields ethylene cyanohydrin. Further reaction with methyl alcohol and sulfuric acid then gives methyl acrylate.

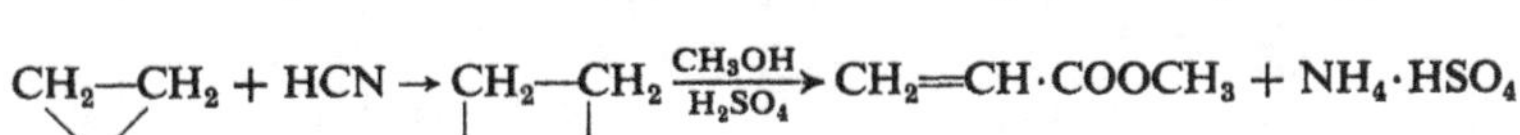

Ethylene oxide, Hydrogen cyanide, Ethylene cyanohydrin, Methyl acrylate, Ammonium bisulfate

(*b*) The following process, starting from acetylene, is due to Reppe in Germany.

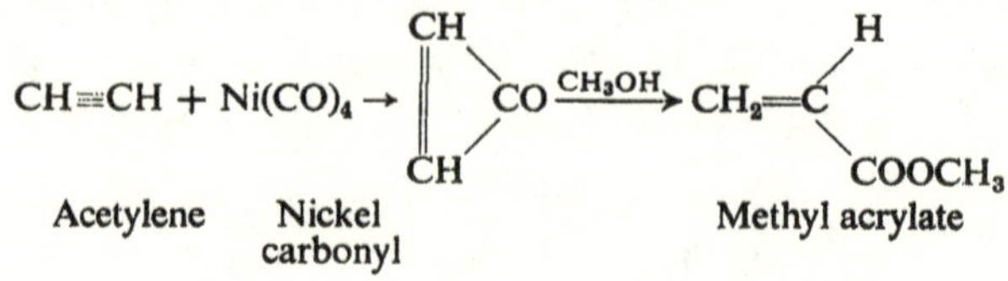

(*c*) A method was worked out in the U.S.A., utilizing lactic acid wastes (after precipitation of casein) from milk.

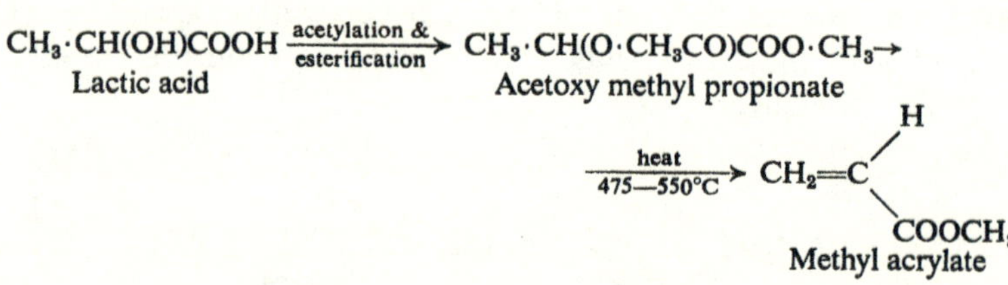

(*d*) A more recent process is based on the oxidation of propylene.

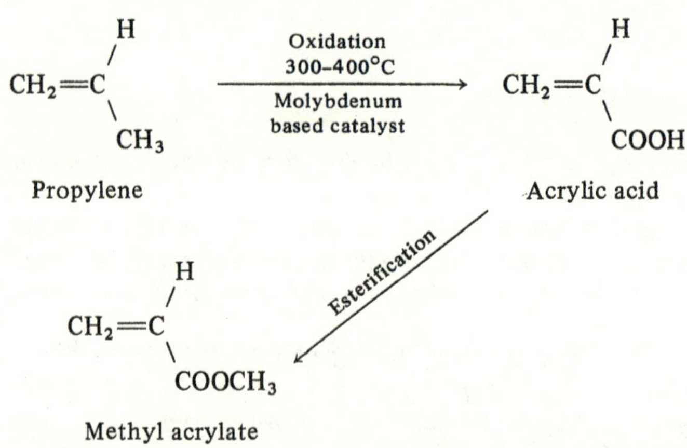

The higher acrylic esters, such as ethyl acrylate, may be produced by processes (*a*) and (*b*) above, by substitution of the appropriate higher alcohol for methyl alcohol.

Preparation of Methacrylates

Methyl methacrylate, the most important of the methacrylic esters, is a liquid boiling at about 101°C. Methods of preparation are given below.

(*a*) The first commercial preparation of methyl methacrylate was as follows.

Acetone was reacted with hydrogen cyanide, or sodium cyanide, in the presence of potassium hydroxide. The acetone-cyanohydrin so formed was then reacted, first with sulfuric acid, and then with methyl alcohol and water.

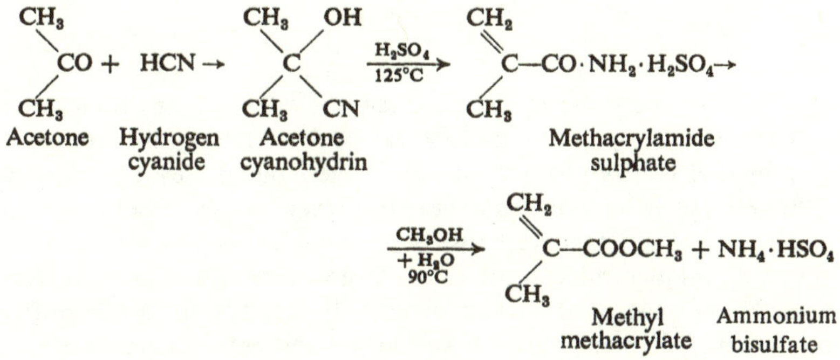

(*b*) A process utilizing acetylene is the following:

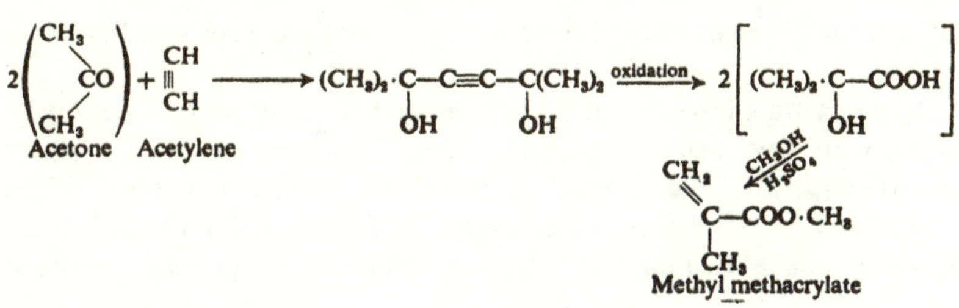

Higher esters may again be prepared by substitution of the appropriate higher alcohol for methyl alcohol.

Polymerization

Five basic methods can be used for the polymerization of the acrylates and methacrylates. These are:

(*a*) Bulk polymerization.
(*b*) Solution polymerization.
(*c*) Emulsion polymerization.
(*d*) Suspension polymerization.
(*e*) Granulation polymerization.

(*a*) *Bulk Polymerization.* This process is the one used for the manufacture of cast sheet, rods, and tubes and can be illustrated by the method of making methyl methacrylate sheet. Since the monomer usually contains an inhibitor (such as hydroquinone), to prevent polymerization during handling and storage, this must first be removed by washing with alkali or by distillation under reduced pressure. A monomer-soluble catalyst, such as benzoyl peroxide or azoisobutyronitrile, is added to the inhibitor-free monomer and polymerization allowed to proceed between glass plates in ovens kept at medium temperatures. In practice, there are several problems to be overcome in order to produce flawless sheets. Allowance must be made during polymerization for removing the heat of reaction and for the shrinkage of 20 percent in volume which occurs. Another problem arises from the very low viscosity of methyl methacrylate monomer which tends to lead to leakage from the glass-plate cells. The following account of a typical casting process should serve to show how these problems have been overcome.

Before pouring into the cells, the monomer is thickened into a syrup by heating with, say, one half percent benzoyl peroxide for a few minutes, with stirring. It is then cooled to room temperature and deaerated by vacuumization. After filtering through a cloth, it is ready for use. If the syrup is not to be used immediately, it can safely be stored under refrigeration.

The shrinkage problem is solved in an ingenious manner by causing the glass plates, between which polymerization takes place, to move together during the polymerization process. This is done by making up cells of glass plates separated by polyvinyl chloride, or some other similarly compressible gaskets. When filled, the cells are placed in an oven at about 45°C for 12-40 hours (according to the thickness of the sheet). The air in the oven is kept well circulated in order to dissipate the heat generated during polymerization. Towards the end of polymerization, the viscosity becomes extremely high thus preventing free circulation of the small amount of monomer still remaining. The temperature is raised, therefore to about 90°C for a short time in order to cure the sheet completely. After removal from the cell, the sheets are annealed to remove residual stresses, by hanging vertically in an oven, and heating to about 140°C for a period of time (dependent on the thickness of the sheet), following this by slow cooling.

(*b*) *Solution Polymerization.* This method is of particular value where the resultant product is desired for use in solution form, as in adhesives or lacquers. The molecular weight is usually lower than is attainable by the bulk process. The monomer is added to an organic solvent in which both the monomer and polymer are soluble. Typical solvents which have been used for acrylic esters are toluene, acetone, and ethyl acetate. The catalyst used must also be soluble in the solvent, and organic peroxides such as benzoyl peroxide and lauryl peroxide have been found suitable. Solution polymerization is carried out in stainless steel or glass-lined reaction vessels fitted with agitators and cooling jackets.

(*c*) *Emulsion Polymerization.* The emulsifier and a water-soluble catalyst are first added to the water, followed by the monomer which is dispersed by agitation. Polymerization is then carried out by heating. Typical water-soluble catalysts suitable for this reaction include hydrogen peroxide, ammonium persulfate, and potassium persulfate. Emulsifiers may be anionic or nonionic surface active agents but the nonionics are favored because of their greater stability. Acrylic emulsions so formed are used for leather-coatings and in the manufacture of water-based paints.

(*d*) *Suspension Polymerization.* This method is very similar to emulsion polymerization described above and utilizes similar equipment. The monomer is added to water and broken into small droplets by agitation. A protective colloid is also added in order to prevent the possible coalescence of these droplets when polymerization is partially completed. Typical protective colloids include polyvinyl alcohol and sodium polyacrylate. In suspension polymerization, the reaction takes place in the monomer phase so that monomer-soluble catalysts must be used, the most common being benzoyl peroxide. The water acts as a heat transfer medium and the temperature of the bath remains constant. The product is obtained in the form of small beads which are screened, washed, and dried.

(*e*) *Granulation Polymerization.* This process is one often used for the production of molding powders and is a variant of the suspension technique. Only a small amount of water is used, together with some lubricant and either a water or monomer-soluble catalyst. Polymerization is carried out in a heavy duty mixer with a sigma blade agitator. Continual kneading of the material during polymerization results in the formation of fluffy, granular particles. No emulsifiers or protective colloids are added and the product is thus extremely pure.

Properties

A wide range of properties exists, depending on the type of monomer or monomers used. In general, for both polyacrylates and polymethacrylates, the lower esters are tougher, harder, and have higher softening points than the higher esters. For example, polymethyl methacrylate has a softening point of about 125°C and a tensile strength of 9000 $lb/in.^2$, while polybutyl methacrylate softens at 33°C and has a tensile strength of 1000 $lb/in.^2$.

Polymethyl methacrylate has excellent transparency, with a light transmission of about 92 percent. It also has extremely good resistance to weathering which has led to its extensive outdoor application. Polymethyl methacrylate does not have a sharp melting point but softens gradually over a fairly wide range of temperatures. At about 150-170°C the polymer is sufficiently rubbery to be capable of shaping. In general, the properties of cast sheet are better than extruded sheet, being harder with a higher tensile strength and a better optical surface.

Polymethyl methacrylate is soluble in most ketones, esters, and aromatic and chlorinated hydrocarbons, and is swollen by contact with ether. It is resistant, however, to attack by alkalis, most dilute acids and aqueous solutions of inorganic salts.

Electrical properties are excellent, a particular virtue of polymethyl methacrylate being its nontracking properties.

Some of the physical and chemical properties of polymethyl methacrylate are shown in Table 13.1.

Table 13.1

PHYSICAL AND CHEMICAL PROPERTIES
OF POLYMETHYL METHACRYLATE

Property	*Units*	*Cast*	*Molded*
Specific gravity		1.17–1.20	1.17–1.20
Refractive index		1.48–1.50	1.49
Tensile strength	lb/in.^{2}	8000–11,600	7000–11,000
Elongation at break	%	3–7	3–10
Modulus of elasticity	lb/in.^{2}	$3.5–4.5 \times 10^5$	4.5×10^5
Impact strength	ft lb/in. notch	0.4–0.5	0.3–0.5
Hardness	Rockwell Scale M	80–100	85–105
Water absorption	% in 24 h	0.3–0.4	0.3–0.4
Vol. resistivity	ohm/cm	$>10^{15}$	$>10^{14}$
Power factor (10^6 cycles)		0.02–0.03	0.02–0.03
Effect of weak acids		Negligible	Negligible
Effect of weak alkalis		Negligible	Negligible
Effect of strong acids		Attacked by oxidizing acids and sulfuric acid	
Effect of strong alkalis		Attacked	Attacked
Effect of organic solvents		Resistant to alcohols at room temperature. Soluble in esters, ketones, and aromatic chlorinated hydrocarbons	

The most important differences between polymethyl methacrylate and polymethyl acrylate are as follows:

(*a*) The methacrylate is more stable to aging and to discoloration than the acrylate.

(*b*) The methacrylate has the lower water absorption of the two.

(*c*) The acidity of the methacrylate is nil whereas the acrylate has a slight acidity.

(*d*) The methacrylate is resistant to alkaline hydrolysis whereas the acrylate hydrolyzes to give the sodium salt of polyacrylic acid, which is water soluble.

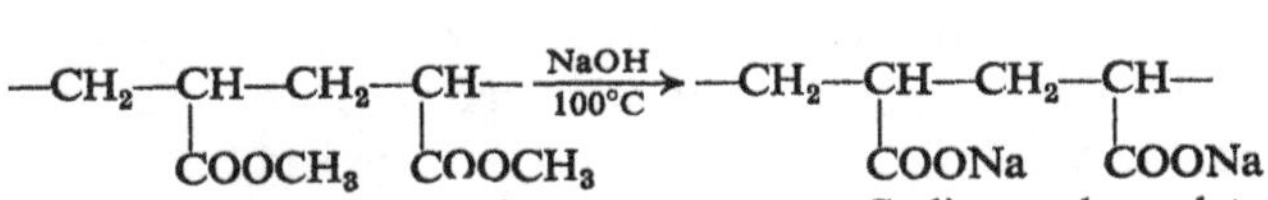

Polymethyl acrylate — Sodium polyacrylate

Many other esters are known and a wide variety of materials has been obtained by copolymerization but these are outside the scope of this book. However, mention may be made of polyethoxyethyl methacrylate which is obtainable as a soft, flexible film, and polybutyl methacrylate which is intermediate in properties between the polyethoxyethyl methacrylate and polymethyl methacrylate (which is hard and brittle).

Processing Techniques

Polymethyl methacrylate can be machined with conventional wood or metal working machinery provided the tools are sharp and cleanly ground. Frictional overheating must be avoided in order to prevent gumming up of the tools. One method of cooling is to blow compressed air on the work; this has the additional advantage of removing swarf, etc. After machining operations have been completed, the work can be polished by means of a rotating head fitted with chamois or flannel. If the finish is too rough after machining, it will be necessary to sand and buff the work before final polishing can be carried out. It is also desirable to finish with an annealing operation to relieve any stresses induced by the machining or by the heating and cooling undergone during the finishing operation.

Polymethyl methacrylate sheet can readily be formed when heated to 150-170°C, when it is soft and pliable. If no stretching is required, the material is simply draped over the required form and allowed to cool. If stretching is necessary, matched molds are used or else the sheet is made to conform to the surface of a single mold by means of air pressure or vacuum. Only simple molds are required as a rule, and these may be made from wood, metal, plaster, etc. Vacuum-forming of polymethyl methacrylate and other plastic sheets is considered in some detail in Chapter 28.

Polymethyl methacrylate molding powders can be injection molded, extruded, or compression molded. The extrusion process obviously offers another means of manufacturing sheet, especially where very large sizes are concerned. Widths of up to five or six feet are available with, of course, a practically unlimited choice of lengths. Much developmental work has been carried out on the extrusion of acrylics and it is now possible to produce sheet with uniform thickness, freedom from residual stresses, an even gloss, and freedom from surface defects. Air-quenching is used, and the cooling should not be too rapid otherwise freeze-off lines appear across the sheet. The recommended temperature of the first set of take-off rolls is between 38 and 55°C. The rolls must also be clean and free from any sort of surface imperfections. Conventional machines and molds are used in injection molding but it should be noted that the molding powder has a tendency to pick up moisture, and predrying for two to three hours in tray ovens at about 85-95°C is recommended just prior to molding.

Articles can be fabricated in polymethyl methacrylate by means of cements. The three main methods of joining polymethyl methacrylate to itself are by straight solvents (chloroform, ethylene chloride, etc.) or by solutions of the polymer in either the monomer or a solvent.

Emulsions and solutions of acrylate and methacrylate can be made up into paints and adhesives, or used for the coating of textiles, paper, leather, etc. Among the desirable properties which can be imparted to textiles by acrylic emulsions are resistance to bacterial action or mildew, and grease and oil. Acrylic emulsions are also fast to washing and do not yellow with age. Acrylic emulsion paints exhibit good scrub resistance, are resistant to freezing and, again, do not yellow with age.

Applications

One of the best-known uses for polymethyl methacrylate sheeting is in glazing. During WW II it was extensively used on aircraft for wind-shields, cockpit covers, gunners' blisters, etc. Similar applications have also been found in the motor-boat industry, namely, wind-shields and canopies. The use of acrylic sheeting in buildings is also growing. In addition to the replacement of glass in conventional glazing, polymethyl methacrylate sheet lends itself to the use of molded skylight domes, which are less liable to breakage, are airtight and rain-proof. Another form of roof glazing increasing in popularity is corrugated polymethyl methacrylate sheet. Indoor building uses include room dividers, doors, and shower cubicles.

A further large market for polymethyl methacrylate sheet includes outdoor signs and other display materials where its optical properties lend themselves to some interesting lighting effects.

In the domestic field, the development of sanitary ware and baths manufactured from acrylic sheet provides color at no extra cost. It has greatly increased the market for colored baths compared with that which existed when only expensive enameled cast-iron baths were available. The injection-molding field, too, has benefited from the color possibilities of acrylics, a good example being telephones.

Industrial uses of polymethyl methacrylate are also numerous and include instrument panels, safety shields, inspection windows, protective goggles, and water baths. It has also been used to construct various demonstration models of equipment, the clarity of the material allowing complete visibility of all working parts.

Medical uses of the acrylics are many, the most interesting being the manufacture of artificial eyes, contact lenses, and artificial dentures. The latter utilizes a slurry of a polymer in a monomer. The polymer powder is mixed with the monomer and made into a dough with a catalyst and an activator. The dough is pressed into a plaster mold which has been coated with a release agent, and polymerization is carried out at low temperatures.

The resistance to breaking of polymethyl methacrylate is also used to advantage in the manufacture of so-called unbreakable watch crystals and the faces of delicate instruments. Similarly, it is extensively used for lenses (torches, magnifying lenses, etc.) and bulkhead covers for lighting units.

Acrylic polymers in the form of emulsions and solutions are used extensively in the treatment of textiles, paper, and leather. In the treatment of textiles, for instance, acrylic resins are added to improve abrasion and flex resistance. They are also added to give improved crease resistance to fabrics already treated with a thermosetting resin. Up to about 5 percent of the acrylic polymer is usually recommended. In leather, acrylic emulsions are used for the base coating, the emulsion serving as a binder for the pigment and forming a strong bond with the surface of the tanned hide.

Paints based on acrylic emulsions are now available in the U.S.A., as well as the better known polyvinyl acetate and butadiene/styrene based products. The acrylic emulsion paints are claimed to possess better scrub resistance and to be applicable over damp surfaces. Another advantage is their resistance to yellowing on aging.

A wide range of adhesives is available using various acrylate monomers and polymers. One extremely important property in certain applications is their clarity. Adhesives based on monomers alone are usually polymerized *in situ* to form excellent bonds. Polymer-based adhesives, on the other hand, can be solvent based (air drying), hot melt, or pressure sensitive types.

The versatility of acrylic emulsions is also shown by their use in floor waxes. The polymers normally used for this purpose are based on methyl methacrylate, methyl and ethyl acrylates, and methacrylic acid. The acrylic emulsions are usually added to the wax or wax/polyethylene emulsion and are said to give clear, hard, and glossy films with excellent toughness and rigidity.

Reference may also be made here to the use of methacrylate monomer as a cross-linking agent for polyesters (*see* Chapter 5).

MISCELLANEOUS ACRYLICS

Acrylic Rubbers

The main class of acrylic rubbers is formed by the copolymerization of acrylonitrile and butadiene. The production of both of these monomers has already been described in Chapter 11. The normal method used for copolymerizing butadiene and acrylonitrile is based on emulsion techniques. The latex so produced can be used directly for coating or impregnation or it can be coagulated by precipitation with a salt solution, followed by washing and drying. This produces a dry crumb which is then compounded and milled into slab form.

Basically acrylonitrile/butadiene rubbers are resistant to solvents and oils but the extent of this is dependent on the acrylonitrile content of the rubber. This, of course, also holds true for other properties. An increase in the acrylonitrile content produces an increase in oil and solvent resistance, tensile strength, and hardness, together with a corresponding decrease in solubility, resilience and low-temperature impact strength. Acrylonitrile/butadiene rubbers have low permeability to gases and a better resistance to chemicals than that of natural rubber. Resistance to heat, abrasion, and creep is also good.

Acrylonitrile/butadiene rubber latices are used for coating industrial gloves, various metal items, tarpaulins, and electrical tapes. They also find outlets in caulking and gasketing materials. These rubbers are discussed in more detail in Chapter 21.

Acrylic rubbers based on ethyl acrylate, have also been produced. One example of this type is, Lactoprene EV, which was originally developed by the U.S. Department of Agriculture. It is produced by the emulsion polymerization of ethyl acrylate (95 percent) and α-chloroethyl vinyl ether (five percent). It possess good oil resistance, flex life, and resistance to degradation by heat and aging. Lactoprene, and other rubbers based on ethyl acrylate are easily cross-linked (vulcanized) using such systems as amines, metal oxides, dioximes and sulfur, plus accelerators.

Fluoroacrylic elastomers are also known but are best treated in relation to the other fluorocarbon polymers (*see* Chapter 14).

Acrylic Fibers

These are based on acrylonitrile and have made an important contribution to the science of man-made fibers. Research was first started on the possible use of polyacrylonitrile as a fiber by Du Pont, in the early years of WW II, and considerable success has since been achieved. High molecular weight polymers of acrylonitrile are necessary for the production of fibers. There are two possible routes for fiber formation, depending on whether the polymer is melted or dissolved in a suitable solvent. Whichever method is used, the resultant viscous liquid is extruded through a die with very fine holes. At this stage, the fiber is amorphous in form and is too weak to be of value. The next stage, therefore, is to stretch the fiber in order to orient the molecules along the axis of the fiber and produce a degree of crystallization. This is accompanied by a great increase in tensile strength, the fiber then being ready for normal processes such as spinning and felting. It can also be used in monofilament form.

Dyeing 100 percent acrylonitrile presents problems because of its chemical inertness. This has led to the production of various copolymers in order to improve dye reception. Other methods used to improve dyeability include opening the fiber to facilitate diffusion of the dyestuff, and the use of anionic dye receptors.

In general, polyacrylonitrile fibers have good resistance to chemicals, heat and moisture, while their resistance to aging and sunlight is outstanding. In addition to the extensive use of acrylic fibers in the field of clothing, they are used for filter cloths, blankets, awnings, and ropes. Acrylic fibers have also been used in the manufacture of excellent imitation furs.

MISCELLANEOUS COPOLYMERS

Styrene-Methyl Methacrylic Copolymer

This was developed to improve the ease of molding of polymethyl methacrylate while still retaining its good resistance to outdoor weathering, its light stability and its heat resistance. An example of the type of molding possible with this copolymer is a street-lamp globe developed in the U.S.A. The globe was injection-molded in two halves and solvent cemented with toluene. The diameter of the larger part was 16 inches and it was 16½ inches high. Shatter resistance was found to be much greater than that of similar glass globes and the total efficiency, in terms of lighting performance, was similar. The diffusion properties, however, were significantly better in the plastic globe.

Ethylene-Methyl Acrylate Copolymer

This material has a very high degree of flexibility, coupled with good low-temperature characteristics (usable down to $-100°C$). Typical applications include flexible conduit for industrial use. This tubing is constructed by extruding a continuous strip, roughly S-shaped in cross section, which is then spirally wound and heat sealed.

Lopac

Lopac is the trade name for a material made by Monsanto Chemical Company Limited and is a terpolymer. The basic monomer is methacrylonitrile and this is copolymerized with about 5 percent styrene and 5 percent methyl styrene. It is a hard and rather brittle polymer but it has a high resistance to creep and is an excellent barrier to gases, especially carbon dioxide. It also possesses good clarity. The terpolymer was developed mainly as a bottle blowing material for the carbonated drinks market where these properties are important.

Chemical resistance to aqueous liquids is good but the material is soluble in acetone, methyl ethyl ketone, and methylene chloride.

Barex

Barex is the trade name for a material made by Vistron Division of Standard Oil of Ohio. It is made by copolymerizing a 75:25 mixture of acrylonitrile and methyl acrylate in the presence of a small amount of a butadiene/acrylonitrile elastomer. This material was also originally developed as a bottle blowing material for carbonated drinks. It has good clarity, excellent gas barrier properties, and high resistance to creep. In addition, it has good impact strength and is insoluble in most organic solvents.

It is also produced in the form of film and sheet and when laminated with other materials such as low density polyethylene it is suitable for thermoforming containers for cheese and meat. In such applications it competes with nylon/polyvinylidene chloride laminates by virtue of its low gas permeability, its resistance to uv light and its toughness.

In semi-rigid form such laminates are finding increasing use in disposable medical device packaging. The Barex can be sterilized either by ethylene oxide or by gamma radiation. Laminations with aluminum foil are also used, in applications ranging from food packaging to oil-drilling core wraps.

Blow molding applications include injection blow molded bottles for ink correction fluid, nail enamel and other cosmetics. Larger bottles, produced by extrusion blow molding have been used for agricultural chemicals, fuel additives and difficult to package household chemicals. Extrusion stretch blow molding is also being used for bottles of ½ liter or above. Such bottles have improved impact strength and so can be made with lower wall thickness.

14.

FLUOROCARBON POLYMERS

The family of fluorocarbon polymers is relatively new although the industry based on organic fluorochemicals has now been in existence for over thirty years. It stems from the refrigeration industry where, due to a number of fatal accidents, a nonflammable nontoxic low boiling liquid refrigerant became an imperative need. Investigations began in the U.S.A. around the 1930's, special attention being given to compounds of carbon fluorine, and chlorine, of which dichlorodifluoro methane (CCl_2F_2) showed most promise. Success came fairly quickly and in 1931 the Kinetic Chemical Corporation began commercial production of a range of colorless odorless, nontoxic and nonflammable fluorocarbons, named freons. Once such materials were industrially available other applications were then investigated. The use of these materials as aerosol propellants was then developed and soon accounted for a high proportion of fluorocarbon output. More recently fluorocarbons, or more accurately, chlorofluorocarbons (CFCs) have been implicated in the gradual erosion of Earth's ozone layer which protects us from the sun's rays. The use of CFCs is, therefore, being phased out in many countries.

It has been reported that the polymers based on the fluorocarbons were discovered by accident in the laboratories of E. I. Du Pont de Nemours. The polymer originally developed was derived from tetrafluoroethylene although commercial development of PTFE (polytetrafluoroethylene) as the polymer is now known, did not come until some years later. In WW II it became essential to find materials with extreme chemical resistance which could be put to good use in carrying highly corrosive products used in the preparation of Uranium 235. PTFE was successfully used for such applications and after the war it was released to industry for general use since then it has become established as a high performance polymer.

Many other fluorine-containing polymers have been developed, including polytrifluorochlorethylene (PTFCE), polyvinyl fluoride (PVF), and fluorinated ethylene propylene (FEP). These materials may be regarded as derivatives of TFE which is now renowned for its ability on polymerization to yield polymers with extreme chemical resistance, combined with excellent dielectric properties and wide working temperatures.

Specialized industries such as the aircraft (including missiles and rocket), automobile and chemical, are continually searching for elastomers combining the good physical properties exhibited by the conventional organic elastomers with the high thermal stability usually associated with inorganic materials. These requirements have been met, to a large extent, by the fluorine-containing elastomers. Typical fluorine-containing elastomers are the vinylidene fluoride/hexafluoro propylene copolymers, fluorosilicones, fluoroacrylates, and certain other fluorine substituted elastomers.

PTFE was the first of this family to be commercially exploited, and much of the developmental work on other fluorocarbons was based on knowledge gained from this polymer.

POLYTETRAFLUOROETHYENE

Raw Materials

The monomer, tetrafluoroethylene must be of very high purity for polymerization. It is a gas at atmospheric pressure and is obtained by pyrolysis at about 800° of chlorodifluoromethane. The monomer must be virtually free from oxygen and from compounds containing hydrogen or chlorine since such impurities can prevent polymerization or reduce molecular weight to an unacceptable degree. It is therefore essential that during the production of the monomer it undergoes an efficient scrubbing and distillation process.

$$\underset{\text{(Fluor-spar)}}{CaF_2} + \underset{\text{(sulfuric acid)}}{H_2SO_4} \rightarrow CaSO_4 + 2HF$$

$$2HF + CHCl_3 \rightarrow 2HCl + \underset{\text{(Chlorodifluoromethane Arcton 22*)}}{CHClF_2}$$

$$2CHClF_2 \xrightarrow[800^\circ C]{\text{Pyrolysis}} 2HCl + \underset{\text{(Tetrafluoroethylene)}}{C_2F_4}$$

* Manufactured by Imperial Chemical Industires, Ltd., General Chemicals Division.

Polymerization

Tetrafluoroethylene is polymerized under pressure in the presence of excess water, which acts as a medium for the removal of the large amount of heat evolved during the extremely fast polymerization. Polymerization is initiated by a water soluble catalyst capable of generating free radicals, e.g., peroxides.

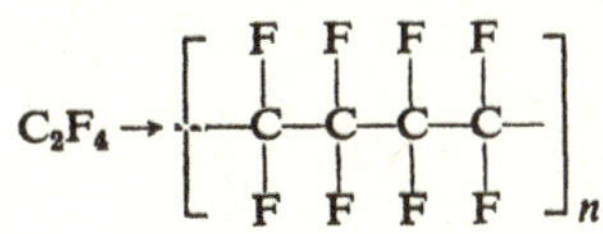

Depending on the specific polymerization conditions used the polymer can be obtained as an ivory-white granular powder or as an aqueous dispersion. The powder must first be washed and dried ready for sintering and disintegration, whereas aqueous dispersion may be either coagulated and dried or stabilized with surface active agents.

Structure

PTFE is a fully fluorinated polymer whose chemical structure resembles that of polyethylene described in Chapter 9. Such resemblance is reflected in its mechanical properties and dielectric behavior at room temperature. Before undergoing any fabrication or processing the polymer is a highly crystalline material (crystallinity above 94 percent) having a molecular weight in the range 5×10^5 to 5×10^6. The polymer molecules may be regarded as unbranched stiff chains in which the fluorine atoms are closely packed round the main carbon chain. It is due to the relative size of the fluorine and the carbon atoms that close packing is achieved. The fluorine/carbon and carbon/carbon bonds are among the strongest known in organic compounds, and many of the remarkable properties of PTFE can be related to the strength and tightness of this molecular structure.

Processing Techniques

PTFE is a thermoplastic polymer by definition but it cannot be processed by conventional thermoplastic techniques such as injection molding and melt extrusion. This is largely due to the fact that it has so high a

Table 14.1

COMPARISON OF PROPERTIES OF FLUOROCARBON POLYMERS WITH SOME OTHER POLYMERS

	PTFE	*PTFCE*	*Polyethylene*	*Polystyrene*	*Nylon*	*Silicone elastomers*
Density (g/cm^3)	2.2	2.1	0.95	1.05	1.14	1.4–2.0
Tensile strength ($lb/in.^2$)	2000	5000	1850	7000	$>$10,000	200–1100
Elongation (%)	50–400	20–80	550	3	80–100	75–200
Impact strength (ft lb)	4	3.5	$>$3.0	0.3	2.5	–
Heat resistance (°C)	230–300	$>$PTFE	100	95	150	150–260
Volume resistivity (ohm-cm)	10^{19}–10^{21}	10^{18}	ca. 10^{19}	10^{19}	10^{13}	10^{12}
Dielectric strength (kV/cm)	200	600	400	400	160	150–225
Dielectric constant	2.0	2.5	2.3	2.5	4.0	3
Power factor (tan δ)	0.0001	0.0002	0.0002	0.0002	0.02	0.003
Chemical resistance: Acids	Resistant	(N.B. inferior to PTFE)	Resistant	Attacked	Attacked	Resistant (swollen although not attacked chemically)
Alkalis	Resistant	Resistant (N.B. inferior to PTFE)	Resistant	Resistant	Resistant	Resistant (swollen although not attacked chemically)
Solvents	Resistant	Resistant (N.B. inferior to PTFE)	Resistant	Attacked	Resistant to most common solvents	Fairly resistant

melt viscosity that it cannot be made to flow by conventional means and, to add to the difficulty, at present it cannot be plasticized. The powder forms cannot be processed by thermosetting techniques, but must be compacted initially into a preform or a billet. They are produced in the cold, then heated to above 327°C, either with or without pressure, so that the individual particles coalesce. This final process is known as sintering, after which the shaped moldings can be machined to various sizes.

Properties

This polymer is smooth and waxy to the touch. It is normally white or mottled bluey grey in color depending on processing conditions. As mentioned previously this polymer possesses outstanding mechanical, electrical, and chemical characteristics. Table 14.1 shows the more important properties of PTFE and gives, for comparison, average figures for the properties of some other polymeric materials.

POLYMONOCHLORTRIFLUOROETHYLENE (PCTFE)

This material was originally produced in Germany around 1934, also in the U.S.A. The monomer is obtained from hexachloroethane as follows:

$$\underset{\text{Hexachloroethane}}{CCl_3CCl_3} \xrightarrow[\text{HF}]{\text{Anhydrous}} CCl_2FCClF_2 \xrightarrow[\text{boiling ethyl alcohol}]{\text{Zinc}} CClF{=}CF_2$$

The method of polymerization is similar to that used for PTFE although it is somewhat more difficult to control. The polymerization products range from mobile oils to horny chips. The chlorine atom in the chain produces a certain amount of electrical unbalance and also reduces regularity, which impedes crystallinity. The close packing of the chain is also reduced and the polar nature of this group results in a higher interchain attraction. On balance the material is still crystalline but less than PTFE, and the reduction in close packing produces a more flexible polymer. Like PTFE this polymer is also nonflammable but its chemical resistance and thermal stability are somewhat inferior, although it is more transparent; when shock-cooling PCTFE from the melt it is almost glass clear.

PCTFE may be processed on conventional plastics machinery with some difficulty. It is essential, however, for the processing machinery to be made from hard metals because of the high processing temperatures required. Any dirt present in the machinery would undoubtedly carbonize and affect the properties of the molding.

PCTFE has a wide range of uses as a packaging film. Among its properties are the highest water vapor barrier properties of any commercially available film, a high resistance to oxygen transmission, and a high resistance to a range of chemicals. It is a flexible film and retains its flexibility down to temperatures of around −195°C. It also has a high temperature of use, with a softening point between 185 and 250°C, according to grade and crystallinity. The film can be laminated to foil, paper, or to a range of other plastics films. It can also be vacuum metallized.

Its main packaging use, either alone or laminated, is for the strip or blister packaging of hygroscopic pharmaceutical tablets and capsules, where a high degree of protection from water vapor is essential. In the U.S.A., the film has also been used in the critical area of packaging aerospace hardware and valves. With other packaging materials there is the risk of particles being abraded from the package and these represent a potential explosion hazard when in contact with the liquid oxygen used as rocket fuel. PCTFE film has proved to be stable under impact in such conditions.

POLYVINYL FLUORIDE (PVF)

The monomer vinyl fluoride has been known for over fifty years, having been first prepared by Swarts in 1901. At one time it was considered impossible to polymerize but, fairly recently, the difficulties have been overcome and it is now produced on a commercial basis in the U.S.A.

Vinyl fluoride is a gas (b.p. −72°C at atmospheric pressure) and is prepared by the action of hydrogen fluoride on acetylene in the presence of a suitable catalyst.

As with TFE it is essential that the monomer is purified before polymerization, which is normally carried out at 85°C, under 300 atm. pressure,

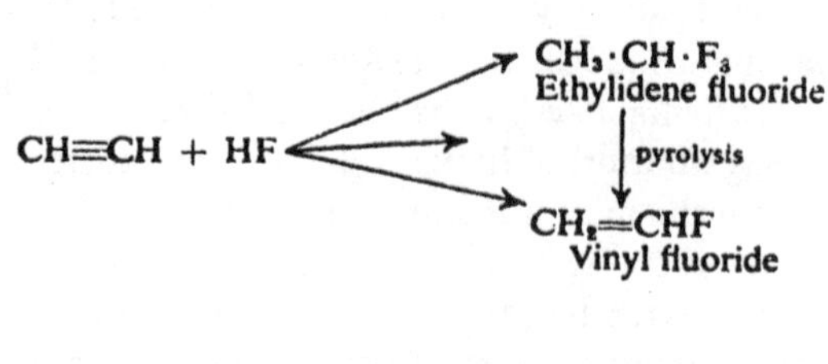

using a peroxide catalyst such as benzoyl peroxide.

Polyvinyl fluoride is a high melting, chemically inert substance with a specific gravity of 1.39. It can be fabricated into tough, weather-resistant films, fibers, and moldings.

POLYVINYLIDENE FLUORIDE

Polyvinylidene fluoride is melt-processible and can be injection or compression molded and screw extruded. It has a lower specific gravity (about 1.8) than that of PTFE and PTFCE and this, coupled with its good processing characteristics, gives a combination of economy with excellent physical and chemical properties.

It is stiffer and has a greater resistance to creep than has PTFE but its chemical resistance, lubricity, antistick properties and electrical characteristics are inferior. Its useful temperature range is from −68° to 150°C.

ETHYLENE CHLORTRIFLUOROETHYLENE COPOLYMER

This is a strong material with a high impact resistance and a wide temperature range. It is readily melt processible.

It has good electrical properties, including a low dissipation factor (0.0008 to 0.015 depending on temperature and frequency) and a volume resistivity of 10^{15} ohm cm. It also exhibits excellent ac loss properties.

ETHYLENE TETRAFLUOROETHYLENE COPOLYMER

This copolymer is closely related to PTFE but is readily processed by injection molding and extrusion. It is a tough polymer with excellent flex life, impact strength, and abrasion resistance.

It has a continuous use temperature of 150°C but can be used intermittently up to 200°C depending on load, exposure time and environment. Weather resistance is excellent and the material is inert to most chemicals, including solvents.

Electrical properties are excellent, including a low dissipation factor (0.0006), a high dielectric strength and high resistivity.

POLYPERFLUOROALKOXY (PFA)

PFA resins combine the ease and economy of conventional melt processing techniques with a high temperature performance in the range of polymers such as PTFE.

In general they have resistance to practically all chemicals, antistick properties, low coefficient of friction and excellent electrical properties. They also have a wide temperature range and excellent weather resistance. Their strength and stiffness at high operating temperatures are comparable to those of PTFE and their creep resistance is superior.

Applications include liners for pumps, valves, roll covers, pipe and fittings, and electrical wire insulation.

FLUORINATED ETHYLENE PROPYLENE COPOLYMER (FEP)

The commercial availability of this copolymer was first announced in the U.S.A. early in 1960. It is closely related to the older TFE polymer manufactured by copolymerizing tetrafluoro ethylene with hexafluoro propylene. The FEP polymers have the fully fluorinated structure and as such possess virtually all the properties for which PTFE is noted. They also exhibit a melt viscosity low enough for conventional thermoplastic processing and can be injection, extruded, thermoformed, compression and transfer molded. Such a factor eliminates the machining of components from block, thereby reducing production costs of a variety of items which could not previously be molded in PTFE.

FLUOROCARBON ELASTOMERS

Vinylidene Fluoride/Chlorotrifluoroethylene Copolymers

These copolymers are commonly known under the name of the Kel-F elastomers. They are produced by copolymerizing TFE with vinylidene fluoride, which results in hydrogen atoms being introduced into the chain and produces polymers that are no longer rigid but flexible. The following structure is obtained:

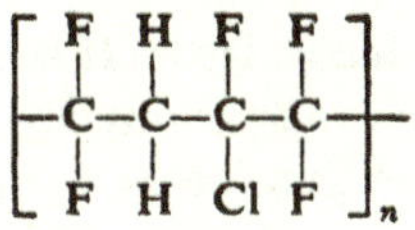

Various grades are available including Kel-F 5500 which is a 50/50 copolymer of vinylidene fluoride and chlorotrifluoroethylene. This elastomer possesses excellent resistance to strong oxidizing agents, such as nitric and sulfuric acid, and is also resistant to hydrocarbon fuels and ozone. In addition, it has excellent flame-retardant properties. Although it would seem to be an excellent all-round material, unfortunately it has disadvantages which include poor resistance to compression-set at elevated temperatures, and fabricating difficulties when producing such items as o-rings and gaskets.

Another grade is Kel-F 3700, which is a copolymer of 30 percent chlorotrifluoroethylene and 70 percent vinylidene fluoride. This particular elastomer possesses marginally better low-temperature flexibility and resistance to hydrocarbon fuels than Kel-F 5500. The Kel-F elastomers may, in general, be cured by peroxides or organic amines. To obtain optimum physical properties they must be given a post cure in addition to the normal press cure. This group of elastomers can be molded, extruded, and calendered with normal rubber processing machinery.

Vinylidene Fluoride/Hexafluoropropylene Copolymers

Important members of this family, which have aroused considerable interest in the last few years, are the Vitons.* They exhibit a combination of heat and oil resistance obtainable from very few elastomeric materials. Basically they are copolymers of vinylidene fluoride and hexafluoropropylene, prepared by an emulsion technique at temperatures up to 100°C, using a 'redox' initiation system (persulfate/bisulfite combination). The elastomer is obtained initially as a latex which is coagulated normally by adding sodium chloride or hydrochloric acid. The resulting product is a white powder which, on drying and pressing, becomes clear.

* Trade name of E. I. Du Pont de Nemours & Co. Inc.

Early attempts to convert the base gum to a useful cross-linked product, using a peroxide curing system, were not completely successful. It was not until elastomers were produced by an amine curing system that the full potential of the family was fully recognized. It is generally agreed that the cross-linking takes place through the preliminary elimination of hydrogen fluoride from the polyvinylidene fluoride segments. With the liberation of the hydrogen fluoride it is obvious that an acid acceptor must be added, and magnesium oxide has proved suitable.

The air-aging and oil-resistant properties of these elastomers are excellent. They combine excellent heat stability with a resistance to a wide variety of chemicals and petroleum fuels.

Fluorel** is another copolymer of vinylidene fluoride and hexafluoropropylene. The base material is cured to a suitable elastomer by organic amines, peroxides, or controlled dosages of irradiation. The most suitable curing agent is probably hexamethylene diamine carbamate. As with the previously discussed elastomer the actual compound must be stabilized during curing, and it has been found necessary to add magnesium oxide, although, as alternatives, zinc oxide or dibasic lead phosphite may be used. This elastomer, as are the Vitons, is suitably compounded on conventional rubber machinery and fabricated by the normal extrusion and calendering methods.

The vinylidene fluoride/hexafluoropropylene copolymers are of particular interest due to their stability to most lubricants and hydraulic fluids used in the aircraft industry. They also exhibit a high degree of resistance to ozone which, to the aircraft industry, is extremely important.

FLUORO ACRYLATE ELASTOMERS

The most common member of this group is poly 1, 1-dihydroperfluorobutyl acrylate, which can be represented as follows:

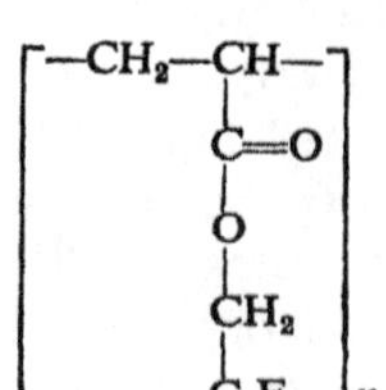

** Trade name of Minnesota Mining & Manufacturing Co. Ltd.

This was given the trade name of Poly FBA and exhibits excellent resistance to many fuels, lubricants, solvents, chemicals, and ozone, plus very good stability at high temperatures. Unfortunately, this elastomer has a brittle point in the region of $-25°C$ and its usefulness is limited to $-20°C$, although it retains its elasticity down to the brittle point. Subsequent development of superior special-purpose high-temperature elastomers has resulted in the rapid decline of the usefulness of the fluoro acrylates.

FLUORINATED SILICONES

It has been stated previously that silicone rubber resembles natural and synthetic elastomers in many ways. The deficiencies of silicone elastomers include their relatively poor resistance to aircraft fuels, hydraulic fluids, and certain solvents. It was with this in mind that developmental work was initiated to introduce fluorine containing substituent groups into the polysiloxane polymers. This has resulted in fluorinated silicone elastomers which possess superior resistance to the swelling effects of many solvents becoming commercially available. These elastomers are cured by peroxides, and it is essential that they are given a post cure in addition to the normal press cure.

MISCELLANEOUS FLUORINE ELASTOMERS

Fluoro polyesters which have shown potentialities as elastomers have been developed in the U.S.A. One such material was hexafluoropentylene adipate which could be cross-linked to a tough elastomeric material. It was discovered that crystallization occurred at moderately low temperatures, and this seriously limited its usefulness. Work was carried out on other fluoro polyesters but with so many other fluorocarbon elastomers available the work remained of academic interest only.

Other fluorocarbon polymers of interest include the addition product of trifluoronitrosomethane and tetrafluoroethylene. This is an interesting semiorganic polymeric substance which may one day find some use as an elastomer. It has been reported that elastomers with molecular weights in excess of 1×10^6 which possess good chemical, nonflammability and low temperature resistance characteristics have been produced.

APPLICATIONS

It is obvious from the properties described that PTFE is attractive to many industries because of its wide working temperature range (from −100°C to +300°C). In this respect and coupled with its resistance to exposure and chemical attack PTFE is superior to any organic polymer at present available. These properties, together with excellent electrical insulating properties and its unusually low coefficient of friction make it of particular interest to the electrical, chemical and engineering industries and whenever the problem of handling sticky substances is encountered.

Until the introduction of certain of the fluorocarbon polymers the use of plastics in engineering and chemical plant was limited by their relatively low softening points (in comparison with those of metals and ceramics). Other deficiencies were their inability to withstand prolonged contact with certain organic materials such as solvents and oils, and in some cases by the deterioration of their mechanical properties and color on exposure to extreme weather conditions. The introduction and development of polytetrafluoroethylene has drastically changed this state of affairs. Although the fluorocarbons are relatively high-priced polymers, their remarkable combination of properties makes them invaluable for certain applications that were hitherto unavailable to plastics. In the electrical industry PTFE has been used for such applications as valve bases, crystal holders, plugs and sockets, cable and wire coverings, cable spacers, and transformer terminals, to mention but a few. It is the high working temperature of PTFE that makes it so attractive as an insulating material, and in the electrical applications it has the advantage over ceramics, which are by comparison, brittle, difficult to mold to close tolerances, and moisture absorbent.

PTFE, because of its resistance to chemical attack is virtually indestructible and, as such, the uses for this polymer in the chemical industry are numerous. They include gaskets in a wide variety of forms, packings, valve seats, washers, tubes, seals, flexible couplings, diaphragms, and filters.

The range of applications where a polymer of low coefficient of friction would be required is so wide that it is impracticable to give details of all uses. However, among the more important uses of PTFE for nonstick applications the following may be mentioned: covered rollers for handling dough, toffee, size, and other such sticky materials, piston rings, metal bearings for high-speed unlubricated applications and as a sheet material for lining hoppers, chutes, grids, etc., in machinery handling sticky products.

PTFCE is unaffected by most corrosive chemicals and organic solvents and therefore is also used in the chemical industry. This polymer is used for electrical fittings where equipment and installations must resist corrosive chemicals, humidity, and abrasion. Like PTFE this polymer has a low coefficient of friction and nothing will adhere readily to its surface. Materials such as dough, toffee, glue, paint, ink, rubber and synthetic resins readily wipe off or can be washed away. These factors plus the polymer's nonporousness make it ideally suitable for machine coverings where nonstick applications are necessary.

Polyvinyl fluoride exhibits extreme resistance to solvents and weathering. Injection molding is somewhat difficult and therefore certain applications are limited. However, film is prepared from this polymer and it has been suggested that it has broad possiblities in the building industry. Such uses would be as a surfacing material for prefabricated metal structures and wooden panels. A service life (maintenance free) of 20 years outdoor exposure is estimated. Other suggested applications include industrial and decorative tapes, outdoor signs, and grease-resistant laminates.

Ease of fabrication, excellent dielectric properties, and high temperature resistance make fluorinated ethylene propylene polymers ideal for use in electical and electronic components. Applications include coil forms, insulators, hermetically sealed terminals, and injection-molded sockets. It is also widely used as a high-temperature insulating and jacking material for wires and cables.

The fluorocarbon elastomers have all found extensive use in many industries, in particular the aircraft industry, where the many stringent requirements are not met by conventional organic elastomers. These applications include o-rings, gaskets, seals, wire coverings, and linings resistant to oils and fuels. The chemical industry has also utilized the outstanding properties offered by these elastomers for tank and pipe linings.

SUGGESTED FURTHER READING

Fluorocarbons. (1958) Merritt A. Rudner (New York: Reinhold).

The Character of Fluon; The Processing of Fluon (Bulletins 2–5); and *The Properties and Applications of Fluon.* Technical Bulletins issued by I.C.I. Ltd.

'Teflon' Fluorocarbon Resins. Technical Bulletin issued by E. I. Du Pont de Nemours & Co. Inc.

'Some Newer Polymers of Current Interest,' Part III–'Fluorinated Elastomer.' (1960) D. C. Blackley. *Applied Plastics,* August.

Film, Fluoropolymers. Article by A. B. Robertson and K. R. Habermann in Encyclopedia of Packaging Technology. Edited by M. Bakker and D. Eckroth (1986). New York: John Wiley and Sons.

15.

THERMOPLASTIC POLYESTERS

As was mentioned in Chapter 5, the general term, polyester, is used to describe a wide range of materials derived from a condensation reaction between a polyhydric alcohol and a polybasic acid, and in that chapter we dealt with the thermosetting polyesters.

In this chapter we shall look at the manufacture, properties and applications of thermoplastic polyesters, in particular, poly(ethylene terephthalate) or PET. During W.H. Carother's classical work on polymer formation he investigated a large number of polyesters. He recognized the ability of certain linear polymers to form fibers, and patents were subsequently taken out defining the conditions for fiber-forming polyesters. In particular, fibers formed from dibasic acids and glycols were mentioned. His basic work was later substantiated by two British scientists, Whinfield and Dickson of Calico Printers who, in 1941, found that the reaction between terephthalic acid and ethylene glycol produced a resinous mass that could be drawn into fibers. The patent rights were then acquired by Imperial Chemical Industries (ICI) and Du Pont who eventually produced the now world-famous fibers Terylene* and Dacron**, respectively.

Around 1966, poly(ethylene terephthalate) was developed as an injection molding resin while later, in 1969, came the introduction of poly(butylene terephthalate) and poly(methylene terephthalate). In the 1970s poly(ethylene terephthalate)—PET—was first used in the stretch blow molding of beverage bottles. This market has grown from 200 million in 1978 to 5 billion (5,000 million) in 1985 in the USA alone. Similar growth has occurred in many other countries.

*Registered trade mark of Imperial Chemical Industries, plc.
**Registered trade mark of E. Du Pont de Nemours.

FIBERS

As mentioned earlier the first uses for linear polyesters were as fibers (Terylene and Dacron). PET itself is made by two different routes, namely, from terephthalic acid (TPA) or from dimethyl terephthalate (DMT), both of which are reacted with ethylene glycol (EG) to give PET.

(a) Terephthalic acid $C_6H_4(COOH)_2$, is prepared by the oxidation of p-xylene.

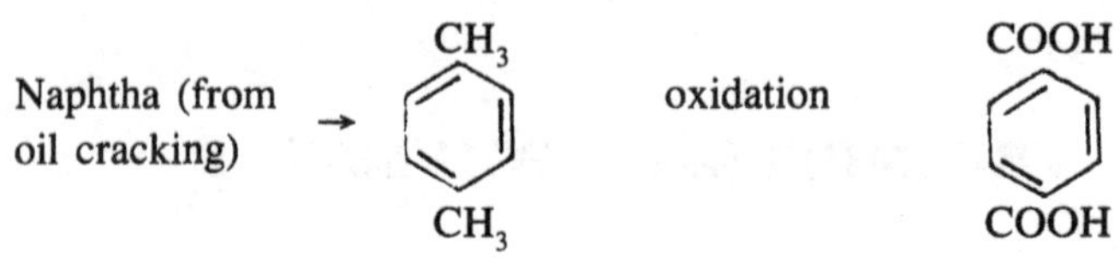

p-Xylene Terephthalic acid

(b) Dimethyl terephthalate, $C_6H_4(COO.CH_3)_2$, is prepared by reacting terephthalic acid with methyl alcohol in the presence of sulphuric acid.

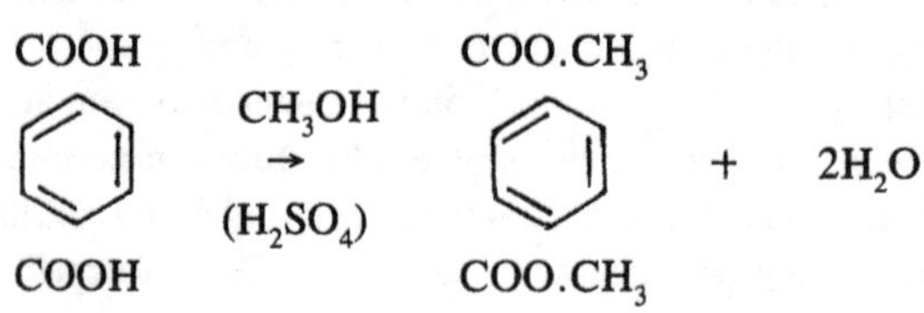

Terephthalic acid Dimethyl terephthalate

(c) Ethylene glycol, $(CH_2OH)_2$, is prepared by oxidation and hydrolysis of ethylene.

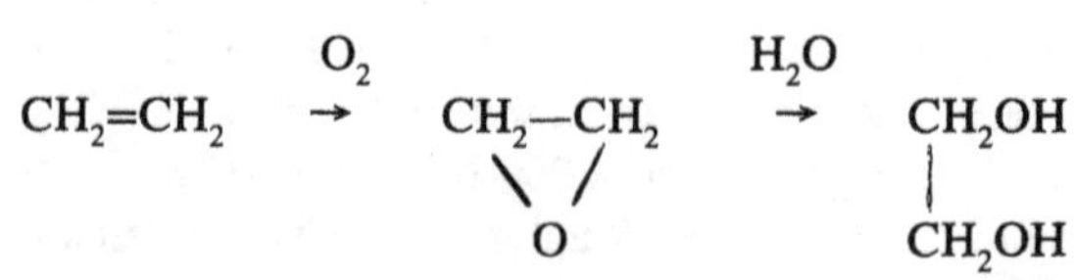

Ethylene Ethylene oxide Ethylene glycol

Manufacture of PET

(i) From TPA (by esterification)

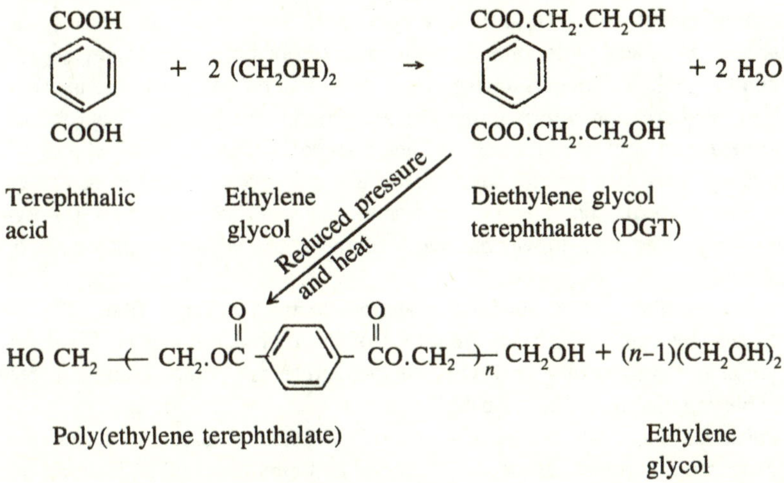

(ii) From DMT (by ester interchange)

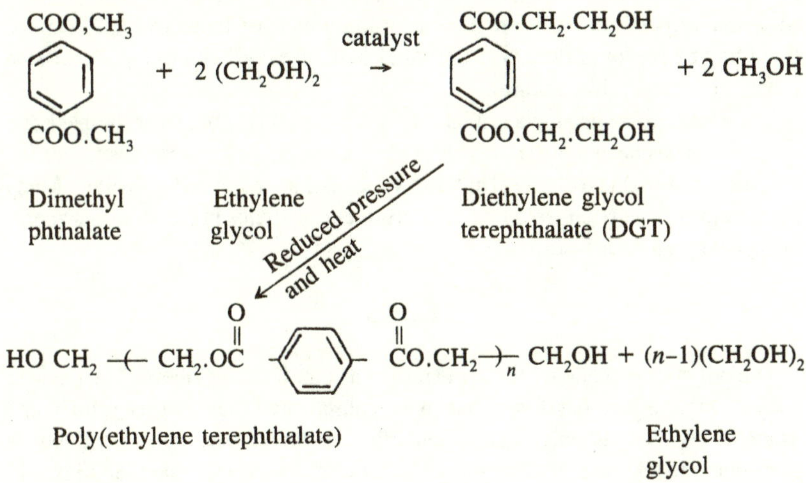

The ester interchange reaction can be catalyzed by materials such as sodium methoxide and is normally carried out at temperatures between 190 and 280 °C.

The polymer is obtained in the form of chips which must first be dried. The fibers are produced by a melt spinning process which consists essentially of pumping molten polymer through a spinneret. A spinneret may briefly be described as a metal plate containing a number of fine holes. (see Figure 30.13) When the molten filaments emerge from the spinneret and meet cold air they solidify and are then collected together and wound on a bobbin. This product is termed spun yarn. It is unorientated and does not possess the required properties characteristic of textiles, therefore it is mechanically stretched or drawn to 4-5 times its original length to develop the characteristic properties of polyester fibers. The fully drawn continuous filament yarn is then wound on a special bobbin.

Polyester fiber is also obtained in another form, viz. staple fiber. This is produced by combining many hundreds of the extruded undrawn filaments together in the form of a tow, which is then stretched, crimped, and cut into suitable lengths and, finally, baled.

Polyester fibers will, in general, resist hydrocarbon solvents, chlorinated hydrocarbons, ketones, esters, concentrated hydrochloric acid, and certain dilute acids. They are, however, decomposed by alkalies, oxidizing agents and, to a greater extent, by alcoholic alkalies.

The uses of polyester fiber are probably best known in the fields of textiles and carpets but there are also a large number of industrial applications including conveyor belting, vee-belting, hoses, tire cord, strapping and coated fabrics for protective clothing.

These particular products are based on woven fabrics but over the past few years there has been a large growth in the area of polyester nonwoven fabrics. A typical material is Reemay (Du Pont) which is a spun bonded polyester. Many applications have been found for this product especially in the shoe, garment, and packaging industries.

FILMS

The polymer is made by the same means as that used for fibers. The process for making the film itself is rather more complicated than that used for some other films, such as polyethylene and PVC. It is extruded as normal, but is then immediately quenched to prevent it from crystallizing, and then biaxially oriented by stretching it at a temperature above 80 °C. This has the effect of improving the mechanical properties and reducing brittleness. If the film is subsequently reheated to above 80 °C it will tend to contract considerably. A third manufacturing stage is added, therefore, to improve heat stability. The oriented film is firmly held to prevent contraction and is then heated to about 200 °C. This causes a considerable amount of crystallization without appreciably affecting the orientation and, when subsequently cooled, the film is stable dimensionally up to 200 °C.

Properties

Polyester films are tough, strong, and have excellent clarity. Slip characteristics are poor unless slip additives are incorporated, and these tend to make the film somewhat hazy. Heat sealing is difficult because of shrinkage and embrittlement which occurs at the heat seal. Adhesives can be used but the best results are obtained by laminating the polyester with low density polyethylene film and using the excellent heat seal properties of the latter.

Polyester film has a high resistance to tearing, a high fold endurance, and is also resistant to abrasion. The impact strength is good and remains so down to about −70 °C.

Polyester film is a fairly good barrier to water vapor and a good barrier to gases and odors. It has good electrical insulation properties including high breakdown strength and good dielectric properties. The volume resistivity is also extremely high.

Applications

Shrinkable grades of polyester film have been used for the vacuum wrapping of cooked meat products where a low oxygen permeability is essential. Polyester bags have also been used for long term storage of frozen poultry.

Packaging uses for the heat stable film include boil-in-bag-packs (when laminated to low density polyethylene) and the overwrapping of oily products. Laminates of polyester with polyethylene or PVDC are also used for vacuum or gas flushed packs of coffee.

A fast growing market for PET film is for 'ovenable' trays for frozen food and prepared meals. The trays are thermoformed from cast PET film and crystallized. This heat-sets the tray and prevents deformation during cooking and serving. The main advantages of PET for this application are suitability for microwave and conventional hot air ovens, light weight and appearance.

There are also a large number of non-packaging uses for polyester films. These include magnetic recording tapes, pressure sensitive tapes and cable wrapping. The use of polyester film for recording tapes allows longer playing time per given reel size because it can be used in thinner gauges due to its high strength. There are many other electrical uses such as the manufacture of small capacitors. Polyester film's higher breakdown strength means that smaller capacitors can be made for a given voltage when it is used instead of paper as the dielectric. Vacuum metallization can lead to even further miniaturization. Electric motors, too, can be insulated with a single layer of polyester film. Its dimensional stability, allied to its insulation value, makes it suitable as a base for flexible circuits when laminated to copper foil.

The surface of polyester film can be roughened mechanically, or modified by lacquering, so as to accept pencil and ink marking. It is then used in the manufacture of drafting materials for original drawings or for tracing. Polyester film is particularly suitable because of its durability, high tear resistance, and dimensional stability. It has also been used for carbon 'papers' and for electrical typewriter ribbons where a clean release and a sharp impression of the type is important.

Polyester films have a number of decorative uses, especially when metallized by vacuum deposition methods. Metallized polyester film is laminated to PVC, for instance, and used in the manufacture of handbags, shoes and decorative trim for cars, etc. Lacquered metallized film is also slit into very narrow strips (about 1 mm wide) and then used as textile threads.

INJECTION MOLDING

A range of thermoplastic polymers have been used in injection molding including poly(butylene terephthalate)—PBT—and poly(tetramethylene terephthalate)—PTMT. PBT is made by substituting 1,4-butanediol for ethylene glycol.

Thermoplastic polyesters for molding applications have certain key properties: dimensional stability, extremely low moisture absorption, chemical resistance, excellent surface lubricity and resistance to wear, high fatigue endurance limit, heat deflection temperatures to 215 °C, excellent mechanical properties, stable electrical properties in a wide range of environments, and outstanding rigidity.

The physical properties of a commercial range of thermoplastic polyester molding materials are given in Table 15.1.

STRETCH BLOW MOLDING

The melt polycondensation process described under the heading, FIBERS, produces amorphous PET. This product is not suitable for the molding of food containers for a number of reasons, one of which is the inherently high acetaldehyde level which can affect the taste of some foods. There are other impurities, too, that might promote degradation during processing, and amorphous polyester resin tends to fuse and form lumps in the drying hopper. The polymer is upgraded, therefore, by crystallizing the chips to avoid subsequent sticking, and then dried to reduce hydrolysis at high processing temperatures. The process can be a batch one (high temperature under vacuum) or continuous (high temperatures in a nitrogen or dry air stream).

Oriented PET has a high tensile strength and is well able to resist the pressures generated by normal carbonated soft drinks (50-60 psi). Gas barrier properties are superior to those of PVC, polycarbonate, and the polyolefins but are not as good as those of polyacrylonitrile. It is also a good barrier to water vapor. The softening point is high, the material melting over the range of 243–270 °C. It has a high thermal and thermo-oxidative stability, is resistant to many chemicals and has good sparkle and optical clarity (comparable to glass). The polymer must be dried to a moisture content of 20 ppm before molding to minimize hydrolytic degradation.

Although gas barrier properties are good there are areas where improvements in carbon dioxide retention and resistance to oxygen ingress would open up large areas of business. It is known that PET bottles of 1-liter or above can retain carbon dioxide to an acceptable extent (not more than 15% lost over about 4 months). As bottle size is reduced, the surface area/volume ratio increases and makes this limit difficult to achieve. In addition, beer is affected by the presence of traces of oxygen. Ingress of oxygen in small bottles of PET is sufficient to make them unacceptable for beer bottling. Better gas barrier properties can be achieved by:

(a) modifying the PET polymer to give inherently better barrier properties
(b) using thicker walled bottles
(c) producing a multi-layered bottle via dip coating, co-injection molding or coextrusion.

The first solution is the neatest and work has already been carried out. Using thicker walled bottles suffers from the disadvantage that carbon dioxide is lost, not only by permeation but by solution in the bottle wall. There is an equilibrium between the two that limits the thickness that can be employed. As far as multi-layered bottles are concerned, an in-line co-injection stretch blow machine has been developed using, for example, PET/acrylonitrile copolymer bottles. Dip coating, using PVDC lattices, has also been used on commercial bottles.

Resin viscosity is important as it directly controls the minimum or natural

Table 15.1

TYPICAL PROPERTIES OF VALOX THERMOPLASTIC POLYESTER

Property	*Units*	*Test Method*	*"Valox" 310 (unreinforced)*	*"Valox" 310-SEO (unreinforced and UL 94V-0 recognized)*	*"Valox" 420 (glass-reinforced)*	*"Valox" 420-SEO (glass reinforced and UL 94V 0 recognized)*
Specific gravity (73°F)		D792	1.31	1.41	1.52	1.58
Specific volume	cu. in./lb		21.1	19.6	18.1	17.5
Water absorption (73°F)	%	D570				
24 h			0.08	0.08	0.06	0.07
equilibrium			0.34	0.38	0.26	0.25
Mold shrinkage	in./in./ x 10^{-3}		17–23	17–23	4–6	4–6
Tensile strength	psi	D638	8,000	8,900	17,300	17,000
Elongation at break	%	D638	300	100	5	5
Flexural strength	psi	D790	12,800	14,700	27,500	27,000
Flexural modulus	psi	D790	340,000	380,000	1,200,000	1,200,000
Compressive strength (10% def.)	psi	D695	13,000	14,500	18,000	18,000
Shear strength	psi	D732	7,700	7,700	8,900	9,000
Izod impact strength	ft lb/in.	D256				
notched			1.2	0.9	2.2	1.8
unnotched			No break	No break	15	15
Gardner impact	ft lb	Falling Dart	40–45	35–40		
Chemical resistance			Excellent	Excellent	Excellent	Excellent
Deformation under load (122°F, 2000 psi)	%	D621	0.17	0.15	0.12	0.15
Creep (73°)	%	D674	1.1(1000 psi)	1.1(1000 psi)	0.44(4000 psi)	0.44(4000 psi)
Fatigue endurance limit (10^7 cycles)	psi	D671	2,850	3,500	5,000	5,000

Property	Units	Test				
Coefficient of friction (self		D1894	0.17	0.16	0.16	0.16
metal)			0.13	0.14	0.14	0.14
Rockwell Hardness (R scale)		D785	117	120	118	119
Tabor Abrasion (CS-17 wheel, 1000 g load)	mg/1000 cycles	D1044	6.5	10.0	9.0	11.0
Heat deflection temp.	°F	D648				
66 psi			310	325	420	420
264 psi			130	160	415	415
Coefficient of thermal expansion	in./in./°F x 10^{-5}	D696	5.3	4.7	3.3	3.5
Thermal conductivity	BTU/h/ft^2/°F/in.	C177	1.1	1.2	1.3	1.3
Flammability		UL94	94HB	94V-O	94HB	94V O
	in./min.	D635	1.0	0.0	1.3	0.0
Volume resistivity	ohm. cm X 10^{16}	D257	4.0	4.0	3.2	3.4
Dielectric strength,	volts/mil	D149				
short time, 1/16" thickness			590	750	750	750
short time, >1/10" thickness						
73°F, 50% R.H.			400	375	475	425
150°F, 50% R.H.			400	375	470	420
250°F, 50% R.H.			400	340	450	415
73°F, 100% R.H.			400	375	475	425
73°F, after immersion in 130°F water			400	375	440	425
Dielectric constant		D150				
100 HZ			3.3	3.3	3.8	3.8
10^6 HZ			3.1	3.1	3.7	3.7
Dissipation factor		D150				
100 HZ			0.002	0.003	0.002	0.002
10^6 HZ			0.02	0.02	0.02	0.02
Arc resistance	s	D495	190	63	130	80
Hot wire ignition, 1/16"	s	UL55	16	20	63	34
U.L. temperature index	°C(°F) @ 30 mils	UL guide				
elec. & mech. properties with impact		QMFZ2	120(248)	120(248)	140(284)	130(266)
mech. prop. w/o impact			140(284)	140(284)	140(284)	140(284)

draw ratio that has to be applied to obtain a nearly uniform wall thickness. Resins of relatively low viscosity (intrinsic viscosity less than 0.75) may be used for stretch blow molding of large bottles (2 liters or more) but for smaller bottles (1.5 liters or less) high viscosities are preferred. Melt viscosity can be increased, without affecting the intrinsic viscosity, by chemical means and these are being developed. However, altering the chemistry of PET can adversely affect other inherent properties such as mechanical and gas barrier properties.

Orientation of PET induces crystallinity, fixing the polymer chains into ordered positions. When the bottles are heated to temperatures of around 78 °C (the glass transition temperature of PET) the polymer chains vibrate and the bottles tend to distort. PET bottles are normally unsuitable, therefore, for steam sterilization or for hot filling above 75 °C. Annealing at temperatures above 140 °C overcomes this but there are problems such as loss or clarity, and of some mechanical strength in the thicker areas of the bottle. Special bottle blowing equipment has been developed using hot molds to heat-set the bottles and such bottles are said to be hot fillable at temperatures up to 85 °C. It is possible to produce bottles stable at 100 °C but higher cycle times are necessary to allow equilibrium to be reached and this increases the cost.

The catalyst system used in the polymerization of PET has an effect on bottle clarity. Catalyst systems used are usually based on antimony, germanium or a mixture of antimony/germanium containing compounds. All produce very clear bottles but germanium also adds an extra 'sparkle' to the bottle.

Although PET is an inert polymer, degradation of the molten polymer leads to the formation of acetaldehyde. This is not toxic in the amounts likely to be found in PET bottles but it can taint the taste of certain critical drinks such as colas. Much work has been carried out, and is continuing, to reduce acetaldehyde levels in the finished PET bottles. Processing temperatures are limited by a lower figure in terms of bottle clarity, and an upper figure in terms of acetaldehyde production. The specification set by cola manufacturers is 3 micrograms per liter. Taking into account other restraints the ideal processing temperature range is said to be between 277 and 282 °C. Residence time is equally important, acetaldehyde generation being almost directly proportional to melt residence time.

COPOLYMERS

Copolymers can be made by using more than one dibasic acid and/or more than one glycol. One dibasic acid used as a comonomer is isophthalic acid. This is prepared from m-xylene and gives resins with higher than normal molecular weights. These resins exhibit superior strength characreristics, toughness and increased heat resistance. The use of isophthalic acid also reduces the rate and degree of crystallization. This, in turn, widens the processing range of the blow molding equipment.

Glycols available as comonomers include neopentyl glycol, cyclohexane dimethanol, and diethylene glycol.

SUGGESTED FURTHER READING

'Characteristics of PET Polymers for Bottle Manufacture'. W.Wick. SPE International Seminar on PET Bottles, held on 20-21 May 1980 in Antwerp (pp 2.1-2.20)

'Principles of Manufacture and Fundamental Properties of PET'. J.M. Leigeois. ibid (pp 8.1-8.16)

Polyesters: Volume 1. Saturated Polymers (1965) I. Goodman and J.A. Rhys. London. (Plastics and Rubber Institute Monograph)

16.

HIGH-PERFORMANCE THERMOPLASTICS

In recent years, the light and nonferrous metal industries have met with keen competition from plastics, particularly in the building and automotive industries. Some of the advantages of plastics, as compared with some of the better-known metals, are their excellent resistance to corrosion, nonrequirement of highly skilled methods of finishing, and comparatively low price per unit volume. Aluminum, in particular, has been rivaled by polyethylene, the modifed polystyrenes and polyvinyl chloride. The introduction and development of other thermoplastics, dealt with in this chapter, has made available materials which are suitable for high performance applications that hitherto have been restricted to metals.

POLYACETALS

Polymers based on formaldehyde have been known since 1859 when they were first described by Butlerov. Formaldehyde is a very reactive material yet the initial polymers obtained from it were of low molecular weight and melted with rapid decomposition at about 150 °C. Until recently, therefore, these polymers have been of little commercial use and were originally largely regarded as mere research curiosities. It has been suggested earlier that thermoplastic polymers produced from aldehydes are of no importance for such applications as fibers, surface coatings, plastics or rubbers due to their tendency to depolymerize rapidly. On polymerizing, this material may take one of the following forms:

Table 16.1

PROPERTIES OF POLYACETALS AND POLYCARBONATES

Test	*Method A.S.T.M.*	*Polycarbonate*	*Polyacetal homopolymer*
Specific gravity	D 792	1.2	1.425
Tensile strength (lb/in.^{2})	D 638	9000	10,000
Elongation at break (%)	D 638	60–100	15–75
Flexural strength (lb/in.^{2})	D 790	11,000	14,000
Impact strength (ft lb/in. of notch)	D 256	12–16	1–2
Thermal expansion (10^{-5} per °C)	D 696	7	8
Heat distortion temp. (66 lb/in.^{2}) °F	D 648	280–290	338
Volume resistivity (ohm-cm)	D 257	2×10^{16}	6×10^{14}
Dielectric strength (V/mil)	D 149	400	450
Dielectric constant (10^6 c/s)	D 150	3	3.7
Power factor (10^6 c/s)	D 150	0.01	0.004
Inflammability	D 635	self-extin guishing	burns very slowly
Water absorption % (24 h)	D 570	0.3	0.12
Mold shrinkage in./in.		0.006	0.025
Chemical Resistance			
Ketones		A**	R*
Water		R	R
Aldehydes		A	R
Esters		A	R
Aromatic hydrocarbons		A	R
Weak organic bases (aniline)		A	R
Ammonia		A	A
Weak acids		R	R
Strong acids		A	A
Weak alkalis		R	R
Strong alkalis		A	A

*R = resistant
**A = attacked

(i) Polyoxymethylene or polymethylene glycols

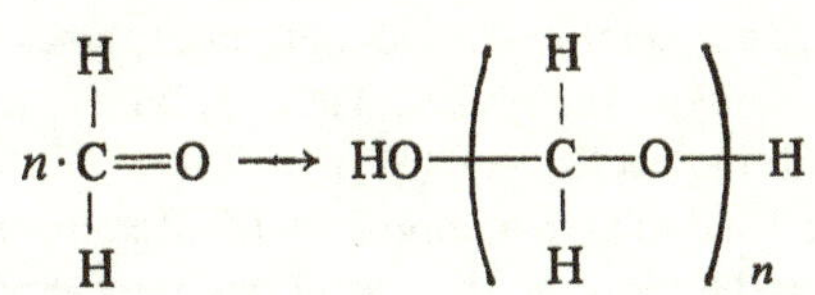

(ii) Sugar-type polymers

If formaldehyde is polymerized under mildly alkaline conditions, polyhydroxy compounds are formed:

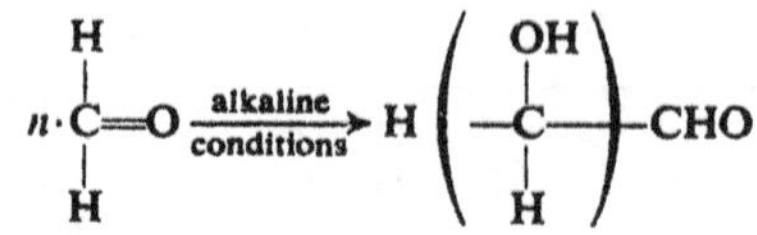

Most of the oxygen atoms are lateral to the main backbone in sugar-type polymers; they are of low molecular weight and are thought to have little or no industrial applications.

A great deal of research work has been carried out on these polymers throughout the past years, and early as 1920 Staudinger and coworkers produced formaldehyde polymers which exhibited plastic-like properties. It was established that these polymers were not of the sugar-type but consisted essentially of molecular chains with repeating oxymethylene groups. Unfortunately, although insoluble in cold solvents, they decomposed on heating. The fault of decomposition on exposure to heat has now been overcome and during the last twenty years a large chemical company, E. I. Du Pont de Nemours, has been trying to correct this drawback. Methods have now been devised for polymerizing formaldehyde to give high molecular weight polymers possessing high thermal stability. Either nucleophilic or electrophilic agents are used as initiators in forming industrial polymers.

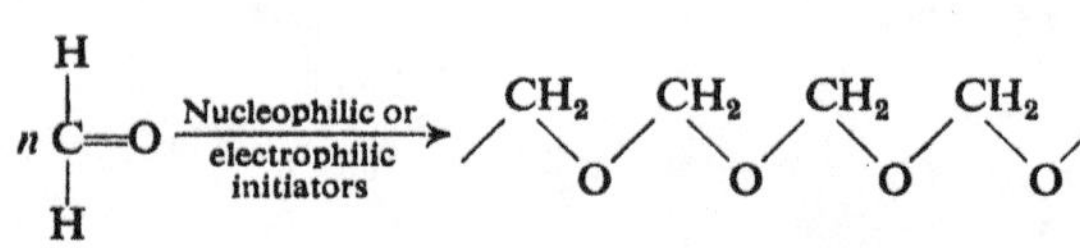

Polyformaldehyde
(mol. wt approx 20 000)

These are high molecular weight stable polymers with good mechanical strength. This latter property results from the ability of the oxymethylene ($-CH_2O$) chains to pack together into a highly ordered crystalline configuration as the polymer moves from the molten to the solid state.

The types of initiators used are: simple acids and bases, organic acids, Lewis acids, amines, organo-metallic compounds, phosphines, arsines stilbenes, or carbonyls of iron, cobalt, and nickel. These polymers may be regarded basically as linear, with molecular weights in excess of 20,000. The end hydroxyl group may be esterified to give a polymer with acetate end groups, which give it greater thermal and hydrolytic stability. For a typical polymerization method for formaldehyde, the reaction medium can be hexane and the initiator may be triphenyl phosphine.

A later development in this field was the introduction by the Celanese Corporation of America of a copolymer having formaldehyde as the major constituent and a cyclic ether such as ethylene oxide as the minor one. Another variation is the polymerization of trioxane, a cyclic trimer of formaldehyde, to give acetal resins. Such a polymer has a highly crystalline structure and the C—C double bonds which are randomly distributed in the polymer chain from the comonomer help to stabilize the polymer against thermal oxidative and acidic attack. If these copolymers are heated to high temperatures, oxidation occurs and the depolymerization which generally results in such a reaction is cut short at the —C—C— bond which both account for the high level of resistance of these polymers to attack by strong alkalis.

These polymers are known as acetal resins and are manufactured in the U.S.A. under the trade names of Delrin* for the homopolymer and Celcon** for the copolymer. Their unique properties are regarded as a bridge between plastics and metals, having, in fact, replaced metals in a large number of important mechanical structural applications. The metal-like properties exhibited by these polymers make them high performance materials, with high strength and rigidity, excellent dimensional stability, resilience, and toughness.

Properties

The properties of polyformaldehyde homopolymer are tabulated in Table 16.1 and the copolymer in Table 16.2 from which it can be seen that the family is one of the toughest and most resilient of the known thermoplastics. The load-bearing characteristics of the acetal polymers, in both compression and tension, are impressive compared with conventional thermoplastics. Other outstanding properties are their resistance to repeated impact and high abrasion resistance. The physical properties are virtually unaffected by continuous service at 80 °C or intermittent use at 120 °C and it remains dimensionally stable over a wide range of operating temperatures. These characteristics make the polyacetals attractive for a wide variety of industrial applications.

They are excellent dielectric materials and retain their electrical characteristics under conditions of high humidity, factors which, in combination with their high mechanical properties, make them of particular value for electrical applications where high strength, resistance to abrasion, dimensional stability and

* Trademark of E. I. Du Pont de Nemours & Co. Inc.

** Trade name of Celanese Corp. of America.

Table 16.2

PROPERTIES OF HIGH PERFORMANCE THERMOPLASTICS

		UNREINFORCED RESINS					
Property	*ASTM Method*	*Acetal Copolymer*	*Acetal Homopolymer*	*Styrene Modified PPO*	*Polycarbonate*	*G.P. and H.S. Nylon 66*	*Thermoplastic Polyester*
Specific gravity (g/cm^3)	D 792	1.410	1.425	1.06	1.20	1.14	1.31
Tensile strength: @ yield (psi)	D 638	8800	10,000	9600	9000	8500*	8000
Elongation: @ break (%)	D 638	60	25	25	110	300*	300
Tensile modulus (psi $\times 10^{-5}$)	D 638	4.10	5.20	3.55	3.45	–	–
Flexural strength (psi)	D 790	13,000	14,100	13,500	13,500	6000*	12,000
Flexural modulus (psi $\times 10^{-5}$)	D 790	3.75	4.10	3.60	3.40	1.85*	3.30
Fatigue endurance limit (psi/no. of cycles)	D 671	4200/ 10MM	5000/ 1MM	2500/ 2MM	1000/ 2.5MM	–	2850/ 10MM
Compressive stress, psi: @ 10% Def.	D 695	16,000	18,000	16,400	12,500	–	13,000
Rockwell Hardness (M)	D 785	80	94	78	70	61*	78
Notched Izod Impact (ft-lb/in. of notch)	D 256	1.3	1.4	5.0	14	2.0*	1.0
Tensile Impact (ft-lb/in.^{2})	D 1822	70	100	120	–	475*	–
Shear strength (psi)	D 732	7700	9500	10,500	10,000	8200*	7700
Water absorption: 24 h Immersion (%)	D 570	0.22	0.26	–	0.15	1.50	0.09
Tabor abrasion: 1000 g load, Cs-17 wheel (mg/1000 cycles)	D 1044	14	20	20	10	5*	6.5
Coefficient of dynamic friction against:	D 1894						
Steel		0.15	0.10–0.30	0.35–0.40	–	0.15	0.13
Brass		0.15	–	0.38–0.49	–	0.11	0.13
Aluminum		0.15	–	–	–	0.12	0.13
Self		0.35	–	0.24–0.30	–	0.24	0.17
Mold shrinkage:	–						
Average (mils/in.)		20	20	5–7	6	16	20
Flow Dir. (mils/in.)		22	22	–	–	–	23
Transverse Dir. (mils/in.)		18	18	–	–	–	17

*Value obtained on specimens containing equilibrium moisture content at 50 percent R.H. (73°F).

good high temperature performances are required. The polyacetals also exhibit excellent resistance to organic solvents, although resistance to strong acids, bases, and oxidizing agents is somewhat limited for the homopolymer. They are also unaffected by ultraviolet light and are nontoxic.

The Celcon resins are self-lubricating with good resistance to abrasion and hot water exposure. They are highly resistant to gasolene, motor oils, and greases even at elevated temperatures. Specimens of this copolymer can be immersed in boiling 50 percent sodium hydroxide solution and other such bases for many months and show no change. However with strong acids prolonged contact should be avoided to prevent breakdown.

The polyacetals offer a combination of properties which make them important as engineering plastics. These include:

1. High rigidity combined with good short-term and long-term mechanical strength.
2. Good performance in mechanical applications combined with stability of properties and dimensions under changing ambient conditions.
3. Excellent resistance to organic chemicals and water combined with a wide operating temperature.

Processing

The acetal polymers may be fabricated with the same ease as is normally expected with the conventional thermoplastics, with consequent cost advantages over metals. The normal molding range is between 200–230 °C and the best performance is achieved on machines with fast ram speeds and high available pressure. It has been reported that one-piece integral shapes produced on injection molding do not normally require finishing and this, of course minimizes the number of rejects. Rod, tube, and sheeting can be extruded from the polyacetals on conventional equipment. Extruders with a high length to diameter ratio (20:1) are recommended, along with a metering screw with a constant-pitch feed section. Other methods of forming, such as blow molding, may be used and both thin-walled and thick-walled containers may be produced. Thick-walled containers as required for aerosols are now manufactured fom these polymers, and many are already on the market, replacing the traditional metal ones. Molded components can be machined quite simply and they may also be welded, painted, metallized, and printed on.

Applications

Typical uses that have already been found for this versatile family of plastic include ring bearings and cams, venturi-tubes, automobile instrumental housings, door-knobs, pump and air impellors, hinges, shoe heels, mechanical toys, and window catches. Their use as aerosol containers is now established and other miscellaneous uses include pen and pencil parts, bottle closures, and business machine housings.

These polymers are sold in the medium price range and because of the exceptional resistance to hot water exposure they are being widely used for plumbing fittings. Added to this characteristic is a high level of corrosion resistance with a smooth self-lubricating surface which obviously resists the build-up of scale and thus makes it eminently more suitable than metals for plumbing. Further still, this family of polymers costs less than brass on a cents per cubic inch basis and can be processed into finished parts by injection molding. There has been a continued expansion of applications for these polymers over the last few years. As stated previously, much of the market for these materials has been the replacement of metals of both the ferrous and the nonferrous types. Particularly die-cast zinc, aluminium, and brass have been replaced because of improved properties at reduced costs. Many other uses have been established as well as in the plumbing industry, particularly in the all-consuming automotive industry. The plumbing applications include such items as ball-cocks, shower heads, valves, pumps, and faucets.

POLYCARBONATES

These polymers may be regarded as linear polyesters of carbonic acid. As early as 1898 attempts were made by A. Einhorn to prepare cyclic carbonates from the reaction of hydroquinone with phosgene in an alkaline medium, but his work was not particularly successful in this respect since he obtained a product which was insoluble, and infusible. This material was probably a polycarbonate and the forerunner of the polymers now known today. He also reacted phosgene with resorcinol and obtained a polymer with a melting point around 200 °C.

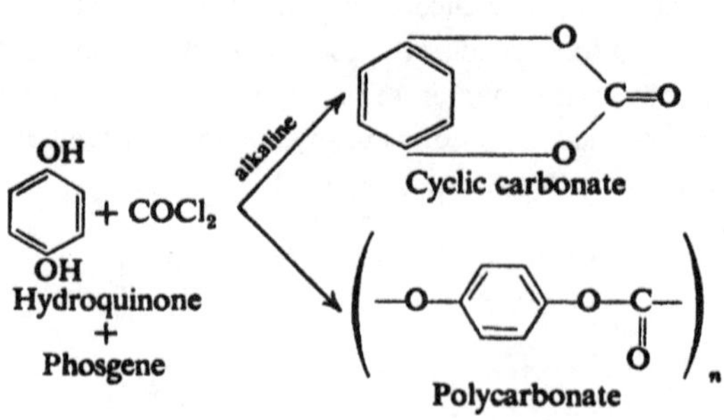

Bischoft and von Hedenstrom repeated Einhorn's work early in the twentieth century, preparing the same materials by means of an ester exchange reaction. Further research was carried out both in Germany and the U.S.A. but it was not until 1956 that the polycarbonates were first reported as useful plastic materials. The polymers of commercial importance now manufactured are produced from bisphenol A and a diphenyl carbonate. The preparation of bisphenol A has been described in Chapter 6.

Commercially available materials include Makrolon (Bayer-Germany), Lexan (General Electric—U.S.A.), and Merlon (Mobay—U.S.A.).

Structure

H. Schnell in his 1976 Swinburne award address to the Plastics and Rubber Institute in the U.K., discussed and described the structure of a bisphenol A polycarbonate. He found that this polymer is capable of crystallization but it is greatly impeded. Moldings can be produced from this polymer which are either (a) x-ray amorphous, (b) x-ray amorphous and oriented, (c) crystalline, or (d) oriented and crystalline.

Manufacture

There are at present three possible processes used in manufacturing polycarbonates:

(i) Ester-interchange. This process involves the heating of an aromatic dihydroxy compound such as bisphenol A with a metacarbonate such as diphenyl carbonate. The reaction is carried out by heating the reactants to 200–300 °C, at the same time reducing the pressure to 20–30 mm Hg to remove the phenol, and finally raising the temperature to 290–300 °C. The process is a simple ester interchange which may be catalyzed by a basic material such as sodium hydroxide. On completion of polymerization the resultant polymer is extruded from the reaction kettle under an inert gas. The reaction must be stopped when the polymer is sufficiently liquid to be extruded from the resin kettle.

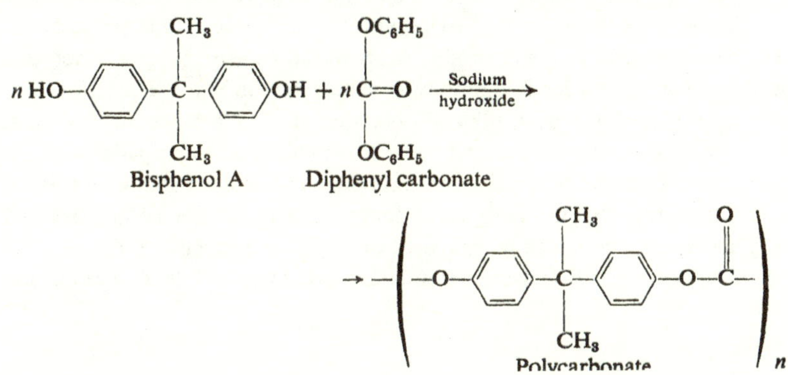

(ii) Phosgenation Process. This process resembles the original method used by Einhorn in preparing the first known polycarbonate. Basically, it consists of reacting the disodium salt of bisphenol A with phosgene, when a condensation reaction takes place. The bisphenol A is suspended or dissolved in sodium hydroxide and the reaction is carried out in the presence of a solvent such as methylene chloride or pyridene which is capable of dissolving the phosgene as well as the polycarbonate which is formed. This is followed by an addition of an equimolar quantity of phosgene at 20–30 °C. The resultant product is a solution of low molecular weight polycarbonate which may then be converted to a higher molecular weight by continual stirring or by the addition of catalysts such as tertiary amines ot quaternary ammonium bases.

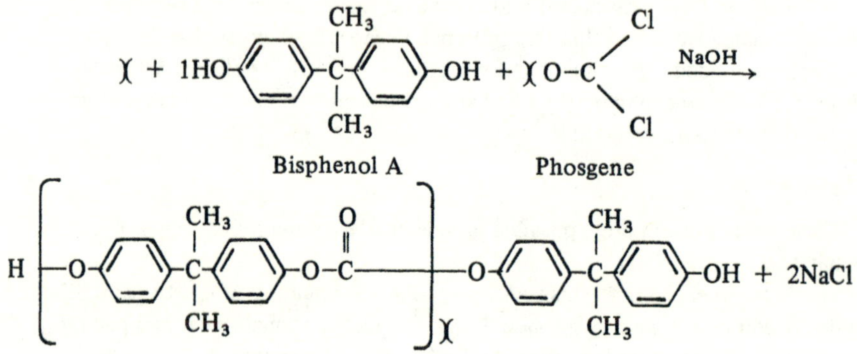

There are at least two mols of pyridine per mol of bisphenol A.

(iii) Interfacial Polycondensation Process. This method has achieved great significance but only recently became a practical proposition when it was discovered that the reaction could be catalyzed by the addition of tertiary amines. This reaction must be kept to a minimum to prevent the saponification of the phosgene. The viscous solution so formed must be washed with water and a good recovery of the solvent must be made to be an economic process.

Of the methods of polymer production it would appear by comparison that the phosgenation process is the more advantageous in that the reaction proceeds quickly with the possibility of obtaining very high molecular weights. Disadvantages include the necessity to recover solvents, and separation of inorganic salts from the polycarbonate solutions. The ester interchange method has the advantage that the inorganic salts at this stage can be completely and easily separated, and it also produces an easily formed granular polymer. The main disadvantage is that the reaction is carried out at high temperatures and in vacuum, which leads to the formation of unwanted side reactions. It is difficult to choose between the two methods of production but all commercial processes known today start from bisphenol A.

Properties

The properties of the polycarbonates are listed in Table 16.1.

A polycarbonate does not have a sharp melting point but begins to soften above the glass-transition temperature of 149 °C. It then goes into a rubbery state until between 200–220 °C, when it passes into the truly thermoplastic state. These materials have fairly high molecular weights and this coupled with high melt viscosity imposes a limitation as to what molecular weight types can be commercially exploited. Their excellent mechanical properties make it possible to mold high-precision structural articles with a combination of hard, rigid, and resilient characteristics, they have good thermal properties, and they have excellent electrical insulating properties, as well as transparency. Other useful properties include low water absorption, the ability to self-extinguish and fair chemical resistance to aqueous solutions. Their resistance to organic bases such as ammonia, however, is limited. The polymers are readily soluble in chlorinated hydrocarbons, are partially soluble in aromatic hydrocarbons and ketones, and are decomposed by strongly alkaline substances.

The outstanding features of thse polymers are their clarity and high impact strength as well as high thermal stability, which is an important fact in processing; this is, of course, only true if moisture is absent.

Processing

Before processing polycarbonates it is essential for the base polymers to be absolutely dry. Although they possess relatively high melt viscosity they can still be processed on all conventional types of molding machines and may be injection molded at temperatures between 240–300 °C. Extrusion can also be carried out satisfactorily at this temperature and good results have been obtained with single, flighted screws of hardened steel with equal pitch and increasing core diameter. Rod, pipe, sheeting, and film can be easily processed by extrusion. Coatings may be cast from solution or applied by fluidized-bed techniques. Molded parts are easily finished by such operations as drilling, turning, milling, planing, sawing, punching, and polishing. It can also be metallized, printed, and heat-sealed.

Applications

This material has found many uses as a film and molding material, and the potential industries include electrical, building, automobile, and lighting. Applications include tail lights for cars, parts for x-ray equipment, holders for fluorescent tubes, bowls and measuring cups, to mention but a few, and household utensils which are unbreakable and easy to clean. In the medium price range the polycarbonates compete with better known thermoplastics and with some metals, and glass.

Unlike many thermoplastics the parts made from a polycarbonate resin are 100 percent dishwasher-proof. In the kitchen the transparency of glass is a very important feature but the hazards make it potentially dangerous. The impact strength of polycarbonates is fifty times greater than glass which means a dramatic reduction in breakages with the added advantage of safety. Combining identical parts made of glass and polycarbonate, it can be seen that the plastic weighs one-third of its counterpart. Processing is also easier with the plastic although the transparency of light transmission is better (91 percent), the polycarbonate has a figure of 89 percent and this difference is of very little significane.

A typical vacuum cleaner made of polycarbonate is 50 lighter than its metal counterpart. It offers strength to weight ratios equal to die cast aluminum, twice that of zinc with four times greater impact strength than either metal. Whereas metals are conductors the polycarbonates have high dielectric strength.

The field of packaging is also opening up new applications to polycarbonate. Coextrusion makes posssible the use of polycarbonate in small quantities so that high performance can be obtained at reasonable cost. Autoclavable medical trays and medical blister packs are two applications which benefit from polycarbonate's clarity and heat resistance. Multilayer coextrusion blow moldings based on polycarbonate offer the possibility of producing clear and strong substitutes for glass bottles. Possible structures include PC/tie layer/PP and PC/tie layer/ PET/tie layer/PC.

Polysulfones

Another thermoplastic which saw its introduction onto the market in the 1960's is the family of resins based on polysulfone. These polymers were developed as a result of the work carried out by the Union Carbide Corporation in the preparation of new aromatic polyethers by nucleophilic aromatic substitution. The synthesis involves the condensation of a dialkali metal salt of a dihydric phenol with an 'activated' or negatively substituted dihalide or anhydrous dipolar aprotic solvent at elevated temperatures. The most commercially successful of these materials from Union Carbide can be prepared by this method and is based on the reaction of the disodium salt of bisphenol A [sodium salt of 2,2 bis (4-hydroxyphenyl) propane] with 4,4′-dichlorodiphenyl sulfone in dimethyl sulfoxide. The sodium phenoxide end groups react with methyl chloride to terminate the polymerization. This enables the molecular weight of the polymer to be controlled and at the same time assisting in thermal stability.

From the following reaction it can be seen that the diaryl sulfone grouping is a feature of the molecular chain structure.

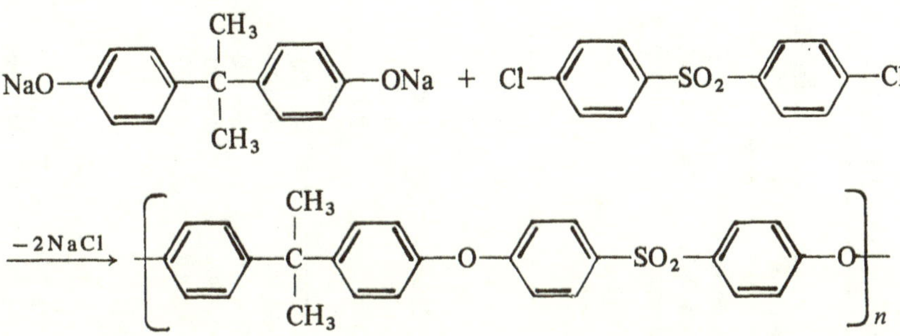

This is a highly resonating structure and the sulfone group tends to attract electrons from the phenyl rings. Oxygen atoms are in the para position to the sulfone groups which enhances resonance and this tying-up of the electrons imparts good resistance to oxidation of the polymer chain.

The bond strength is increased and the grouping is spacially fixed into a planar configuration which is due to the high degree of resonance. This leads to good thermal stability and the chain exhibits rigidity even at relatively high temperatures.

Properties

Currently this family of polymers have been given the trade name of Udel by Union Carbide. They are a family of tough, rigid, high strength thermoplastics which maintain their properties over a wide temperature range from −65 °C to above 150 °C. The deflection temperature of this resin is 175 °C at 264 psi and 180 °C at 66 psi. Aging these resins for long periods at 150–200 °C has little or no effect on their electrical or physical properties. Even down to temperatures as low as −65 °C there is little deterioration in the properties of these

Table 16.3

TYPICAL PROPERTIES OF UDEL® POLYSULFONE

Property	*ASTM Methods*	*Value* P-1700	P-1710	P-1720	P-3500	P-6050
Appearance in Natural Form		Natural 11 or transparent light amber. Available in opaque and transparent colors.	Natural 15 or light beige opaque.	Natural 13 or light beige translucent.	Natural 11 or transparent light amber. Available in opaque and transparent colors.	Natural 15 or opaque off-white.
General						
Melt flow, at 650°F (343.3°C), 44 psi, g/10 min	D 1238	6.5	6.5	6.5	3.5	3.0–7.0[(a)]
Density (g/cm³)	D 1505	1.24	1.25	1.25	1.24	1.40
Mold shrinkage (in./in.)	D 955	0.007	0.007	0.007	0.007	0.005
Water absorption (% in 24 h)	D 570	0.3	0.3	0.3	0.3	0.26
Mechanical						
Tensile strength at yield (psi)	D 638	10,200	10,200	10,000	10,200	13,000
Tensile modulus (psi)	D 638	360,000	360,000	360,000	360,000	720,000
Tensile elongation at break (%)	D 638	50–100	50–100	50–100	50–100	4.0
Flexural strength (psi)	D 790	15,400	15,400	15,400	15,400	20,000
Flexural modulus (psi)	D 790	390,000	390,000	390,000	390,000	750,000
Izod impact (at 72°F, 18-in. specimen, ft-lb/in. notch)	D 256	1.3	1.3	1.3	1.3	1.5
Izod impact, at −40°F (1/8 in. specimen, ft-lb/in. notch)	D 256	1.2	1.2	1.2	1.2	–
Tensile impact, ft-lb/in.²	D 1822	200	200	160	200	43
Rockwell hardness	D 785	M69 (R120)	M69 (R120)	M69 (R120)	M69 (R120)	M75

Property	ASTM					
Heat deflection temp., at 264 psi (°F)	D 648	345	345	345	345	355
Coefficient of linear thermal expansion, in./in./°F	D 696	3.1×10^{-5}	3.1×10^{-5}	3.1×10^{-5}	3.1×10^{-5}	2.5×10^{-5}
Flammability (*)						
average time of burning (ATB), seconds	D 635	5	5	<5	5	10
average extent of burning (AEB)	D 635	0.4 in. (10 mm.)	0.4 in. (10 mm.)	<0.2 in. (<5 mm.)	0.4 in. (10 mm.)	<0.2 in. (<5 mm.)
	UL (File No. E-36098A)	← 1/16-in. thickness: 94V-2 →		0.040-in. thickness; 94V-1	–	–
		← 1/8-in. thickness: 94V-2 →				
		← 1/4-in. thickness: 94V-O →		1/16-in. and thicker: 94V-O		
Oxygen index rating	D 2863T	30	30	32	30	–
Thermal conductivity Btu/h/ft²/°F/in.	C 177	1.8	1.8	1.8	1.8	1.85
Electrical						
Dielectric strength, 1/8-in. specimen, S/T, v/mil	D 149	425	425	370	425	425
Arc resistance, s (tungsten electrodes)	D 495	122	122	60	122	–
Volume resistivity (at 72°F, ohm-cm)	D 257	5×10^{16}	5×10^{16}	5×10^{16}	5×10^{16}	5×10^{15}
Dielectric constant (at 72°F, 60 Hz-1 MHZ)	D 150	3.07–3.03	3.07–3.03	3.15–3.06	3.07–3.03	3.43–3.34
Dissipation factor (at 72°F, 60 Hz-1 MHZ)	D 150	0.0008–0.0034	0.0008–0.0034	0.0008–0.0056	0.0008–0.0034	0.011–0.005

(a) at 707°F (375°C).

* NOTE: This numerical flame spread rating is not intended to reflect hazards presented by this or any other material under actual fire conditions.

Table 16.4

EFFECT OF COMMON CHEMICALS ON UDEL© POLYSULFONE

Property	*P-1700*	*P-1710*	*P 1720*	*P-3500*	*P-6050*
Inorganic acids			No effect		
Alkalies			No effect		
Alcohols			No effect		
Esters			Partly soluble, swells		
Ketones			Partly soluble, swells		
Aliphatic hydrocarbons			No effect		
Aromatic hydrocarbons			Partly soluble, swells		
Chlorinated hydrocarbons			Dissolves		

resins. The electrical properties of polysulfones are good and are maintained over a wide temperature and frequency range. Added to these excellent characteristics is the fact that such resins also have low creep and cold flow properties. These materials differ from other amorphous thermoplastics in that their rheology is Newtonian and is not particularly shear sensitive. The glass transition temperature is 175 °C.

Polysulfone resins exhibit a high degree of resistance to mineral acids, alkalis, and salt solutions. They are also resistant to detergents and hydro-carbon oils even at high temperatures under moderate stress. Polar organic solvents such as ketones, chlorinated hydrocarbons, and aromatic hydro-carbons attack these polymers.

Processing

These resins must be thoroughly dried before any melt processing is attempted. While the presence of water has no effect on the overall characteristics, it causes bubbling in the finished molding.

These materials are moldable in most commercial injection molding machines. Mold temperatures are about 95 °C for the moldings and 160 °C for thicker components with stock temperatures around 340–400 °C. They have been extruded as sheet and thin film, pipe, tubing, and other profiles as well as wire insulation. The temperature in extrusion is higher than normal for most thermoplastics and typically within the range of 300–360 °C which is the extrusion melt temperature. Typical properties are given in Tables 16.3 and 16.4.

Applications

The Udel polysulfones can be used in a wide variety of applications because of a good balance of mechanical, electrical, and thermal properties. They have good dimensional stability and offers the user many opportunities. These include: electrical-electronic (circuit carriers, terminal blocks, brush holders), domestic appliances (coffee makers, parts for cameras), automotive and aircraft (cabin interior products).

Polyphenylene Oxide

This polymer was introduced to the market in the middle 1960's by General Electric Company, U.S.A. under the trade name Noryl. It was made by a patented process from the oxidative coupling of phenolic monomers and currently the manaufacture consists of the oxidation of 2,6-xylenol using a copper catalyst. The basic phenylene oxide structure is as shown:

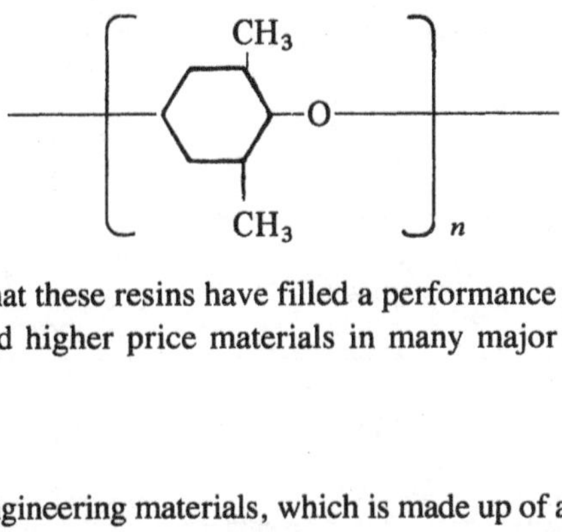

It is claimed that these resins have filled a performance gap between commodity plastics and higher price materials in many major markets.

Properties

This family of engineering materials, which is made up of at least eight different grades, is characterized by outstanding dimensional stability at elevated temperatures with a broad temperature range for use (−5° to above 150°C). It has a low coefficient of expansion with a low creep figure, a high hydrolytic stability, and excellent dielectric properties over a wide range of frequencies and temperatures.

Unlike many other thermoplastics, repeated autoclave cycling of hot and cold water or prolonged exposure to steam have little effect on the properties of these resins. They are an ideal choice for applications under load or elevated temperatures because of the heat resistance under high levels of stress. The properties of an unfilled polymer N300 and a filled RN300 are given in Table 16.5 and a range of properties of a modified polymer can be found in Table 16.2.

Table 16.5

PHYSICAL PROPERTIES OF POLYPHENYLENE RESINS

Properties		*N300*	*RN300*
Specific gravity	73°F	1.06	1.26
Water absorption	24 h; 73°F; %	0.006	0.005
Heat deflection temperature	°F@264 psi °F@ 66 psi	300 315	300 315
Thermal conductivity	Btu/h/ft²/°F/in.	1.10	1.15
Flammability	1/16″	UL 94 V-O	UL 94 V-O; V-5
Tensile strength (psi)	73°F 150°F 200°F 250°F	11,000 8,600 7,500 5,400	12,000 10,000 8,700 7,500
Elongation at break	%	20	15
Flexural strength (psi)	73°F 150°F 200°F 250°F	15,100 12,200 9,900 7,300	17,500 14,700 12,700 10,500
Flexural modulus (psi)	73°F 150°F 200°F 250°F	350,000 325,000 305,000 275,000	575,000 540,000 520,000 500,000
Izod impact Notched–ft lb/in. 1/8″ Specimen Unnotched–ft lb/in.	 73°F –40°F 73°F	 10.0 3.2 No Break	 3.0 1.5 45.0
Gardner impact in.-lb	73°F	300+	250
Mold shrinkage in./in. y 10^3		.005–.007	.002–.004

Polyphenylene oxide resins dissolve or soften in many halogenated or aromatic hydrocarbons. They are resistant to hot and cold water, detergents, and weak and strong acids and bases.

Processing

All types of injection molding machines can be used with these resins although screw types are preferred over the plunger or combination presses because of the rapid uniform heating of the resin. General purpose screws with *L/D* ratios up to 24/1 and compression ratios 2/1 to 3/1 are satisfactory although vinyl screws and tips are not recommended.

Typical mold temperatures range from 85–105 °C and the cylinder temperatures, 255–300 °C. These resins can be painted with a variety of attractive finishes such as alkyd, acrylic, vinyl, or epoxy base paint. Both dry offset and silk screen printing can be used to achieve exact durable image reproduction on these resins. Other processing techniques which can be used successfully are metallizing and electroplating.

Applications

These resins have found increasing use in several major areas; these include appliances (parts for cameras and projectors), electrical-electronic components (switches and terminals), furniture (institutional, such as nurseries or hospitals where stringent conditions are required), automobile, business machines, and components for the handling of liquids such as valves and pumps.

Poly(etherimide)

This polymer was also developed by General Electric Company, U.S.A. It was introduced commercially in 1982 under the trade name Ultem. It is an amorphous thermoplastic and is said to combine the high performance usually associated with exotic speciality plastics, and the excellent processing characteristics typical of engineering plastics.

Properties

Poly(etherimide) combines a number of properties that are important to the design engineer. It has high heat resistance, for example, coupled with excellent strength and flexural modulus. It is inherently flame resistant and has low smoke evolution. Its continuous use temperature rating is 170 °C.

Poly(etherimide) is an excellent electrical insulator, with a dielectric strength of 242kV/mm at 1.6 mm and a high volume resistivity (6.7×10^{15} ohm-m).

Chemically, it is unaffected by most hydrocarbons although it is soluble in partially halogenated hydrocarbons such as methylene chloride and trichloroethane. It is resistant to mineral acids, mineral salt solutions and dilute bases (pH less than 9).

Processing

Ultem resin is ideal for precision injection molding of intricate, close tolerance parts. It is also said to be easily processed by extrusion, blow molding and structural foam techniques, requiring no special equipment. For optimum properties and part appearance, Ultem resin must be thoroughly dried before processing (to a moisture content of less than 0.05%.) Typical mold temperatures range from 65 to 175 °C with cylinder temperatures of from 340 to 425 °C.

Applications

Applications for Ultem resins include a high voltage circuit breaker housing and a reflector in a medical lamp. In the latter, the risk of breakage during operating procedures is reduced and the resin provides crucial infra-red transparency and high heat resistance (up to 200 °C) under load.

Other actual or suggested uses are high temperature connectors, electric motor parts, integrated circuits, roller bearing cage and gears, car under-the-hood applications and knife-fuses.

Poly(etheretherketone) — PEEK

This is one of the latest high performance thermoplastics and is suitable for injection molding and extrusion. It was developed initially as a coating and insulation material for high performace wiring and is also showing promise in injection molding, monofilaments, fibers, films and composites.

PEEK has a number of high performance properties such as high melting temperatures (334 °C), high service temperature (200 °C) and high temperature cut-through and abrasion resistance. All this is combined with low flammability and fume emmission, and chemical and solvent resistance.

Glass fiber and carbon fiber reinforced grades are also available, offering heat distortion resistance at temperatures up to 300 °C.

17.

HEAT-RESISTANT THERMOPLASTICS

Today's technology is making increasing demands for materials which can be fairly readily processed but which can also stand up to high temperatures, not only without melting but with the maintainance of adequate mechanical and electrical properties. Supersonic aircraft, particularly those operating at Mach-2 or higher, need materials that can survive temperatures of 300°C for the construction of windshields, radomes, and door seals. Mach-3 or Mach-4 could demand materials that maintain adequate strength at temperatures that are as high as 1000°C for short periods. Aerospace uses. too, need extremely high performance materials, while in the electrical field the discovery that electric motors function more efficiently at higher operating temperatures has led to the need for insulation with greater thermal stability.

These, and a great many other requirements for high temperature uses have stimulated extensive research programs to discover plastics with greater heat resistance and a number of materials have been developed with melting points up to about 600°C. Many of these can survive temperatures well above 300°C and some can retain their form and strength (as films or fibers) for periods of up to several months.

Apart from finding chemical structures with an increased resistance to bond disruption by heat, improvements can also be made by increasing the stiffness of the polymer chain so that the rigidity of the polymer is retained at higher temperatures of use. The replacement of carbon atoms in the main polymer chain by inorganic atoms or groupings is one way of producing structures with greater heat resistance; those materials are dealt with in Chapter 22. Another method is to incorporate the aromatic ring structures in the polymer chain. Not only do they increase the rigidity of the polymer chains but they also possess intrinsically stronger bonds that are more resistant to heat. The addition of side chains at regular intervals is another way of increasing heat resistance because the side chains tend to interlock mechanically and thereby reduce the tendency of the molecules to slide over one another when heat energy is applied. The addition of bulky side chains also helps the polymer retain its mechanical properties at elevated temperatures because the very bulk of these side chains prevents the polymer chain from folding easily.

If heat resistant thermoplastics are taken to be those that are melt processable and have a continuous service temperature of 150°C or more, then of the thermoplastics already discussed in earlier chapters only the fluorocarbon polymers (Chapter 14) can really be classified as heat resistant. Other thermoplastics that have been developed and meet the needs for high temperature performance are discussed below.

AROMATIC POLYAMIDES

These are polymers in which aromatic rings have been incorporated into typical polyamide chains giving structures similar to the one below.

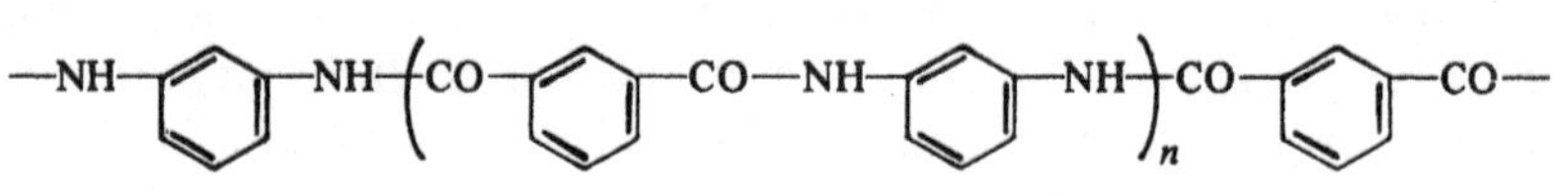

Fibers made from aromatic polyamides have been used in applications involving exposure to temperatures of up to 300°C. The retention of mechanical strength is good and examples of 50 percent retention of room temperature strength at temperatures of 285°C are known. Although aromatic polyamide fibers are not completely nonflammable, ignition occurs only after prolonged exposure to a direct flame and does not continue after removal of the flame. One successful use is in flame protective clothing while another is in conveyor belts for carrying articles through high temperature ovens.

In electrical applications, aromatic polyamides are used for the insulation of motors and generators operating at temperatures of up to 220°C. The success of aromatic polyamides in such applications can be judged from the fact that failure in these motors often occurs because of failure of some metal component rather than the aromatic polyamide insulation.

POLY (AMIDE-IMIDE)

Amide-imide polymers are based on the combination of trimellitic anhydride with aromatic diamines. The general chemical structure may be represented as follows.

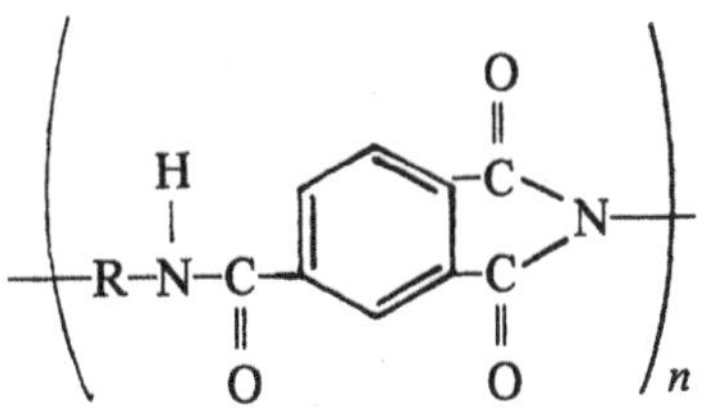

Amide-imide polymers are available as solutions for use as high-temperature wire enamels, high performance decorative coatings, laminates, and adhesives. They are also obtainable in powder or pellet form and can be injection molded, compression molded, or extruded on standard commercial equipment. The solutions referred to above are applied to the appropriate substrate and then thermally cured to produce insoluble heat-resistant coatings. As wire enamels, amide-imide polymers have received an IEEE 57 thermal stability rating of 225°C for 20,000 hours and exhibit good resistance to thermoplastic flow, heat shock, and burnout. Decorative coatings can be produced in a variety of colors and exhibit good resistance to high temperatures while adhesives give high-temperature lap shear strengths.

Solid parts can be fabricated with properties such as high tensile, compression, and flexural strength and these properties are maintained at high temperatures—of the order of 260°C. Impact resistance is also high, as is resistance to creep. The development of optimum physical properties in fabricated parts requires post-mold curing at 260°C for 24 hours. Sprues and runners intended for reprocessing should be removed prior to the post-curing process.

Chemically, poly (amide-imide) is virtually unaffected by aliphatic, aromatic, and halogenated hydrocarbons and to most acid and base solutions. It is, however, attacked by some acid and caustic systems at high temperatures.

Uses for molded parts include valves, bearings, pumps, and other functional parts formerly the domain of metals.

POLYIMIDES

In polyimides, the polymer chain is a rigid one consisting of heterocyclic rings bracketed with aromatic rings, as in the following example.

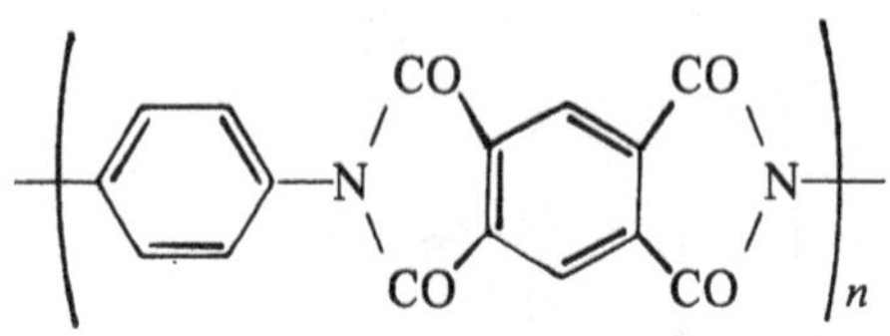

This molecule is made by first synthesizing long chain precursor molecules in which the heterocyclic rings are not closed. These precursors are soluble in certain solvents and so can be used for impregnation of glass fabrics or for bonding of laminates. The temperature is then increased to drive off the solvents and complete the heterocyclic ring closure.

The polyimide structure shown is typical but many variations are possible. Commercially available products are normally based on the reaction between aromatic dianhydrides and aromatic diamines. Melting points range up to about 600°C and polyimide films have been used in applications where temperatures range from −270°C to 400°C for short term use. For continuous use the recommended upper limits are 260°C in air, and 315°C in an inert atmosphere. Fabrication of a polyimide resin involves molding at high temperatures and pressures into shapes from which parts may be machined. In some cases parts may be formed directly by techniques similar to powder metallurgy processes. Resistance to abrasion is good and a graphite-filled grade is available for bearing applications having constant wear resistance over a wide temperature range. Polyimides are also noted for resistance to electrical and nuclear radiation. They are extremely resistant to organic solvents and to dilute acids but alkalis and concentrated acids have a detrimental effect. Water or steam above the atmospheric boiling point causes hydrolytic degradation.

Polyimide film has been used for flexible printed circuits because of its solder resistance and dimensional stability at high temperatures. The long term use at high temperatures of polyimide glass cloth reinforced composites is illustrated by experiments carried out for an exposure period of 10,000 hours at 260°C where retention of strength was excellent.

Polyimides are also used in electric motor insulation, especially in those used in electric locomotives which generate great surges of power (and hence heat) when starting up. Molded polyimides are used as back-up rings in jet aircraft turbofan engines, gear wheels (operating at high temperatures), and for the vanes of the helium compressor in a gas cooled atomic reactor. Bearing uses have also been found in office equipment, home appliances, and automobiles.

POLYETHERSULFONE

The structure of polyethersulfone consists of phenyl groups linked by thermally stable ether and sulfone groups.

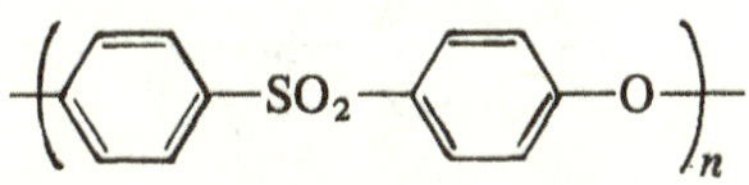

The absence of aliphatic groups, which would be liable to oxidative attack, accounts for the material's resistance to oxidation and its retention of mechanical properties at high temperatures. In particular, the sulfone group confers high temperature performance while the ether linkage aids practical processing by allowing mobility of the polymer chain when in the melt phase. The resin was developed originally for injection molding but it can also be extruded, the films so formed having good dimensional stability and high tear strength (both initiation and propagation).

Polyethersulfone is both tough and strong and retains these properties over a wide range of temperatures. Its tensile strength at 20°C, for example, is 12,000 lb/in.^{2} and this is reduced by less than 50 percent at 180°C. Polyethersulfone also exhibits low creep. A constant stress of 3000 lb/in.^{2} at 20°C for three years produces a strain of one percent, while a stress of 6500 lb/in.^{2} results in a strain of only 2.6 percent, again over three years. Results are also good at elevated temperatures. At 150°C, for example, a stress of 3000 lb/in.^{2} produces a strain of only 2.55 percent after three years. Above 150°C its load bearing properties are reduced but it can still be considered for applications up to 180°C. It remains form-stable to above 200°C and has a heat-deflection temperature of 203°C at a stress of 264 lb/in.^{2}.

Chemical properties include resistance to acids, concentrated alkalis, oils and greases, aliphatic hydrocarbons, and alcohols. Polyethersulfone is attacked by esters, ketones, and aromatic and halogenated hydrocarbons.

Uses for polyethersulfone include door frames and handles in aircraft ovens, bakery oven windows, and sterilizable medical components. Reinforced polyethersulfone composites are also being used in aerospace components, radomes, and automatic transmission housings.

POLYPHENYLENE SULFIDE

Polyphenylene sulfide is a crystalline polymer with a symmetrical, rigid backbone chain of recurring para-substituted benzene rings and sulfur atoms and is produced from *p*-dichlorbenzene and sodium sulfide in a polar solvent.

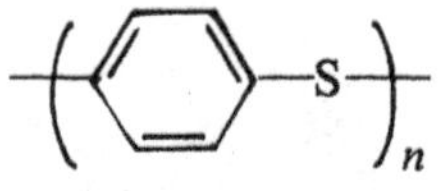

This chemical structure accounts for its high melting point (288°C), good chemical resistance, thermal stability, and fire resistance. Polyphenylene sulfide has a high stiffness and good retention of mechanical properties at elevated temperatures. The unfilled polymer is hard and has high tensile and flexural strengths. These properties are further improved by the incorporation of fillers, particularly glass. Polyphenylene sulfide can be injection or compression molded or applied as a coating, using slurry coating techniques. Coatings require a baking operation of 45 minutes at 355°C.

Chemical resistance is very good and there are no known solvents below 190–215°C. Electrical properties are also excellent, including low dielectric constant, low dissipation factor, and high dielectric strength.

Among the many uses for molded polyphenylene sulfide are aircraft and automobile parts, military hardware, pump impellors, vanes and housings, and chemical etch bath trays. As a coating it is used in valves, etch tanks, and gate valves.

AROMATIC POLYESTER

A range of materials is available, including a homopolyester of repeating *p*-oxybenzoyl units and a series of *p*-oxybenzoyl copolyesters. All of them have a combination of good mechanical, electrical, wear, and high temperature properties.

The oxybenzoyl homopolyester is linear and highly crystalline. It decomposes at 550°C and does not melt below this temperature. When compression sinter molded the homopolyester has a high flexural modulus (above 1,000,000 lb/in.^{2}), a flexural strain of less than 1 percent and a low deformation to load. Sintered blends of around 25 percent homopolyester with PTFE have lower densities and better wear resistance than pure PTFE and can be used at temperatures of up to 315°C. Applications include self-lubricated bearings, bearing races, seals, rotors, and the vanes of processing pumps.

The copolyesters are marketed as compression and injection molding grades. The former is made from isophthalic acid, *p*-hydroxybenzoic acid, and bisphenol, while the injection molding grade is made from terephthalic acid, *p*-hydroxybenzoic acid, and bisphenol (giving poly-*p*-oxybenzoate).

Both materials have a high resistance to radiation, with continuous use temperatures in the range 275-300°C. Electrical properties include a low dielectric constant and a high dielectric strength. They are resistant to a wide range of organic solvents (including benzene, *N*-heptane, ketones, and methanol) and acids but are slightly attacked by alkalis.

Applications for compression molded articles include high temperature self-lubricating bearings, bearing cages, seals, radar windows, and high pressure seals. Injection molded articles have been used in valve seats, high performance aircraft parts, and electrical connector components.

POLYARYL ETHER

This is a high impact material with a heat deflection temperature of 150°C at 264 lb/in.^{2}. Resistance to organic solvents is good, with the exception of esters, ketones, and chlorinated aromatics. It also withstands prolonged immersion in boiling water.

It has a low shrinkage factor that allows the precision molding of void-free parts. It can be solvent welded, chrome plated, and painted without pretreatment although standard priming techniques may be necessary for optimum adhesion and surface appearance. The combination of heat resistance and good paint adhesion makes polyaryl ether a candidate for use in oven-baked, painted exterior automotive trim.

POLYARYL SULFONE

Polyaryl sulfone consists mainly of phenyl and biphenyl groups linked by thermally stable ether and sulfone groups. The absence of aliphatic groups renders it less liable to oxidative attack than polysulfone polymers. It retains its mechanical properties at high temperatures, therefore, while the presence of the ether linkages gives chain flexibility, thus permitting fabrication by normal melt processing techniques.

Polyaryl sulfone has a high heat-deflection temperature (275°C at 264 lb/in.^{2}) and maintains a tensile strength of 4000 lb/in.^{2} and a flexural modulus of 250,000 lb/in.^{2} at 260°C. At normal room temperature polyaryl sulfone is a strong, stiff, and tough material with good electrical properties. It is resistant to a wide range of chemicals including acids, bases, and common solvents. It is unaffected by most fuels, lubricants, hydraulic fluids, and cleaning agents used on, or in the proximity of, electrical components. Polyaryl sulfone is, however, dissolved by highly polar solvents such as *N*, *N*-dimethylformamide and *N*-methylpyrrolidone.

Polyaryl sulfone has been used successfully for electrical connector bodies that are required to meet severe structural and environmental conditions. Other applications include integrated circuit test sockets, high temperature bobbins used in relays and solenoids, and sleeving for high-temperature solderless terminals.

PARYLENE

Parylene is a series of polymers developed for use as coatings or films. Parylene N, for example is poly (*p*-xylylene) and has the structure:

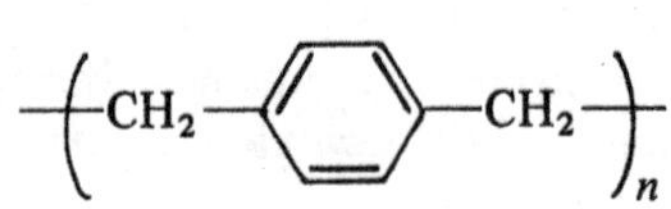

Other commercially available polymers are parylene C, poly (monochloro-*p*-xylylene) and parylene D, which has two chlorine atoms on each benzene ring.

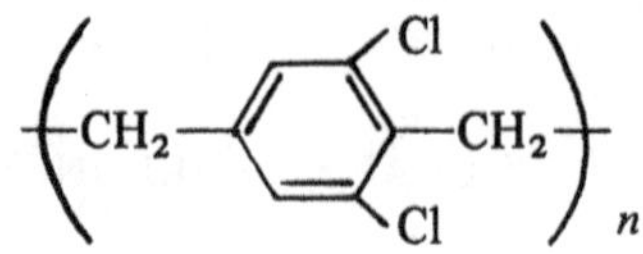

Unlike other materials, parylene is not produced and marketed as a polymer because it is not practical to extrude or mold it as with other plastics. Solvent casting is also out because it is insoluble under conventional conditions. In practice, parylene is produced by a vapor phase deposition process. The process starts with the solid dimer which goes through a two-stage heating process to convert it to the reactive monomer vapor. When passed over objects at room temperature the vapor rapidly coats them with polymer. Coating thickness is controlled by preweighing the amount of dimer to be vaporized. If deposited on a cold condenser it can be stripped off as a very thin film. It is also possible to produce it in particulate form by polymerizing in an aqueous medium.

In general the parylenes are crystalline high melting point polymers, the melting points being 400°C for parylene N and around 300° for parylenes C and D. The short-term (1000 hours) use temperatures are not as high as would be expected, however, being about 95°, 115°, and 130°C for parylenes N, C, and D respectively. This is due to oxidative degradation. It will be seen that although parylene N has the highest melting point, it has the lowest continuous use temperature. This is because the chlorine substituted polymers are more stable to oxidation. In an inert atmosphere the 1000-hour use temperatures are: parylene N, 271°C; parylenes C and D, 279°C. Degradation under these conditions is due mainly to cleavage of the C—C bonds and there is then little difference in the stability of the three polymers.

Low temperature performance is extremely good, as is dimensional stability with respect to humidity changes. However, resistance to weathering is not good and the parylenes are not recommended for extended exposure to sunlight. Chemical resistance is outstanding and the parylenes are insoluble in most organic solvents up to about 150°C. They are also resistant to environmental stress cracking. Moisture vapor permeability varies with the type of polymer. That of parylene D is one of the lowest known (3.9 g/m^2/24 h/mil, 37°C 90 percent R.H.) while parylene C is slightly higher (7.8 g/m^2/24 h/mil). Parylene N's permeability is the highest of the three (24.8 g/m^2/24 h/mil) but is still not much higher than that of low density polyethylene. Gas permeabilities are also low.

Electrical properties are good, those of parylene N being the best. Parylene N has a very low power factor which changes only slightly with frequency. It is, therefore, classed as a primary dielectric. The chlorine in the other two parylenes increases both the dielectric constant and the power factor.

The main fields of use for parylene depend on the fact that it can be produced as very thin coatings (down to 0.004 mil) with good adhesion and high purity and uniformity. Coatings can be produced on almost any substrate that can resist exposure to a relatively mild vacuum. Because this is a low stress condition, even delicate substrates such as paper or easily corrodible metals can be coated in this way.

Parylene N's particular balance of properties makes it very suitable for electronic applications. It has been used, for example, in capacitors, its advantages being a low power factor, a good frequency response, and a high operating temperature. The thinness of the films plus the excellent electrical properties also allows a reduction in size—down to about a fifth of the size of a polystyrene capacitor of equivalent rating. Other parylene insulated or encapsulated electronic devices, such as resistors, switches, and transistors show similar advantages.

Interesting experiments with parylene C point to its possible use as a coating for other plastics materials in order to reduce their permeability. Polyethylene bottles coated internally with 0.25 mil thick parylene C have been filled with lemon oil for three years and have lost only two grams through permeation whereas lemon oil passes extremely rapidly through uncoated polyethylene. Parylene is, however, expensive and can only be used where its special properties are fully justified.

One of the most important applications of parylene is the protection of circuit boards or modules from hostile environments. Uniform, pinhole-free coatings are possible, with a high degree of conformity, such that even fine lead wires and their sharp snipped-off ends are fully coated and protected.

One other different application is worth noting. This is the use of deposited parylene film to procure surface replicas. One example is the coating of single crystals of polyethylene, followed by solution of the polyethylene and shadowing of the surface replica. It has then proved possible to observe details, such as screw dislocations in the crystals, by electron microscopy. The height of such dislocations is only 0.0004 mil (0.01 micron).

POLYBENZIMIDAZOLE

The molecular structure of this polymer is similar in some ways to that of polyimide but here the heterocyclic ring contains two nitrogen atoms.

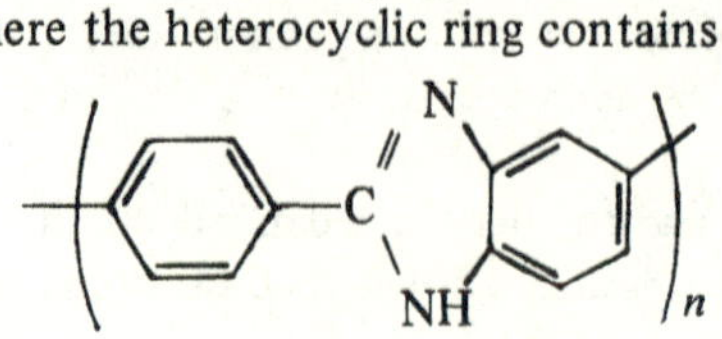

It is unlike other polyaromatic heterocyclic polymers so far considered inasmuch as it is soluble in the more common solvents thus allowing more flexibility in processing. The melting point is around 590°C. Possible uses include aircraft skin coatings and other aerospace applications.

POLYBENZTHIAZOLE

The heterocyclic ring in this polymer contains a sulfur atom as well as a nitrogen atom.

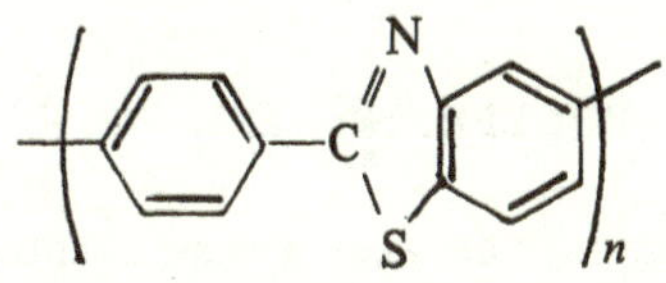

The melting point is a little higher than that of polybenzimidazole (600°C) with applications in the fields of adhesives and laminating agents.

POLYBENZOXAZOLE

Replacement of the sulfur atom in the previous polymer by oxygen gives polybenzoxazole.

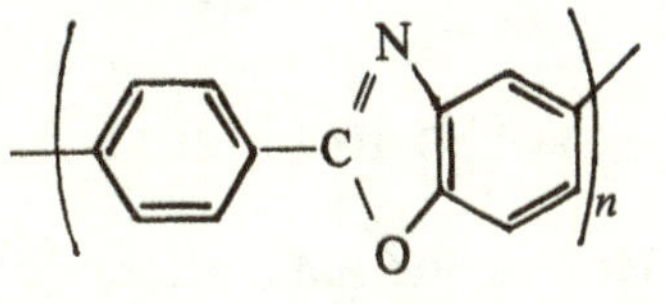

The melting point is again 600°C and the uses are similar to those of the thiazole polymer.

POLYOXADIAZOLE

This has a structure different from the preceding polyaromatic heterocyclic polymers since the aromatic and heterocyclic rings are joined only by single bonds. The heterocyclic ring is also different as it contains an oxygen atom and two nitrogen atoms.

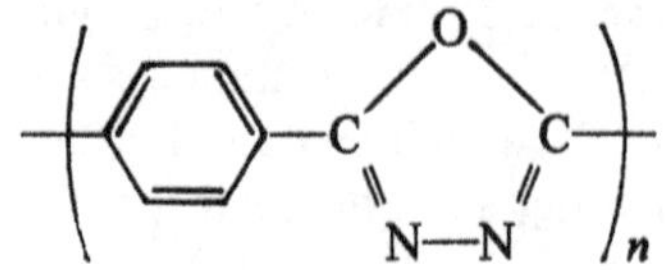

The melting point is lower than those of polymers in which the rings are joined at two points. The melting point is above 450°C but fibers made from the polymer disintegrate after about one hour at 450°C, in air, and films become brittle after two hours. The polymer has been used as a fiber, as a laminating agent, and in coatings.

POLYPHENYLTRIAZOLE

The links in this polymer between aromatic and heterocyclic rings are also single bonds. The melting point is slightly higher, around 480°C, and the main suggested use is as a fiber.

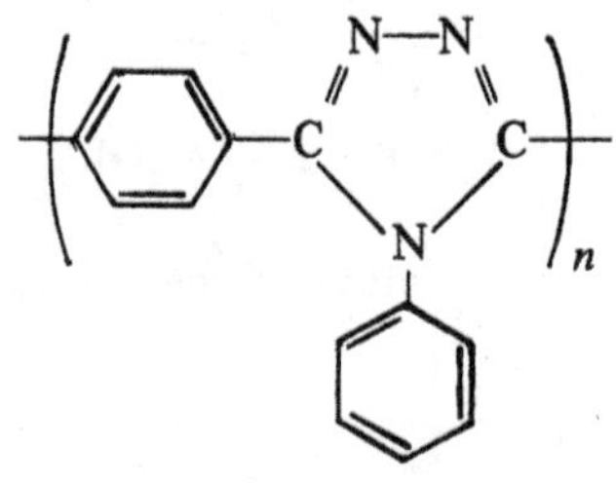

POLYTHIADIAZOLE

This polymer also has aromatic and heterocyclic rings joined by single bonds and this is reflected in its melting point of just over 450°C. The heterocyclic ring contains a sulfur atom and two nitrogen atoms.

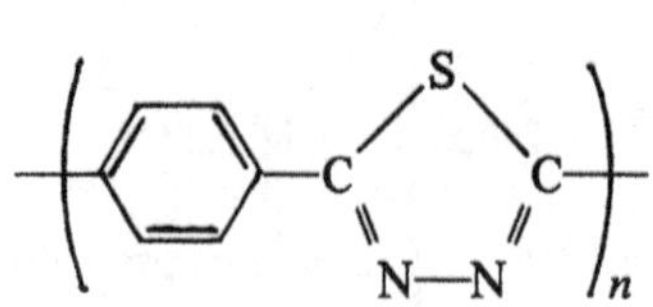

The main uses would appear to be in the fields of adhesives and laminating agents.

LADDER POLYMERS

One weakness with the polyaromatic heterocyclic polymers dealt with earlier is the single bond which occurs at intervals. If the whole molecule consisted of rings joined together at two points the molecule would have a ladder-like structure. This would not only improve the overall rigidity (and hence the retention of mechanical properties at high temperatures) but it would also mean that the molecule would not be broken simply by rupture of a single bond (Fig. 17.1). It could, indeed, survive the rupture of two bonds provided that they were not between the same pair of 'rungs' (Figs. 17.2 and 17.3).

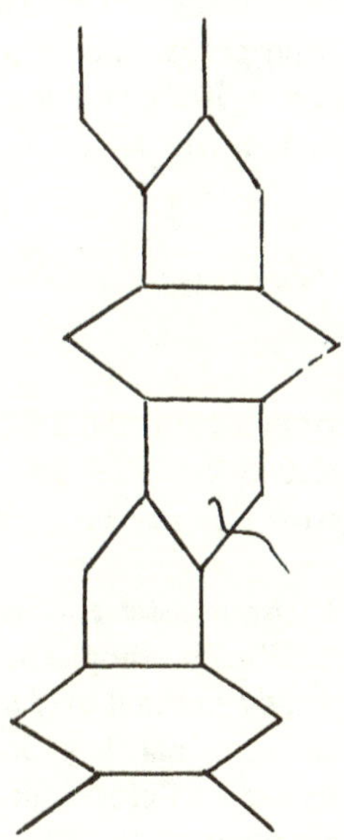

Fig. 17.1 No breakage of molecule

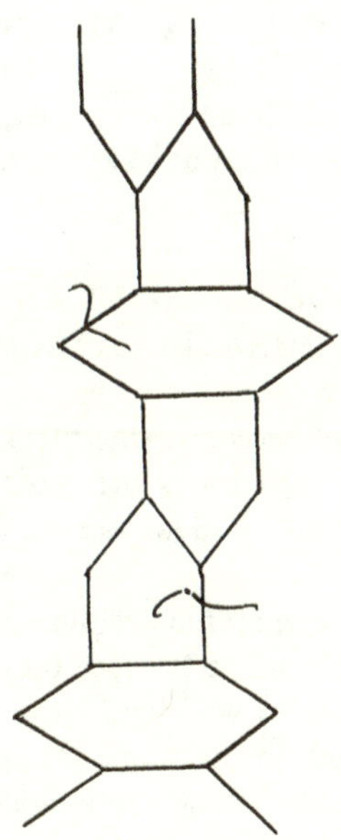

Fig. 17.2 No breakage of molecule

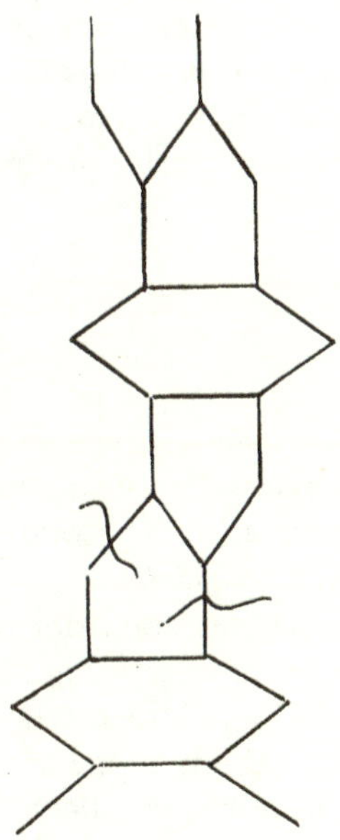

Fig. 17.3 Breakage of molecule

One example of such an approach is poly (imidazopyrrolone) which has the repeating structure shown below.

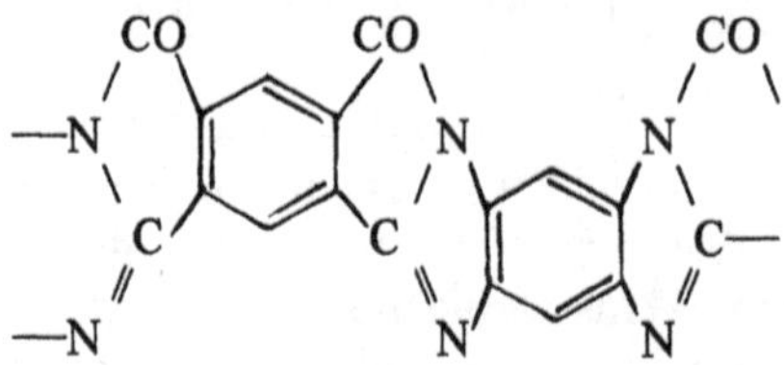

Ladder polymers such as this are usually made by a two stage process. A soluble prepolymer is first prepared and this is then turned into a ladder polymer by ring closure. Using this technique, ladder polymers have been made in the form of films, coatings, and fibers. Heat stability of these types of polymers is good. Ladder polymers have been prepared that are stable in air up to 500°C and in an inert atmosphere up to 600°C but commercially viable production methods have yet to be developed.

POLYPHENYLQUINOXALINE AND ACETYLENE–TERMINATED OLIGOMERS

These high temperature resistant polymers are formed from the polycondensation of aromatic tetramines and aromatic bisbenzils. Such polymers exhibit a combination of high thermal oxidative stability coupled with toughness.

During fabrication these materials remain truly thermoplastic and as such can be a distinct disadvantage because they can exhibit thermoplastic yield or failure at temperatures well below the limit set by thermal oxidative stability. One approach to overcome the disadvantage of plastic yield is to react the quinoxaline with oligomers which are capable of cross-linking without the release of volatile by-products. The resultant system thermosets with properties similar to the polyimides.

A typical synthesis of such thermosetting systems involves the reaction of benzil end-capped phenylquinoxaline oligomers and 3 (3,4-diaminophenoxy phenylacetylene). The benzil end-capped oligomers are prepared by the dropwide addition of a solution of diamino benzidine in *m*-cresol to a solution containing a twofold excess of the appropriate aromatic bisbenzil dissolved in *m*-cresol containing a trace of acetic acid.

Variations can be made in the polymer structure by reacting different aromatic bisbenzils. The general reactions are as follows:

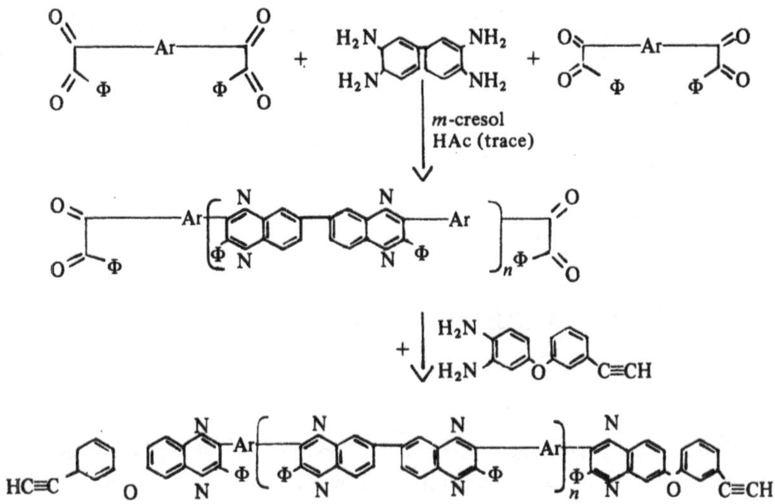

Claims have already been made for the use of this family of polymers as structural adhesives at temperatures close to 300°C.

Section C Natural Polymers and Derivatives

18.

POLYMERS OF NATURAL ORIGIN

We have already pointed out, in the chapter on raw materials, that many polymeric materials occur naturally. Some of these are utilized directly, with little or no chemical treatment except that needed in the separation and purification processess, while others are chemically altered to give materials with little resemblance to the originals. In this chapter those natural polymers which are used without substantial alteration to their chemical make-up, are considered; while in the following chapter chemical derivatives of natural polymers, such as cellulose acetate and ethyl cellulose are discussed.

Regenerated cellulose is also included in the next chapter because although the processess undergone in its manufacture are really separation and purification, it seemed more logical for it to take its place among the other variations on the theme of cellulose.

BITUMENS

Historoy of Development. There is evidence that an appreciation of the thermoplastic character of natural biumens has existed since very early times. In this connection we may cite the discovery of a water tank, dated about 3000 B.C., in the Valley of the Indus. The tank was fairly large (about 39 feet × 23 feet × 8 feet) with a three-foot thick wall, and was water-proofed with a one-inch thick bitumen layer, followed by futher brickwork.

The existence of the Trinidad Asphalt Lake was mentioned by Sir Walter Raleigh after one of his voyages, round about 1595, while in 1870 deposits of bitumen were discovered in Utah in the U.S.A. No commercial exploitation of these deposits occurred for many years until S. H. Gilson began marketing the material under the name of Gilsonite. At the end of the nineteenth century fresh impetus was given to the use of asphalt by the development of motor vehicles with pneumatic tires, and a subsequent demand for good roads.

General Nomenclature

The terms bitumen, pitch, asphalt, etc., are used very loosely but those most generally accepted are:

(a) Bitumens. Naturally occurring complex hydrocarbon mixtures related to petroleum oils. Gilsonite is an example of this type.

(b) Asphalts. These are bitumens occurring naturally but with a heavy mineral loading, as in Trinidad Lake asphalt.

(c) Petroleum Still Residues. These are sometimes termed 'artificial bitumen.'

(d) Pitches. The term 'pitch' is applied to distillery residues from many industrial processes as, for example, coal tar, wood tar, stearine, whale oil.

Processing Techniques

Bituminous plastics can be molded and because of their cheapness can often be considered for the manufacture of large articles such as water tanks and battery boxes which would not be economically possible in synthetics.

The molding compounds employed are varied in character but in general they may contain some, or all, of the following ingredients.

(i) Bituminous base
(ii) Linseed or tung oil plus a dryer
(iii) Wax (paraffin or carnauba)
(iv) Coumarone resins
(v) Soft pitch (viscous tars used as processing aids)

(vi) Asbestos fibers (these must be of the acid resisting type)
(vii) Mineral fillers (these may be chalk, china clay, slate dust, etc.).

The drying oils are incorporated to give some degree of heat hardening character to the composition. Another means of achieving this is to add sulfur but this is not feasible when the molding is for use in electrical appliances, since sulfur attacks copper. The composition is heated in an open trough mixer to 150–200°C with oil or gas heating.

Flash or semi-positive molds are normally used and are cored for steam or water heating. The mold is heated to about 100°C, a slab of pre-heated dough is dropped in, and the press is closed. After chilling to about 50°C the press is opened and the molding is ejected. Pressures are of the order of ½–1 ton/in.^{2} and the complete cycle takes about four minutes. This process has been called 'cold molding' since the temperatures used are relatively low compared with those used for synthetic polymers, but the term is more properly used to describe the cold compression of the composition (with fillers, drying oils, etc.) followed by ejection in the 'green' state. These 'green' moldings are finally hardened by air drying or by stoving.

Properties

Chemical resistance is generally excellent and there is a wide range of physical properties to choose from. Electrical properties are also good. Colors, of course, are dark although not always black, but this can be combated to a certain extent by using heavy loadings of earth pigments. Brittleness is also a problem, but this can be eased by addition of asbestos, rubber, drying oils, and high molecular weight synthetic resins.

Applications

The use of asphalts in road-making compositions has already been mentioned and this remains one of the most important applications, especially in the U.K. Bitumen is also used as a water-vapor-proof seal for covering cold-store insulation, and as a water-vapor barrier in many laminations, e.g., in multiwall paper sacks. A similar application is as a water-resistant impregnating agent for materials such as roofing felts.

In molded form its water-resistant properties are made use of in the manufacture of water tanks, while its acid-resistant and good electrical properties are utilized in battery boxes.

SHELLAC

History of Development. Shellac, like bitumen, is of great antiquity and, together with bitumen, was the only significant thermoplastic until the commercial advent of celluloid and ebonite in 1870. Because of its geographical origin (India and Burma) its use was limited for many years to the Orient, although it was mentioned by Marco Polo in the year 1290. It was long recognized as an extremely good lacquer for woodwork, but in 1895 it found another important outlet when Emil Berliner developed the disc-type of gramophone record.

Origin and Preparation

Shellac is the protective coating produced by the lac bug (*Tachardia lacca*) which attaches itself to the softwood sap trees of India. The sap is absorbed through the lac bug's stinger and exuded as a gum known as stick-lac. The gum is deposited over the eggs which the bug lays on the tree, and also covers the bug itself. The exudation hardens and kills the bug but acts as an efficient protective coating for the eggs until these hatch out about six months later. The young bugs break through the shell and eventually the whole process is repeated.

The lac is scraped from the twigs with knives, then ground and extracted with water or dilute soda solution to remove the coloring matter (lac dye). The residue is dried and is then known as seed lac. The seed lac is melted in a sausage-shaped bag suspended over a charcoal fire, and the lac squeezed out. The product is either stretched into a large sheet and then broken into 'flake shellac' or poured into circular molds, when it is known as 'button lac.' Where a colorless shellac is required, to give a clear coating over natural wood, the product is bleached by dissolving it in hot-soda solution, treating with chlorine or sulfurous acid and precipitating with acid. The precipitated 'bleached shellac' is then collected, washed, and 'pulled' under water into sticks.

Composition

As collected, the lac contains about 6 percent wax, 6.5 percent of a red coloring matter (laccaic acid), 70-85 percent resin, and a small amount of insect remains and vegetable debris.

The resin itself is not a chemical entity but consists of several different compounds of polyhydroxy acids, the chief of which are aleuritic acid (trihydroxy palmitic acid) and shellolic acid ($C_{13}H_{16}(OH)_2(COOH)_2$). Shellac has a relatively high molecular weight (1000–1500) for a natural resin and this gives it a certain toughness and elasticity.

Properties

Shellac has a softening point of 75–85 °C and an acid value of 60–70. It is soluble in alcohol, giving solutions with good adhesive properties, but insoluble in ketones, esters, glycerol, and hydrocarbons. It is also soluble in carbonate solutions, being precipitated unchanged by the addition of acid.

Its resistance to water absoption is poor, which gives it a tendency to 'bloom' in damp conditions. Shellac has a low dielectric constant which, when allied to its adhesive and binding properties, has led to its use with fillers, such as flaked mica or asbestos, to produce electrical insulators and forms of coils. Another outstanding advantage of electrical insulators based on shellac is their excellent arc resistance.

Applications

As we have already seen, shellac is an excellent lacquer for woodwork because of the hardness and brilliancy of its films. Shellac varnishes are slow drying, but drying times can be greatly reduced by incorporating cellulose nitrate. Treatment with sulfurous acid produces aqueous dispersions which can then be used as a basis for oil-bound distempers, the stiffening of paper and felt (as in felt hats), quick-drying water-based paints, or for the formulation of wax polishes.

Shellac can also be compression molded, and in this form has played an outstanding part in the history of sound reproduction. Its choice for the manufacture of gramophone records was due to the fact that it faithfully records all impressions made on it, remaining hard and dimensionally stable. About 22 percent lac is mixed with cotton flock, carbon black and mineral fillers, then mixed on rollers, pre-pressed and cooled. A weighed amount is then treated at 80–90 °C under a pressure of ½-¾ ton/in.^{2} followed by cooling to 30 °C prior to extraction from the mold.

The above refers, of course, to the original 78 r.p.m. records. These have since been replaced by 45 r.p.m. (extended play) and 33 r.p.m. (long play) records whic are made from vinyl chloride/vinyl acetate copolymers.

ROSIN

Historical Development. Rosin is one of the constituents of the oleoresins exuded by pine trees. This oleoresin is also a source of turpentine, the recovery of which from pine trees is mentioned in Pliny and Herodotus. Essential oil of turpentine was also said to have been known to Marcus Graecus, the inventor of 'Greek FIre.'' Pine and fir resins were also well-known to the Ancients.

Up to the eighteenth century, rosin supplies were obtained mainly from France, but America is now by far the most important source of supply, producing about two-thirds of the world's total.

Origin and Preparation

Oleoresin is a normal physiological product of pine but the amount produced is greatly increased by injury to the tree. In the cup and gutter method, a blaze or groove is cut in the tree and metal or earthenware cups are attached to the trunk at the bottom of the blaze. The flow of resin is directed into the cup by one or two strips of galvanized iron known as gutters.

The crude oleoresin is sent to distilleries where it is steam distilled. The distillate is turpentine, which is then rectified and later used in medicine as an external counterirritant and rubefacient. The molten rosin left in the still may contain bark, pine needles, twigs, and other foreign matter which are removed by screening, before running into barrels for shipment. Rosin can also be obtained from old pine stumps by distillation.

Composition

Rosin consists essentially of the anhydride of abietic acid, which is present in a number of isomeric forms. Abietic acid is a cyclic acid having the structural formula:

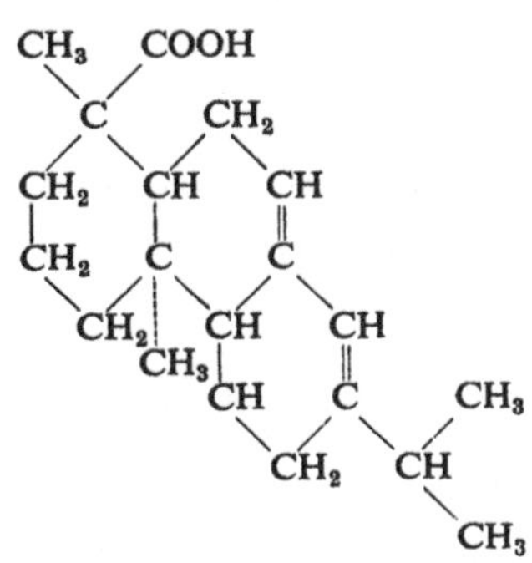

In addition rosin contains a resene, colophenic acids, and traces of a volatile oil.

Properties

Rosin, as it is obtained from pine oleoresin, is ruby red in color, but by purification or special treatment with filter clays such as fullers earth a wide range of resins with colors ranging from very dark to water-white can be obtained. Rosin is brittle and is easily powdered. At about 100°C it fuses gradually, while at high temperatures it burns with a smoky flame, leaving a very small residue of ash (approximately 0.1 percent). Rosin is insoluble in water but soluble in alcohol, ether, benzene, and carbon disulfide.

Applications

Rosin is very widely used as a sizing agent in paper-making, in the form of a resin soap from which the rosin (as the aluminum salt) is precipitated on to the paper fibers by adding aluminum hydroxide. Another large outlet for rosin is in the manufacture of yellow laundry soap.

Other metallic salts of rosin, such as those of cobalt and manganese, are used to promote the drying of oils in paints or are added to oils in the manufacture of grease.

Unmodified rosin is used, in combination with drying oil, in the manufacture of linoleum and as a soldering flux, while a more romantic use is for rosening violin bows.

19.

DERIVATIVES OF NATURAL POLYMERS

The materials discussed in the following pages are all derivatives of the same natural polymer, namely, cellulose; such materials as cyclized rubber, chlorinated rubber, and rubber hydrochloride are considered in a separate chapter in the section on rubber.

Cellulose is the most widely available chemical raw material since it is the chief constituent of nearly all forms of plant life. Not all plants, however, are suitable commercial sources of cellulose, the two most important materials in current use being cotton linters and wood pulp.

Cotton linters are the short fuzz fibers remaining on the cotton seed after the long fibers have been removed by the process known as ginning, and are a particularly pure form of cellulose. A final chemical purification is still necessary, however, and this usually consists of treatment by caustic soda solution followed by bleaching.

The sulfite process is used for isolating cellulose from wood, the lignin and other noncellulosic constituents being removed by heating in an acid calcium-magnesium bisulfite liquor. Sulfite wood pulp is extensively used in the manufacture of paper but as a raw material for plastics a further purification, consisting of treatment with caustic soda, is necessary.

Cellulose is a carbohydrate with the chemical structure shown below:

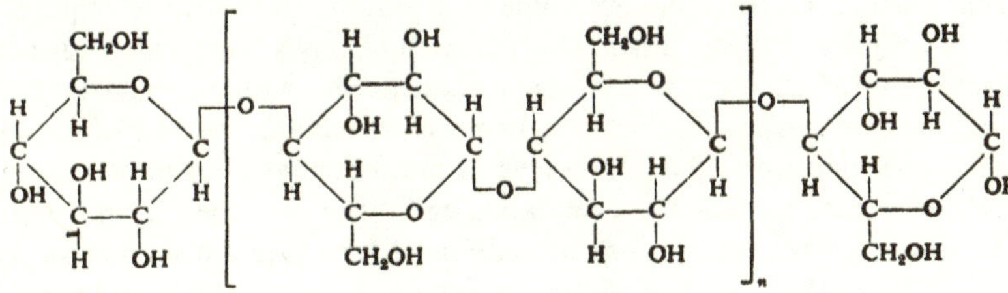

The important thing to note about this formula is that each six carbon-atom (glucose) units of the cellulose molecule contains three hydroxyl groups, and it is at these groups that substitution occurs during the manufacture of the various cellulose plastics.

Experiments have shown that the terminal unit contains four hydroxyl groups instead of the three contained by the interchain units, while other work indicates that one of the terminal groups can exhibit aldehydic properties. This can be explained by a molecular rearrangement to give the configuration shown below:

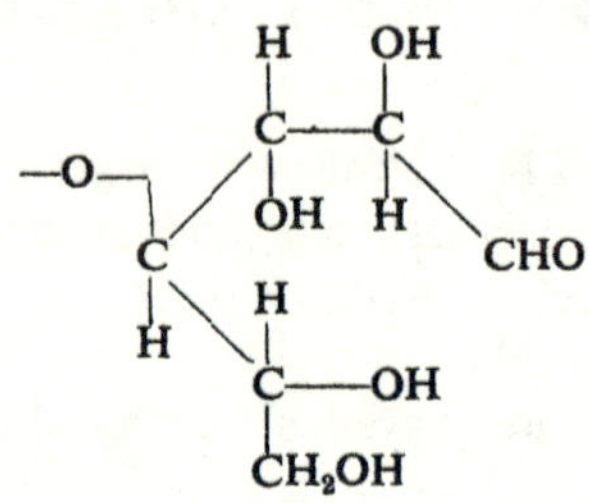

The physical structure of cellulose is also important both in considering its properties and in the manufacture of its derivatives. In general, we can visualize natural cellulose as consisting mainly of long, fine fibrils of a largely crystalline character. At first sight one might wonder why a material consisting of long-chain molecules with no cross-linking is not thermoplastic but this is easily explained by the enormous possibilities for hydrogen bonding which exist with such a high proportion of hydroxyl groups present in the molecule. The application of heat, therefore, leads to degradation (charring) before the point is reached at which the hydrogen bonding could be overcome sufficiently to allow relative movement between the molecules. The insolubility of cellulose is also due to hydrogen bonding which must, therefore, be overcome if a soluble derivative is required. This is achieved by substituting sufficient hydroxyls by other groups to prevent the reformation of hydrogen bonds. Substitution must be uniform if local hydrogen bonding is to be prevented and in the case of acetylation, particularly, the cellulose is given a pretreatment to open up the cellulosic structure so that the acetylating agent can penetrate it uniformly and rapidly. Two main types of cellulose derivatives are commercially available, the esters and the ethers. In esterification, the hydroxyl groups are replaced by ester groups such as acetate and nitrate, while in etherification, the hydrogen atom of the hydroxyl group is replaced by groupings such as ethyl (C_2H_5), methyl (CH_3), and benzyl ($C_6H_5 \cdot CH_2$), the oxygen atom of the hydroxyl acting as a bridge.

CELLULOSE NITRATE

Historical Development. Cellulose nitrate has a special place in the history books because it was the first synthetic polymer and the one on which the present-day plastics industry is founded. Although the official birth of the plastics industry dates from the day that Alexander Parkes exhibited Parkesine at the Great Exhibition of 1862, the nitration of cellulose to cellulose nitrate was carried out as far back as 1832 by Professor Henri Bracconot in France. Six years later a fellow countryman, Pelouze, put forward the suggestion that the reaction between nitric acid and cotton could be utilized to make explosives, but it was left to Professor Schoenbein, of Basle University, to isolate cellulose nitrate from the reaction product of paper and mixed nitric and sulfuric acids. Following further work by Pelouze, an American chemist, Maynard, found that cellulose nitrate was soluble in a mixture of ethyl alcohol and ether, giving a solution which he named collodion. Collodion was later used as a liquid plaster for wounds. Apart from this, no commercial use was proposed for cellulose nitrate until the time of Alexander Parkes.

Parkes's original samples on show at the 1862 Exhibition were composed of cellulose nitrate and caster oil but in 1865 he patented the use of camphor as a plasticizer for cellulose nitrate and no one has yet discovered a better. When reporting his discoveries to the Royal Society of Arts, Parkes put forward an extensive list of possible uses for this material, including knife handles, combs, brush-backs, walking sticks, buttons, buckles, and photographic baths. The Parkesine Company was formed in 1866 to operate Parkes's process but had only a brief life, mainly due to lack of interest and commercial instinct by the inventor. Meanwhile, in the U.S.A. the brothers J. W. and I. S. Hyatt had successfully established a company called the Celluloid Manufacturing Company and their trade-name Celluloid has since become the general descriptive term for cellulose nitrate. After the failure of the Parkesine Company, manufacture of celluloid was continued in the U.K. by Daniel Spill who was formerly Parkes's assistant. The final result of Spill's endeavors was the formation of the British Xylonite Company in 1877 in association with L. P. Merriam.

In addition to its use as a molding composition, and in the form of film and sheet, cellulose nitrate enjoyed a brief existence as a substitute for silk. The first patent was taken out by Chardonnet in 1884 and full-scale production was achieved by 1891. The fiber was not denitrated and was explosively inflammable. In addition to this major disadvantage, other and cheaper processes were developed and Chardonnet artificial silk was short-lived.

Raw Materials

The raw material for the basic cellulose nitrate is cellulose prepared from cotton linters. Other raw materials directly concerned in the manufacture of celluloid are camphor and ethyl alcohol. Naturally occurring camphor is produced in Formosa but the compound normally used in the U.S.A. and Europe is synthesized from oil of turpentine.

Cotton Linters. Although cotton-seed fiber is the purest form of cellulose to be found in nature some kind of purification is still necessary. Bales of compressed linters from the cotton-seed mills are run through a breaker, which consists of a rotating cylinder with bronze spikes, followed by blowing up to an air separator at the top of the building. This eliminates any heavy impurities. A conveyer then carries the linters to the digester where they are treated under a pressure of between 30 and 100 $lb/in.^2$ with a 3 percent solution of caustic soda. The linters are thoroughly wetted with three percent caustic soda solution while on the conveyor immediately prior to their entry into the digester. Digestion takes two to six hours and is designed to remove seed hull and other noncellulosic matter by solution in the caustic soda. The pure cellulose can then be obtained by washing out the solution of impurities. The washing stage is extremely thorough but the color of the linters is still not quite good enough, being greyish or light yellow. Multistage bleaching by gaseous chlorine and sodium or calcium hypochlorite is then carried out followed by washing and drying. A multistage process is used in order to avoid degradation of the cellulose by too severe a reaction.

Manufacture of Cellulose Nitrate

The purified and bleached linters are dried to a moisture content of about 0.5 percent in order to prevent excessive dilution of the nitrating acids with consequent irregular nitration and degradation due to heat when the concentrated acids are diluted in the interior of the fibers. A typical composition for the nitrating acid is nitric acid 25 percent, sulfuric acid 55 percent, water 20 percent, and a charge of 1200 lb of acid to 30 lb of dried linters is used. The mixture is stirred in a stainless-steel tank for about 20 minutes, the temperature being strictly controlled to within plus or minus 5° of 100°F. The batch size may seem small compared with other, similar, processes but this is to minimize the risk of large explosions. At the completion of the nitration process the contents of the vessel are discharged through a valve at the bottom of the tank into a centrifuge where spent acid is removed. One centrifuge usually serves four nitrating tanks. From the centrifuge the cellulose nitrate is dropped into a 'drowning' tank where the remaining acid is diluted with a large enough volume of water to prevent a rise in temperature and to ensure complete stoppage of the nitration reaction. The resultant slurry is pumped to blending tanks to accumulate. While in the blending tank, it is refluxed for 10–16 hours to remove all traces of the unstable cellulose sulfate which could cause instability of the cellulose nitrate on storage. The dilute acid is then removed and the cellulose nitrate washed and dried.

Structure

The physical structure of cellulose nitrate is similar to that of the original linters but there is a harsher feel due to a straightening-out of the convolutions in the original cotton. Chemically, the basic structure of the cellulose molecule is retained but with, on average, two out of the three hydroxyl groups per glucose unit replaced by nitrate ($O \cdot NO_2$) groups.

Processing Techniques

Cellulose nitrate cannot be used in injection molding machines because of its sensitivity to high temperatures, neither is it compression molded since its thermoplasticity at relatively low temperatures makes it necessary to chill the mold before removing the molded article. Celluloid can be extruded and is obtainable in the form of rod, tube, or shaped profiles. It can also be cast into a thin film or pressed into large blocks, from which thicker sheet can be obtained by slicing.

Once the material has been obtained in sheet form, the way is open to fabrication by various molding processes. At a temperature of about 110°C cellulose sheet can be molded into complex shapes which are retained on cooling. Simple shapes can be molded by heating suitable blanks between pieces of thick flannel on steam-heated hot-plates and transferring these to a mold consisting of a hollow form and corresponding plunger. Deep drawing is accomplished by placing a preheated blank between two plates held together under light pressure. These plates have openings through which a shaped plunger is passed to draw the sheet with it against the frictional resistance of the pressure between the plates. After setting, the shape is taken off the plunger. Hollow articles such as toy ducks and dolls are made by blowing steam or heated compressed air through a nozzle between two sheets of celluloid held together at the edges by pressure in a heated mold. The material is forced out to the shape of the mold and becomes welded together at the edges. After the mold is cooled, the article is extracted.

Articles can also be fabricated from celluloid by means of a cement made by mixing equal volumes of acetone and amyl acetate. Another use for celluloid is as a covering material. Metal tubes, for instance, are covered by taking cellulose tubing, softening it in boiling water, stretching it on an oversize mandrel, then cooling and sliding it over the object to be covered. Subsequent immersion of the tube and the celluloid in hot water causes the celluloid to shrink to a tight fit.

Sheet. The cellulose nitrate as prepared above has a brownish yellow color which is removed by bleaching before being used for the production of transparent or delicately colored celluloid. After bleaching and washing, the cellulose nitrate is treated with an antichlor such as sodium sulfite or sodium thiosulfate followed by more washing and the removal of as much water as possible by centrifuging. It will have become obvious by now that a great deal of water is used in the manufacture of cellulose nitrate and, in fact, about 35,000 gallons of water are used for every ton of cellulose nitrate produced. This introduces another problem, that of contamination by metallic ions. Iron, particularly, adversely affects the color and stability of cellulose nitrate and the iron content must be kept below 0.03 ppm. This necessitates extremely strict control and treatment of the water supply.

The material is next dehydrated either by spinning with 85 percent ethyl alcohol or by pressing out a wet cake through which alcohol is forced at a pressure of 250 $lb/in.^2$. The alcohol-wet cellulose nitrate is converted to celluloid by mixing it with camphor in the proportion of 240 lb of cellulose nitrate to 80 lb of camphor, for one hour at 50°C. In the initial stages steam is circulated through the jacket of the kneader but the reaction is exothermic and the steam is soon shut off. The fibers soon swell and eventually the fibrous form disappears and a doughy mass is obtained. Stabilizers, such as zinc acetate or urea, and any necessary coloring agent are usually added at this point. When blending is complete the dough is filtered through calico cloth under hydraulic pressure, transferred to a mixer, and kneaded in vacuo to reduce the alcohol content to about 25 percent. The dough is now rolled, first on fast running, heated rollers, then on comparatively large, slow-running cool rolls or calenders. The condition of the 'hide' which is produced on these rollers must be closely watched, and the production of a satisfactory sheet is still largely an art. The alcohol content at this stage of the proceedings has been reduced to about 12-16 percent. The hides are about ½ in. thick and are cut to sheets of the same dimensions as the block press. A number of these sheets are pressed together under a pressure of 350 lb/in.^{2}. The press is heated with hot water to a temperature of 93°C and the temperature and pressure are maintained for about 4-5 hours. Subsequent cooling is carried out slowly in order to avoid the formation of solvent bubbles, and the finished blocks are conditioned in a cool room for 24 hours to allow the attainment of a uniform temperature inside the blocks.

Slicing is carried out by securing the celluloid block to a moving table which passes backward and forward under a knife inclined at an angle of 30 degrees to the block surface. Sheet thicknesses of from 0.005 in. to one inch can be obtained in this way. The alcohol content is, by now, down to about 12-14 percent and this is finally removed by seasoning the sheet. This is accompanied by shrinkage for which allowance must be made in the original calculation of sheet thickness. Initially, seasoning is carried out by hanging the sheets for 24 hours in a room maintained at a temperature between 27-32°C through which air is circulated continuously and then passed through a solvent recovery system. Seasoning is completed by laying the sheets on a slatted floor over hot-water pipes maintained at 50°C.

Finally the sheets are polished by pressing between heated steel plates at a pressure of 2000 lb/in.^{2} for 10-15 minutes, dip polishing by immersion in a solvent bath, or buffing with an abrasive such as tripoli powder.

Rods and Tubes. Cellulose nitrate can be extruded at a temperature of 85°C and a pressure of 2000 lb/in.^{2}, but since it is soft and easily deformed as it leaves the nozzle it has to be supported, usually by providing a conveyor belt take-off. Celluloid extrusions are seasoned in the same way as sliced sheet.

Film. The manufacture of celluloid film is essentially a continuous casting process in which cellulose nitrate solution is deposited on a smooth surface, the film being removed later as a transparent gel after partial evaporation of the solvent. The film is then passed through a heater chamber to remove the last traces of solvent, and finally reeled up.

The casting solution is prepared from cellulose nitrate having a higher degree of substitution and the percentage of added camphor is lower in order to give a tougher and more water-resistant material. The solvent consists of a 70 : 30 mixture of ether and alcohol and the solution is prepared in paddle mixers. After eight to ten hours of mixing, which results in a smooth, viscous 'dope,' the product is filtered through cellulose wadding and deaerated in vacuo at a temperature just under the boiling point of the solvent.

Film casting is carried out on an endless belt, a smooth surface being obtained by depositing a layer of hard gelatin on a copper support. After being stripped from the casting surface the film is seasoned in a heated drying cabinet where it passes over a zig-zag pattern of rollers. After a final passage over chilled rollers to cool the film it is wound into rolls (Fig. 19.1).

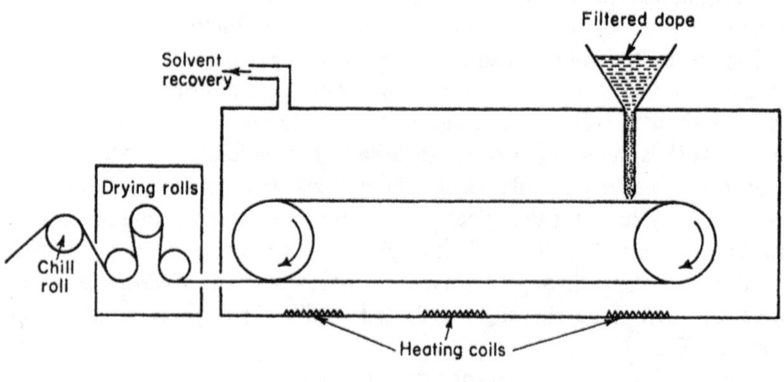

Fig. 19.1 Endless belt method of casting cellulose nitrate film

Properties

Although celluloid is classed as a thermoplastic there is, in fact, an appreciable amount of degradation each time the temperature is raised to the softening point. In the natural, i.e., unpigmented state, celluloid is a clear, tough, rigid solid with a slight yellow tint. This basic color does not, however, interfere with its colorability, and celluloid can be obtained in an unlimited range of color effects. It has a specific gravity of 1.38 and a refractive index of 1.5. It has a low water sensitivity, with a consequent good dimensional stability, and is resistant to dilute acids and alkalis. It is very inflammable, burning with a hot, white intense flame although it does not explode unless in the form of dust. It tends to discolor on aging and becomes brittle due to loss of camphor. It is soluble in methanol, ether/alcohol mixtures, esters, and hydrocarbons. It should also be noted that certain solvents such as chloroform do not dissolve cellulose nitrate but do extract the camphor, thus leading to adverse changes in the mechanical properties of the celluloid. Some typical physical properties of a general purpose grade of celluloid are given in Table 19.1.

Table 19.1

PHYSICAL PROPERTIES OF CELLULOID

Specific gravity	1.35–1.70
Refractive index (n_D 20°C)	1.5
Tensile strength (lb/in.^{2})	3–10 $\times 10^3$
Elongation at break (%)	25–35
Compressive strength (lb/in.^{2})	2–3 $\times 10^4$
Flexural strength (lb/in.^{2})	6–14 $\times 10^3$
Impact strength (Izod-ft lb/in. notch)	2–8
Coefficient of thermal expansion (per °C)	6–16 $\times 10^{-5}$
Softening point (°C)	80–90
Volume resistivity (ohm-cm)	2–30 $\times 10^{10}$
Dielectric strength (volts/mil)	600
Dielectric constant (10^6 cycles)	6.2
Power factor (10^6 cycles)	0.07–0.10

Applications

As we have already seen, celluloid has many valuable properties (toughness, rigidity, water resistance, etc.), is easily fabricated, and can be obtained in a wide range of colors. Since it is also relatively cheap it still has a wide range of applications in spite of competition from the newer thermoplastics. Its biggest disadvantage is its inflammability but, on the other hand, the fire hazard arises mainly when celluloid is stored in bulk so that the domestic user is little affected.

Among the main applications for celluloid are tooth-brush handles, knife handles, combs, brush backs, spectacle and handbag frames, and photographic film. Table tennis balls and billiard balls also provide outlets for celluloid as do bicycle mudguards and the coverings for pumps and handlebars.

The continued use of celluloid for table tennis balls is due to the fact that it satisfies the stringent requirements for height of bounce laid down in the regulations of the game.

Finally, no mention of the applications of cellulose nitrate would be complete without at least a brief reference to cellulose nitrate lacquers. These first appeared on the market in the 1920's but they were not used to any great extent until taken up by the automobile industry where they enabled tremendous reductions to be made in overall painting time.

CELLULOSE ACETATE

Historical Development. Cellulose acetate was first reported by Schutzenberger in 1865 when he obtained it by heating cotton in sealed tubes with acetic anhydride at 130–140°C. In 1879, Franchimont deduced from earlier experiments that esterification proceeded more smoothly if a dehydrating agent was present and discovered that the use of sulfuric acid in this role rendered the application of external heat unnecessary. Commercial development may be considered as having its beginning in 1894, when Cross and Bevan published a patent which outlined a practical manufacturing method. As in the case of Parkes and cellulose nitrate, Cross and Bevan were insufficiently interested in the commercial exploitation of this particular discovery but many other chemists had now entered the field and a steady stream of patents began to appear. Extensive use of cellulose acetate was retarded, however, by its insolubility in any but expensive or toxic solvents such as chloroform or tetrachloroethane. This problem was solved in 1905 by Miles, an American, who showed that a mild acid hydrolysis of the normally produced acetylation product gave a cellulose acetate (secondary cellulose acetate) which was insoluble in chloroform but soluble in acetone. From then on, progress was steady. In 1909 it was used for safety film, while WW I stimulated a demand for fire proof dopes for aeroplane wings, and in 1916 a factory was set up at Spondon, Derby under the direction of the Dreyfus brothers from Switzerland. At the end of the war other outlets had to be found for large quantities of cellulose acetate and this stimulated research into applications such as acetate rayon and cellulose acetate film and sheet. In 1919, Eichengrun first successfully molded cellulose acetate and this laid the foundation of the injection molding side of the plastics industry. Since WW II when cellulose acetate became in short supply, many markets have been lost to polystyrene, but although it is no longer the preeminent thermoplastic it still has many important applications.

Raw Materials

The subject of cellulose has already been dealt with in the section on cellulose nitrate but a few brief notes on the other raw materials utilized in the manufacture of cellulose acetate are given below.

Acetic Acid. Pure acetic acid melts at 16.6 °C and has a boiling point of 118.1°C. The most important commercial process for the production of acetic acid is by the oxidation of acetaldehyde which itself can be obtained either from acetylene or ethyl alcohol. In some of the most modern plants ethyl alcohol is converted to acetaldehyde and then to acetic acid continuously as a two-stage process. The ethyl alcohol is passed over a metallic silver or copper catalyst at 300–350°C in the presence of air. The resultant acetaldehyde is oxidized by air or oxygen in the presence of cobalt or manganese acetates at a temperature of 40–80°C and pressures of 50–100 lb/in.^{2}.

Acetic Anhydride. Acetic anhydride is a clear, colorless liquid boiling at 139.5°C and freezing at −73.1°C. One method of manufacture extensively used in the U.K. and U.S.A. is based on the cracking of acetone at 650–800°C to give ketene and methane, the ketene being absorbed in acetic acid.

$$\underset{\text{Acetone}}{(CH_3)_2CO} \longrightarrow \underset{\text{Ketene}}{CH_2{:}CO} + \underset{\text{Methane}}{CH_4}$$

$$\underset{\text{Ketene}}{CH_2{:}CO} + \underset{\text{Acetic Acid}}{CH_3COOH} \longrightarrow \underset{\text{Acetic Anhydride}}{(CH_3CO)_2O}$$

Methylene Chloride. Methylene chloride is a clear, colorless liquid boiling at 39.5°C and freezing at −96.7°C. It is prepared commercially by direct chlorination of methane with chlorine at 380–390°C. The resultant gases are washed to remove hydrochloric acid, condensed, and fractionally distilled. Methylene chloride constitutes about 35 percent of the condensate.

Manufacture of Cellulose Acetate

(a) Secondary Cellulose Acetate. The cotton linters or sulfite wood pulp are usually given a pretreatment with glacial acetic acid to open up the cellulosic structure. This enables the acetylating agent to penetrate the fibers quickly and uniformly, thus allowing acetylation to proceed simultaneously throughout the fiber instead of just on the surface. This also has the effect of reducing acetylation time considerably with a consequent saving in overall costs.

The suggested use of sulfuric acid as a dehydrating agent by Franchimont in 1879 has already been mentioned and it is an interesting fact that although over 80 years have passed since then, sulfuric acid is still almost universally employed.

A typical continuous acetylation process is the one used by I. G. Farbenindustrie and this will be briefly described. The pretreated cellulose is charged into the acetylizers, which consist of horizontal cylinders with bronze stirring paddles on a horizontal shaft, and this is followed by the reaction mixture in two or more portions. The reaction mixture consists of one part sulfuric acid, 300 parts acetic anhydride, and 400 parts methylene chloride per 100 parts of cellulose. Cooling water is circulated through an outer jacket and the temperature of the mixture is controlled at 50°C. The reaction takes about five to six hours, and the contents of the acetylizers are then pumped into the ripening and precipitating vessel which is a vertical cylinder fitted with a powerful stirrer. Water and sulfuric acid at 15-20°C are then added in the proportion of 50 parts water and 6.5 parts of concentrated sulfuric acid per 100 parts of cellulose, and agitation is continued until the required degree of hydrolysis is reached. Twenty parts of a 30 percent sodium acetate solution (per 100 parts cellulose) are then added to prevent further deacetylation.

The methylene chloride is then distilled off and recovered and the temperature of the residue raised to 80°C to maintain it in a fluid state. While at this temperature, and with careful stirring, a dilute solution of acetic acid (between 5 and 12 percent) is added at the rate of 800 parts acetic acid solution to 100 parts cellulose. The resultant slurry is discharged from the bottom of the vessel via a cutter which breaks up any large lumps which may have formed. The cellulose acetate now undergoes extensive washing by means of a continuous countercurrent operation followed by drying, first by a centrifuge and later by vacuum dryers.

(b) Cellulose Triacetate. The production of cellulose triacetate in fibrous form is achieved by carrying out the acetylation in nonsolvent

media so that the original fibrous form of the cellulose is preserved. Benzene, toluene and xylene have all been suggested, with perchloric acid and its salts replacing the sulfuric acid as catalyst. The change in catalyst was suggested in order to prevent the retention of sulfur residues, which lead to loss of stability. The removal of sulfur residues is not, of course, such a problem in the solution processes used to prepare the secondary cellulose acetate. Later work, however, has largely overcome the difficulties inherent in the nonsolvent media acetylation process.

Processing Techniques

Cellulose acetate can be made into sheet by similar techniques to those used for cellulose nitrate. Cellulose acetate flake, solvent, plasticizer and, if required, colorants are mixed together in a Werner-Pfleiderer type kneader, and the resultant viscous mixture is filtered through screens, under pressure. After filtration the composition is passed to a pair of heated rollers, and rolling is carried out until the solvent content is reduced to about ten percent. The mass is cut and removed from the rolls as a hide, about 1 in. thick, and a number of these hides are then pressed into a block at about 95°C and a pressure of about one ton/in.^{2}. After several hours the block is slowly cooled, still under pressure, and transferred to the slicing machine.

Sheet and film can also be produced by extrusion, either from flat dies or from circular dies, with subsequent slitting. No solvents are used but plasticizers are incorporated, together with dyestuffs if a colored sheet is required. Thin film is also manufactured by solvent casting on to endless bands, as described for cellulose nitrate.

Cellulose acetate can also be injection molded, extruded as rod, and tube, etc., and blow molded into bottles.

Cellulose acetate fiber (acetate rayon) is manufactured by dissolving the cellulose acetate in acetone and forcing the solution through spinnerets. A typical spinneret contains about 60 fine jets 0.03 mm in diameter, but this depends on the type of yarn required. The streams of viscose solution emerging from the jets are forced downwards against a countercurrent of warm air which evaporates the acetone leaving the fibers to be wound up on bobbins.

Properties

Cellulose acetate has a specific gravity of 1.32 and decomposes at 240°C.

On burning, the odor of acetic acid is noticeable. It is sensitive to water absorption and hence is not dimensionally stable. In this respect the triacetate is superior.

The properties of cellulose acetate vary with the degree of acetylation, and the degree and type of plasticizer used, but, in general, it is a hard, tough material with good electrical properties when dry, combining good insulating properties with good corrosion and arc resistance.

Tensile strengths vary from about 5000 to 11,000 lb/in.^{2} with impact strengths from 1-2 ft lb/in. notch but mechanical properties, generally, are reduced by moisture pick-up. The effect of aging or sunlight is slight provided the sulfur content has been effectively reduced after acetylation.

Weak acids and alkalis have only a slight effect but strong acids and alkalis cause decomposition. Secondary cellulose acetate is soluble in acetone and 80/20 acetone/water, but insoluble in methylene chloride, whereas the triacetate is insoluble in acetone and acetone/water but soluble in methylene chloride.

One of the main differences between cellulose nitrate and cellulose acetate is flammability, the latter being virtually nonflammable.

The properties of cellulose acetate which are of most interest in the form of film are crystal-clear transparency and a fair degree of permeability to water vapor and gases. The permeability decreases, however, with degree of acetylation.

In fiber form, cellulose acetate confers on textiles such properties as softness and suppleness, crease resistance and a lower degree of moisture absorption than wool, silk, or viscose rayon.

Applications

As an injection molding material, cellulose acetate has the advantage of flexibility in formulation by virtue of the range of properties available when changing the molecular weight, degree of acetylation, or the amount and type of plasticizer added. In general, it is used for its combination of hardness and toughness coupled with good moldability, but it is susceptible to dimensional changes in moist atmospheres. The chief applications for cellulose acetate are toys, combs, cutlery handles, spectacle frames, handles for domestic equipment, and certain automotive and electrical components.

Cellulose acetate extruded tubing is used in the manufacture of fountain pens, bicycle pumps, etc., and, to some extent, in the covering of handrails, although in this application it has been largely superseded by

PVC.

Cellulose acetate film has many uses, particularly in the fields of packaging and display. The fact that it permits the passage of water vapor and gases while remaining impervious to liquid water makes it particularly useful where a 'breathing' film is required as, for instance, in the packaging of certain fruits and vegetables. It is also used for the manufacture of window cartons, the cellulose acetate being sealed to the cardboard by suitable adhesives. One disadvantage of cellulose acetate as a wrapping material is the fact that it is difficult to heat-seal, but this has been overcome by coating with compositions based on cellulose nitrate, a polyamide, and plasticizers. Cellulose acetate film has also found use as a glazing material for greenhouses, chicken coops and cold frames since it is reasonably transparent to uv light (about 85 percent transmission through 10 mil film—transmission of visible light is about 93 percent through 10 mil film). Reinforcement is sometimes given to the sheet by placing cotton scrim or wire mesh between two layers of film and laminating in a hot press, or by passing the reinforcement through a cellulose acetate casting solution. Cellulose triacetate film is widely used for cinematography and x-ray film although experiencing competition from the physically tougher polyethylene terephthalate.

A thicker film of cellulose acetate (between five and 10 mil) is used for the manufacture of rigid containers either by cutting, creasing, and welding (solvent or HF) by spiral winding, or by pressing and deep drawing. Vacuum forming is also a suitable technique for the creation of various shaped containers. Such rigid containers are used for the display of flowers, especially orchids, roses, etc., and for fancy assortments of sweets or chocolates.

Acetate film is also widely used in lamination for greetings cards, magazine covers, and price tickets, etc., or in combination with paper, aluminum foil, or rubber hydrochloride in the manufacture of sachets for shampoos, etc. In the latter case, the acetate film is used as the outer layer and carries the printing on the sachet. The film is reverse printed and is thus sandwiched between the paper of aluminum foil and the cellulose acetate, giving good appearance with complete protection against scuffing for the printing.

Cellulose acetate sheet can be used as a dielectric in capacitors and for electrical tape while among the miscellaneous nonelectrical uses are various types of lamp shades, typewriter ribbon, and (laminated to fabric) collars and cuffs.

The use of cellulose acetate as a fiber for textile manufacture has already been mentioned and one should also be reminded here of its use as a lacquer, as in its early use as an aircraft 'dope.'

CELLULOSE ACETATE/BUTYRATE (C.A.B.)

Manufacture

Cellulose acetate/butyrate is manufactured in the U.S.A. and in Germany but not in the U.K. The manufacture is similar to that used for cellulose acetate, a typical process being as follows:

One hundred kilograms of cellulose (from cotton linters or wood pulp) are stirred with 400 kg of methylene chloride, 250 kg of mixed acetic and butyric anhydrides together with 1 kg of sulfuric acid as a catalyst. The reaction is continued for 4 hours and the solvent removed by distillation. The residue is 'drowned' in a large volume of water and centrifuged. After neutralizing the residual acidity the product is finally washed and dried.

Properties

Cellulose acetate/butyrate is somewhat lighter than cellulose acetate, having a specific gravity of 1.21 (cellulose acetate–1.32). It has a higher softening point than cellulose acetate and a better resistance to acids. Although it is discolored by oxidizing acids such as 10 percent nitric, it resists 10 percent HCl and 30 percent H_2SO_4. It can be made quite nonflammable by the addition of trichloro-ethyl phosphate but on storage it generates the objectionable odor of butyric acid (the smell is similar to that of rancid butter). The moisture absorption of cellulose acetate/butyrate is less than that of cellulose acetate and this, in combination with the fact that cellulose acetate/butyrate needs less plasticizing to attain any particular performance, renders it more dimensionally stable and gives it a better weathering resistance than cellulose acetate. It also has good resistance to petrol and oil.

As an injection molding material cellulose acetate/butyrate is preferred to cellulose acetate because less pressure is needed to mold the former while the resultant molding can be obtained harder and with increased heat resistance.

Applications

In the U.S.A., cellulose acetate/butyrate has been used for injection molding large items such as telephone housings and it is also used for car steering wheels.

As vacuum formed sheet it has found widespread use, in the U.S.A. particularly, for outdoor signs where its rigidity and weatherability are the important characteristics.

Cellulose acetate/butyrate is easily extruded and pipes made of it have been used for carrying water and oil.

Many of the above applications are being challenged by other thermoplastics but one interesting application is particularly well adapted to cellulose acetate/butyrate. This is the protection of tools and gages, etc., by strippable coatings. The plasticized cellulose acetate/butyrate formulation is heated and the object dip-coated. When the tool is required for use the coating is easily removed by stripping.

CELLULOSE PROPIONATE

The manufacturing methods and equipment used for cellulose propionate are similar to those used for cellulose acetate, the resultant product being dimensionally stable, hard without being brittle and possessing good flexural strength and high impact strength. It has a good color range and an extremely good surface finish when molded and, like cellulose acetate/butyrate, it resists petrol and weak acids and alkalis. It has a slightly lower softening point but a higher impact strength than cellulose acetate/butyrate. In the amount of plasticizing necessary to obtain equivalent flow during molding, cellulose propionate lies between cellulose acetate and cellulose acetate/butyrate.

Main applications so far have been in the fields of portable radio sets, housings for small domestic appliances, and telephone hand sets.

REGENERATED CELLULOSE

Historical Development. The fact that cellulose could be dissolved in an ammoniacal solution of copper oxide, and regenerated by dilution with water or weak acids and bases was discovered by Schweitzer in 1857 and formed the basis of his cuprammonium rayon process. In 1892, another means of producing regenerated cellulose was discovered by Cross, Bevan, and Beadle. This process, which is known as the xanthate process, was the foundation stone for the cellulose film and viscose rayon industries. A method for coagulation and regeneration of xanthated cellulose was patented by Stearn in 1898, but the person chiefly responsible for the development of equipment for the continuous production of cellulose film was Brandenberger to whom patents were first issued in 1911. The uses for cellulose film were at first rather restricted because it had a high transmission rate for water vapor, but, in 1927, processes were developed for moisture-proofing the film, and the demand greatly increased.

In the field of fiber, the cuprammonium rayon is now produced on a minor scale only, for high-grade fabrics, because of its high production costs, and viscose rayon now holds the field.

Manufacture

Regenerated Cellulose Film. Cellulose, in the form of cotton linters or sulfite wood pulp, is steeped in a 17-21 percent aqueous solution of sodium hydroxide for about an hour at room temperature. The excess alkali is pressed out and the alkali-cellulose is shredded in a Werner-Pfleiderer type mixer and allowed to age at room temperature for about two to three days. During this time, the alkali-cellulose absorbs oxygen from the air and the length of the cellulose chain is shortened, this being necessary to reduce the viscosity of the resultant xanthate solution.

The aged alkali-cellulose is next fed to xanthating churns or drums which rotate slowly on a horizontal axis while carbon disulfide is sprayed on the pulp. The amount of carbon disulfide added is usually in the range of 30-40 percent, reckoned on the dry weight of cellulose. The xanthating reaction is exothermic, and cooling of the churns is necessary. After about three hours, the sodium cellulose xanthate is discharged into a tank where it is dispersed by stirring with a dilute sodium hydroxide solution. The dispersion takes about two to six hours stirring at room temperature and the resultant solution (six to eight percent cellulose and six to seven percent caustic soda) is known as 'viscose.' The color of the solution is a bright orange due to the presence of sodium trithiocarbonate (Na_2CS_3). The basic reaction in the manufacture of viscose can be represented by the following equation where R represents the cellulose molecule.

$$R{\cdot}OH + NaOH \rightarrow R{\cdot}ONa + H_2O$$

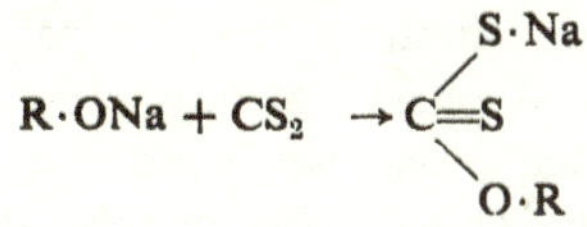

It should be understood, however, that the actual number of xanthate groups is of the order of one per two glucose units in the cellulose molecule.

The viscose is filtered and is then passed into a vessel where it is allowed to ripen for a period of from four to five days. During this ripening a slow hydrolysis takes place, which is followed by measurements of viscosity. If allowed to ripen too far, the viscose forms a gel which is useless. Air bubbles in the solution are removed by a vacuum process prior to delivery to the casting hopper. The thickness of the foil is regulated by varying the opening at the bottom of the hopper where the viscose is delivered to the casting drum or endless band which revolves in the first of a series of tanks.

The first tank contains the coagulating solution consisting of 10–15 percent of sulfuric acid and 10–20 percent sodium sulfite at a temperature of 35–40°C. This bath not only coagulates the viscose but initiates the regeneration back to cellulose. The coagulated film is now led through a series of tanks containing guide rolls and various solutions designed to complete the regeneration (sulfuric acid/sodium sulfate at a lower concentration than the first), wash out the carbon disulfide, hydrogen sulfide, sulfuric acid, and sodium sulfate (hot water at 90°C), complete the removal of sulfur (sodium hydroxide/sodium sulfide) and bleach the now transparent but still slightly colored film (sodium hypochlorite or sodium peroxide). The film is next run through a bath containing glycerol or ethylene glycol, which acts as a plasticizer and confers flexibility on the film, after which it is passed through a drying oven and reeled up.

At this stage we have plain cellulose film which, as we mentioned earlier, is permeable to moisture vapor. In addition, this plain film is nonheat-sealable because cellulose itself is not a thermoplastic so that a subsequent coating treatment must be given if moisture-vapor proofness or heat-sealability are required. A typical coating consists of a dibutyl phthalate plasticized cellulose nitrate, together with waxes and natural resins, dissolved in volatile organic solvents. The film is passed through the coating bath, dried to evaporate the solvent, then passed through a high humidity chamber to restore flexibility to the film. Film of this type is referred to as moisture-proof cellulose film. Coatings of this type are prone to float off in wet conditions but most films now incorporate urea formaldehyde to give an anchored coating that resists wet conditions. Cellulose film can also be coated with polyvinylidene chloride copolymers to give extra good moisture-vapor barrier properties. Such films are usually referred to as polymer coated cellulose films.

Viscose Rayon. The viscose solution is prepared as for the regenerated cellulose film. It is then fed to spinnerettes which contain a large number of fine holes through which the viscose is forced into the coagulating bath by metering gear pumps. The composition of the coagulating bath is slightly different from that used for film and the temperature is a little higher (40-55°C) since here coagulation and regeneration must proceed simultaneously. In addition to sulfuric acid and sodium sulfate a little zinc sulfate is added, together with a wetting agent.

If continuous filament is required, the spinnerettes contain up to a 1000 holes each of diameter 0.002-0.010 inches, while for the production of staple fiber large spinnerettes with 3000 to 15,000 holes are used. In the continuous process for spinning rayon, a single strand is passed through a series of baths, as described under 'Regenerated Cellulose Film,' and is then wound up on bobbins.

Staple fiber is spun from short lengths of rayon in the same manner as wool and cotton (which are, of course, obtained naturally in the form of short lengths). To produce rayon staple, the filaments from a group of spinnerettes are drawn into a continuous loose (untwisted) rope, known as 'tow,' and cut either immediately or after de-acidification. The cut fiber is then desulfurized, bleached, and dried.

Properties

Regenerated Cellulose Film. Regenerated cellulose film has a light transmission equal to that of glass in the visible region (about 90 percent) and greater than glass in the uv region.

Dry cellulose film is practically impermeable to the permanent gases but becomes more permeable when wet, while as already indicated, in the uncoated state cellulose film has a very high moisture vapor permeability, but this is very much reduced by coating, especially with polyvinylidene chloride copolymers (*see* Table 19.2).

Table 19.2

MOISTURE VAPOR PERMEABILITY OF VARIOUS GRADES OF REGENERATED CELLULOSE FILM ($g/24\ h/100\ in.^2/mil$ at 100°F 95% R.H.)

Plain cellulose film	15.0
Moisture-proof cellulose film	1.0
Polymer coated cellulose film	0.7

Other properties, such as tear strength, tensile strength, etc., are given in Table 19.3 but it should be noted that these properties are dependent on

Table 19.3

PHYSICAL PROPERTIES OF REGENERATED CELLULOSE FILM (ONE MIL THICK)

Tensile strength (of 1 in. strip)–machine direction	15.8 kg
transverse direction	11.6 kg
Elongation–machine direction	14.3 percent
transverse direction	26.2 percent
Tear resistance–machine direction	2.4 g
transverse direction	4.4 g

the ambient humidity, a typical example being a film with a tear strength (in grams) of 3, at 7 percent R.H., 4½ at 35 percent R.H. and 8½, at 60 percent R.H. Another point to be noted with regenerated cellulose film is that the moisture vapor resistance of coated films is reduced on creasing because of damage to the coating.

Applications

Before dealing with the applications of regenerated cellulose film it should be noted that there are over a hundred different types of film, varying in factors such as weight per unit area, type of coating, color, etc. In the U.S.A. the three major manufacturers have their own designations for the different films except that the letter M always denotes moistureproof grades and the letter S denotes heat sealing grades. In Europe there has been a great deal of standardization using codes made up of four elements, written in the following order:

1. A code number indicating the approximate weight in grams of ten square meters of film.
2. One or more letters denoting the basic type of film, such as P for

plain (nonmoistureproof).

3. A letter denoting special characteristics, such as C for colored.

4. A number used as a suffix where an additional description is required to cover special features of manufacture and certain end-uses.

The letters denoting the basic type of film are given below:

P	This is the basic film, nonmoistureproof and nonheat sealable.
M	Nitrocellulose-coated on both sides. Moistureproof. Coating is heat sealable only when letter S is included.
S	Coating is heat sealable.
QM	Nitrocellulose-coated on both sides but less moistureproof than M.
PS	Nitrocellulose-coated on both sides but less moistureproof than QM.
DM	Nitrocellulose-coated on one side.
MXXT/A	Copolymer (polyvinylidene chloride type) coated by aqueous dispersion on both sides.
MXXT/S	Copolymer coated by solvent process on both sides.
B	Opaque.
C	Colored (name of color follows).
F	For twist wrapping (more flexible).

Nonmoisture-proof Films. These films are used where protection from dust or grease is required but where mold growth may be a problem. For instance, meat pies, cakes with a high moisture content and little or no fruit or sugar, and cakes containing fondants, all require a highly permeable film. If heat sealability is required then QMS grades would be used rather than P. Nonmoisture-proof film is also suitable for chocolate-coated sweets and is the most generally used film for wrapping fresh sausages.

P film is also used as the base for pressure-sensitive tapes.

Moisture-proof Films. The use of moisture-proof grades is a must when dealing with hygroscopic products and such grades of regenerated cellulose film have been used for packaging biscuits, sugar confectionery, potato crisps, shelled nuts, dried fruit, bread, and for the over-wrapping of cartons of sweets and pharmaceuticals.

Single-sided Film. This type of film (DMS) is used for packaging fresh meat with the uncoated side in contact with the meat. The moistening of the uncoated film raises its permeability to gases, including oxygen, and gives just the right combination of conditions to keep the fresh color of the meat.

Single-sided film is also used when extrusion coating with polyethylene is required, the polyethylene having a more secure key on the base cellulose film than to the cellulose nitrate coating. Such cellulose/polyethylene combinations are used for vacuumized pouches of bacon, cheese, coffee, etc.

ETHYL CELLULOSE

Manufacture

Ethyl cellulose is the most widely manufactured of the cellulose ethers and methods for its production were patented in 1912 almost simultaneously by Dreyfus, Lilienfeld, and Leuchs. Commercial production was first achieved in the U.S.A. in 1935. The standard commercially available ethyl cellulose has been substituted to the extent of about two and a half ethoxyl groups per glucose unit (out of a maximum of three).

The first step in the production of ethyl cellulose is the formation of alkali cellulose by treatment of linters or pulp with 50 percent caustic soda solution. The excess of alkali is pressed out, the alkali cellulose broken up to a 'crumb' and dried in a vacuum oven to a moisture content of between 10-25 percent. The 'crumb' is then conveyed to reactors, where ethyl chloride is added (under pressure, since ethyl chloride boils at 12°C) and the temperature raised to between 80-150°C for periods varying from 6-24 hours. By varying temperature, reaction time, pressure, and the ratios of raw materials, it is possible to control the ethoxy content and the viscosity (a measure of molecular weight) of the resultant ethyl cellulose.

When the reaction is judged to be complete, excess ethyl chloride and the ethyl alcohol and ether formed as by-products are distilled off and the ethyl cellulose is thoroughly washed until free from salt and caustic soda. Drying is carried out in a tray drier at 90°C.

Properties

Ethyl cellulose is a tough and flexible thermoplastic with a specific gravity of 1.14-1.20 and, is thus, the lightest of the cellulose plastics. The material containing 2.4-2.5 ethoxy groups per glucose unit is soluble in a wide range of solvents including benzene; toluene; xylene; ethyl and butyl acetates; acetone; methyl, ethyl, and butyl alcohols; ethylene dichloride; and carbon tetrachloride. It is also compatible with a wide range of plasticizers, resins, and waxes, thus allowing a great deal of flexibility in formulation to give specific properties. Moisture absorption is slightly less than cellulose acetate but higher than cellulose nitrate, and it is extremely resistant to alkalis. Ethyl cellulose is not readily flammable and has good electrical insulating properties. These are particularly valuable for low-temperature uses where flexibility is important, since ethyl cellulose retains its toughness and flexibility down to about −40°C. The softening point is between 60–80°C.

Applications

Ethyl cellulose has been used in injection-molded parts requiring high impact strength at low temperatures or where resistance to alkalis is important. It is widely used in coating compositions with other resins, where it is added to improve toughness and flexibility, and in solvent-based strippable coatings. Ethyl cellulose is also used in adhesive compositions, giving low-temperature toughness and control of tackiness and softness at room temperature. Ethyl cellulose can also be prepared as a film which is flexible and tough at low temperatures.

BENZYL CELLULOSE

Benzyl cellulose was manufactured in Europe in the early 1930's and thus predated the manufacture of ethyl cellulose. Benzyl cellulose, however, has certain disadvantages, notably a lower softening point than ethyl cellulose and instability to heat and light. Benzyl cellulose has never achieved commercial success partly because of its drawbacks and partly because ethyl chloride is more freely available at a more economic price than benzyl chloride.

The production of benzyl cellulose is similar to that of ethyl cellulose except that it is not necessary to react under pressure since benzyl chloride has a higher boiling point (175°C). Benzyl cellulose is relatively nonflammable, is resistant to water, acids, and alkalis and is flexible at low temperatures.

METHYL CELLULOSE

Methyl cellulose is prepared in an identical fashion to ethyl cellulose, using methyl chloride or methyl sulfate instead of ethyl chloride. Only in the washing of methyl chloride is there any difference of approach, hot water being used because methyl cellulose is soluble in cold water. Methyl cellulose is used as a thickening agent and emulsifier in various cosmetic creams, and in adhesives. It is also edible, so that it can be used in salad dressings and other thickened food products. It is a useful thickening agent for water-based paints and has been used as a size for textiles and in the manufacture of grease-proof paper.

SODIUM CARBOXY METHYL CELLULOSE

This particular ether is prepared from alkali cellulose by reaction with the sodium salt of chloracetic acid in a Sigma blade mixer. When the reaction is completed the excess of sodium hydroxide is neutralized and the water-soluble product precipitated with alcohol and washed with aqueous methyl alcohol.

The degree of substitution achieved commercially is usually about 0.7 ether groups per glucose unit, and this type of sodium carboxy methyl cellulose is water soluble. It is extensively used as a stabilizer and thickening agent in ice cream, synthetic creams, cosmetic creams, wall-paper adhesives, and in water-based paints. It is also used in detergent and soap formulations where its function is to strengthen the foams produced and prevent redeposition of the dirt which was originally loosened by the detergent.

20.

NATURAL AND MODIFIED RUBBERS

The birth of natural rubber dates back to the fifteenth century when Columbus first reported that he had been intrigued on his journey through the New World to see natives playing with a ball that actually bounced.

The *Hevea brasiliensis* tree is the main source of natural rubber and its existence in the Amazon basin was first reported by Charles de la Condamine to the French Academy in 1736. He described to them the native methods of obtaining rubber from this tree and also methods that were used to convert it into useful products. La Condamine named this material 'cahutchu,' the word being of Indian origin, and from this has come the more common word 'cauoatchouc.' The actual derivation of the word 'rubber' is attributed to John Priestley, the chemist who discovered oxygen in 1774, who successfully used this material for rubbing out unwanted pencil marks.

In 1876, Sir Henry Wickham collected many thousands of seeds used for growing rubber trees in South America, and shipped them to Kew Gardens. Many of these rubber-tree seeds grew into healthy young plants some of which were sent to Ceylon. A few years later, Henry Ridley, who was at that time a director of Singapore Gardens, against strong opposition planted rubber trees in Malaya. Today, thanks to his action, this country is the leading producer of raw natural rubber.

The uses of natural rubber for certain applications were limited for some time owing to the fact that it deteriorated badly in the presence of oxygen. Fortunately, chemicals were developed for inhibiting this action, and in 1924 the first commercial antioxidant became available, since when many other similar materials have been used successfully.

Raw Materials

Raw natural rubbers are obtainable from a wide variety of trees and plants found in tropical surroundings. The *Hevea* tree is by far the most important source, due largely to its high yield. The rubber is obtained as a milky latex which comes not from the sap, as many people believe, but from the latex capillary vessels found between the bark and the wood of the tree. These vessels are cut by making an incision in the bark of a tree and allowing the latex to drip out slowly, the process being commonly known as 'tapping'. The actual process consists of tapping the tree, allowing the latex to collect in an aluminum cup, then straining and diluting followed by coagulation with formic or acetic acid. The coagulated material is then washed and thoroughly dried. It may then be converted into any one of the following forms:

(a) Pale Crepe. In this process the coagulated material is first washed and pressed between tightly closed rolls until a thin crepe is obtained. This is then allowed to dry in the air, a light-colored rubber with a slow cure rate being obtained.

(b) Smoked Sheet. Smoked sheet is obtained by pressing the coagulated rubber between rolls, washing and finally drying in an atmosphere of smoke and hot air. A brown rubber results which cures faster than crepe rubber.

(c) Blankets. The coagulated rubber may be kept for some weeks on the plantation where it undergoes a maturing cycle. This essentially is decomposition of the nonrubber substances by bacterial attack. The natural rubber is subsequently passed through rolls to form a thick crepe which is then air dried. The rubber is a dark color and cures fairly quickly.

(d) Concentrated Latex. The latex may be concentrated by centrifuging and then stabilized by adding ammonia as a preservative.

Structure

Chemically natural rubber is largely *cis* 1,4 polyisoprene and the hydrocarbon has the empirical formula C_5H_8. It is a linear high molecular weight polymer of the general formula

$$\left[-CH_2-\overset{\overset{\displaystyle CH_3}{|}}{C}=CH\cdot CH_2- \right]_n$$

The naturally occurring polyisoprenes are in the predominantly *cis* form of natural rubber or the *trans* form of gutta percha. The geometric isomers are as follows:

(*a*) *cis*-isomer (natural rubber)

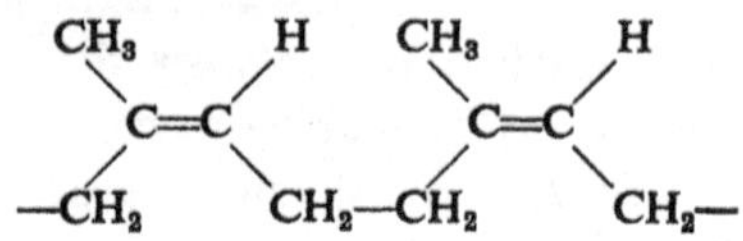

(*b*) *trans*-isomer (gutta percha, balata and chicle)

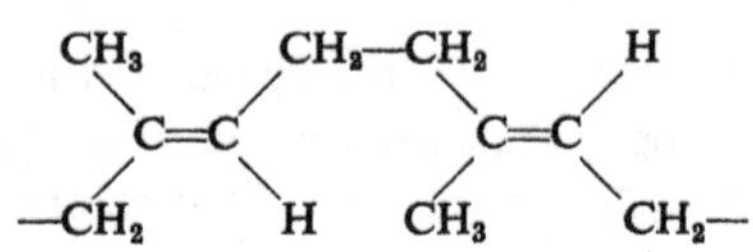

Compounding Processes and Ingredients

Before the raw natural rubber is processed, it is essential to compound it because in the raw state it is technically and economically unsuitable for direct use. To make the natural rubber suitable for compounding it must undergo a process known as 'mastication,' which consists of kneading the rubber between warm rolls until it becomes thermoplastic by breaking down the polymer chains. The rubber is then soft enough to enable the compounding ingredients to be added.

Mastication

Raw natural rubber consists of long chains as follows:

Fig. 20.1

There are basically two forms of mastication:

(a) Physical Chain Scission. This effect is achieved by the high shearing forces obtained by passing rubber through a two-roll mill.

(b) Oxidative Chain Scission.. This method can best be accomplished by masticating the rubber in an internal mixer such as the Banbury.

The resultant masticated rubber consists of short chains and the rubber also develops greater optical homogeneity.

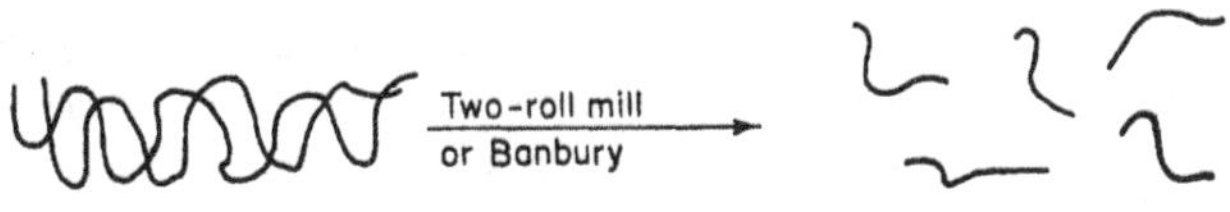

Fig. 20.2

Peptizers

These speed up the break-down in the process of mastication. They are chemicals, the most important of which are the aromatic hydrazines (e.g., phenylhydrazine), aromatic mercaptans (e.g., β-naphthyl mercaptan), and aromatic thiophenols (e.g., pentachlorothiophenol). Their main function is to block off the broken ends of the polymer chain, thereby preventing the short chains from recombining.

Softeners

Softeners are added to rubber compounds to facilitate the addition and dispersion of the other solid compounding ingredients as well as to enhance the processing properties for extrusion, calendering, etc. Tack is another property achieved by adding softeners, and is essential for many applications such as tire building.

Numerous materials are used as softeners in the rubber industry, including vegetable and petroleum oils, waxes, bitumens, pine tars, and various resins. Coumarone-indene is one such resin; it improves the adhesive tackiness of the rubber, thus making it suitable for such applications as adhesive plasters and tapes.

Softeners used as processing aids for the uncured rubber have also been referred to as lubricants, plasticizers, tackifiers, and dispersing agents.

Another important softener is known as 'factice' which may be prepared by heating drying oils with sulfur or sulfur monochloride. The product varies from a soft to a hard rubbery material, depending upon conditions used. It is now used not only as a low density resilient filler but also as a processing aid and softener for rubber. Factice has also been shown to have some curing action when added to rubber compounds.

Vulcanization

The process of vulcanization improves the rubbers' mechanical strength and elasticity, and reduces its sensitivity to changes of temperature. The discovery and development of the process of vulcanization is discussed in the section on 'Hard Rubber'; it can be briefly defined as the treatment of a raw rubber which decreases its plasticity but retains its elasticity. The process is schematically represented in Fig. 20.3.

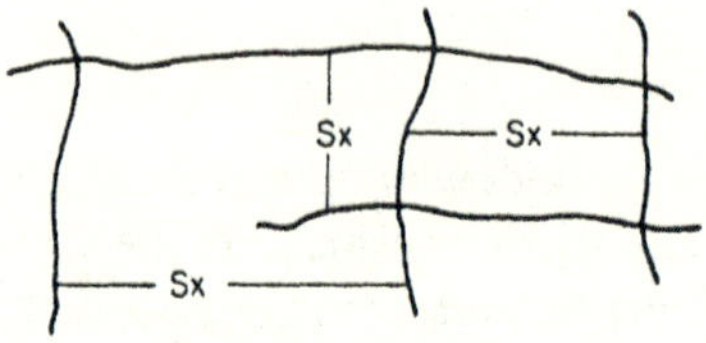

Fig. 20.3

There are various theories of vulcanization but most authoritative bodies agree that it is a cross-linking process. A three-dimensional structure is obtained by cross-linking the individual polymer molecules at those points of the molecule which react with the vulcanizing agent.

The vulcanizing process with sulfur alone takes some time to carry out and, therefore, to speed up the reaction, accelerators and activators are mixed with the rubber compound. The accelerators and activators are chemicals whose main functions are to reduce the time and temperature necessary to obtain a rubber possessing the desired physical properties.

The most popular and widely used vulcanizing agent is sulfur, although many other materials such as selenium, tellurium, sulfur monochloride, thiuram disulfide, or some alkyl phenol sulfides may be used. As little as 0.2 percent of sulfur causes vulcanization, although it is normal to use 2-3 percent to give a flexible rubber, and as much as 30 percent for a hard rubber.

Accelerators

These materials are chemicals and, as the name implies, serve to accelerate the rate of vulcanization. The original accelerators were inorganic oxides and salts such as lead carbonate, lime, magnesia, zinc oxide, and litharge. Lead carbonate was mentioned in the original patent of Goodyear on vulcanization.

The work carried out on the inorganic accelerators was followed by the discovery, by George Oenslager, in 1906, that aniline and other organic materials such as hexamethylene tetramine speeded up the rate of vulcanization. Aniline, however, is toxic and after a few years it was replaced by *p*-aminodiphenylaniline and diphenylguanidine. These developments were then followed by the introduction, in 1925, of the thiazoles, of which mercaptobenzthiazole (MBT) has become one of the most well-known and probably the most widely used. New classes that have found use since are the thiurams, their derivatives the carbamates, and the xanthates.

Accelerators can therefore be conveniently classified in the following general chemical groups:

(*a*) Aldehyde-amine reaction products (e.g., aniline and hexamethylene tetramine)
(*b*) Guanidine derivatives (e.g., diphenylguanidine and diortho-tolyl guanidine)
(*c*) Thiazoles and their derivatives (e.g., mercaptobenzthiazole)
(*d*) Thiurams and dithiocarbamates (e.g., tetramethylthiuram disulfide and zinc dimethylthiocarbamate)
(*e*) Xanthates (e.g., zinc isobutyl xanthate and sodium isopropyl xanthate)
(*f*) Sulfonamides (e.g., *N*-cyclohexyl benzothiazole-2-sulfonamide and N-tert-buryl benzothiazole-2-sulfonamide)
(*g*) Thioureas (e.g., thiocarbanilide)

Activators

Both organic and inorganic types may be used to activate the accelerators mentioned above without any apparent effect on the final properties. The majority of accelerators require both zinc oxide and a fatty acid such as stearic acid to develop optimum final physical properties. The amount may vary between 2-5 parts by weight per hundred parts of rubber (p.h.r.), zinc oxide, and 1-3 p.h.r. stearic acid. The activators should always be added at the beginning of the compounding process because of the difficulty in dispersing them evenly throughout the rubber mix.

Antioxidants

The main function of an antioxidant is to protect the rubber against decomposition by oxidation and thereby improve the resistance to aging. Antioxidants therefore maintain the physical properties of the rubbers while in service and constant use. The types used for rubbers are normally oxidizable derivates of aromatic amines and certain phenols. These materials are described more fully in Chapter 23. However, in this particular chapter mention is given to the general classification into three main groups: (i) Nonstaining–these are also non-discoloring for use in white or light colored stocks. Typical materials are 2,2′-methylene-bis (4-methyl-6 tertiary butyl phenol), fortified alkylated mixed phenols, alkylated bis phenol and straight alkylated phenols. (ii) Semistaining–these are used for light colored applications but not where minimum discoloration is a requisite. Typical materials are octylated diphenylamine and polymerized 1,2 dehydro-2,2,3-trimethylquinoline. (iii) Staining–these are not used for critical applications and find use in tire stocks and other black furnace compounds. A typical material is the reaction product of diphenylamine and acetone.

Retarders can also be used with thiazoles or activated thiazole type accelerators to prevent scorching during processing. One of the more common materials is a modified phthalic anhydride.

Ozone attacks rubber, but antiozonants can be added to improve its resistance; antiozonants are alkyl substituted amines. A typical product is N,N′-bis (1,4-dimethylpentyl-*p*-phenylene diamine) and these materials also offer fair protection against oxidative degradation.

Fillers

Raw, uncompounded rubber is deficient in mechanical strength but certain inorganic compounds improve its physical properties. The requirements of a suitable filler are discussed later in the book, but they may be classified in two broad groups:

(a) Reinforcing Fillers. This group includes carbon black, furnace and channel black, magnesium carbonate, china clay, hard clay, hydrated silica, and aluminum and calcium silicates.

(b) Nonreinforcing Fillers. Fillers of this group impart little or no improvement to the physical properties of a rubber. They do, however, perform a useful function in improving processing properties, hardness, and chemical resistance. They may also act merely as diluents, and as such reduce the cost of the final molding. Known also as inert fillers, they include whiting, barytes, talc, kieselguhr, and slate dust.

Processes and Processing Techniques

Natural rubber and compounded latex are processed and fabricated by many of the techniques used for plastics, which are, in fact, basically modifications of those developed originally for rubber. Such processes include compression and transfer molding, extrusion, calendering, and even injection molding, and since these are described fully elsewhere in the book no further elaboration is made here. It must be pointed out, however, that processing conditions used for plastics cannot, without discretion, simply be transposed to rubbers. The schematic representation in Fig. 20.4 gives a general picture of the processes natural rubber undergoes before becoming a sound useful technical and commercial product.

Properties

Physical. The properties of natural rubber are largely dependent on the degree of cure and type of additives and filler reinforcement used. To list properties of all the possible different natural rubber mixes is beyond the scope of this book but as a general guide some of the typical physical properties of natural rubber in comparison with synthetic rubbers are listed in Table 21.1.

Chemical. Natural rubber, in general, has good resistance to alkalis as well as to dilute acids. The resistance to most concentrated acids is not good and resistance to aromatic and aliphatic hydrocarbon solvents is poor. Natural rubber is unaffected by concentrated hydrochloric acid, although it is decomposed by oxidizing acids such as concentrated nitric acid.

Applications

The uses to which natural rubber, both in a solid form or as a latex, has been put are so numerous and widespread that enumeration would be difficult. Briefly, the uses for solid rubber may be summarized as follows: rubber types, gaskets, seals, o-rings, footwear, cable covering, adhesives,

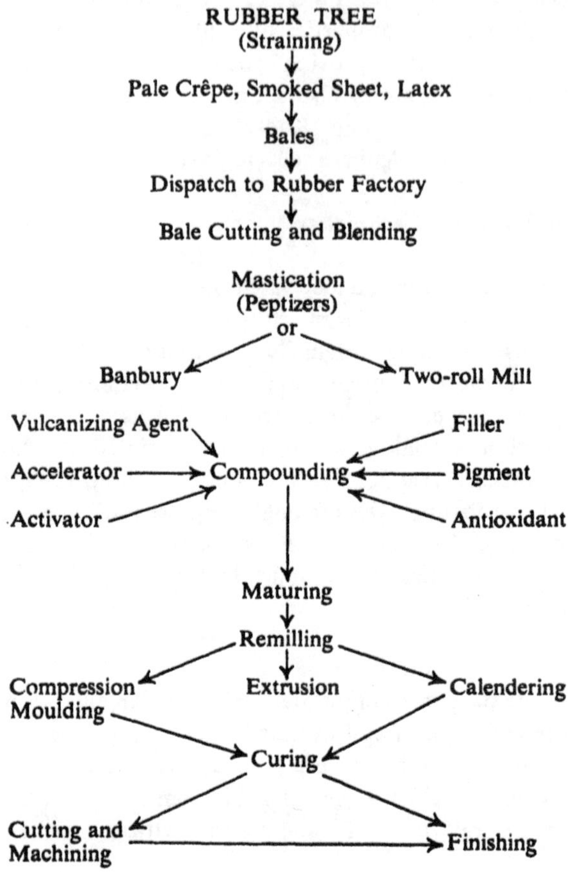

Fig. 20.4 Processes involved in converting raw rubber into finished molded articles

belting, hose, proofing, flooring, etc. The uses for the latex form includes seating (cushioning), coatings, paints, and adhesives.

Before leaving this section, mention must be made of reclaim rubber. This name is used to describe scrap vulcanized rubber that has been subjected to chemical and mechanical breakdown. These processes make the rubber soft and plastic and thereby capable of further processing. The advantages of incorporating such a material in rubber compounds are obvious, since the reclaim rubber contains nearly all the ingredients of the original rubber compound plus the added factor that it is cheaper than the raw rubber. This makes it an extremely valuable extender, and a large industry has grown around this material.

MODIFIED RUBBERS

Hard Rubber

The terms hard rubber, ebonite, and vulcanite are all used to describe the product obtained by heating a mixture of rubber with up to 30 percent by weight of sulfur to a temperature of 120–160°C until the reaction is virtually complete.

It has been stated previously that rubber in the raw uncompounded state tends to lose its strength on continuous use and is adversely affected by temperature changes, and this is presumably why no large-scale use was made of rubber until 1839. Fortunately, at this time two pioneers of the rubber industry, Thomas Hancock (Britain) and Charles Goodyear (U.S.A.), both discovered that raw rubber could be hardened by heating with sulfur, thereby improving the physical properties of the rubber. Hancock termed this process 'vulcanization,' derived from Vulcan (Roman god of fire and volcanoes). He based the name on the fact that both heat and sulfur are present in volcanoes. In the strictest sense, the term vulcanization is not correct as it may be effected without heat or the addition of sulfur.

General

The methods of processing and compounding hard rubbers are essentially the same as those used for natural rubbers. The most popular filler is ebonite dust, which is obtained from scrap ebonite or by grinding specially prepared blocks, although coal dust, certain clays, whitings, and even carbon black have been used for this purpose. It has already been mentioned that the speed of chemical reaction between natural rubber and a vulcanizing agent can be increased by the addition of accelerators and activators. This is, of course, also true for hard rubber, and the most effective ones have been established as diphenylguanidine, butyraldehyde-aniline, zinc dimethylthiocarbamate, and tetraethylthiuram disulfide. The amount varies between 1–3 p.h.r.

In compounding hard rubber it must be pointed out that although metallic oxides are suitable as activators for soft natural rubber, other effects are obtained when used in conjunction with hard rubber.

Mixing

Hard rubber mixes are normally prepared by using a two-roll mill or an internal mixer (Banbury). If the two-roll mill method is preferred it is normal to band the raw rubber on the front roll at a temperature of 40-60°C. The compounding ingredients may be then added in the following order.

(i) Peptizers
(ii) Fillers (ebonite dust)
(iii) Processing aids
(iv) Accelerators
(v) Antioxidants (if required)
(vi) Sulfur

The actual order of addition is sometimes a matter of personal choice.

Properties

The physical and chemical properties of hard rubber are given in Table 20.1.

Table 20.1

PHYSICAL AND CHEMICAL PROPERTIES OF HARD RUBBER (ACCORDING TO B.S. 903)

Test	*Property*
S.G.	1.13–1.18
Tensile strength (lb/in.^{2})	7000–9000
Elongation at break (%)	3–10
Cross-breaking strength (lb/in.^{2})	12–16,000
Impact strength (ft lb/inch notch)	0.5
Dielectric constant ($f = 10^6$ c/s)	3
Power factor (10^6 c/s)	0.5–0.8
Volume resistivity (ohm-cm)	2–3 $\times 10^{16}$
Chemical Resistance	
Hydrochloric acid	Resistant
Dilute sulfuric acid	Resistant
Concentrated sulfuric acid	Attacked
Organic acids	Resistant
Sodium hydroxide	Resistant
Aromatic hydrocarbons	Attacked
Aliphatic hydrocarbons	Attacked
Esters	Resistant
Ketones	Resistant

Limitation and Applications

Unfortunately, hard rubber has poor light stability and degrades on prolonged exposure to light. The relatively long cure times necessary are not attractive to industry and other materials are now preferred for certain specific applications. Hard rubber has a high shrinkage value and the better the grade the more prone it is to shrink, which limits its use for precision requirements.

Uses found for hard rubber include: battery boxes, although in this application hard rubber is losing ground to certain other polymers such as styrene/butadiene and PVC. There is still, however, some use for hard rubber as battery separators. It has also found some use in general chemical plant equipment and as linings for industrial chemical plant such as containers and structural units. Expanded ebonite has been used extensively for thermal insulation especially at very low temperatures although this material is now receiving severe competition from cork, expanded polystyrene, and polyurethane foams. Miscellaneous uses include combs, mouthpieces for musical instruments, and barrels for pipes.

Cyclized Rubber

In the latter part of the eighteenth century Leonhardi (1791) treated rubber with sulfuric acid and observed that the material became hard and brittle. It was not until some time later, when many of the classical theories of organic chemistry had been developed, that Harries (1910–1919) suggested cyclizing rubber by treating it in a solution of benzene with sulfuric acid. The mechanism of isomerizing rubber has been examined by many scientists, and the structure shown below has been supported by M. Gordon (1951)

Structure

Natural rubber may be modified by heating with sulfuric acid, chlorostannic acids, metallic halides, or phenols. A reduction in unsaturation occurs and is attributed to cyclization.

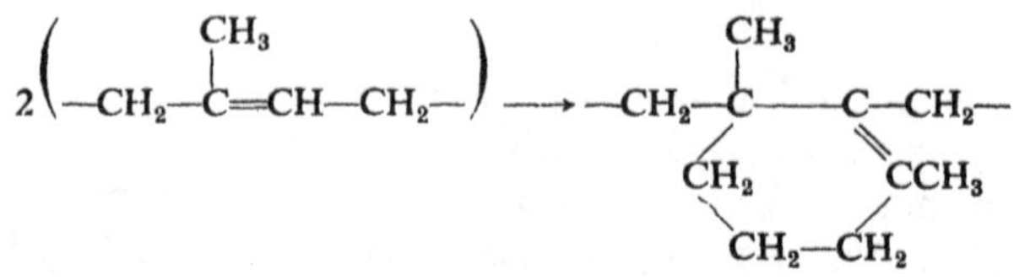

Fig. 20.5

Properties

The properties of cyclized rubber are given in Table 20.2.

Table 20.2

PHYSICAL PROPERTIES OF CYCLIZED AND CHLORINATED RUBBER

Test	*Cyclized rubber*	*Chlorinated rubber (properties dependent upon chlorine content)*
	Typical	*Typical*
S.G.	1.0–1.05	1.06–1.07
Hardness–Shore	85–90	70–80
Tensile strength (lb/in.^{2})	3000–4000	4000–4500
Elongation at break (%)	low	2–10
Flexural strength (lb/in.^{2})	6000–8000	10,000–14,000
Impact strength Izod (ft lb/in. notch)	3	1–2
Water absorption (%)	<1.0	<1.0
Water permeability	Highly impermeable	Highly impermeable
Dielectric constant	2.5	2.5–3.5
Power factor (1 Mc/s)	0.002	0.006
Volume resistivity (ohm-cm)	10^{16}	10^{13}–10^{14}

Applications

Cyclized rubber originally found use as a compounding ingredient in shoe-soling but has now been displaced by the more popular high-styrene resins. It is also used as an adhesive for rubber to metal bonding, and as a moisture-proof coating for paper and for paints and varnishes.

Chlorinated Rubber

The action of chlorine on rubber was investigated in 1880 by W. Roxburgh. Industrial applications were first suggested for these derivatives in 1881 but it was not until 1930 that the excellent oil-resistance of these rubbers was utilized.

Preparation and Structure

(a) Chlorination. Chlorination is carried out by substitution and addition of the halogen to a latex solution. The percentage of chlorine can be as high as 72 but to obtain maximum stability the polymers should not contain more than 68 percent of chlorine. Commercial polymers usually contain between 64-66 percent chlorine. Alloprene, which is manufactured by Imperial Chemical Industries Ltd., contains approximately 65 percent chlorine in its combined state and has the empirical formula $C_{10}H_{11}Cl_7$. The molecular weight varies according to the different grades and is from 5000 upwards.

The structure may be represented as follows:

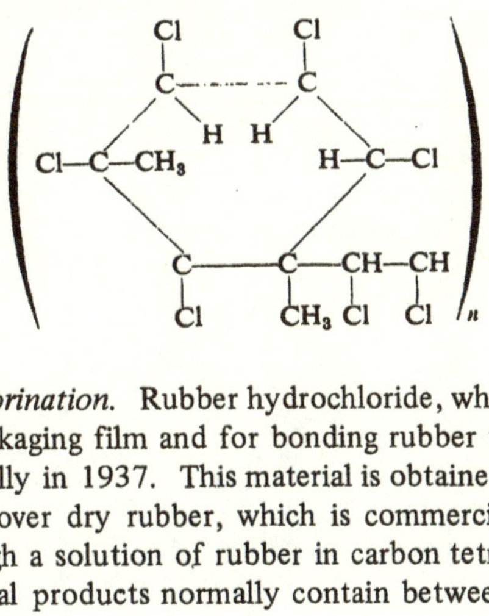

(b) Hydrochlorination. Rubber hydrochloride, which is being used extensively as a packaging film and for bonding rubber to metal, was introduced commercially in 1937. This material is obtained by passing dry hydrogen chloride over dry rubber, which is commercially impractical, or better still through a solution of rubber in carbon tetrachloride or in benzene. Commercial products normally contain between 25-32 percent of chlorine.

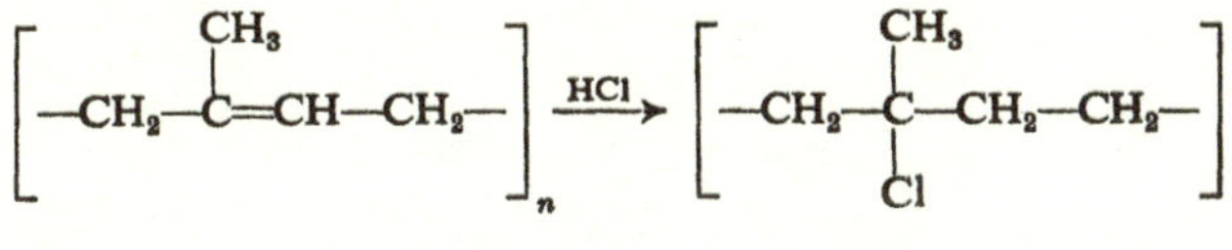

Properties and Applications

The physical properties of chlorinated rubber are given in Table 20.2.

Rubber hydrochloride has found wide use as a protective film (grease resistant) both alone and as a laminate to paper and aluminum foil, etc. Chlorinated rubber has been successfully used for bonding rubber to metal and in formulating electrical insulating lacquers.

SUGGESTED FURTHER READING

Alloprene. Technical Bulletin issued by Imperial Chemical Industries Ltd.

Rubber Technology and Manufacture. (1971) Ed. C. M. Blow (London: Butterworths).

21.

SYNTHETIC RUBBERS

The history of synthetic rubber can be traced back to 1926, when Faraday provided what may be considered the foundations for most of the future developmental work by establishing the empirical formula, C_5H_8, for natural rubber. The exact structure of natural rubber has been the subject of much controversy, nevertheless Faraday's work has been of great value to all the scientists who carried on his work in their attempts to produce synthetic rubbers similar and possibly better in certain specific properties than natural rubber.

The next step in this history was the success of Greville Williams who, in 1860, isolated isoprene by the dry distillation of natural rubber. This was followed by the work of Bouchardat, who, in 1879, obtained a mass resembling natural rubber by heating isoprene with hydrochloric acid in a sealed tube. The fact that isoprene could be polymerized, and seemingly had an industrial potential, was first noted by an English chemist named Tilden, in 1884. It was not until some time later, however, that Euler, in 1897, made isoprene synthetically and thus, in the chemical sense, completed the synthesis of rubber. The first commercial production of synthetic rubber was begun in Germany during WW I. The Germans produced methyl isoprene on a limited commercial scale, but the quality was poor, nevertheless they used it for lack of anything better. After the war, however, natural rubber again became available and production of the synthetic rubber was abandoned for the time being, although it was realized that something better than natural rubber was required for certain applications. This led to further developmental work which resulted in the production, in 1932, of Buna N (acrylonitrile/butadiene) and Buna S (styrene/butadiene). About this same period considerable work was being carried out on synthetic elastomers in the U.S.A., and as a result of the labors of W. H. Carothers and coworkers, in 1931 a new synthetic elastomer was developed. This resulted in an American company, E. I. du Pont de Nemours, manufacturing a new elastomer based on chloroprene, which they named Duprene, a name that they changed later to Neoprene. This has now been accepted as the generic name for this particular group of synthetic rubbers. Polysulfides were also developed around the early 1930's but it was not until 1935 that polysulfide elastomers became commercially available.

Nitrile rubbers date back to the discovery of Tschunkur and Bock who, in 1933, discovered that butadiene and acrylonitrile could be combined to give a copolymer which was capable of undergoing vulcanization to a rubber-like material. However, the first commercial production of this family of rubbers was not started until 1937, when rubbers were sold in Germany under the trade names Buna N and Perbunan.

The last 25 years have produced many new synthetic rubbers including butyl, silicone, fluorinated silicone, fluoroacrylic, fluorinated ethylenes, chlorosulfonated polyethylene and solid polyurethane rubbers. It was not until fairly recently, however, that an elastomer resembling natural rubber was produced synthetically. This material is *cis* 1,4 polyisoprene; polybutadiene has now also been produced by similar techniques. The conception of these synthetic elastomers was achieved largely as a result of the rapid developments in the polymer field and in particular to the work of Professors Natta and Ziegler who have made possible the synthesis of isoprene rubber by using stereospecific catalysts, mentioned in Chapter 9. An even later development can be attributed to such catalysts: the discovery of ethylene/propylene rubbers. This has also led to the development of a completely new class of rubbers which are termed thermoplastic rubbers.

Synthetic rubbers must not be classed as mere substitutes for natural rubber; they are rubbers in their own right. Their compounding is basically the same as for natural rubber to which reference should be made (Chapter 20).

Polyurethane rubbers are described in Chapter 8.

Nomenclature

Various abbreviations have been used in the terminology of synthetic rubbers, and the American Society for Testing Materials (A.S.T.M.) has made an attempt to standardize them as follows:

ASTM D 1418	*Trade name or accepted name*	*Chemical type*
IIR	Butyl	Isobutylene/isoprene
SBR	GRS, Buna S	Styrene/butadiene
NBR	Nitrile, Buna N	Acrylonitrile-butadiene
CR	Neoprene	Chloroprene
IR	Synthetic natural	Synthetic polyisoprene
NR	Natural rubber	Natural polyisoprene
BR	Cis-polybutadiene	Polybutadiene
T	Thiokol	Polysulfides
EPM	EPR	Ethylene propylene
EPDM	EPT	Ethylene propylene terpolymer
CSM	Hypalon	Chlorsulfonated polyethylene
SIL	Silicone	Polysiloxane
U	Urethane	Polyurethane-reaction product of a diisocyanate and a polyalkylene ether glycol
FLU	Viton	Fluorinated hydrocarbon
ABR	Acrylics	Acrylate butadiene
COX	Carboxylic; Hycar	Modified butadiene/acrylonitrile with carboxylic groups
CO	Hydrin	Homopolymer of epichlorhydrin with ethylene oxide

At present no abbreviations have been approved for ethylene/propylene rubbers but it is likely that EPR will be used for the saturated rubbers and EPT for the sulfur curable terpolymers.

POLYISOBUTYLENE

Liquid polymers of isobutylene were described, in 1873, by Butlerov and Gorianov. The actual polymerization of isobutylene to form solid polymers did not follow until some sixty years later. The development of solid polyisobutylene rubber was a result of a joint research scheme carried out by Standard Oil Development Co. (U.S.A.) and I. G. Farbenindustrie (Germany).

Preparation and Structure

Polyisobutylene is a by-product of the petroleum industry and is also obtained from natural gas. The lower members are liquids of low molecular weight and the higher members, high molecular weight solids.

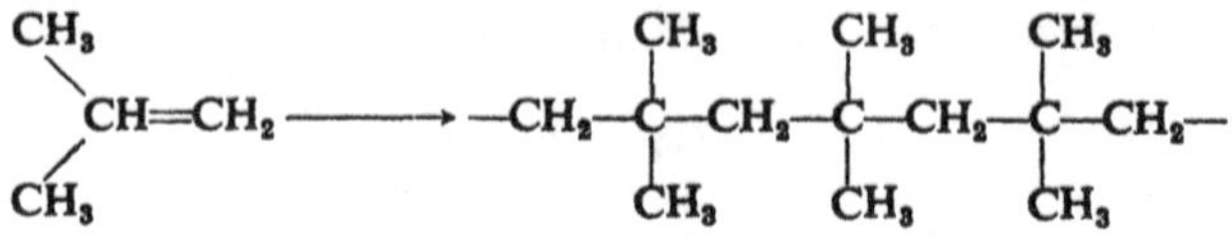

Properties and Applications

The physical properties of polyisobutylene are somewhat similar to those of natural rubber although this synthetic elastomer is saturated and, consequently, fairly resistant to oxidizing agents, with little inherent strength.

The main use of this rubber was in the modification and compounding of natural rubber, although this is no longer true. It is miscible in all proportions with natural rubber, and is also added to polyethylene to function as a plasticizer and to inhibit stress cracking. Other applications include its use as a pressure-sensitive adhesive, damp-course protector, caulking material, and as a synthetic lubricant.

BUTYL RUBBER

Butyl rubber is a generic term for copolymers of isobutylene and one to two percent of a diolefine. The diolefines in most common use are isoprene and butadiene. Butyl rubbers are available in several degrees of unsaturation.

Polymerization and Structure

Isobutylene polymerizes readily but, unfortunately, on polymerization it is fully saturated and cannot therefore be cured. However, by introducing a diolefine, unsaturation is produced and permits a curing process to take place.

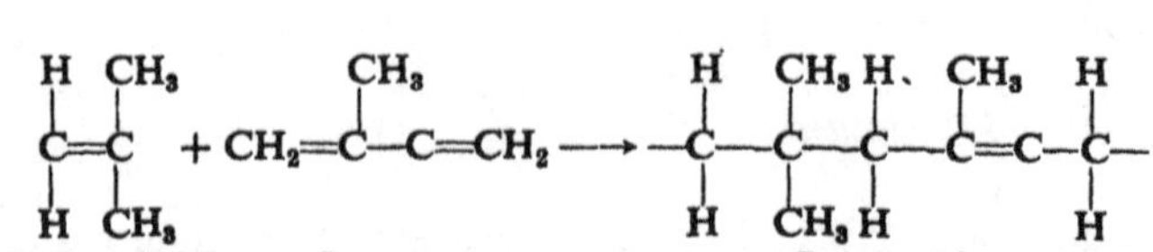

Isobutylene Isoprene Butyl rubber

Manufacture

Butyl rubber is not generally prepared by the popular emulsion polymerization technique but by a solution technique using a Friedel Craft catalyst. The isobutylene and isoprene are first mixed in the required proportions with an inert diluent such as methyl chloride. The resultant reactants are cooled to around −80°C and a catalyst such as aluminum chloride is quickly mixed in with the monomers, whereupon the polymer forms almost at once. The reaction is exothermic and it is essential that the temperature of the reaction is kept to around 80°C. A small quantity of zinc stearate may then be added, to prevent agglomeration, together with an antioxidant. The polymer is discharged as a dispersion of very fine crumbs in the methyl chloride and is in the form of a slurry. The diluent must then be stripped off to leave behind the rubber in the form of a crumb, which is finally filtered and dried. The resultant rubber may then be compounded and is suitable for curing with conventional vulcanizing agents such as sulfur, or by para quinone dioxime and red lead. The rate of cure of the butyl rubber can be increased by chlorination. This produces further active sites for vulcanization; there is approximately one chlorine atom per double bond site. The same curing agents as mentioned previously are still used to effect the actual curing.

As with natural rubber it is essential to use fillers such as carbon black or fine silica to obtain suitable physical properties.

Properties

To provide consistent values of physical properties suitable for correlation is not always easy. It is essential to quote exact details of the formulation and compounding methods, in particular, type and quantity of filler, type and length of cure, etc. There are a great many combinations of different mixes and to provide figures for all synthetic rubbers would be a mammoth task beyond the scope of this book. Table 21.1, however, gives a general classification of the synthetic rubbers in respect to certain physical tests.

One specific characteristic of butyl rubber is essential to mention; the fact that this synthetic elastomer cannot be used in blends with natural rubber.

Applications

The main uses of butyl rubber include car inner-tubes, although this

Table 21.1

PHYSICAL PROPERTIES OF SYNTHETIC RUBBERS

(Natural Rubber is Included for General Information)

Properties		*Natural rubber*	*SBR or Buna S Butadiene styrene*	*Butyl isobutylene isoprene*	*Thiokol organic polysulfide*	*Nitrile (Buna N) butadiene acrylonitrile*	*Neoprene poly-chloroprene*
Tensile strength (lb/in.^{2})	*Pure gum*	>3000	<1000	>1500	–	<1000	>3000
	Black loaded stocks	>3000	>2000	>2000	–	>2000	>3000
Hardness range (Shore Durom. A)		30–90	40–90	40–75	35–80	40–95	40–95
Specific gravity (base material)		0.92	0.94	0.92	1.34	1.00	1.23
Vulcanizing properties		Excellent	Excellent	Good	Fair	Excellent	Excellent
Adhesion to metals		Excellent	Excellent	Good	Poor	Excellent	Excellent
Adhesion to fabrics		Excellent	Good	Good	Fair	Good	Excellent
Tear resistance		Good	Fair	Good	Poor	Fair	Good
Abrasion resistance		Excellent	Good to excellent	Good	Poor	Good	Excellent

Compression set		Good	Good	Fair	Poor	Good	Fair to good
Rebound	*Cold*	Excellent	Good	Bad	Fair	Good	Very good
	Hot	Excellent	Good	Very good	Fair	Good	Very good
Dielectric strength		Excellent	Excellent	Excellent	Fair	Poor	Good
Electrical insulation		Good to excellent	Good to excellent	Good to excellent	Fair to good	Poor	Fair to good
Permeability to gases		Fair	Fair	Very low	Low	Fair	Low
Acid resistance	*Dilute*	Fair to good	Fair to good	Excellent	Fair	Good	Excellent
	Concentrated	Fair to good	Fair to good	Excellent	Fair	Good	Good
Solvent resistance	*Aliphatic hydrocarbons*	Poor	Poor	Poor	Excellent	Excellent	Good
	Aromatic hydrocarbons	Poor	Poor	Poor	Good	Good	Fair
	Oxygenated (ketones, etc.)	Good	Good	Good	Good	Poor	Poor
	Lacquer solvents	Poor	Poor	Poor	Good	Fair	Poor

specific use has lost considerable ground with the introduction of the tubeless tire. An application which is becoming of increasing importance is its use as a mastic. These mastics are used for glazing and sealing multistory buildings, in particular those of curtain-wall construction. An important use of butyl rubber is for sheathing electric insulating cables. Miscellaneous uses include tank linings, air bags for tire curing, and rubber cements.

Chlorobutyl

In 1960 another type of butyl rubber was introduced by the Standard Oil Co., U.S.A., and this modified butyl contained approximately 1.2 weight percent chlorine. The claims for such a rubber are that it possesses greater vulcanization flexibility with improved cure compatibility with other more conventional rubbers.

The rubber is produced by introducing a continuous stream of chlorine gas into a solution of butyl rubber in hexane. For each chlorine molecule reacted one molecule of hydrogen chloride is evolved and one atom of chlorine appears in the polymer chain.

Curing

Because of the presence of both olefinic unsaturation and reactive chlorine in this polymer chain there is the facility for a great variety of curing techniques to be exploited. Sulfur cure is possible but this is of minor significance and there are two other main types of cure which essentially work with the chlorobutyl but not with the more conventional butyl rubber. One reaction uses zinc oxide and is more commonly recognized with the neoprene rubbers and the other is a bis-alkylation mechanism. This latter process is usually carried out with primary amines and the reaction is as follows:

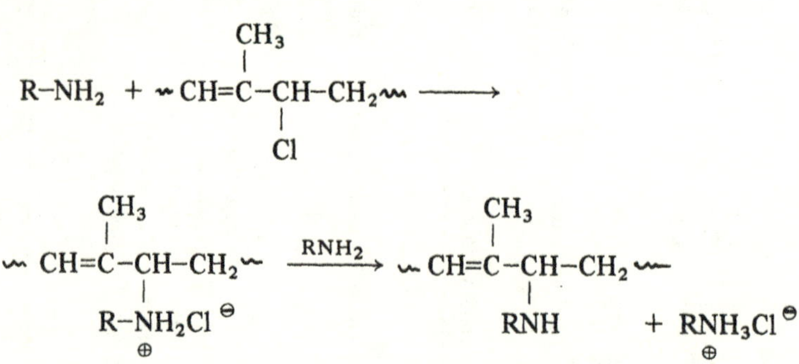

Properties

The chlorobutyl rubber exhibits all the excellent properties associated with the more established butyl types. It does have the added advantage of the ability to be cross-linked in a variety of ways to produce highly heat stable cross-links. Because this rubber can be cured with nontoxic materials such as zinc oxide, it has the ability to be used in contact with foodstuffs. The addition of the chlorine in the polymer chain improves the adhesion to many polar substances.

Applications

They are used in the tire industry for inner liners for tubeless tires, tire side walls, and heat resistant track inner tubes. Other applications include hoses (for automotive use), adhesives and sealants, conveyor belts and gaskets in the aircraft and automobile industry.

STYRENE/BUTADIENE (SBR)

A styrene/butadiene rubber was developed and produced in Germany during the period 1930-35 and called Buna S. The polymerization process used was rather lengthy, and when some years later, in the U.S.A., it was decided to produce a rubber having approximately the same composition as Buna S, a different process was therefore used. The original German process took weeks to achieve a high degree of conversion but there are new processes used today which take only a matter of hours. There are basically two types of emulsion polymerization used in its preparation: one is carried out at 0°C, the product being called 'cold-rubber'; the other takes place at 50-60°C and is referred to as 'hot-rubber.'

The normal method for preparing SBR is to use an emulsion polymerization technique which involves the chemical linking or copolymerization of high purity butadiene and styrene into long-chain molecules of high molecular weight. The preparation of the monomers, styrene, and butadiene, as well as acrylonitrile, has been described in Chapter 11.

Manufacture

The styrene and butadiene are first emulsified in an immiscible liquid such as water by using fatty acid soaps, resin acid soaps, or a mixture of the two. They are then charged to the reaction tanks, the styrene/butadiene ratio being adjusted to give a polymer normally containing about 23-26 percent bound styrene. The chain reaction is started by means of a free-radical initiator, maintained by an activator, and is generally referred to as a 'Redox' system. This system may consist of an activator, a solution of inorganic salts including the reducing agent, and an oxidant, normally an organic hydroperoxide. The chain length of the polymer during the reaction is controlled by means of a modifier, usually a mercaptan. At the required degree of conversion of monomers to polymers the reaction is terminated with a 'short-stop' (chain stopper). This is an essential step if a high level of physical properties are required in the final rubber. If a short stop is omitted and the polymerization allowed to continue to near completion an increase in the branching and crosslinking occurs, making processing difficult or near impossible.

On completion of the polymerization process, the unreacted styrene and butadiene are stripped from the reaction mixture. An antioxidant or stabilizer is added to the latex formed at this stage to prevent degradation of the rubber during drying and storage. It is also possible at this stage to oil-extend the latex, and normally 37.5 parts of oil are added. In manufacturing oil-extended styrene-butadiene rubber, the effect of the oil is compensated by polymerizing the rubber hydrocarbon to a higher degree than normal. The resulting elastomer is easily processed and gives excellent physical properties.

The rubber particles are then coagulated and the coagulum is filtered, washed, dried, and compressed into standard bales. The bales may be wrapped in polyethylene film to prevent adhesion of the rubber to the multi-ply paper bags or containers in which they are then packed.

Polymerization and Structure

$$\underset{\text{Butadiene}}{CH_2{=}CH{-}CH{=}CH_2} + \underset{\text{Styrene}}{CH_2{=}\underset{\displaystyle C_6H_5}{\overset{|}{CH}}} \longrightarrow$$

$$-CH_2{-}CH{=}CH{-}CH_2{-}\left(CH_2{-}\underset{\displaystyle C_6H_5}{\overset{|}{CH}}\right)_n$$

(*N.B.* Molecular weight dependent on styrene/butadiene ratio)

Properties and Applications

Numerous grades are produced by the manufacturers of SBR rubber, in many countries throughout the world. They may be compounded for a wide variety of applications including household moldings, sports goods, toys, wire, and cable sheathing, battery boxes, adhesives, shoe soles, and white sidewall tires.

BUTADIENE/ACRYLONITRILE (NITRILE RUBBERS)

Rubbers produced from copolymerizing butadiene and acrylonitrile are commonly referred to as 'nitrile' rubbers, containing between 15-40 percent of acrylonitrile. The most significant property of these rubbers is their excellent oil and aromatic solvent resistance, which is governed to a large extent by the acrylonitrile content. The higher this value, the better the oil resistance although this is accompanied by a reduction in the rubber's low temperature flexibility characteristics.

Manufacture

Nitrile rubbers may be prepared by either a hot or cold emulsion polymerization in a similar manner described previously for SBR.

Polymerization and Structure

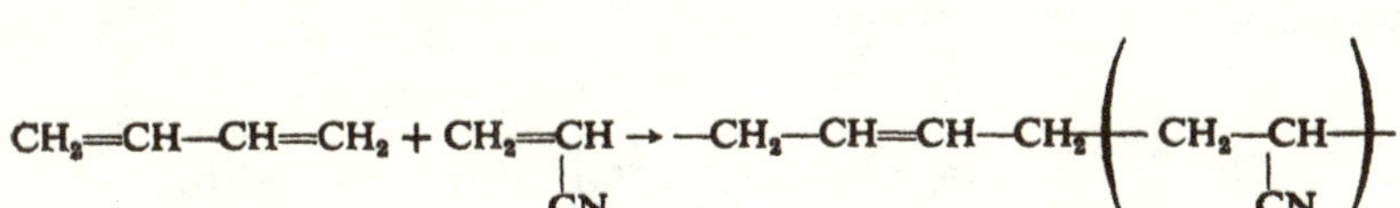

Butadiene Acrylonitrile Butadiene/acrylonitrile copolymer

Properties and Applications

The properties of nitrile rubber are given in Table 21.1. In addition to their outstanding oil resistance nitrile rubbers exhibit excellent abrasion resistance and compression-set characteristics. They can be worked on conventional rubber machinery but operations are slower since they are less thermoplastic than natural rubber. It has been established in practice that thorough mastication on a cold mill is necessary.

The use of nitrile rubbers as oil seals is extensive while other applications include flexible fuel tanks, oil-resistant hoses, and printing rollers. These rubbers are polar by nature and are, therefore, used extensively as adhesives for the bonding of PVC to PVC, PVC to metal, and shoe soles to their uppers. The nitrile rubbers may also be used in latex form for a wide variety of applications in the leather, paper, and textile-coating industries.

POLYCHLOROPRENE (NEOPRENE)

Polychloroprene was discovered as a result of the work carried out by Nieuwland on the polymerization of acetylene. He found that on passing acetylene into a concentrated solution of cuprous and ammonium chlorides, and allowing the subsequent reactants to stand, liquid polymers were formed.

$$CH{\equiv}CH \xrightarrow[\text{chlorides at } 10°C]{\text{Cuprous and ammonium}} CH{\equiv}C{-}CH{=}CH_2$$

Monovinylacetylene
(plus divinyl acetylene)

The divinyl acetylene is removed by separation and the monovinylacetylene is then passed into an aqueous hydrochloric acid solution containing cuprous and ammonium chlorides.

$$CH{\equiv}C{-}CH{=}CH_2 + HCl \xrightarrow[\text{and ammonium chlorides } 30°C]{\text{Cuprous}} CH_2{=}\overset{\displaystyle Cl}{\overset{|}{C}}{-}CH{=}CH_2$$

Chloroprene

Manufacture and Structure

Chloroprene can be polymerized by the emulsion technique, the polymerization occurring quite rapidly under controlled conditions. The chloroprene is first emulsified in water and subsequently polymerized at 40-50°C, using potassium persulfate as an initiator. The solid polychloroprene is obtained from the latex in a similar manner to that for natural rubber by acid coagulation or by multivalent salt addition to the latex.

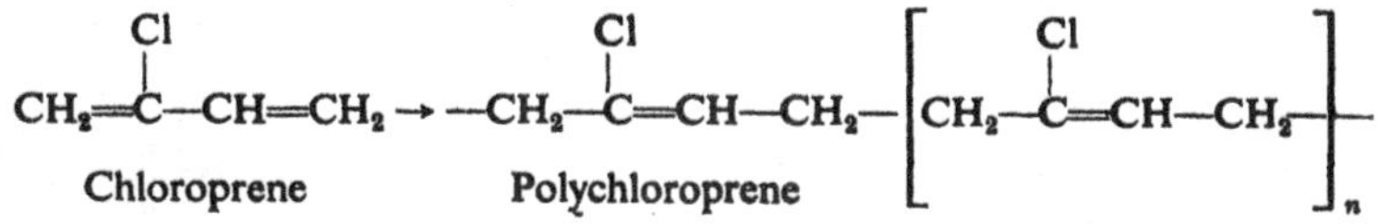

The chloroprene units mostly polymerize in the 1 : 4 position but there is occasional 1 : 2 addition. The manufacturing process and many end uses are represented diagrammatically in Fig. 21.1.

Curing. The actual mechanism of vulcanizing neoprene is not widely agreed upon but in general can be considered to consist of bridging and chain extension, the 1 : 2 additions in the polymerization providing active chlorine atoms to react with metallic oxides such as zinc oxide.

Properties

Neoprene possesses many of the excellent physical characteristics of natural rubber, viz. toughness and resilience, coupled with resistance to oils, greases, and paraffinic solvents. It is also stable towards sunlight and the oxidizing effect of ozone and heat-aging. Neoprene also possesses outstanding abrasion resistance and other advantageous properties which include flame and chemical resistance.

Typical physical properties for neoprene are given in Table 21.1.

Applications

Neoprene has been developed for uses where natural rubber gives poor service due to certain limitations in its physical properties. The good oil and abrasion resistance of neoprene has led to its widespread use as a gasketing material in the aircraft and automobile industries. Whether used in tension, compression, or shear these rubbers display good deformation under load and are therefore ideally suitable for use as gaskets, packings, and engine mounts. This rubber has also found some use for bearing pads in bridge construction, its function being to compensate for thermal and mechanical beam movements. Protective clothing may be prepared from neoprene compounds by calendering the rubber on to various textiles. Other coatings uses include conveyor belts and giant collapsible containers. Neoprene, as a base material, has also become very important in the field

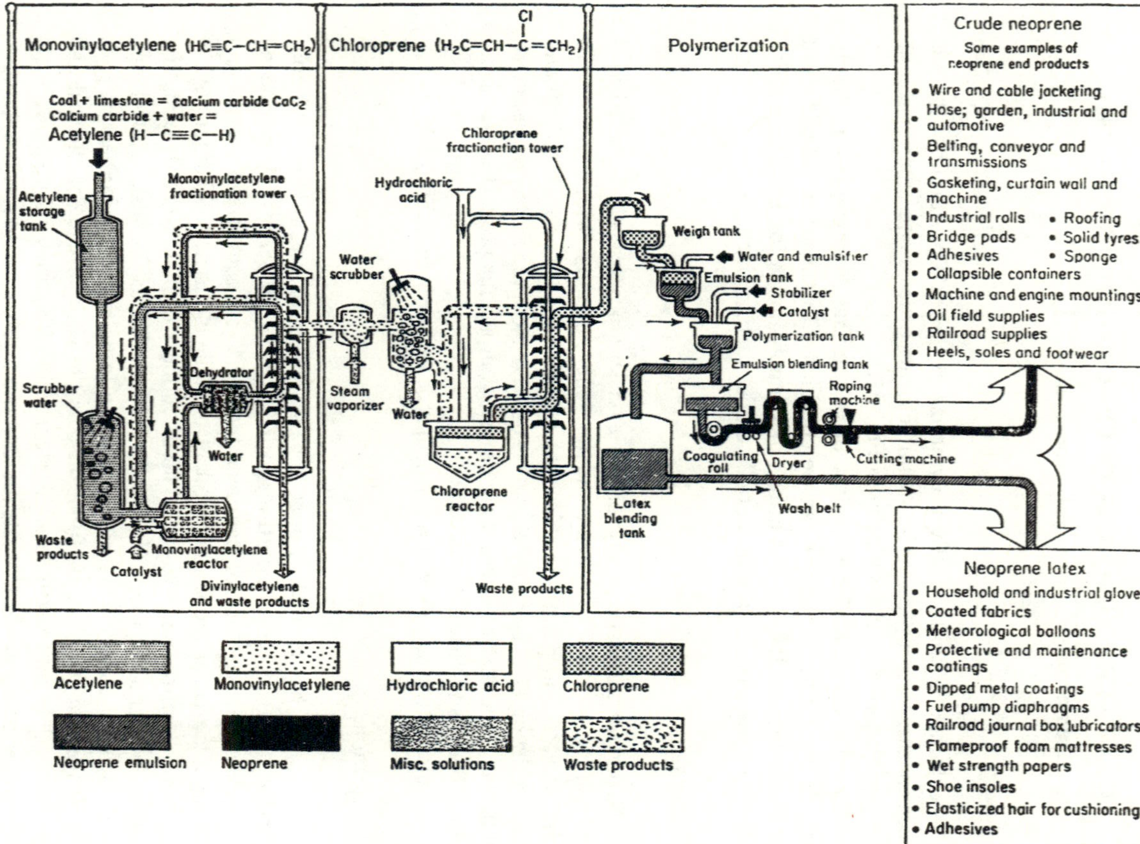
Monovinylacetylene (HC≡C–CH=CH2)
Chloroprene (H2C=CH–C(Cl)=CH2)
Polymerization
Coal + limestone = calcium carbide CaC2
Calcium carbide + water =
Acetylene (H–C≡C–H)
Acetylene storage tank
Monovinylacetylene fractionation tower
Scrubber water
Dehydrator
Water
Waste products
Catalyst
Monovinylacetylene reactor
Divinylacetylene and waste products
Water scrubber
Hydrochloric acid
Chloroprene fractionation tower
Steam vaporizer
Water
Chloroprene reactor
Waste products
Weigh tank
Water and emulsifier
Emulsion tank
Stabilizer
Catalyst
Polymerization tank
Emulsion blending tank
Roping machine
Coagulating roll
Dryer
Cutting machine
Wash belt
Latex blending tank
Crude neoprene
Some examples of neoprene end products
• Wire and cable jacketing
• Hose; garden, industrial and automotive
• Belting, conveyor and transmissions
• Gasketing, curtain wall and machine
• Industrial rolls • Roofing
• Bridge pads • Solid tyres
• Adhesives • Sponge
• Collapsible containers
• Machine and engine mountings
• Oil field supplies
• Railroad supplies
• Heels, soles and footwear
Neoprene latex
• Household and industrial gloves
• Coated fabrics
• Meteorological balloons
• Protective and maintenance
• coatings
• Dipped metal coatings
• Fuel pump diaphragms
• Railroad journal box lubricators
• Flameproof foam mattresses
• Wet strength papers
• Shoe insoles
• Elasticized hair for cushioning
• Adhesives
Acetylene
Monovinylacetylene
Hydrochloric acid
Chloroprene
Neoprene emulsion
Neoprene
Misc. solutions
Waste products

Fig. 21.1

of synthetic adhesives and rubber cements.

POLYSULFIDES (THIOKOL)

Although the basic reaction of an alkyl ester with an alkali sulfide was known early in the nineteenth century, little significance was attached to it until the period 1925-27. In 1926, J. Baer observed that a polymeric, rubberlike substance was obtained by reacting an alkali sulfide with methylene chloride and other dihalides. During the same period another chemist, J. Patrick (1927), who was not principally interested in producing polymers, was working to develop a new antifreeze by reacting ethylene chloride and sodium polysulfide in an aqueous medium. His work resulted, however, not in a new antifreeze liquid but a yellow elastomeric solid. Patrick followed up his accidental discovery by taking out a patent for the process and in 1932 the first commercial production of polysulfide rubbers began in the U.S.A. These materials are commonly referred to as Thiokols, being manufactured by the Thiokol Corporation of U.S.A.

Manufacture

The commercial process consists of adding a dihalide slowly to an aqueous solution of sodium polysulfide in the presence of a dispersing agent such as magnesium oxide. The form of the final product is controlled either by adding excess polysulfide, which produces a solid rubber, or by adding excess dihalide, which produces a viscous liquid suitable for casting. The polymerization is exothermic and takes between two to six hours to reach completion at 70°C.

Unlike the polymerization of most other synthetic rubbers, this reaction is basically one of condensation and not one of addition. The polymer is obtained in suspension of latex form, from which it can be washed free of sodium chloride and any excess sodium polysulfide present. The polymer is then finally dried and packed ready for shipment.

Various grades of polysulfide are available, depending on the nature and proportion of the base raw material used initially. A well-known type is Thiokol B which is made from bis- (2 chloroethyl) ether and a specific type of sodium polysulfide. This rubber may be cured by using trichloropropane in the initial reaction, which then allows a degree of cross-linking. The liquid polysulfides are generally prepared by reacting bis-chloroethyl formal with a solution of polysulfide containing emulsifying agents.

Polymerization and Structure

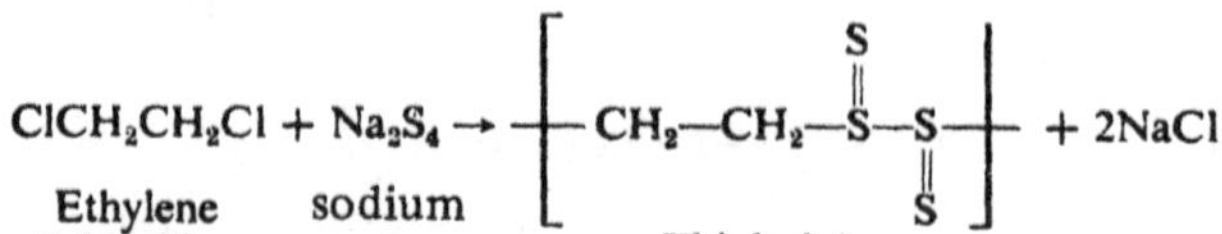

Ethylene dichloride   sodium polysulfide   Thiokol A

Other specific types are:

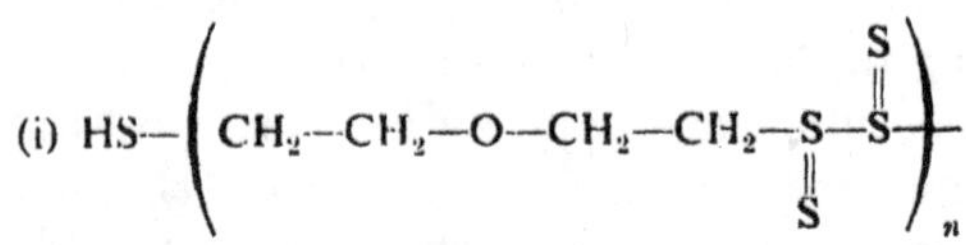

Thiokol B

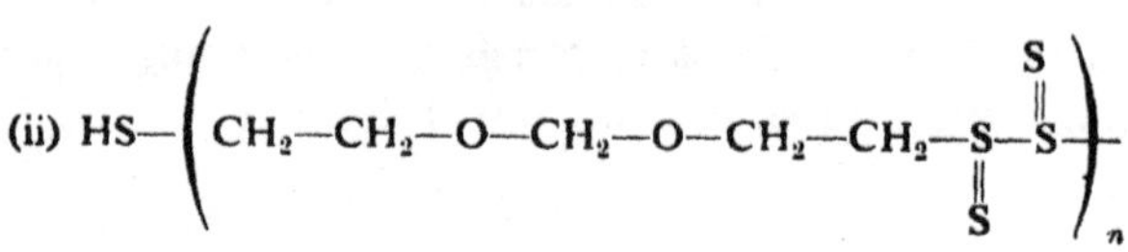

Thiokol ST

(Note terminal mercaptan groups.)

Curing

Polysulfide rubbers may be cured by metallic oxides such as zinc oxide or lead peroxide. The mechanism for curing can, in general, be considered as being effected by the condensing of the terminal mercaptan groups with oxidizing agents.

Properties

The uncompounded polysulfide rubbers generally possess low mechanical strength, but this may be improved by filler addition, etc. The main advantages of polysulfide rubbers are their outstanding resistance to petrol and a variety of oils, greases, and solvents. They also have a very low permeability to gases and vapors. The main disadvantage of this family is their characteristic obnoxious odor.

Applications

Only a relatively small amount of polysulfide elastomers are produced today although they are used for such diverse applications as printing rollers, paint sprays and petrol hoses, cable-coverings and gaskets. The low permeability to gases and vapors makes them ideally suitable for coating balloon fabrics and gas-proof diaphragms.

The liquid rubbers such as Thiokol LP 2 and LP 32 have a particular advantage in that they do not contract on curing and are thus finding extensive use as casting and caulking materials (sealing). Typical sealing applications include aircraft landing strips, windows, and integral fuel tanks for aircraft. They are also used for sealing ships' decks as well as in buildings for curtain-wall structures. As discussed in Chapter 6, they are used as a flexibilizer for epoxy resins to give a tough and resilient end product.

CIS 1 : 4 POLYISOPRENE (IR) AND BUTADIENE (BR)

Until the work of the two eminent scientists, Ziegler and Natta, on stereospecific catalysts, all reported attempts to polymerize isoprene and butadiene yielded polymers containing at least two isomers which were inferior in properties to that of the natural product. Natta discovered that by suitable choice of a stereospecific catalyst, it was possible to obtain polymers with predominantly *cis* or *trans* configuration or even to obtain isotactic or amorphous polymers with 1 : 2 linkages.

Structure

There are two main groups of *cis*-polyisoprene produced at present, depending on the type of catalyst used. One group is based on a lithium or lithium alkyl-type of catalyst and gives high molecular weight polymers with a *cis* 1,4 content in the region of 92 percent. The other group uses a Ziegler-type catalyst and gives polymers with a higher *cis* content in the region of 96 percent. The *cis* polybutadiene can also be manufactured with varying *cis* contents from 92–97 percent, although there is also a third group manufactured, with a *cis* content as low as 40–50 percent, which is normally blended with natural rubber in equal proportions.

Properties

The specific gravity of IR is equal to that of natural rubber (0.92) and BR is slightly lower (0.91). Both IR and BR have low ash contents, being less than one percent in both cases. IR is very light in color and is normally stabilized with nonstaining antioxidants because of the requirements of many of its anticipated end uses. BR is also inherently very light colored and both rubbers may be extended with oil. The tensile strengths of both IR and BR in the raw uncompounded state are inferior to NR but superior to SBR. However, with IR, the elongation is higher than that for NR and it is being found in practice that IR and BR are generally required to be blended with NR.

Applications

Similarity in properties has permitted consumers to use IR to supplement NR in a wide variety of applications, including tire treads (not as yet important), carcasses, and side walls. In nontire applications the light color, excellent flow characteristics, and low ash content of IR are distinct advantages. Such applications include hot-water bottles, bathing caps, and molded footwear. The IR compounded stocks flow very easily, a property that is especially attractive in molded products where good surface finish is important. These rubbers show excellent potential for injection molding even in the conventional machines used for thermoplastics.

It is thought, as a result of preliminary investigations which have shown that the tread vulcanizates have an attractive combination of high resilience, low heat build-up, and outstanding abrasion resistance, that BR rubbers have promising uses for tires especially in the heavy duty field. Other potential markets include belting, resilient mountings, solid tires, flooring, and for improving the impact strength of polystyrene (i.e. toughened polystyrene).

Polybutadiene

On writing this section it was very difficult to assess where the new polybutadiene resins should be placed in the context of this book. Whatever decision the authors came to, a good case could be made to place it elsewhere. It was decided to insert it immediately after the new Butadiene rubbers (BR) although it belonged in the relevant section on thermosetting resins as well.

A relatively new type of material is the polybutadiene thermosetting resins which include up to 90 percent of vinyl functionality. There are two main types available, the low molecular weight oils which range in functionality up to 50 percent and the very viscous (semi-solids) with up to 90 percent vinyl content. With the lower vinyl types, high *cis* or *trans* structure varieties are commercially available.

The low molecular weight polymers are regarded as prepolymers or resin intermediates. They are all cured using peroxide catalysts which can take place at room temperature or at temperatures up to 160°C. The resultant cross-linked structure can be regarded as polyethylene with a cross-link at every other carbon atom in the main chain and can be suitably classed as polyvinyl ethylene.

Properties and Processing

These resins, when cured satisfactorily, exhibit good dielectric and chemical resistance. In addition they have a heat deflection temperature far above that of thermoplastics such as polyethylene.

The resins are generally mixed with high amounts of reinforcing fillers and the mixing takes place in dough-type equipment. The compound so formed can be transfer or injection molded.

Uses

These resins are finding growing applications especially in requirements where dielectric qualities coupled with high temperature performance, corrosion resistance, quick cure with good stability are required. They are used in fabric coating as well as finding some use in printed circuit work.

ETHYLENE/PROPYLENE RUBBERS

These rubbers are one of the newest additions to the rapidly growing list of synthetic rubbers available today. Such rubbers are produced from relatively low-cost monomers that yield elastomers with low specific gravity (0.87). This added to their outstanding ozone and weather resistance make them attractive to a wide variety of industries.

Polymerization and Structure

These rubbers are made basically by copolymerizing ethylene and propylene in a hydrocarbon solvent, using a heterogeneous catalyst of the Ziegler type based on an aluminum alkyl and and titanium tetrachloride. There are two main types at present available: one is the completely saturated copolymer of ethylene and propylene distributed at random (EPR); the other a terpolymer of ethylene, propylene, and a nonconjugated diene such as dicyclopentadiene (EPT) which provides double bonds capable of undergoing cross-linking. Commercial polymers contain between 40–60 percent of ethylene.

(i) EPM

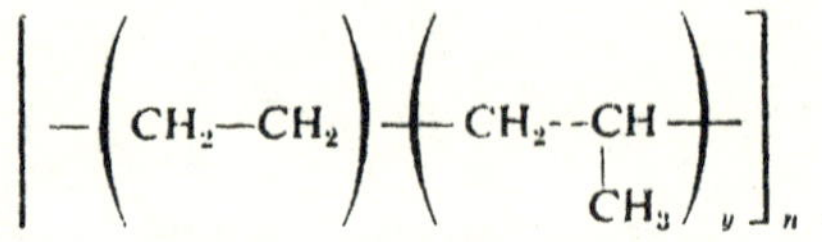

(ii) EPDM

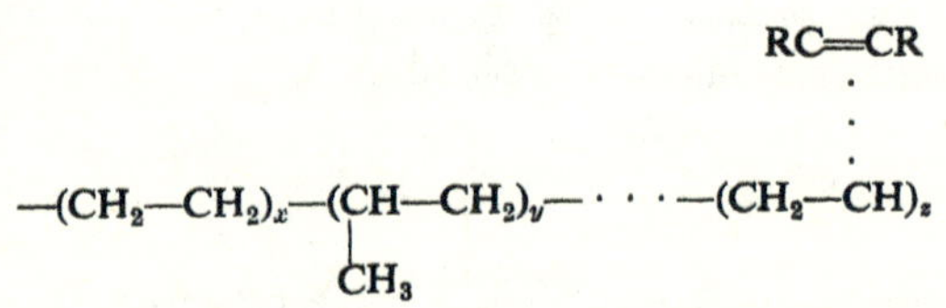

Curing

The unsaturated elastomers are cured by a free radical peroxide-type catalyst such as dicumyl peroxide, in addition to sulfur. Such catalysts are fairly expensive and reports have been made that the end-products possess an obnoxious odor. This has led, to some extent, to the introduction of the diene type (EPT) which are suitable for the conventional sulfur cure. It has been found in practice that zinc oxide, sulfur, and an organic accelerator are necessary for curing if sound useful elastomers are to be obtained. The curing characteristics are similar to butyl since relatively few double bonds are present.

Properties

The ethylene propylene rubbers are characterized by excellent resistance to aging, ozone and chemicals, good electrical and dynamic properties and satisfactory tensile and abrasion resistance. An unfortunate disadvantage is the fact that these rubbers are incompatible with all other rubbers, and they suffer from a lack of tack.

Applications

The good ozone and weather resistance of these new rubbers make them ideally suitable for outdoor use on vehicles, etc., for window strips and weather-proofing profiles. They combine good weathering properties with heat and insulating resistance and, therefore, are ideally suitable as a cable covering medium. Other possible outlets include tank linings, gaskets, seals, diaphragms, sports goods, footwear, proofing, sheets and hose, conveyor belting, also floor tiling. The use for tires will not be high until the poor adhesion of the rubber to the tire fabric can be overcome.

Thermoplastic Rubbers

Over recent years we have seen the introduction of thermoplastic rubbers which have the unique advantage of exhibiting many of the end-use properties characterized by the cross-linked rubbers, yet having the rapid processing advantages normally associated with thermoplastics. They may be regarded as combining properties of high performance rubbers and plastics.

These new rubbers meet the ASTM definition of a rubber or elastomer which states 'a material which at room temperature can be stretched repeatedly to at least twice its original length and, upon release of the stress, returns immediately with force to its approximate original length.'

The majority of commercially available thermoplastic rubbers are based on polyolefin or block polymer technology. In simple terms they can be considered as a polymer structure with a rubbery center segment and two glassy or crystalline plastic end blocks (A—B—A), e.g.,

♦♦♦●●●●●●♦♦♦

- ♦ is the A monomer units in the plastic end blocks
- ● is the B monomer units of the rubbery middle segments

More than two monomers can be used (A—B—C) as well as branched $\begin{array}{c} \text{A} \\ | \\ \text{(A—B—A)} \end{array}$ or repeating units (A—B—A—B—A).

Another interesting type is referred to as radial teleblock polymers with the following structure

$$(\text{A—A—A—A—B—B—B—B—B—B})_n\text{X}$$

where A equals a styrene glassy segment, B is the rubbery diene and X is the coupling agent.

Commercial polymers of this type are available and included in this classification are Solprene 406, 411, 414 which are butadiene-styrene radial teleblock copolymers, and the Solprene 418 is an 85 percent isoprene and 15 percent block copolymer with properties as shown in Table 21.2 and 21.3.

Table 21.2

SOLPRENE BUTADIENE-STYRENE PLASTOMERS

Solprene plastomer	*Butadiene/styrene ratio*	*Molecular weight, Mw*
406	60 : 40	280,000
411	70 : 30	300,000
414	60 : 40	160,000
416X	70 : 30	150,000
417X	80 : 20	180,000

Table 21.3

TYPICAL PROPERTIES OF SOLPRENE 418 PLASTOMER

Property	*Solprene 418 plastomer*
Structure	radial teleblock
Isoprene/styrene content	85/15
Average molecular weight, Mw	300,000
Melt flow @ 356°F (180°C) (5 kg)	1.5
Toluene solution viscosity (cP) 25 wt. % @ 77°F (25°C)	2200
Tensile strength* (psi)	2300
Elongation* (%)	1200

* Compression-molded slab, 80-mil thickness.

The polymers mentioned earlier with a soft segment (B) are often based on butadiene, isoprene, ethylene-propylene, and polyether ester repeating units while polystyrene is generally preferred as the hard plastic glassy segment although crystallizable polyesters can also be used.

It is considered that the rubbery structure is cross-linked by the thermoplastic segments but in actual fact a reinforced network structure is formed which does behave in the same manner as a conventional cross-linked type. When the polymer is heated through the glass transition temperature of the thermoplastic phase, the plastic block softens and subsequently plasticizes the rubber segments allowing the whole system to be processed on conventional plastics machinery. On cooling, the network structure is reformed to exhibit characteristics associated with cross-linked rubbers.

Chemical Types

Commercial material available include block copolymers of polystyrene (S) with butadiene (B) or polyisoprene (I). The S–B–S and S–I–S are called Kraton (Shell Chemical Company) and Solprene (Phillips Petroleum Company). The former rubber was developed by Shell in the middle 1960's but by 1970 there were several other manufacturers with commercial grades. A more recent development by this company was the introduction of a thermoplastic elastomer with an EPR as the center block and is designated Kraton G.

In 1972 Du Pont introduced a thermoplastic polyester elastomer made by reacting a dicarboxylic acid such as terephthalic acid with a polyglycol such as polytetramethylene glycol and a short chain diol with the terephthalate forms crystalline short, hard segments in the linear copolyester chain. The softer rubber amorphous phase contains the long polyether segments which are bound by the hard segments. The network so formed serves to cross-link the polymer and on heating this is broken down allowing the elastomer to be processed on conventional thermoplastic machinery.

Properties

The thermoplastic elastomers are anistropic, i.e., properties which are dependent on orientation. The thermal stability of both the olefinic and the styrenic types is good at normal processing conditions. The latter types have better elongation and recovery although at the expense of hardness and rigidity. The thermoplastic polyesters have high tensile and tear strength and in general they have properties somewhat superior to the other types of thermoplastic elastomers. They have the highest heat resistance and can be used at temperatures up to 300°F whereas other types are lower.

The electrical properties of the copolyesters are poor and they are not good insulators whereas the styrenic and olefinic types are better.

The chemical resistance of these thermoplastic elastomers is of course dependent on the composition of the blocks. The styrenic types, as would be anticipated, have poor chemical resistance and are soluble in ketones, esters, and aromatic solvents whereas the olefinic saturated types have improved chemical resistance. These latter types also have improved oral resistance but still not as good as the copolyester materials. These copolyesters have good resistance to solvents including both aromatic and aliphatic hydrocarbons.

Processing and Applications

These materials can all be processed on conventional thermoplastic machines and readily lend themselves to injection molding.

The styrenic and olefinic types are finding uses as adhesives, coextrusions, coatings, and polymer modification. The copolyesters are being used because of their high mechanical strength and excellent resistance to hydraulic fluids as a hose material. The high strength to weight ratio and resistance to flex fatigue have led to its use as a material for power transmission belting. Other uses include flexible couplings, pipes, cushions, and small gears.

Miscellaneous Rubbers

There are several other rubbers which are demanding interest but as yet have to grow in use to any significant tonnages. These include the rubbers based on epichlorhydrin and the copolymers based on epichlorhydrin and ethylene oxide. These rubbers are recognized as having excellent resistance to swelling in oils and fuels in addition to having good chemical resistance to a wide variety of solvents. The properties of these rubbers appear to lie somewhere between those of the neoprene and nitrile types.

It is claimed that the ethylene-vinyl acetate elastomers are characterized by excellent resistance to ozone, weathering, light, and color stability. The character of this type of copolymer is essentially determined by the ethylene to vinyl acetate ratio. Such elastomers with a low acetate content are crystalline at room temperature, while those with a higher acetate content are softer, more amorphous and more easily vulcanized. At present, a commercial grade is now available containing about 45 percent of vinyl acetate.

The polyacrylate and fluorinated rubbers are described under the relevant chapter headings elsewhere in the book.

Before closing this chapter, mention must be made of the carboxylic rubbers which are becoming more and more important in adhesive applications. In general, they have been prepared principally by the copolymerization of olefins and diolefins with acrylic acid type monomers. The adhesive characteristics of a rubber are altered considerably by the addition of very small amounts of carboxyl groups to the polymer chain. Various copolymers exist commercially such as butadiene with acrylic or methacrylic acid, carboxylic butadiene-acrylonitrile and more recently the liquid carboxylic types (CRLP). These latter materials are capable of undergoing cross-linking with epoxide groups and aliphatic hydroxyl groups. Most of these so-called rubbers exist as aqueous dispersions, solutions, or even solids which make them eminently suitable for hot melt applications.

Another member of this family of carboxylic rubbers is an even later development which is based upon carboxylated vinylidene chloride/butadiene latices. Commercial polymers have been introduced by Dow Chemical Company which contain equal parts of butadiene and vinylidene chloride with about 36 percent of chlorine. A further product in this family contains 60 percent of vinylidene chloride with 40 percent of chlorine. These latices are having good oil and chemical resistance and are comparable in properties to the styrene-butadiene types with better chemical resistance than the nitriles.

Addition of these latices to adhesive formulations imparts a high level of ignition deterrence. Generally these materials do not require curing agents to give ultimate properties but improvements can be achieved by cross-linking with heavy metal salts or melamine formaldehyde which react with the carboxyl groups on the polymer chain. The main area of application involves the laminating of two substances together, where ignition deterrence is required. Typical applications are foil to paper, foil to scrim to paper, and foil or vinyl film to fiberglass, in loose or dense duct board for insulation purposes.

SUGGESTED FURTHER READING

Polymer Chemistry of Synthetic Elastomers. (1968) J. P. Kennedy and Eric G. M. Tornquist (New York: Interscience).

Plastics and Synthetic Rubbers. (1970) A. J. Gait and E. G. Hancock (New York: Pergamon).

Rubber Technology. (1973) Edited by M. Morton (New York: Van Nostrand).
Block and Graft Copolymers. (1973) Ed. R. J. Ceresa (New York: Wiley).

Section E Inorganic Polymers

22.

INORGANIC AND SEMIORGANIC POLYMERS

The title of this chapter could well be 'Polymers of the Future,' for while considerable research and developmental work is being carried out on such polymers only small quantities are, at the time of writing, being produced on a commercial scale.

The considerable interest shown to polymers exhibiting resistance to high temperatures is largely a result of the increasing need for space exploration and missile defence. Polymers available at present for these applications are based upon a backbone of carbon-carbon bonds. Unfortunately, as we are now well aware, these polymers have certain limitations, including the susceptibility to oxidation and relatively low heat resistance. It has been with these limitations in mind that an extensive search has been made for polymers which possess the high thermal stability exhibited by certain of the naturally occurring inorganic materials such as silica and asbestos. Fluorocarbon and silicone polymers have gone a long way to meet these needs, but there are still requirements where operating temperatures are of the order of 400°C plus.

The inorganic type of polymers so far developed have in many cases met this thermal stability requirement although they have been found lacking in some other characteristic and, therefore, commercial exploitation has been somewhat slow. If previous experience of the polymer industry is anything to go by, large-scale production of inorganic polymers is probably many years away. This statement is substantiated particularly in the case of polystyrene, which has taken 100 years from the time of being first discovered to that of being commercially exploited. This state of affairs is also true of such polymers as polycarbonates and polyformaldehydes which were also discovered long before they were commercially exploited. The urgent need for high-temperature polymers may induce polymer manufacturers to devote considerable effort to develop these very specialized products.

The first consideration in producing inorganic polymers is the ability of various elements to join by means of a covalent bond. The formation of such polymers may be said to be restricted to elements of low molecular weight and, in practice, it has been reported that phosphorous, nitrogen, boron, silicon, and sulfur are the chief elements worth examining.

Table 22.1 lists the elements in the Periodic Table which are known to form covalent bonds and are thus possibly suitable for use in polymer formation. It is now proposed to discuss certain of these elements in the same sequence as they appear in the table. The semiorganic polymers based on silicon, aluminum, and titanium are also briefly mentioned.

Table 22.1

SECTION OF PERIODIC TABLE–
(ELEMENTS CAPABLE OF POLYMER FORMATION)

Group III	*Group IV*	*Group V*	*Group VI*
Boron	Carbon	Nitrogen	Oxygen
Aluminum	Silicon	Phosphorus	Sulfur
	Tellurium, Germanium	Vanadium, Arsenic	Selenium
	Zirconium, Tin	Antimony	
	Lead	Bismuth	

BORON POLYMERS

Considerable research and developmental work on boron polymers has been carried out in the U.S.A. and in England. Dr. W. Gerrard and co-workers at the Northern Polytechnic, London N.7, have been responsible for much of the fundamental work in incorporating boron in polymer chains. Many boron polymers have been prepared and reported in literature although no extensive industrial use has yet to be reported.

It is now a well-known fact that polymers based on the B—N system are stable and may vary in consistency from viscous liquids to rigid solids. One such material is phenyl-butyl borimide which is said to be extremely stable up to 350°C.

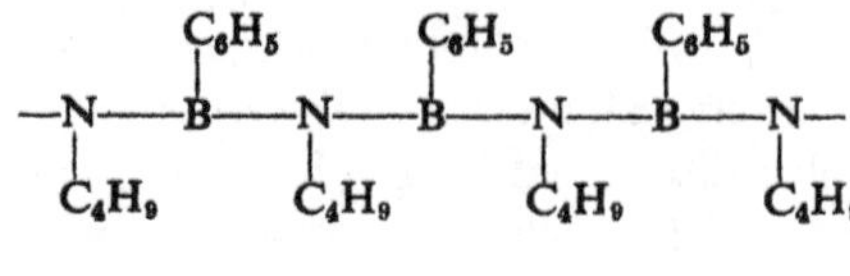

Other interesting polymers reported include the cyclic borazole and boron-phosphorus compounds. Boronazole derivatives may also yield polymers by splitting off amino groups. They are of no commercial importance as yet but investigations are still being carried out to determine their suitability for replacing conventional polymers at very high temperatures.

It is probably true to say of many of the boron polymers that although their heat stability may be superior to that of the carbon containing polymers the susceptibility of the bonds to oxidation and hydrolysis limits, to a certain extent, their potentialities as useful industrial materials.

ALUMINUM POLYMERS

Aluminum form polymers which must be regarded as semiorganic rather than purely inorganic. Their polymer formation is analogous to the silicones, and an interesting reaction to look at is that between triethyl hydroxy silane and aluminum in the presence of a mercuric chloride catalyst, which yields an alumino siloxane.

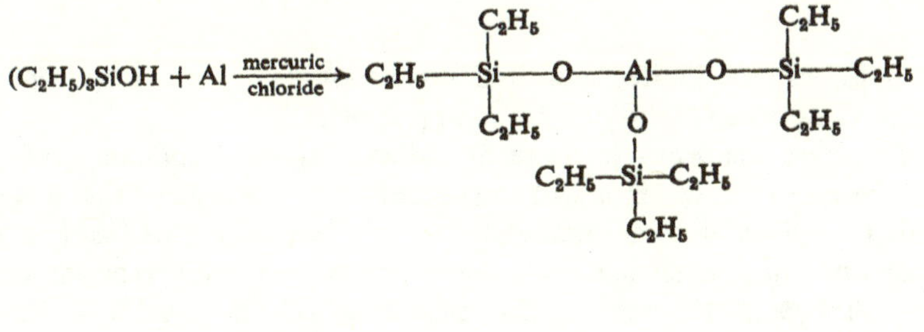

This polymer is based on an unusual backbone structure of —Si—O—Al— and is a crystalline material. Further reactions have been developed such as the simultaneous alkaline hydrolysis of dimethyldichlorsilane with aluminum chloride to give polyaluminotetra-(dimethylphenyl-disiloxane). This is a polymer in which the aluminum is linked to the silicon through oxygen, with the organic groups attached. The important characteristics of these materials is that although they are brittle solids they are, however, soluble in organic solvents such as benzene, alcohol, and acetone. They are, therfore, potentially suitable as lacquers, producing films exhibiting excellent resistance to temperatures as high as 400°C.

SILICON POLYMERS

The semiorganic polymers based on silicon, known as the silicones, have been discussed fully in Chapter 7. Their use in industry is now firmly established and no further mention is necessary here.

TITANIUM POLYMERS

Successful attempts have been made in preparing polymers in which the backbone consists of silicon, titanium, and oxygen. These polymers must therefore also be regarded as semiorganic, and they are obtained by hydrolysis of mixtures of organo-silicon compounds such as alkyl (or aryl) chlorsilanes and the butyl esters of ortho titanic acid. The polymeric structure is as follows.

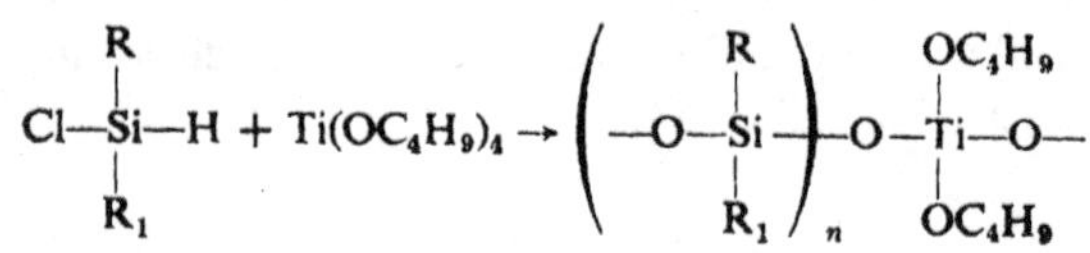

They have found limited industrial applications; small uses include drying oil accelerators and modifiers for resin polymers.

Polymers analogous to those of the semiorganic aluminum type may also be prepared, which is to be expected when one realized that it is possible to replace silcon by titanium in certain silicates. This fact has been established in practice and dodecamethyltitanoxy tetrasiloxane was found to hydrolyze in the presence of acid catalysts to form polymers of varying molecular weight. The lower molecular weight polymers so formed are liquids and the higher ones solid; the linear products are also soluble in various solvents.

TIN POLYMERS

This material exhibits two valency states and, like carbon and silicon, an atom of tin has four electrons in the valency shell. In the stannous form all four electrons are utilized in covalent bonds while in the stannic form the electrons are employed for valency purposes. Polymers of both these materials are known; but it is the stannic types which appear the most promising and may be regarded as semiorganic polymers analogous to the silicones.

The formation of a polydimethyl stannoxane has been reported but was found to be rather unstable on heating, although the phenyl derivative was found to be more stable. An example of such a reaction is a dialkyl tin dichloride which hydrolyzes to form a dialkyl tin dio which, in turn, may be dehydrated to form a polyalkyl stannoxane polymer.

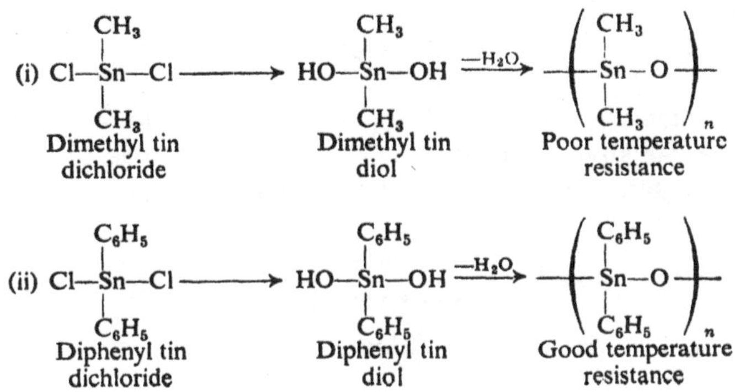

Another group of semiorganic tin polymers that has aroused interest due to its rubber-like characteristics is obtained by the reaction of methacrylic acid with tributyl tin oxide and dibutyl tin oxide. The monomers so formed may be tributyl tin methacrylate or dibutyl tin dimethacrylate.

$$(C_4H_9)_3Sn{-}O{-}Sn(C_4H_9)_3 + 2CH_2{=}C(CH_3)COOH \rightarrow$$

Tributyl tin oxide — Methacrylic acid

$$\longrightarrow 2(C_4H_9)_3Sn{-}O{-}COC(CH_3){=}CH_2$$

Tributyl tin methacrylate

$$(C_4H_9)_2Sn{=}O + 2CH_2{=}C(CH_3)COOH \rightarrow$$

Dibutyl tin oxide — Methacrylic acid

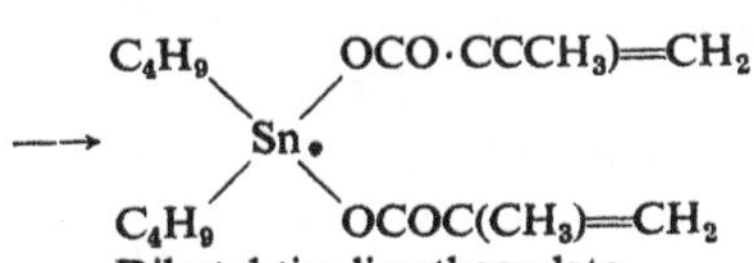

Dibutyl tin dimethacrylate

The monomers may be polymerized using an emulsion technique at 50°C with a potassium persulfate initiator. The resulting polymers may be cross-linked by heating at 250°C with a dibutyl tin oxide, dibutyl tin dihydride, and sulfur. The polymers possess good heat stability but their mechanical properties as yet must be considered poor. Undoubtedly these polymers will be improved when such elastomers may become commercially and technically competent materials.

LEAD POLYMERS

The semiorganic polymers of lead, analogous to those of tin, titanium, and silicon are known to exist although of little commercial importance as yet. The suggested structure of these polymers is as follows:

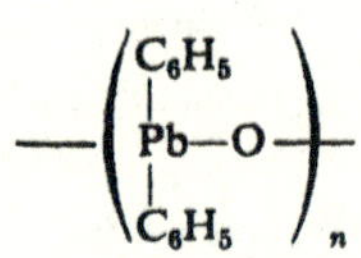

PHOSPHORUS–NITROGEN POLYMERS

The best-known member of the purely inorganic polymers is phosphonitrilic chloride. This monomer was discovered by Stokes in the latter part of the nineteenth century and is prepared by reacting ammonium chloride with phosphorus pentachloride at 120-125°C in boiling tetrachloroethane.

$$PCl_5 + NH_4Cl \xrightarrow[125°C]{120°-} PNCl_2 + 4HCl$$

The pure chloride may then be separated from the reaction mixture to provide a convenient starting point for the preparation of other derivatives. In 1924, Schenck and Rimer found that on heating this crystalline material to between 250-300°C a rubber-like polymer was produced. This material, like natural rubber, exhibited a sharp x-ray diffraction pattern when held in tension. The suggested polymeric structure is as follows.

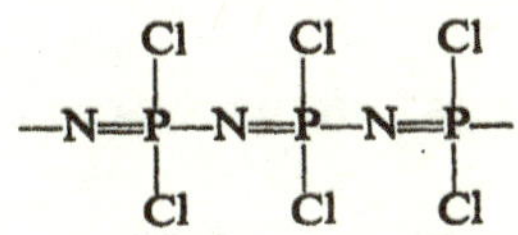

The phosphonitrilic chloride monomer may also undergo polymerization to yield polymers of low molecular weight and exhibiting a cyclic structure differentiating them from the higher molecular weight linear types previously mentioned. Such cyclic polymers may also be converted to linear polymers by reacting with phosphorus pentachloride at 350°C.

These inorganic polymers are chemically inert and have high thermal stability although they have a point of weakness, their hydrolytic instability. They react with water to give a white powder which does not exhibit any elastomeric properties. The hydrolysis has been suggested to take the following form although the complete picture is somewhat more complex.

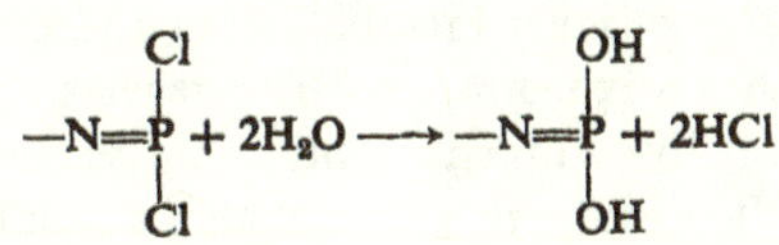

Any attempts to improve the hydrolytic stability have hitherto been unsuccessful although work has been carried out in replacing the chlorine atoms with various other groups, such as the phenyl and ethyl, in an attempt to achieve this result.

The linear phosphonitrilic chlorides are more reactive than the cyclic members, and because they retain the P—N skeleton they show promise as the basis of a completely inorganic type of polymer. Such polymers are no longer mere research curiosities, for already the trimer and tetramer may be obtained commercially. One chemical company in Great Britain is now offering a phosphorus-containing inorganic type of polymer, and numerous patents have been taken out for these polymers in the U.S.A. for use in the flame treatment of textiles, etc.

Polymers in which the —N═P— chain structure is preserved but the Cl atoms replaced by organic radicals have also been prepared. These materials have aroused a great deal of interest as heat-resistant plastics as well as for fire-proofing certain materials. The degree of heat-resistance and flame-proofing characteristics is governed to a large extent by the amount of the organic component in the polymer chain.

ARSENIC POLYMERS

Arsenic polymers at present have found little or no industrial use. They exhibit excellent heat stability, and it has been found that arsenic in the backbone, when cross-linked in the para position, give polymers with softening temperatures that are so high they cannot be formed on conventional plastics machinery. However, if the phenyl nuclei are cross-linked in the ortho or meta position the softening temperatures of such polymers are in the region of 300°C.

SULFUR AND SELENIUM

It is well known by students of chemistry that if sulfur is heated to 250°C for some time, on subsequent chilling, a rubbery mass of 'plastic sulfur' is obtained. This product may be drawn and cooled to produce a fibrous material which, on reheating to 70°C, regains its elasticity. Similarly selenium, which is like sulfur in many of its reactions, may be supercooled from the molten state to give a material exhibiting elastomeric properties over a temperature range of 70 to 75°C. These polymers are considered to be linear as follows.

(i) —S—S—S—S—S— (ii) —Se—Se—Se—Se—Se—

Derivatives of sulfur may also be polymerized, and sulfur trioxide polymerizes in an atmosphere of sulfur dioxide to give a flexible and essentially fibrous material which is believed to have the following structure.

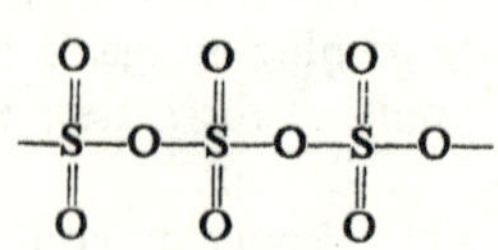

(A cyclized form is also considered to exist.)

Sulfur-nitrogen derivatives have also aroused a certain amount of interest. If cyclic tetrasulfur tetranitride $(S_2N_2)_2$ is heated at low pressures at a temperature of about 300°C the ring is broken and a solid is formed having the empirical formula S_2N_2, which produces a polymer of unknown molecular weight. Such materials form fiber-like crystals with a metallic sheen and, unexpectedly it is a semiconductor.

In general, it is true to say that both sulfur and selenium have not to date found any great commercial use as polymers.

CARBON DIOXIDE

Carbon dioxide has been found to have limited use in synthesis, particularly for polymers, despite the fact that it is a renewable source in abundant supplies. Exceptions include the preparation of urea from carbon dioxide and ammonia and the preparation of hydroxybenzoic acid from carbon dioxide and phenol.

High molecular weight polycarbonates have been synthesized from carbon dioxide and epoxides.

Copolymers with ethylene, propylene, and 2-butene oxide have also been produced. These materials can be molded at 150°C but are subject to thermal decomposition at higher temperatures and are readily hydrolyzed.

Polyureas have also been produced from carbon dioxide and aromatic diamines with molecular weights in the region of 50,000.

23.

PLASTICIZERS, STABILIZERS, AND RELATED ADDITIVES

Plasticizers are regarded by many people in the polymer industry as relatively new materials but they have been known for some hundreds of years although probably not specifically recognized as such. Early reports of the uses of plasticizers can be traced back to the Red Indians who used a mixture of bear's grease and spruce gum to seal the joints in their birch-bark canoes. Other early uses include the plasticizing of pitch with certain oils for caulking ships to render them waterproof.

Probably the earliest use of a plasticizer for plastics was initiated by Parkes who, in 1868, added camphor to cellulose nitrate. This latter material, until this innovation, was a tough inflexible substance, but on adding camphor it became the more flexible material commonly known as celluloid. The family of cellulosics were the major consumers of plasticizers until WW II, when the necessity to find substitutes for natural rubber led to many developments, notably the production of PVC. Plasticization is normally adopted with polymers that possess strong polar forces which tend to make them horny, brittle materials with only a limited range of application and difficult processing characteristics. Polymers such as polyolefins, polystyrenes, polyacrylates, and the thermosets do not, in general, require plasticizers. Over 80 percent of the total plasticizer market is used for PVC which is the main subject for discussion in this chapter although brief aspects of other thermoplastics are also dealt with.

Plasticizers may be defined as nonvolatile solvents added to high polymers to soften and render them plastic or flexible during processing, or to impart permanent flexibility and rubber-like extensibility. The mechanism by which inclusion of plasticizers with PVC results in increased flexibility is attributed to a reduction of the intermolecular forces, i.e., the plasticizer acts as a lubricating agent to allow the macromolecules to slide over one another freely, or by the solvation of the polymer. Unplasticized PVC can be regarded as a three-dimensional gel formed by the attachments between molecules at active centers, i.e., at points of high polarity. Plasticization can be looked upon as a reduction of polymer-polymer unions by creating polymer-plasticizer unions in their place. One important criterion for a material to be considered as a suitable plasticizer is that it should possess a fair degree of polarity in order to effectively solvate the polymer dipoles. This statement can be somewhat contradicted if we look at dimethyl oxalate, which is composed entirely of polar groups but, in practice, is a poor plasticizer for PVC. It has been concluded that as PVC contains both polar and aliphatic groups the plasticizer must bear some relation to the base polymer. It has also been established in practice that a good plasticizer should possess, in addition to a polar group, one of the following: a polarizable nonpolar group (e.g., triphenyl phosphate), a nonpolarizable group (e.g., dialkyl sebacate) or both (e.g., dialkyl phthalate). Therefore, to sum up, it can be said that the kinds of chemical groups present are of fundamental importance in determining the suitability of a material as a plasticizer. Polar groups are essential but other groups must also be present to modify their effects and to provide overall polar characteristics similar to those of PVC itself. The amounts of the various groups present are also of fundamental importance in order to achieve the desired balance of polar forces. Certain disadvantages are unfortunately exhibited by many plasticizers and this has led to another approach which helps to overcome such shortcomings as exudation, extractability, and loss through evaporation of the plasticizer. We have now seen that plasticization consists essentially of the separation of the macromolecules so that the polar forces cannot come into action. This effect can also be achieved by synthesizing a polymer structure with appropriate side chains spaced at intervals along the polymer chain. The side chains so formed would then keep the macromolecules apart, and this is achieved in practice by copolymerization of the monomeric vinyl chloride with vinyl acetate or vinylidene chloride. The copolymerization results in a less regular structure, a reduction in close packing, and attraction of the polar groups.

FUNCTIONS OF A PLASTICIZER

Plasticizers serve three basic functions:

(i) to lower the processing temperature of the polymer to below the decomposition temperature;
(ii) to modify the properties of the finished product (i.e., flexibility or extensibility);
(iii) to modify the processing properties.

GENERAL CLASSIFICATION

The plasticizer field for PVC can be divided into two basic categories:

1. True Plasticizers

These can be further subdivided into

(a) Primary Plasticizers. These are materials which exhibit good compatibility with PVC and can, and often are, used by themselves. They may be regarded as true plasticizers for the polymer.

(b) Secondary Plasticizers. These usually exhibit from fair to good compatibility with PVC but are normally used in conjunction with *(a)*.

2. Extenders

Also referred to as softeners, or even secondary plasticizers, extenders exhibit only limited compatibility with PVC and are used in conjunction with primary plasticizers. They may be regarded as aggregates of molecules in the form of droplets which soften the mix mechanically and, as such, they can be likened to liquid fillers analogous to diluents in surface-coating technology. One basic requirement of these materials is that they must, of course, be reasonably compatible with the polymer.

REQUIREMENTS OF AN IDEAL PLASTICIZER

The properties to be looked for in a plasticizer may be classed under two headings: those that are essential and those that are desirable. Essential properties include free availability, permanency, low volatility and compatibility. Desired properties include low cost, low oder, absence of color, nontoxicity (essential for food contact uses), nonmigratory, nonbleeding, nonflammable, heat and light resistance, and good processing characteristics (e.g., ease of dispersing in resin). One main property which is highly desirable for a plasticizer is its ability to impart to the polymer, and to maintain, the characateristics of an elastomer over the widest possible temperature range. Unfortunately, no single plasticizer embodies all the desired properties mentioned above. Therefore, for any specific application it is necessary to choose the correct plasticizer or combination to suit the requirement by using a general-purpose plasticizer as the basis and building-in other specific requirements with specialized plasticizers.

PLASTICIZATION OF RUBBERS

We have seen in Chapter 21 that rubbers can be plasticized with a variety of materials such as plasticizers, softeners, and processing aids. These may be classified as follows:

(*a*) *Pure softeners or plasticizers.* It must be noted here that, as with the polar thermoplastics, the usefulness of a material as a plasticizer for a given polymer depends on the relative polarities and, therefore, a material suitable as a plasticizer for one polymer may be unsuitable for another. These materials include the mineral oils and the ester plasticizers, the latter type being used primarily with polar rubbers.

(*b*) Materials which aid processing but have no permanent softening action, e.g., coumarone-indene resins, mineral rubber, and phenolic resins.

(*c*) Miscellaneous materials of varying compatibility which are analogous to the extenders and lubricants used in thermoplastics.

TYPES OF PLASTICIZERS

Phthalates

These are probably the most important single group of plasticizers, of which, probably, the most significant are dioctyl, diisooctyl, and mixed (C_7 to C_9) octyl (alphanol) types. This type of plasticizer offers a good balance of all round properties within a medium price structure and therefore is probably the most widely used type. In general, such plasticizers have medium volatility, extraction, migration, moisture resistance, chemical stability, and low temperature flexibility.

The phthalates are prepared mainly from *o*-phthalic anhydride, obtained by air oxidation of naphthalene, though some isophthalates have recently become available. The preparation is effected by esterifying the alcohol with the dibasic acid, using acid catalysts, commonly sulfuric or *p*-toluene sulfonic acid, in an inert atmosphere to minimize color formation. The esterification is carried out in the temperature range 130-200°C for a period between 4-10 hours. Solid plasticizers can also be manufactured by using cyclic alcohols as the esterifying alcohol and a typical plasticizer of this type is dicyclohexyl phthalate.

Phosphates

Tricresyl phosphate (TTP) was the earliest commercial plasticizer for PVC but unfortunately it has rather poor low-temperature characteristics and has been replaced in many applications. This disadvantage can be overcome to some extent by mixing it with certain other phosphate plasticizers such as cresyl diphenyl phosphate.

The phosphates are commonly prepared by reacting coal tar or cresylic acid derived from petroleum and the corresponding alcohol with phosphorus oxychloride or phosphorus pentachloride.

The most widely used of this family are of two general types: triaryl phosphates and alkyl aryl phosphates. The former materials have low volatility and excellent flame retarding efficiency but are relatively poor with respect to low temperature flexibility. The alkyl aryl phosphates are good on low temperature flexibility but at some sacrifice to other properties. The phosphate family is a group of highly compatible plasticizers which are aggresive solvators for most commonly used resins.

Diesters

This group includes the sebacates, azelates, and adipates, of which di (2 ethyl-hexyl) sebacate (DOS), di (2 ethyl-hexyl) azelate (DOZ) and di (2 ethyl-hexyl) adipate (DOA) are typical. These materials are used mainly for their extremely good low temperature properties but normally only in the role of secondary plasticizers due to their limited compatibility with PVC. They are produced by methods similar to those used for the phthalates. The adipates impart excellent low temperature flexibility somewhat better than the phthalates. They are relatively volatile and do not process easily.

Polyesters

This group together with the epoxidized oils are commonly referred to as polymeric plasticizers, although the term is somewhat of a misnomer and 'high molecular weight plasticizers' is to be preferred. There are two major types of polyester plasticizers used. The first have molecular weights in the range of 3000–8000 and are produced as a result of the polycondensation reaction between a dicarboxylic acid, i.e., adipic, azelic, or sebacic and a dihydric alcohol, i.e., propylene or butylene glycol. The end groups of such a material (i.e., OH and COOH) are not modified whereas in the second type the groups are primarily to lower the molecular weight. Monocarboxylic acids or monohydric alcohols can be used, when molecular weights in the range of 850–2000 can be obtained.

In general, these types of plasticizers are incorporated along with primary plasticizers as their compatibility with PVC is often not very good. They are used in compounds primarily to decrease the danger of migration and oil extraction.

Another group of polymeric plasticizers is based on polyurethane chemistry and the resultant plasticizers are colorless and odorless. One commercial product is Ultramoll PU, which is made by Bayer in Germany and contains five percent of PVC and was designed especially for use in plasticizing PVC. This plasticizer has excellent compatibility with the polymer and thus it can be processed in high concentrations. The resultant compound has permanent flexibility and the plasticizer is very resistant to extraction by a wide variety of solvents.

Epoxies

Conventional vegetable oils and alkyl esters of their fatty acids are not sufficiently polar to be compatible with PVC. Epoxidizing the double bonds of the unsaturated acids, as with the triglycerides and certain esters of the fatty acids, remedies this. Thus, with the epoxy plasticizers we have two main types, namely the epoxidized triglycerides produced from soyabean and other oils, and the epoxidized alkyl esters derived from fatty acids such as stearic and oleic. Such epoxidized materials, in addition to this plasticizing action, also exhibit marked stabilizing action against both heat and light.

One method of epoxidizing the double bonds is to add hydrogen peroxide to an acetic acid solution of the unsaturated ester in the presence of a catalyst between 60–70°C.

The most common and best understood method of preparing such materials is the esterification of the epoxy resins with drying oil fatty acids and the subsequent oxidative polymerization of the fatty acid radicals.

Carboxylic acids, which may in fact be of almost any origin, first attack the oxirane ring in preference to the secondary hydroxyls. In this reaction another secondary alcohol grouping is formed:

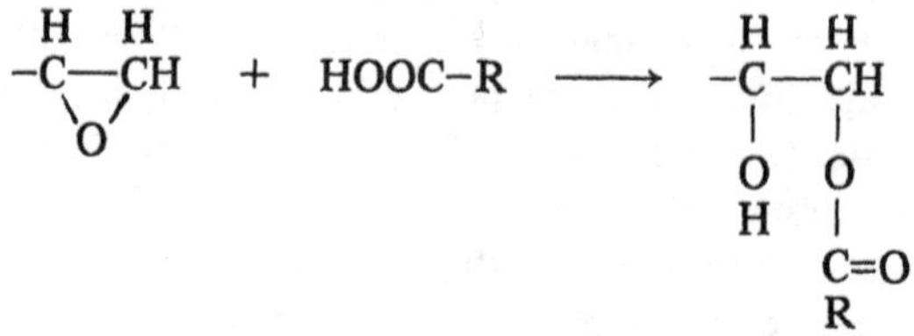

This sequence consisting of the attack on the oxirane ring followed by the generation of an adjacent secondary hydroxyl group may be regarded as another unifying principle for most of the epoxy resins.

Theoretically a complete esterification of all epoxy and alcohol groups is possible:

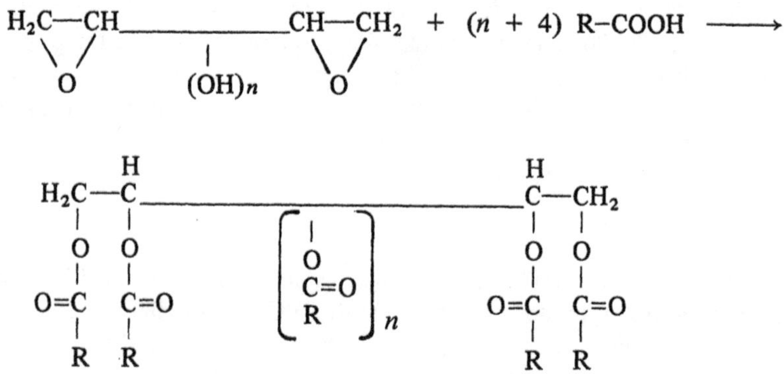

In practice a minimum of 10 percent excess epoxy resin is necessary and usually the excess is much greater.

Nitrile Rubbers

These are used only on a limited scale for applications where specific resistance to petrols and oils combined with low volatility are necessary. They can be vulcanized *in situ* and are compatible in all proportions with PVC.

SULFONAMIDES

This type of plasticizer is generally compatible with resins other than polyvinyl chloride. They are uniquely compatible with some hard to plasticize polymers such as the nylon family. Their high polarity and molecular configuration make them readily adhere to a wide variety of substrates, suggesting they are eminently suitable for such applications as adhesives and surface coatings. Polymers plasticized with these types usually retain a tack free surface which is associated with the unplasticized resin.

The chemical structure of some of the main groups of plasticizers are demonstrated in Table 23.1.

Table 23.1

CHEMICAL STRUCTURE OF PLASTICIZERS

Phthalates	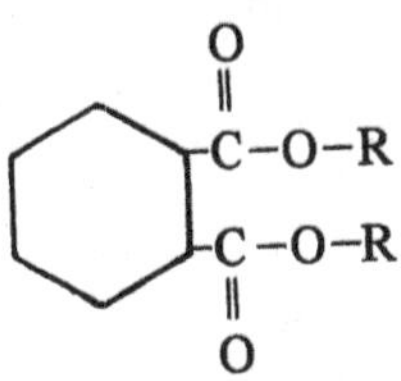
Adipates	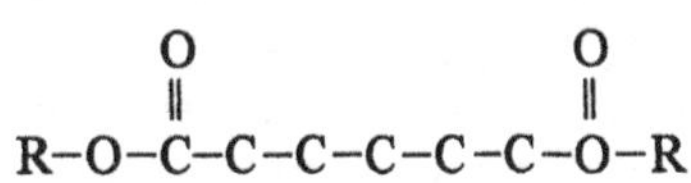
Phosphates	
Sulfonamides	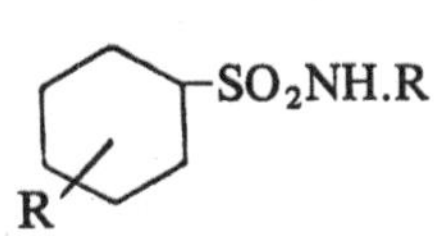

EXTENDERS

These are commonly referred to by other names such as secondary plasticizers or softeners. They are not true plasticizers for the base polymer and as such cannot be used alone. They are used invariably to effect a cost reduction and are normally the cheapest component of a PVC compound apart from the filler. Among the well-known extenders are classed the aromatic hydrocarbons, alkyl-aryl hydrocarbons, hydrogenated alkyl-aryl hydrocarbons, chlorinated paraffins, and special petroleum fractions. Certain of these materials also impart certain improved characteristics, such as the chlorinated paraffins which improve flame resistance. Generally, the use of extenders results in a poor smell and color in the final compound. It is quite usual, therefore, to use conventional fillers rather than extenders to cheapen formulation.

Before leaving this section mention should be made of plasticizers used for thermoplastics other than PVC. Dibutyl phthalate is quite suitable for use with polyvinyl acetate, although with PVC it has been found to be somewhat too volatile. Butyl benzyl phthalate, tricresyl (or tolyl) phosphate (TTP), and diethyl phthalate have all been used satisfactorily with polyvinyl acetate. Cellulose nitrate the 'mother' of thermoplastics has been satisfactorily plasticized with camphor for molding compositions, and for coating applications the following have been used: dibutyl phthalate, dioctyl phthalate, and butyl-benzyl phthalate. For cellulose acetate polymers it is necessary to use polar plasticizers and diethyl, dimethyl phthalate as well as triphenyl phosphate have all been used satisfactorily. In general, it is unnecessary to use plasticizers with polyamides although the flexibility can be increased by incorporating a mixture of *o*- and *p*-ethyl-toluene sulfonamide. Acrylics for molding applications are not, in general, plasticized although for coating applications it has been found necessary and butyl benzyl and dicyclohexyl phthalate have been employed satisfactorily.

STABILIZERS

The majority of polymers known at present suffer some physical or chemical deterioration when subjected to the atmosphere unless previously protected by the discrete use of certain compounding ingredients. This fact is apparent in loss of strength, surface erosion, warping, cracking, loss of transparency, in addition to other phenomena. Certain physical changes are not necessarily inherent disadvantages of the base polymers and such effects as warping, shrinking, and embrittlement may be brought about solely by the loss of plasticizer due to migration. Exposure to alternate high and low humidity may cause swelling and/or shrinkage with hydrophylic materials such as nylon or cellulose acetate. Deteriorative chemical changes in the polymer structure depend largely on the basic design of the polymer. The degradation processes which can effect polymers are as follows: shelf-aging, oxidation (i.e., by metallic catalysts), effect of heat and light, flex cracking (accelerated by ozone), and atmospheric cracking.

Certain specific polymers which are adversely affected by some of the above-mentioned processes follow. It is now well known that when PVC and certain copolymers are heated to high temperatures (above 170°C) they tend to decompose and become discolored. This is largely due to the fact that hydrochloric acid is split off, when the material tends to become brittle and is unsaturated; in addition, it loses many of its electrical properties. The decomposition is believed to begin with the abstraction of a hydrogen atom, by a free radical which can be generated by light, heat, or even shearing. This results in the chain being left in the form of a radical which immediately stabilizes itself by expelling a chlorine atom and forming a double bond. This highly reactive chlorine atom immediately reacts with the nearest hydrogen atom (probably the one adjoining the chain), and the process repeats itself, generating a conjugated polyene structure as follows:

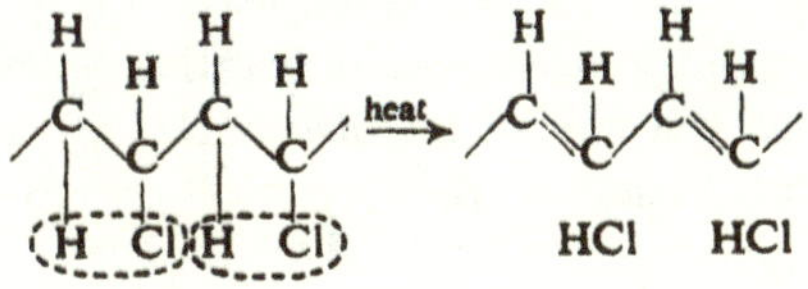

The progress of this 'zipper' reaction is believed to be halted by the branching points and other imperfections in the chain, or by intervention of antioxidants which form a most important part of modern stabilizer systems.

The polyene structure so formed is 'chromophoric' and the orange color formed is partly explained by the fact that the structure so formed resembles that of carotene, the coloring matter of carrots. The color range which occurs when heating PVC at elevated temperatures is yellow, orange, red brown, and finally black, the last color being typical of degraded polymers.

Light stabilization can be distinguished from heat stabilization in that the former provides protection against oxidation whereas the latter effect provides protection solely against heat. Light aging, in comparison to heat, does not cause such severe discoloration but still results in brittleness, loss of rigidity, and decomposition.

It is now evident that with PVC and copolymers in order to obtain useful products it is essential to add materials that prevent degradation. Such materials are commonly referred to as stabilizers and they may be classified in five main groups: (i) acid acceptors, which combine and fix the HCl as it is liberated from the polymer; (ii) antioxidants which prevent any further decomposition due to oxygen; (iii) ultraviolet absorbers which filter out uv light and thus reduce the total energy available to activate the polymer; (iv) materials which react with double bonds, to break up the polyene structure and destroy the color formation; (v) free radical acceptors which terminate or destroy the free radicals which otherwise would initiate and propagate degradation. The five groups may be regarded as requirements of an ideal stabilizer; in addition, the material should also be permanent, compatible, nonleaching, free from odor and taste, nontoxic, and should neither bleed or sweat out of the compound nor be stained by sulfur.

Despite extensive research work to find the ideal stabilizer, as yet no such material has been found. However, many materials now available impart a remarkably high degree of stability to PVC compounds and these are discussed. Before doing so, however, a description must be given of the term 'synergism' which is quite often used in stabilizer systems. This phenomenon can be said to occur when the efficiency of a stabilizing compound is greater than the sum of the efficiencies of each of the constituents which make up the compound.

Types of Stabilizers

Stabilizers may be divided into three main groups: (i) inorganic and organic lead salts; (ii) organic derivatives of various metals of Group II of the Periodic System, i.e., barium, calcium, cadmium, and zinc; (iii) organic derivatives of tetravalent tin. Stabilizers falling into these groups are normally added in varying amounts from 1-3 parts per hundred parts of polymer. Specialized types of the purely organic stabilizers such as diphenyl urea, which gives nontoxic compounds, can be used.

Group 1. Lead compounds were the first really satisfactory materials to be used for vinyl stabilization. They are excellent heat stabilizers and some types also provide light stability. Resins stabilized with these lead compounds have good electrical insulation characteristics, but the following disadvantages limit their general applicability: sulfur staining (black discoloration), inability to obtain a high degree of clarity unless at the expense of heat stability, toxic hazards and poor results given with phosphate plasticizers. However, they are comparatively cheap, and the majority of the previous drawbacks are relatively unimportant in many applications. They therefore account for nearly 70 percent of the total vinyl stabilizer market.

The basic lead content of a lead stabilizer is often taken as a measure of its reactivity and general heat stabilizing efficiency.

The cheapest and most widely used lead stabilizer is basic lead carbonate (white lead). It is normally used in excess but still provides no light stabilizing action at even such high concentrations. Other lead stabilizers used include tribasic lead sulfate, dibasic lead phosphite, dibasic lead phthalate, and tribasic lead maleate. The maleates are dienophils and break up the conjugation of the vinyl system, thereby reducing color formation. Thus, tribasic lead maleate gives excellent long-term color stability to the vinyl compound during processing.

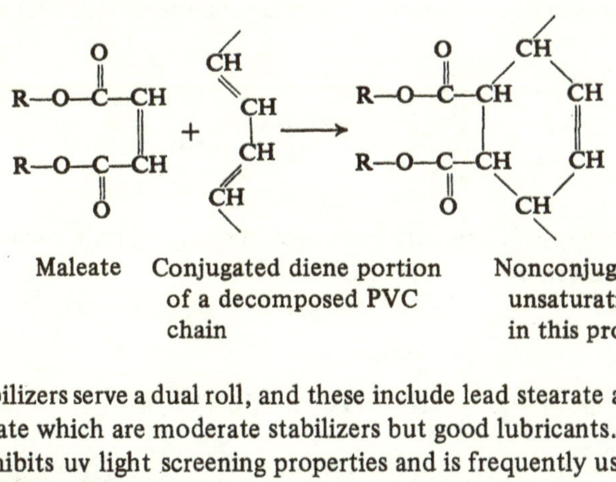

Maleate Conjugated diene portion of a decomposed PVC chain Nonconjugated unsaturation in this product

Some stabilizers serve a dual roll, and these include lead stearate and dibasic lead stearate which are moderate stabilizers but good lubricants. Lead salicylate exhibits uv light screening properties and is frequently used in combination with basic lead stabilizers to improve light stability.

Group 2. We have seen earlier by the definition of synergism, that when certain metallic stabilizers are combined it is often possible to obtain a much higher heat stability for a vinyl compound than by using a single stabilizer type. This effect is of particular value with derivatives of the metals barium, cadmium, zinc, and calcium. An important derivative of these metals are the 'soaps.' The most popular solid soap system is obtained by coprecipitating barium-cadmium laurate, which gives a good general purpose heat stabilizer when combined with some secondary stabilizers such as the epoxides and chelating agents. Other derivatives include barium nonylphenate, barium decanoate, cadmium octoate, zinc laurate, and octoate.

Group 3. The organotins have been claimed to be the best vinyl stabilizers as yet discovered. Their most important advantage is that they produce products with high clarity. They have no effect on paste viscosity and are therefore used for stabilizing plastisol and organosol systems. Unfortunately, they are expensive, which restricts their wider use, and they are therefore limited to certain small specialized applications. Typical materials are dibutyl tin dilaurate, dibutyl tin maleate, dibutyl tin mercaptides, and certain dioctyl tin compounds.

Before leaving this classification it is important to mention the secondary stabilizers of which the epoxy materials are the most important. They have very little influence alone but when used in conjunction with Groups 2 and 3 they have a pronounced effect in improving the heat stability. We have seen in the previous section that there are two main types of epoxides used for plasticizers and these are also used as stabilizers, therefore, we can say that the epoxide or internal group is prone to accept HCl as follows:

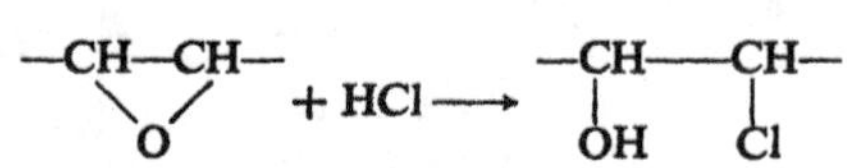

Polyhydric alcohols have also been used in stabilization systems with PVC, especially in conjunction with lead stabilizers.

CHELATING AGENTS

They are used in certain stabilizing systems since they improve the clarity by the fact that they 'chelate' the metal chlorides formed during the reaction of the stabilizer with hydrochloric acid. They are frequently included in barium/cadmium stabilizer systems in a concentration of 1–2 p.h.r. Probably the most common chelating agents used in PVC compounds are the alkyl or aryl organic phosphites such as triphenyl and trioctyl phosphite.

Polyolefins

The problem of stabilization is also necessary with the polyolefins for certain specific applications. It has been shown, with certain members of this family, that if exposed to light and air for 2-3 years they become brittle and useless products; if, however, they are stored in the absence of light at room temperature they remain perfectly satisfactory. Specific experiments carried out on polyethylene which had previously been subjected to such aging conditions showed the following polar groups to be present

$$—CO, —COOH, —OH \text{ and } RCH{=}CH_2R.$$

Protection against such degradative effects can be achieved by adding two percent carbon black or other antioxidants of the nonstaining types which were originally developed for natural rubber, i.e., substituted or hindered phenols or the organic phosphites. It has been reported that an addition of two percent of finely dispersed carbon gives a thirty-fold improvement in the resistance of polyethylene and it is probably for this reason that polyethylene pipes are invariably black. Since carbon black is a pigment, it is also used, aside from its stabilizing qualities, to color the polymer black.

We have seen earlier that there are basically two major types of polyethylenes, low and high density, and both these polymers exhibit a difference in their susceptibility to oxidation. The high-density materials have higher amounts of crystallinity and, consequently, less amorphous regions that the low density and, therefore, they have somewhat better resistance to oxidation. The use of the organic metallic catalyst for the preparation of the high-density polymers necessitates the use of a stabilizer or a chelator to inactivate the metallic salts which could possibly act as a catalyst and speed up oxidation. It would also appear that the amorphous region of the polyethylenes, which have more side chains, are primarily attacked by oxygen.

Polypropylene is more prone to oxidation than the polyethylenes, therefore the overall picture is somewhat more complicated, and polypropylene must also be stabilized against heat degradation during processing. A synergistic mixture of alkylated phenol with certain sulfides has been shown to impart a high degree of stability.

Polystyrene

This polymer has quite good ozone and oxygen resistance although poor weathering properties. When this material is exposed to heat or uv radiation in the presence of oxygen, carbonyl, and hydroxyl groups are formed; further, the uv light attack results in discoloration. Infrared studies and mass spectrometric analysis of the gaseous product obtained on degradation of the base polymer by heat and uv degradation indicated that the degradation involved two different processes, first, a breakdown of thermolabile groups by heat at temperatures above 110°C; second, an oxidation of the polymer by exposure to uv radiation in the presence of oxygen.

ANTIOXIDANTS

We have seen earlier that when a polymer is aged by heat, light, chemical attack, exposure to bacteria or even storage, its properties change, resulting in a loss of certain physical characteristics. This degradative process cannot unfortunately be entirely eliminated, but it can be retarded, and it is presumed obvious that the addition of materials termed antioxidants helps in this respect. It may also help to make the subject clearer if the requirements of an ideal antioxidant are summarized. They must: (i) be permanent; (ii) be capable of inhibiting oxidation under all conditions of storage, processing, and service; (iii) be nonblooming; (iv) not discolor during processing or, later, when the product is exposed to light; (v) not adversely affect any of the characteristic properties of the base polymer; (vi) be nontoxic and odorless.

The properties of a polymer can be seriously affected without the addition of antioxidants. This is deleterious and the following properties can be influenced: melt viscosity, discoloration, clarity, tensile and impact strength.

The oxidation of a polymer is a chain process which is initiated and propagated by a free radical mechanism. This mechanism is basically the same as previously postulated by Bolland and coworkers, which can be illustrated as follows:

Polymer oxidation

1.1 Initiation

$$P \longrightarrow P.$$

1.2 Propagation

$$P. + O_2 \longrightarrow POO.$$

$$POO. + PH \longrightarrow POOH + P.$$

1.3 Chain branching (autocatalysis)

$$POOH \longrightarrow PO. + .OH$$

$$PO.(.OH) + PH \longrightarrow POP(H_2O) + P.$$

1.4 Termination/crosslinking

$$\left.\begin{array}{l} 2P. \\ 2PO. \\ 2POO. \\ P. + POO. \\ P. + PO. \\ PO. + POO. \end{array}\right] \longrightarrow \text{Nonradical products}$$

Inhibition by antioxidant

2.1 By radical termination

$$POO. + AH \longrightarrow POOH + A.$$

$$PO. + AH \longrightarrow POH + A.$$

$$OH. + AH \longrightarrow H_2O + A.$$

2.2 By peroxide decomposer

$$POOH + PD \longrightarrow \text{Stable, nonradical products}$$

Polymer chain scission

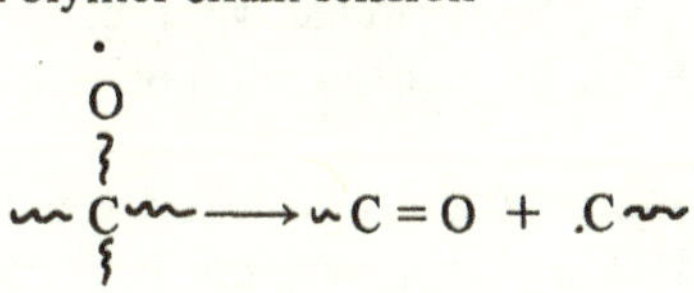

Antioxidants are normally used in commercial liquid stabilized systems to reduce plasticizer oxidation in the plasticized formulations. The organic phosphites have certain antioxidative properties but, as yet, they are not used to any large extent. The most popular types are certain amines and hindered phenols, e.g., 2, 4 dioctyl phenol. Other materials are also used including the condensation products of the previous materials with ketones and aldehydes.

SUGGESTED FURTHER READING

C. R. Beiter, Mod. Plast. 60(10A), 164 (1983)
J. W. Castagno, Plast. Eng. 40(6), 29 (1984)
V. N. Sutan, Plast. Technol. 30(6) 65 (1984)
D. L. Love, Mod. Plast. 61(3), 60 (1984)

24.

FILLERS, COLORANTS, AND SPECIAL ADDITIVES

The science of enhancing the properties and characteristics of a polymer by use of special additives has been known and practiced for many years. The origin of two of the, probably, most important additives, fillers, and colorants, can be traced back to times when polymers were unknown. Industries based on paint and printing ink, for example, were quick to realize improvements that could be achieved by incorporation of such additives.

During the past 50 years, more and more has come to be expected of polymers as regards performance, and many improvements have been achieved solely by the addition of special purpose additives, among these flame-retarders, which have been in use since the early days of cellulosic plastics, and others which by chemical means alter the actual polymer structure. The change in structure, of course, is not a chemical change but, rather, a physical one. The polymer is no longer solid but contains minute, discrete cells which give the polymer a cellular structure. Many changes have taken place in the types of blowing agents used to produce these, and they are discussed later.

It is well known that if rubber is exposed to the weather it loses its elasticity and is said to have 'perished.' This action is accelerated if the rubber is held in tension, the cause probably being due to oxidation and ultraviolet attack. The actual mechanism of the degradation has been the subject of much debate in the past and considerable work has been carried out in trying to find the correct answer.

Space does not permit the exhaustive treatment necessary for describing all these special additives but it is hoped that the reader will consult the reading list at the end of the chapter for further information on each individual additive. These additives are of great importance to the polymer industry, and it must be stressed that to obtain optimum effects it is essential that the correct choice and amount of additive is made.

FILLERS

The use of fillers is recorded in the Bible where references are made to the use of straw in brick-making. Fillers are, of course, not essential for every polymeric compound, but some are vitally necessary for specific end-uses, and many are added solely to reduce costs. They vary widely in availability, cost, composition, and physical and chemical characteristics, and the choice of a filler in the formulation of a molding compound is usually governed by end-use application.

Effects and Requirements of a Filler

Unfortunately, no single known filler meets all the requirements necessary to be classed as the complete filler. The following are some requirements of a potential filler for polymer modification:

(*a*) Low cost.
(*b*) Low specific gravity.
(*c*) Heat stability.
(*d*) Easily and readily obtainable.
(*e*) Neutral–should not be acidic or alkaline.
(*f*) Easy to handle.
(*g*) Low water and oil absorption.
(*h*) Suitable for wetting by the polymer binder.
(*i*) Must not absorb the polymer.
(*j*) Must not prevent flow of the polymer.

In addition to possessing certain of these above properties, the fillers must also confer certain advantages on the final product. These include enhancement of:

(*a*) Heat resistance and thermal stability.
(*b*) Dimensional stability.
(*c*) Mechanical strength, e.g., impact and tensile.
(*d*) Processing techniques.
(*e*) Mold shrinkage and 'after crazing tendencies.'
(*f*) Surface finish of the molding.

(*g*) Thermal expansion.
(*h*) Electrical properties.
(*i*) Chemical resistance.
(*j*) Flow.
(*k*) Special effects such as making the polymer magnetic, conductive, or flame-retardant.
(*l*) Dilution of exotherm as necessary.

Classification

Fillers may be broadly classified as (*1*) inorganic or (*2*) organic. The main fillers are listed in Table 24.1.

1. Inorganic Fillers

These fillers have certain advantages over the organic types but in many respects their uses are complementary to one another. The inorganic fillers are not widely used for modifying thermosetting molding materials but are now used much more in reinforced plastics. The main advantages in comparison with the organic fillers are that they impart better heat, water, and electrical resistance. They also have better dimensional stability and, in some cases, better chemical resistance. The main inorganic fillers are:

(a) Asbestos. Asbestos consists of hydrated metal silicates and is partly related in chemical composition to glass. It is a naturally occurring mineral product and may be found in South Africa, Canada, and Russia. The base material is readily separated into more or less flexible fibers of staple length, varying between 1/16 inch to two inches, although it is not unusual for greater lengths to be used. There are four main types of asbestos occurring in nature of which chrysotile, crocidolite, and amosite are the most important. The first two are used extensively in the polymer industry. Little use has been made of amosite because of the brittleness of the moldings obtained from it. It is, however, used in specialized cases for its acid-resistant characteristics. Chrysotile fibers are long and, thus, ideal for use in laminated plastics, but crocidolite fibers are preferred for general moldings because of the high tensile strength they impart. The moldings produced with asbestos as the filler have excellent heat resistance and low water absorption but they have a high density.

(b) Glass. Glass is an inorganic product of fusion which is cooled to a rigid state without crystallization. It is transparent, hard, and heat-resis-

Table 24.1

PRINCIPAL FILLERS FOR POLYMERS

Type	*Fibrous*	*Nonfibrous*
(1) *Inorganic*	Asbestos Glass fiber Mica	Aluminum powder Iron oxide Titanium dioxide Zinc oxide Antimony oxide Slate powder Barytes Marble dust China clay Calcium carbonate Talc Carbon black (various types) Graphite Silicas Ebonite dust (hard rubber)
(2) *Organic*		
(i) Cellulosic	Cotton fabric, flock, and powder Woodflour, sawdust, and wood shavings Paper and wood pulp Linen fabric and fiber Lignin, sisal	'Microballoons', e.g., P/F and U/F Microspheres
(ii) Noncellulosic	Cork dust Wool and wool felt Cotton seed hulls	
(3) *Miscellaneous*	Walnut shell flour Coconut shell flour Soyaflour and some natural gums	

tant, and has a theoretically high strength as compared with plastics. Fibers produced in glass are of immense importance as a reinforcement for synthetic resins such as epoxies and polyesters. Various forms of fibers are used although it is probably correct to say that terms describe the use of glass fibers in conjunction with resin is reinforcing agent rather than a filler. The fibers are generally used in a cloth or mat form for laminates although use is now made of the fibers for a technique known as dough molding. The various forms of glass fiber are:

1. Continuous filaments–rovings, etc.
2. Staple yarn cloth.
3. Glass mats, needle mat, diamond mat, and coil mat.

(c) Mica. Mica occurs in nature mainly in India and the U.S.A. and, like asbestos, it is a silicate. When used as a filler it imparts excellent electrical properties although, unfortunately, it is not wetted by resins, and homogeneous moldings are therefore difficult to obtain. Use of mica as a filler for polymers is somewhat limited because the base material is fairly costly.

(d) Diatomaceous Silica. This material when used as a filler improves heat resistance and the electrical properties of the polymer without increasing its specific gravity. However, it is an abrasive and will affect the surface of the mold, which is then likely to cause the molding to stick.

2. Organic Fillers

(a) Wood Flour. One of the most common and cheapest fillers, wood flour is used mainly for pheno and aminoplastics. The type of wood used is most important, the best wood flour fillers being obtained from soft woods such as deal and spruce. The wood flour is normally ground to a particle size of a 100 mesh with a moisture content of 6–8 percent. This filler produces moldings with fair strength properties and good moisture resistance, but its electrical properties could be better.

(b) Cotton Flock. This is a short staple cotton fiber which is of little use for spinning due to its relatively small fiber length. It is a pure fiber and is normally light in color with a high bulk factor. It provides moldings with improved impact strength and mold shrinkage. It is more costly, however, than wood flour.

(c) Paper (α-Cellulose). Very important as a filler with the amino plastics and paper is normally used in the form of bleached sulfite pulp. It is practically colorless, with a short fiber and is, therefore, fairly expensive. The use of α-cellulose as a filler, apart from the amino plastics, is somewhat limited.

(d) Chopped Fabric. This type of filler is obtained from cuttings from the garment and related industries. It is, therefore, relatively cheap, but it is difficult to reproduce moldings with it. Moldings of outstanding impact strength are obtained although their water absorption is high.

(e) Walnut-Shell Flour. Moldings with smooth surfaces and medium

gloss are obtained with the filler and their properties are comparable with those obtained from wood flour. The cost however, is higher.

(f) Carbon Black. Channel black has been replaced to a large extent by a furnace black which is a more coarse variety. This latter black consists of rod-like clusters of spherical particles, which can be up to 800 angstrom units.

Carbon blacks have low surface energy and they highlight shielding capacity which makes them very useful for retarding the photo-oxidation of polythene. This type of filler also increases the modulus of elastomers but may decrease the tensile strength and increase the embrittlement temperature of some plastics.

Another use of carbon as a filler is in a fibrous form. Both rayon and polyacrylonitrile can be pyrolyzed to produce carbon fibers. These fibers have found an interesting use as a reinforcing material in conjunction with certain resins in processing technology methods akin to the better known technology associated with polyester glass fiber. The composites so formed have already found uses in the aircraft industry and one particular example is their use as fan blades in the Lockheed Tristar engines.

Properties and Choice of Filler

The choice of a filler plays a very important part in determining the final physical properties of the molding. This is illustrated in Table 24.2, which shows the effect of adding different types of fillers to phenol and urea-formaldehyde resins. This effect is further emphasized in Table 24.3, which lists the physical properties of polyester and epoxy casting resins resulting from using various fillers. In this table it is seen that the calcium-carbonate-filled resins give the best general characteristics and the mica-filled the best electrical properties.

Previous mention (*see* Chapter 5) has been made of a process known as 'dough or gunk' molding, which was originally developed to use waste material obtained from the conventional lay-up technique for fiber-glass reinforcement. It consists essentially of mixing the glass fiber with polyester resins, catalysts, and initiators, pigments, and additional fillers such as calcium carbonate, to form a material with a dough-like consistency. The material is then suitable for molding into a variety of shapes. The term 'gunk' is of American origin and means waste or scrap and this process now consumes a considerable amount of glass fiber as a filler.

Many developments have taken place with fillers for use with polymers, one such being employed for casting resins. These fillers are either low density aggregates or hollow spheres, the actual density of the final product

Table 24.2

SOME TYPICAL PROPERTIES OF THERMOSETS WITH VARIOUS FILLERS

	Unfilled P/F	*Wood flour P/F*	*Asbestos P/F*	*Wood flour U/F*	*α-Cellulose U/F*
Specific gravity 15°C	1.3	1.33	1.7	1.4	1.45
Tensile strength (lb/in.^{2})	5000	7000	5000	7000	7500
Impact strength (ft/lb)	0.1	0.2	1	0.16	0.2
Plastic yield B.S. 771 at 100°C	3	2	2	3	3
Coefficient of linear expansion $\times 10^{-6}$/°C	25	40	25	30	30
Water absorption (168 h) mg B.S. 771	30	90	50	300	300
Volume resistivity (ohm/cm)	1×10^8	1×10^{11}	1×10^9	1×10^{11}	1×10^{12}
Power factor 1 Mc/s	0.2	0.15	0.1	0.3	0.3
Permittivity at 1 Mc/s	6	7	7	6	6

being determined by the amount of low-density filler added. Two types are available, inorganic and organic. The organic low-density fillers are hollow spheres in the form of 'microballoons' and they are used for filling epoxy and polyester resins. The microballoons are tiny nitrogen-filled polymeric spheres, usually produced from urea or phenol-formaldehyde, with a diameter from 0.0002 inch to 0.0035 inch. The inorganic type is based on glass, a material typical of this group being the 'Eccospheres.'

COLORANTS

It has been suggested that one of the reasons for the tremendous growth in the use of polymeric materials is that they can be supplied in almost any color. Extensive experience in color matching is required, for colorants can play a dual role in many polymer formulations, particularly with PVC. The pigments used in PVC paste are added primarily to impart the decora-

Table 24.3

EFFECT OF VARYING TYPE OF FILLER IN CASTING RESINS

Property	*Epoxy*			*Polyester*		
	Unfilled	*Calcium carbonate*	*Mica*	*Unfilled*	*Calcium carbonate*	*Mica*
Specific gravity 15°C	1.2	1.5	1.4	1.25	1.5	1.35
Tensile strength ($lb/in.^2$)	9500	6000	5500	7000	4000	5000
Compressive strength ($lb/in.^2$)	16,000	7500	5500	18,000	18,500	19,500
Coefficient of linear expn. $\times 10^{-6}/°C$	70	55	40	95	84	64
Water absorption (mg)	22	22	22	50	50	50
Dielectric strength (V/mil)	300	350	400	300	270	400
Power factor 1 Mc/s	0.030	0.035	0.035	0.018	0.017	0.02
Dielectric constant 1 Mc/s	4	4.5	4	3	4.75	3.5

tive effect but they may also act as stabilizers. Many pigments produce an adverse effect on some polymers; for example with PVC the volume resistivity is affected. The correct choice of color for individual polymers is essential, Table 24.4 listing the various colorants suitable for different polymers.

Three basic colorants are used for polymers, namely, inorganic pigments, organic pigments and dyes.

Pigments are black, white, colored, metallic or fluorescent solids that are insoluble in, and remain essentially unaffected by, the mediium into which they are dispersed. They are small in particle size, generally in the range of 0.01–1.0 microns in diameter. Pigments produce color by selective absorption of light, but, because they are solids, they also scatter light.

Light scattering may be undesirable in, say, a transparent film but is desirable where opacity is required. There are some organic pigments that are extremely small in particle size and so scatter very little light. They act, therefore, more like dyes. Organic pigments are characterized by high color strength, brightness, low density, high oil absorption, and sensitivity to heat and light. Inorganic pigments are more opaque, less bright and weaker in tint. However, they are more resistant to heat, light, chemical attack, migration and weather-

Table 24.4

COLORANTS FOR POLYMERS

Polymer	*Suitable colorants*
Thermosets	
1. Aminoplastics	Inorganic pigments, organic pigments, or phthalocyanines
2. Phenoplastics	Spirit and hydrocarbon soluble dyes, inorganic and organic pigments
3. Epoxies	Inorganic and organic pigments, also phthalocyanines
4. Polyesters	Inorganic and organic pigments as pastes, also selected hydrocarbon soluble dyes
5. Silicones	Inorganic pigments
Thermoplastics	
1. Polyvinyl chloride and copolymers	Spirit and hydrocarbon soluble dyes. Inorganic pigments (except those with the tendency to bloom and migrate)
2. Cellulosics	Spirit, hydrocarbon soluble dyes, and basic dyes
3. Polyolefins	Organic pigments, phthalocyanine pigments
4. Polymethylmethacrylate	Hydrocarbon soluble dyes, pththalocyanine products
5. Polystyrene	Hydrocarbon soluble dyes, inorganic and organic
6. Polyamides	Inorganic pigments, phthalocyanine, and some anthraquinone pigments
7. Fluorcarbons	Inorganic pigments

ing. They also have higher density, lower cost and less antioxidant effect. One example is molybdate orange for vinyl film.

Inorganic pigments are generally metallic oxides which have been calcined to around 2,000 °C. Typical examples are titanium dioxide, zinc oxide, and antimony trioxide. Inorganic pigments are popular because of their resistance to the higher molding temperatures required by many of the new polymers.

Dyes are intensely colored, solubilized organic substances that are retained by the medium they color, by chemical bonding, absorption or mechanical retention. Dyes produce color by absorption of light without affecting transparency or optical purity. They possess good color strength and are available in a wide range of colors. The specific gravity of dyes is normally low and they are also normally characterized by low plasticizer absorption. However, they possess certain disadvantages including poor heat stability and light resistance. Dyes also tend to bleed out from the polymer and rub off from the surface.

Selection of a colorant for plastics is governed not only by the coloring characteristics of the pigment or dye but also by the resin to be colored—its properties and compatibility—and the method by which it is processed. For example, PVC and its copolymers liberate hydrochloric acid at high temperatures so that acid-sensitive colorants like ultramarine blue and cadmium red must be used with care. In addition, molten nylon acts as a strong acidic reducing agent and can decolorize certain dyes and pigments. Colorants must be heat stable enough to withstand processing temperatures, which are sometimes very high. Colorants may also react with other additives used in the film, such as heat and light stabilizers. As an example, sulfur-containing cadmium pigments react with the nickel-bearing stabilizers used in some film grade polypropylenes.

The requirements of a colorant can be summarized as follows:

(*a*) Low specific gravity.
(*b*) Relatively inexpensive.
(*c*) Heat and light stable.
(*d*) Weather resistant.
(*e*) Low plasticizer and oil absorption.
(*f*) Chemically resistant.
(*g*) Bright and high strength of color.
(*h*) Disperse easily.
(*i*) Resistant to migration.
(*j*) Soluble in suitable solvents.

Methods of Color Addition

An important factor is that of dispersion, to develop color strength and maximum optical properties. Dyes are not normally a problem because of their soluble nature, but pigments need to be properly wetted by the plastic or by suitable vehicles. Poor dispersion can lead to processing difficulties with thin films. Improvements have been made through surface treatment of pigments or through better grinding techniques.

Colorants are normally supplied in one of two forms, as a dried base or as concentration of pigment dispersed in a paste or solid. The thermoplastic polymers may be colored by dry coloring techniques and this is normally carried out in a tumbling machine. The premix thus obtained is sent straight to the injection molding machine. This is a relatively cheap process and has been used successfully for polystyrene, polyolefins, and polyamides. Another method used with thermoplastics quite successfully is the 'masterbatch' technique. In this method the pigment is supplied at a high concentration mixed into a solid polymer, such as PVC, and the technique is to dilute the 'masterbatch'' with further polymer. This has the advantage of being clean and easy to handle although more expensive than dry coloring. Another method that has become extremely popular with some thermosetting resins is that of using high concentrations of prepared colored paste. This technique has been established as the easiest for the fabricator incorporating colorants in polymers. It has been widely adopted for use with epoxy resins, the procedure being to mill the pigment into the resin, the end-product obtained being a reactive part of the overall system.

SPECIAL ADDITIVES

Flame-retarders

In general, the flame-resistance of polymers is good in comparison with such materials as wood, cotton, and other cellulosic derivatives. However, even from the early days of the cellulosic plastics the need to improve the flame- or fire-resistance of polymers has been very much in the minds of the polymer manufacturers. The present-day uses of plastics, rubbers, paints, and fibers for many applications that were hitherto considered impossible has only served further to highlight the limited flame-resistance of many polymers. Typical applications where this problem is of considerable importance include the building and clothing industries.

There are at present many terms used interchangeably to describe the degree of flame resistance and it is hoped that the following suggested definitions help in clarifying this subject.

(*a*) *Fire-resistance.* This is the ability of a material to stop the penetration of flame.
(*b*) *Flame-retardant.* This is the ability of a material to retard the penetration of flame.
(*c*) *Self-extinguishing.* The ability of a material to cease burning immediately on removal of the source of the flame.
(*d*) *Nonburning.* The ability of a material to resist combustion.
(*e*) *Flame-spread.* The spreading of a flame over the surface of a material due to its characteristic properties when in contact with a flame.

Unfortunately, at the present time there is not one polymeric material available that meets all the requirements for complete flame resistance. Fortunately, in many cases certain improvements can be obtained by the addition of flame-retardant chemicals and modification of the polymer chain, although there may be an effective increase in the cost.

Until recent years the most common method of imparting flame resistance to polymers was by chemical additives such as antimony trioxide or phosphorus-containing compounds. Antimony trioxide has successfully been used with polyvinyl chloride and polyethylene. It has, however, been suggested that antimony trioxide does not impart flame-retarding properties unless the polymer to which it is added contains chlorine in the polymer chain or unless a chlorine-containing substance is also added. There are some drawbacks in using antimony trioxide one of which is that the final color of the molded article is affected and may possess a light or dark shade, depending on the grade used. This material can also be used with phosphate plasticizers, and satisfactory results have been obtained with such combinations in PVC mixes. The phosphate plasticizer alone imparts a certain degree of flame-retardancy although it is preferable to mix between two or five parts of antimony trioxide in the compound.

The flammability characteristics of the more common polymers are given in Table 24.5. The main attempts to improve the flame-retardant properties of polymers can be summarized thus:

(*a*) By adding antimony compounds such as antimony oxide and triphenyl stilbene used in conjunction with chlorine releasing agents.
(*b*) By adding monomeric phosphorus containing plasticizers.
(*c*) By treating the polymers with halogen-containing derivatives such as 2,3-dibromopropyl phosphate.
(*d*) By tailoring the polymer chain by such processes as chlorination or bromination.
(*e*) By adding flame-proof low molecular weight inorganic polymers such as phosphonitrilic chloride.

Blowing Agents

A blowing agent may be defined as a compound, organic or inorganic, that releases a gas when heated to a definite minimum temperature. It is, in fact, the decomposition temperature which determines the usefulness of a blowing agent. Many different and diverse materials have been used for blowing agents, ranging from the domestic product baking powder to materials which can be classed as mild explosives. The ideal blowing agent, however, has yet to be marketed. This statement may be better understood if we look at the requirements suggested for an ideal blowing agent.

Table 24.5

INFLAMMABILITY CHARACTERISTICS OF POLYMERS

Material	*Burning rate*	*Contact with a flame*
(*a*) *Rubbers*		
Natural rubber	Burns readily	Melts and decomposes at 200°C
Chlorinated rubber	Burns only on continuous contact with flame	Does not melt fully
Polybutadiene	Burns readily	Decomposes
Polychloroprene	Nonflammable	Decomposes only at elevated temperatures
Polysulfides	Burns readily	Hardens at high temperatures
Polyisobutylene	Burns readily	Decomposes at 200°C
(*b*) *Plastics*		
(i) Thermosets		
Urea-formaldehyde	Does not burn readily	Decomposes at high temps.
Melamine-formaldehyde	Does not burn readily	Decomposes at high temps.
Phenol-formaldehyde	Does not burn readily	Melt and decompose above 150°C
Silicones	Burns slowly	Do not melt
Polyester (cross-linked type)	Burns slowly	Do not melt
Epoxy resins	Burns	Do not melt
(ii) Thermoplastics		
Cellulose nitrate	Burns very quickly	Ignites in air at 160°C
Cellulose acetate	Does not burn readily	Melts above 200°C
Polyamides	Does not burn readily	Decomposes
Polyethylene	Burns readily	Melts between 110–115°C
P.T.F.E.	Burns	Does not melt
Polystyrene	Burns readily	Decomposes around 150°C
Polyvinyl acetate	Burns readily	Decomposes above 200°C
Polyvinyl alcohol	Burns	Decomposes without melting
Polyvinyl acetals	Burns readily	Rapid combustion
Polyvinyl chloride	Does not burn unless continued assistance is given	Decomposes above 150°C
Polyvinyl chloride	Does not burn readily	Melts about 140°C
Polymethacrylate	Flame may be rapidly extinguished	Decomposes around 230–300°C

(*a*) Gas must be released over a definite and short termperature range.
(*b*) The release rate must be controllable, although at a reasonable rate.
(*c*) Gas evolved should be noncorrosive. Nitrogen is normallly preferred although carbon dioxide is used for certain applications.
(*d*) The compounds should disperse readily in the polymer stock and, preferably, dissolve.
(*e*) It should be cheap.
(*f*) Stable in storage.
(*g*) Should not have an unpleasant or obnoxious smell.
(*h*) Nontoxic.
(*i*) Colorless and nonstaining.
(*j*) The compound should not give rise to a large exotherm on decomposition.
(*k*) Should not affect the rate of cure when added to thermosetting polymers.
(*l*) Should work equally well in closed molds and, therefore, not be affected by pressure.

Blowing agents are incorporated in polymer matrices in order to alter the structure of the polymer from that of the solid to the cellular state. Simple inorganic and organic salts such as alkali carbonates (e.g., sodium bicarbonate) have been used but due to their poor dispersability have found limited use. The main materials may be compounds of the following groups: (1) azo compounds, e.g., azobis isobutyronitrile (AZDN); (2) *N*-Nitroso compounds, e.g., N, N^1-dimethyl-N,N^1-dinitroso terephthalimide; (3) sulfonyl hydrazides, e.g., benzene sulfonyl hydrazide.

Many patents for blowing agents have been taken out and the reader should consult the bibliography for further information.

Ultraviolet Absorbers

It has been established that the breakdown of polymeric materials under exposure to ultraviolent rays involves a light-catalyzed free radical mechanism.

This results in rupture of the bond along the polymer chain with the subsequent discoloration and degradation of the physical properties of the polymer. In practice, one or more of the following effects can result from exposure of unstabilized polymers to ultraviolet energy sources: (i) discoloration from the formation of conjugated structures or complex radicals, (ii) reduction in tensile and impact strength, (iii) distortion of initial flexibility, (iv) dimensional change, (v) surface crazing and cracking, (vi) powdering, and (vii) increase in electrical conductivity.

The addition of antioxidants gives some protection already discussed in

Chapter 23. The various polymers, of course, differentiate in their sensitivity to uv light; polymers such as cellulose nitrate deteriorate quite rapidly whereas polymethyl methacrylate is more resistant. This deleterious effect can be overcome by adding compounds which themselves absorb uv light and at the same time resist decomposition. The uv stabilizer addition is incorporated normally at 0.1–0.5 percent of the polymer weight at the point of fabrication or during polymerization.

The requirements of a material used for uv absorption purposes are:

(*a*) Material must absorb uv light in the desired region.
(*b*) Stable to uv light.
(*c*) Stable to processing conditions.
(*d*) Compatible with polymer.
(*e*) Chemically inert.
(*f*) Neutral in color.
(*g*) Nontoxic.

The principal groups of materials successfully used as uv light absorbers are

(i) Metallic Complexes. These include complexed nickel or cobalt salts of substituted phenols, thiobisphenol, alkylphenones, dithiophosphinates and phosponates, e.g., dicyclohexyl dithiophosphinate.

(ii) Substituted 2-hydroxybenzophenones. This is probably the most common and widely used type of which the most popular is 2-hydroxy-4-octoxybenzophenone.

(iii) Benzotriazoles. Most of this family are derivatives of 2-(2′hydroxyphenyl) benzotriazole and the 5′-*t*-octylphenyl) analog. They are very popular with strong intensity, and broad uv absorption.

(iv) Substituted Acrylonitriles. These give good protection to a wide variety of polymers and absorb uv radiation in the range most damaging to the bulk of common polymers. A popular material in this family is ethyl-2-cyano-3,3′-diphenyl acrylate.

(v) Aryl Esters. Aryl salicylates, e.g., p-octyl phenyl salicylate and benzoate are weakly uv absorbing compounds. They are photochemically unstable and rearrange under uv radiation to form more stable products such as 2-hydroxybenzophenones.

(vi) Inorganic Pigments. Most organic pigments absorb uv radiation and act as screening agents. Carbon black is probably the most effective of this group but cannot be used in many applications because of its objectionable color. Other types are titanium dioxide and zinc oxide but have limited use although they have found some application where translucency and opaqueness is of prime importance.

SPECIAL ADDITIVES FOR FILMS

Anti-fogging Agents

Moisture tends to condense as droplets on packaging films, particularly when wrapping fresh produce such as potatoes and tomatoes. This then obstructs the customer's view of the contents. Such droplets can be dispersed by the addition of non-ionic ethoxylates or hydrophilic fatty acid esters, such as glycerol stearate. These act by promoting the deposition of continuous films of moisture, thus improving visibility.

Antistatic Additives

Plastic films are non-conductors of electrical charges. Such charges may be developed by friction between different layers of the plastic itself, between the plastic and the processing equipment or by electro-ionization from dust or radiation. These electrostatic charges can cause problems during processing or during end use.

The two main problems in processing are: (a) adhesion, either between individual particles or layers of the plastic itself, or between the plastics and processing equipment; and (b) discharging to earth by sparking. This can lead to electric shocks to operators, pin holes in film and local overheating. In extreme cases it can cause a fire hazard. End use problems include dust pick-up and adhesion of thin films, coupled with sparking when they are separated.

The two main properties which an effective antistatic agent must have are a degree of electrical conductivity and a marked tendency to migrate to the surface of the film. A wide range of materials has been investigated for antistatic effectiveness, but the most common in actual use are either glycol derivatives or quaternary ammonium compounds. The latter are ionized to some extent in the anhydrous state and hence act as conductors in their own right. Glycol derivatives do have some slight conductivity but their main effect is the sorption of moisture, and it is the moisture layer that acts as the main path to earth for the electro-static charge.

Slip Additives

Thin films, particularly those with smooth surfaces, tend to 'drag' when being pulled over metal surfaces such as the forming plate. Both static and dynamic friction may be important in film handling. Where 'drag' is likely to be a problem, slip additives are incorporated into the resin. Slip additives act as external lubricants and the materials normally used are fatty acid amides.

In addition to solving problems in film handling equipment, slip additives are also useful in applications where sheaves of film are to be dispensed.

UV Screening Agents

UV radiation has a harmful effect on many products, particularly on certain foodstuffs. Where these effects are likely to be significant they can be greatly reduced by the addition of UV screening agents to the film wrapper or sachet. These screening agents act by absorbing or reflecting the UV radiation. The reflected radiation is dispersed harmlessly, but absorbed UV radiation is converted to radiation of a different wavelength, and this might have an effect on the plastic. If the converted wavelength is in the visible or infra-red spectrum then it is relatively harmless, but shorter wavelengths, such as different areas of the UV spectrum, could have a deleterious effect on the film itself. In most food packaging applications, however, the time of exposure to UV is likely to be small and the effect on the plastic should be negligible.

UV screening is achieved either by pigments or by colorless organic compounds that absorb the radiation. The most effective pigment is carbon black whic gives optimum protection at concentrations of about 2–3%. Films intended for outdoor use, particularly in the tropics, are often protected in this way.

SUGGESTED FURTHER READING

'Foaming Agents.' (1977) L. R. Bolduc. *Modern Plastics Encyclopaedia.* (U.S.A.).

UV Absorbers as Stabilizers for Plastics. Geigy Technical Service Bulletin No. 7.

Colorants. (1977) J. Copp, Modern Plastic Encyclopaedia (U.S.A.).

Flame Retardants. (1977) D. P. Wyman, Modern Plastics Encyclopaedia (U.S.A.).

Fillers for Plastics. (1977) R. B. Seymour, Modern Plastics Encyclopaedia (U.S.A.).

Ultraviolet Stabilizers. (1977) Charles S. Abramoff, Modern Plastics Encyclopaedia (U.S.A.).

Coloring of Plastics (1979) Edited by T. B. Webber. New York: Wiley-Interscience.

PART III PROCESSES

Section A Thermosetting

25.

COMPRESSION AND TRANSFER MOLDING

Compression molding is one of the oldest known methods of processing polymers and is still today regarded as a major processing technique. It is used principally for thermosetting plastics although it can be used to process thermoplastics and is still used extensively for many types of rubbers.

Compression molding may be described briefly as the placing of a known weight or volume of molding powder into an open-mold cavity, the mold consisting of two halves, a male and a female part, and normally preheated. On closing the mold, the male half engages with the female, and pressure is applied, normally in the range of 0.5–3 ton/in.^{2} The whole operation is carried out in a compression press, which consists basically of a ram in a cylinder which can be hydraulically operated. The hydraulic power is delivered either by means of a central accumulator system or by separate pumps attached to each press. Heat is also essential with this molding process, and when used in conjunction with pressure the molding powder will soften and flow within the mold cavity and assume the required shape. Correctly designed molds minimize flash and thus little or no finishing operations are usually necessary. It has been seen in previous chapters that thermosetting molding materials undergo a chemical change whereby their structure becomes cross-linked and the material hardens to a permanent shape. To ensure that the curing action is complete it is necessary to maintain the molding pressure for a period.

For complex components which require delicate inserts the compression molding process has been found to have certain limitations, and yet another process has been developed during the past 25 years, which also involves the application of heat and pressure to the molding powder before it is forced into the mold cavity, but which overcomes some of the disadvantages associated with the compression technique. It is commonly called transfer molding. The actual process consists of charging a known amount of molding powder into a heated chamber outside the mold, and when the powder has reached a sufficiently plastic state it is then forced under high pressure through a suitable narrow opening, called a 'gate,' into a closed mold. The technique can be likened to the injection molding of thermoplastics, described in Chapter 27. In both processes the material is 'plasticized' in one chamber and then, while in a fluid state, forced through a 'runner' into a mold cavity.

Much knowledge has been gained by the polymer industry on both compression and transfer molding of thermosets, but in this chapter it is only possible to give an insight of the basic details. Readers wishing to pursue the subject further should consult the reading list for sources of information.

COMPRESSION MOLDS

Compression molds can be divided into three types: (*a*) hand molds, (*b*) semi-automatic and (*c*) automatic. The design of these molds must take into consideration 'venting' for the escape of steam, gas or air produced during the operation. After the initial pressure has been applied it is usual to open the mold slightly to release the gases. This is known as 'breathing.' The molds are designed with either a vertical or horizontal flash and it is convenient to distinguish between the different types as follows (*see* Fig. 25.1).

Flash Type

With this type of mold a slight excess of the molding power is loaded into the mold cavity and, on closing the top and bottom force, the excess powder is forced out and 'flash' is formed. This can be removed quite readily, though it tends to make the process somewhat wasteful. However, the molds are relatively cheap and are suitable for building up into tools containing multiimpressions. Another advantage is that, while to some extent it is wasteful as far as raw materials are concerned, very slight labor costs are necessary in weighing out the powder.

Positive Type

This type of mold is also easy to manufacture, but no allowance is made

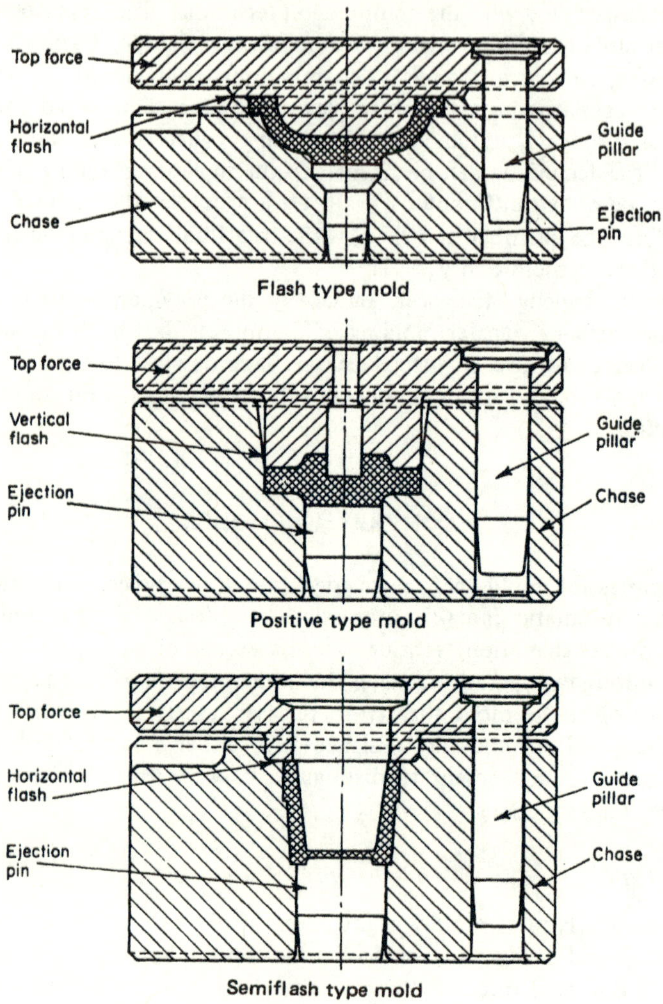

Fig. 25.1

for placing excess powder in the cavity. Should this occur in error, the mold will not close; if however, the correct charge is added only a small amount of flash results, and it can easily be removed. A disadvantage with such a mold is that the gas liberated during the chemical curing action will be trapped, thus resulting in blisters. This type of mold is suitable for molding high bulk factor materials and is used only on a small scale for

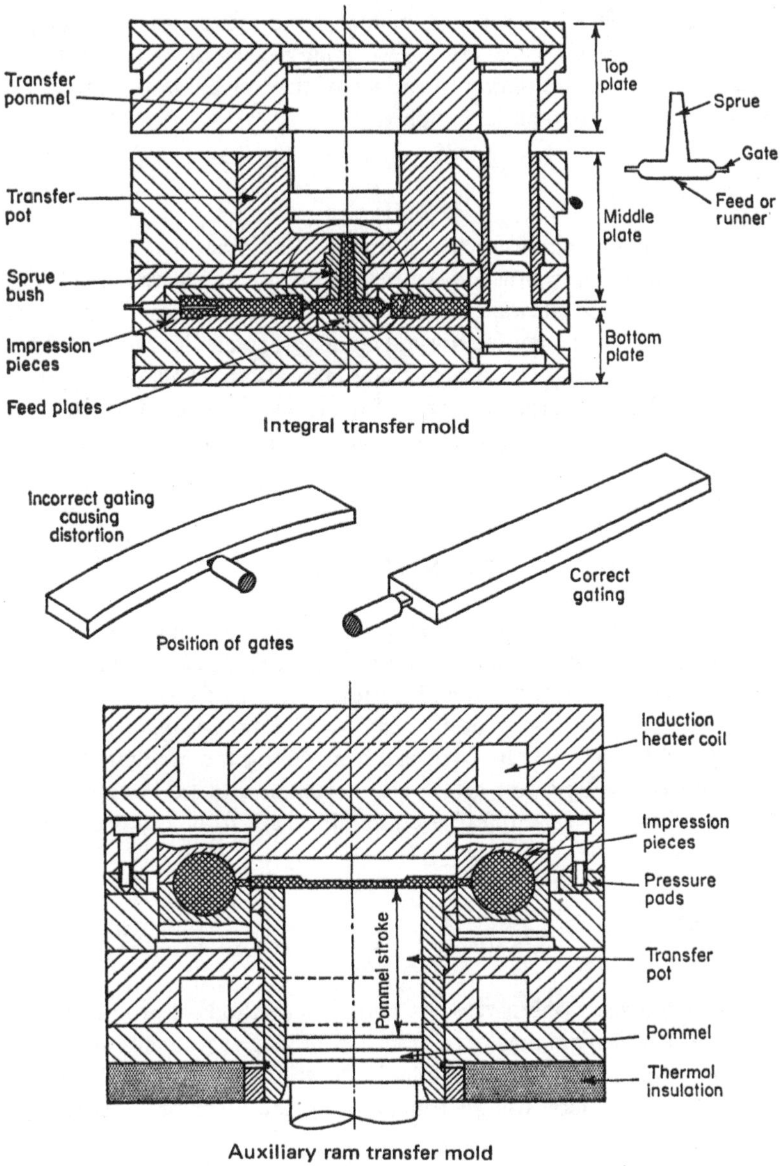
Transfer pommel
Top plate
Sprue
Gate
Feed or runner
Transfer pot
Middle plate
Sprue bush
Impression pieces
Bottom plate
Feed plates
Integral transfer mold
Incorrect gating causing distortion
Correct gating
Position of gates
Induction heater coil
Impression pieces
Pressure pads
Pommel stroke
Transfer pot
Pommel
Thermal insulation
Auxiliary ram transfer mold

Fig. 25.2

molding thermosets. It is, however, used for molding laminated plastics and certain rubber components. Some of the reasons for discarding this type of mold are the necessity for weighing or measuring an exact charge, excessive wear on the sliding fit surface of the top and bottom forces, and the difficulty of ejecting the molding.

Semipositive Type

Such a mold combines certain features of the other two types described above. The molding powder is forced into a complicated mold, at the same time making allowance for excess powder and flash. This type of mold is expensive to manufacture and maintian, and is therefore used where long runs are required. With such a mold it is also possible to get both horizontal and vertical flash. The area of the faces of the mold which come in contact with each other when the mold is closed is generally referred to as the 'land.'

TRANSFER MOLDS

Transfer molding has an advantage over compression technique in that the molding powder is fluid when it enters the mold cavity, which enables slender core pins to be used and intricate sections to be molded. Faster cycles are also possible by using transfer molding, due mainly to the local heating produced from the frictional heat during the molding cycle. This is a result of using narrow gates, and it must be stressed that correct positioning of gates is essential if good moldings are to be obtained (*see* Fig. 25.2).

MANUFACTURE OF MOLDS

Machining and hobbing are the two main methods of producing molds, the most of which are machined from carburizing or normalized steel.

Machining. This is a direct method of cutting out the mold cavity from a solid block of metal using normal cutting and machine tools and techniques.

Hobbing. This is a process in which the mold cavity is formed by forcing a hardened steel master, having the same shape and dimensions as the required molded component, into a soft steel block. The hobbing press must travel at a fairly low speed and each sinking operation normally takes about 70 seconds. The pressure for hobbing ranges from 25-200 ton/in.^{2} and is carried out on a hydraulic press. The cavity formed by hobbing is subsequently trimmed to size, then hardened and, as with the machining techniques, the surface is normally chromium plated.

Other methods are used including spark erosion, electroforming, and casting.

Ejection of Molding

Ejection from the mold can be achieved by using any of the following: ejector pins, sleeves, or stripper plates. The ejector pin method is popular since the pins can be fitted easily and readily replaced. The pins are worked by a common ejector plate or bar located under the mold, which as a rule is operated by a central hydraulic ejector ram. It is necessary for some moldings to remain on the top half of the male part of the tool and when this is required the ejector pins are fitted to the top force.

Heating Systems

The heat required to carry out the molding operation must be sufficient to heat the tool and auxiliary parts to the recommended temperature required for the powder being molded, and to maintain this temperature throughout the whole cycle. Heating is extremely important in the polymer industry and can be used in the following plant: press platens and molds, barrels of injection and extrusion machines, rolls for mills and calenders, resin kettles, mixing equipment, ovens, and drying plant (including preheating, and drying). Heating can even be used in certain cases to facilitate extraction of the molding by making the molding adhere to the required portion of the mold (i.e., by temperature variation in the mold).

Various methods are used to heat the molds, including steam, hot water (high pressure), gas, and electricity (resistance heating, band heaters, low voltage and induction heating). Steam is the preferred method for heating press platens for the compression and transfer technique, although electricity because of its low installation costs and because it is a cleaner method, is also used. The main disadvantage with the latter method is that evenness in heating is not very good and there is tendency to form 'hot' spots. In the manufacture of press platens for steam heating the drilling required for the channels necessary for the passage of steam is a relatively cheap procedure. There are no fire hazards with this method and it is also used for heating mixers and resin kettles. Pressure regulators are normally necessary if close control of the temperature is required. For such processes as injection and extrusion molding, higher temperatures are required than can be conveniently obtained with steam, and electrical heating, generally in the form of cartridge heaters, is preferred.

PRESSES

Presses used for molding thermosets take many different shapes and designs and can be classified as either hand, mechanical or hydraulic types (*see* Fig. 25.3). Hand presses range from 10–100 tons capacity

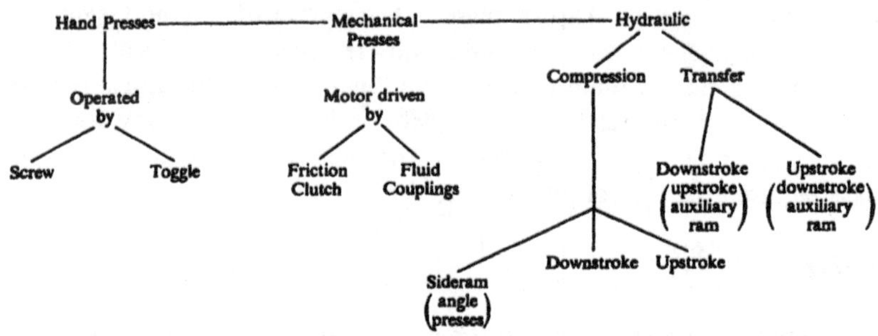

Fig. 25.3 Main types of presses used in compression transfer molding

whereas hydraulic presses are considerably higher (500 tons). Platens are bolted in the press, the space between the platens being called the 'daylight.' Presses with two or more 'daylights' have been developed for greater economy. In most modern molding shops, hydraulic presses are standardized and may be of the upstroke or downstroke variety (*see* Fig. 25.4 and 25.5). The disadvantage of the simple upstroke type is that the return is slow, relying entirely on gravity, although pressure can be applied fairly quickly. Improvements have been sought by using push-back rams. The downstroke-type press incorporating a prefiller, overcomes the disadvantages of the upstroke press, and a high pressure is conserved by prefilling with liquid from a tank (*see* Fig. 25.6). Transfer presses are also used with two independently controlled movements, one for clamping the tool parts together, the other transferring the powder into the cavities. A diagram of a complete unit for a 250-ton press is shown in Fig. 25.7.

The drive for the presses is supplied by single pumps or by central pumping stations. Accumulators are used for storing energy and for supplying the instantaneous pressure demand in excess of the pump delivery.

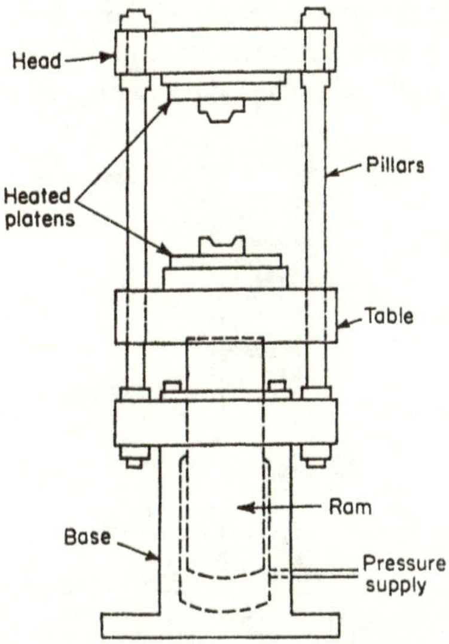

Fig. 25.4 Simple upstroke press

The basic principles of hydraulics are employed, in that liquids are incompressible and transmit pressure equally in all directions. The main fluids used are water and oil. Water is, of course, cheap but causes rust to outside moving parts, whereas oil is more expensive but does not cause corrosion and, in addition, will lubricate the system. The main disadvantage with oil is that it tends to find its way through small orifices and oxidizes, giving rise to sludges. There are two main types of accumulators used, the weight-loaded and air-loaded type. The usual accumulator consists of a single-acting plunger working in a cylinder. The weight of the accumulator (load) divided by the plunger area gives the line pressure, and the plunger area multiplied by the stroke gives the capacity. The main disadvantages of the weight-loaded type is that it is heavy and therefore not very portable. Also, on opening the valve, due to the kinetic energy absorbed in moving the weight, a pressure surge results in the system. However, once this is overcome the pressure given will always be the same. In an attempt to overcome pressure surge and reduce the maintenance normally associated with this method, the air (or gas-loaded) accumulator was introduced. A suitable gas is nitrogen, and the technique used is based on the classical gas law that the product (volume × pressure) of air or a gas at constant temperature is constant. This type is more portable although there is a small pressure loss over the molding cycle.

As with all molding processes attempts have been made to cut down cycle times with compression and transfer molding, to reduce costs and to improve the finished product. Preheating and preforming are two processes

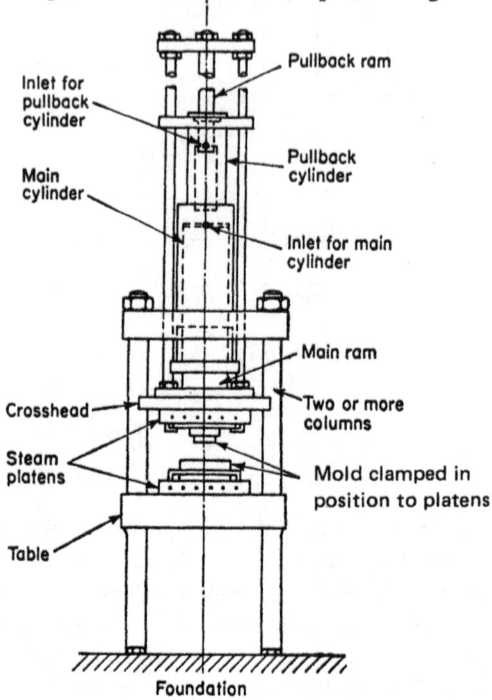

Fig. 25.5 Simple downstroke press

used for such improvements in this form of molding. Preforming is used where the molding is of an awkward shape, i.e., containing both thick and thin sections. The molding powder can be initially compressed into the shape of the molding before being placed in the mold or can be used in the form of pellets which can be placed in awkward spots, where thinning is likely to occur. Pelleting is a preforming process and, as the name implies, consists of compressing the molding powder into pellets of approximately known weight. This process confers many advantages to molding techniques including reduction in bulk factor, avoiding waste, easy measurement and therefore more rapid loading of charge, less pressure than uncompacted material and less risk of contamination. Disadvantages are in obtaining adequate strength of the preform for transmitting purposes and difficulty in obtaining uniform weight. The simple pelleting machines

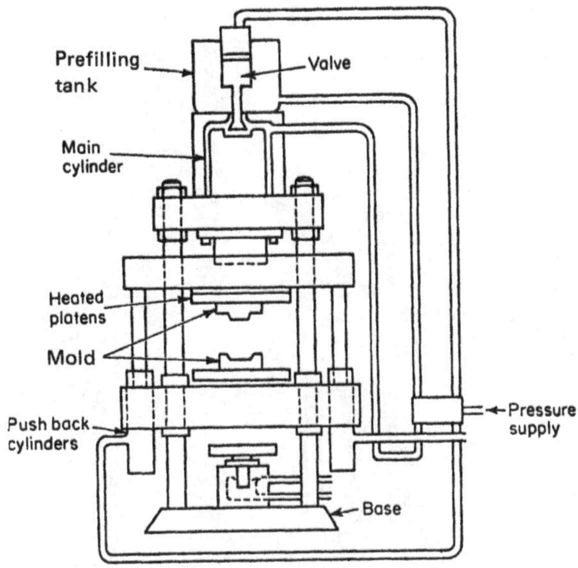

Fig.25.6 Prefilling downstroke press

used until recently worked on the principle of compacting the powder by a short sharp blow rather than with a squeezing action. This led to many difficulties. Horizontal pelleting machines which produce satisfactory pellets and obviate the loss of powder that occurred with other methods are now used. It must be pointed out that while many of these advantages are attractive, the additional cost may not always be justifiable, and with many powders of high bulk factor the process is virtually impossible.

Preheating provides a method for molding relatively thick sections without porosity and the other common faults usually associated with this type of molding. The advantages of the technique include improved flow, reduced flash, lower molding pressures, reduced mold shrinkage and even reduced wear of the molds. The methods of preheating are convection, infrared, radio-frequency and live steam heating. The original method used was a thermostatically controlled gas or electrically heated oven. The quickest, and probably the most efficient method, is the radio-frequency technique, but this is also the most expensive.

1. *main ram*
2. *main cylinder*
3. *return cylinders*
4. *valve in prefilling tank*
7. *piston-type control valve*
8. *pressure gauge*
9. *heated platens*
10. *automatic control of main and transfer ram movements and 'breathing'*
11. *air cylinders operating piston valves*
16. *adjustable delay valve to allow breathing during a compression molding cycle*
17. *ejector gear by hanging bar*
18. *ejector table and stops enabling transfer or ejection*
19. *transfer cylinder*
20. *transfer plunger*
21. *transfer cylinder return rams*

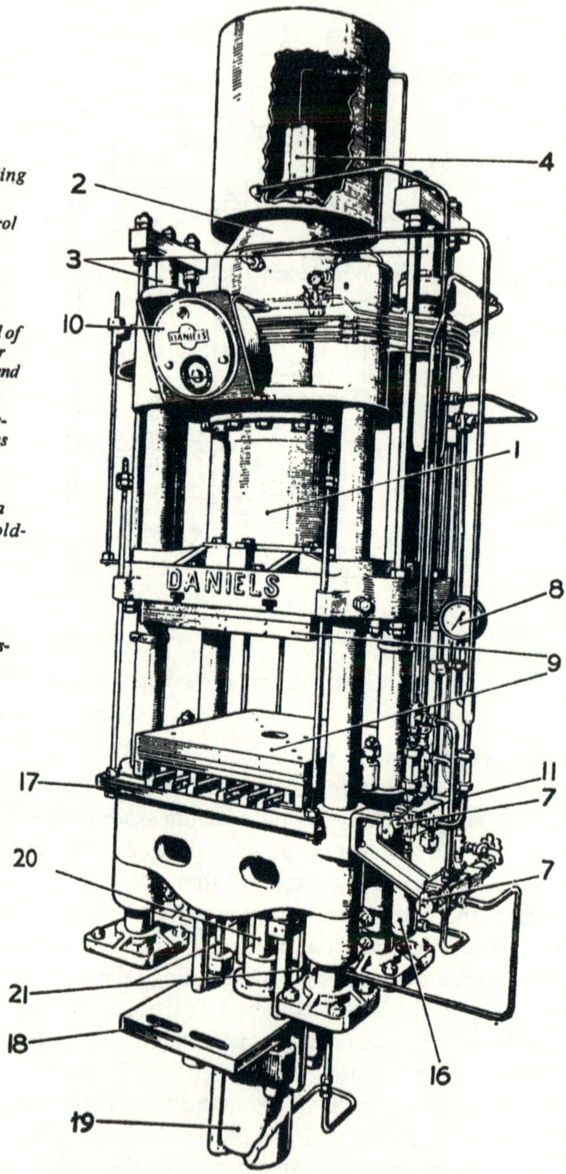

Fig.25.7 Daniel 250-ton transfer molding press

FINISHING

Although mold design takes into consideration the fact that flash must be reduced to a minimum, it still occurs to some extent. Probably one of the best methods of removing flash from thermosetting molding powders is to put them through a process termed 'barrelling.' A barrel in the shape of an octagon is charged with the molding and is allowed to rotate at a speed of 5-50 r.p.m. for 10 minutes to several hours. The moldings are then removed and washed free from dust, etc. Variations include the use of a rubber-lined barrel with steel balls and mineral oxide chips, adding pegs previously impregnated with polish to give a finish to the deflashed molding.

MOLDING FAULTS

The following are some molding faults, reasons, and remedies associated with the molding of thermosets.

Fault	*Possible Causes*	*Remedy*
Short and porous moldings	Insufficient charge of powder	Check charge weight—usually calculated as 5-10 percent in excess of weight of full molding
	Inadequate pressure	Increase pressure or reduce number of impressions (if multiimpression tool) or preheat material
	Lack of air escape	Vent the tool (usually required opposite to the feed in transfer molding)
	Splashing of powder from die in the case of open flash tools	Close press more carefully in final stage to avoid splashing
Soft and porous moldings. Sometimes containing blisters with odor of formaldehyde	Undercure	Increase tool temperature and/or cure time
Blistering accompanied by bleaching and distinct fishy smell	Overcure	Decrease tool temperature and/or cure time until correct cure time is reached

Fault	*Possible Causes*	*Remedy*
Flow marks (parallel to flow of powder) mainly found in articles of deep draw	Trapped gas or air	Use stiffer flow, coarse grind powder. If flow marks persist either 'breath' or 'dwell' before final cure
Burning and discoloration	Extreme overheating	Adjust temperature
Dull moldings	Mold too hot or insufficient pressure. Possibly too much lubricant	Repolish or replate mold
Warping	Irregular cooling, uneven stresses due to widely differing wall thicknesses. Careless ejection	Cool on a jig, modify design. Check method of ejection
Ripple marks	Powder too soft, flows so readily that only low pressures attained esp. in flash type molds	Use a stiffer grade of molding powder
Thick flash	Overfilling mold. Pressure too low. Temperature too low. Too stiff a grade powder	Improve powder measurement or use pellets or preforms. Flash rims and sprue grooves can sometimes be adjusted. Increase pressure or increase temperature or use softer grade of molding powder
Cracks around metal inserts	Material shrinks more than the metal or uneven cooling	Change design or use smaller inserts
Contamination	Molding shop conditions faulty or even packaging of molding powder may be at fault	Check molding shop conditions and check powder against newly opened container

SUGGESTED FURTHER READING

Electrical Properties of Polymeric Materials. (1973) M. E. Baird (London: The Plastics and Rubber Institute).

Moulding of Thermosetting Plastics. (1988. 3rd Ed.) T. Whelan and J. Goff (London: Plastics and Rubber Institute).

Thermosetting Materials and Processing (1994) J. F. Mark (Ed), (London: Longman).

Section B Thermoplastics

26.

EXTRUSION

The basic extrusion process is designed to continuously convert a thermoplastics material into a particular form. Extrusion is an extremely versatile process and the final shapes or forms which can be produced include pipes, films or sheets, fibers, profiles, coatings for paper and other substrates, and coverings for wires and cables. Extruders are also used to feed blow molding machines which produce bottles and other hollow articles. The basic sequence of events is as follows:

(*a*) Plasticization of the granular raw material.
(*b*) Metering of the plasticized product through a die which forms it to the desired shape.
(*c*) Solidification into the desired shape and size.
(*d*) Winding into reels or cutting into units.

Processes (*a*) and (*b*) are carried out in the extruder, while (*c*) and (*d*) are ancillary processes.

The extruder itself (Fig. 26.1) consists fundamentally of an Archimedian screw which revolves inside a close-fitting, heated cylinder or barrel. The thermoplastic granules are fed through a hopper at one end, and carried forward along the cylinder by the action of the screw. As the granules move along the screw, they are melted by contact with the heated walls of the barrel and by the generation of frictional heat in the viscous melt. The final action of the screw is to force the melted polymer through a die which determines its final form.

This apparently simple process is, in fact, far more complicated than it sounds, and highly developed machines and techniques are necessary to achieve the best results. For an understanding of some of the problems involved, a more detailed examination of the various components of an extruder, and of their function, is necessary.

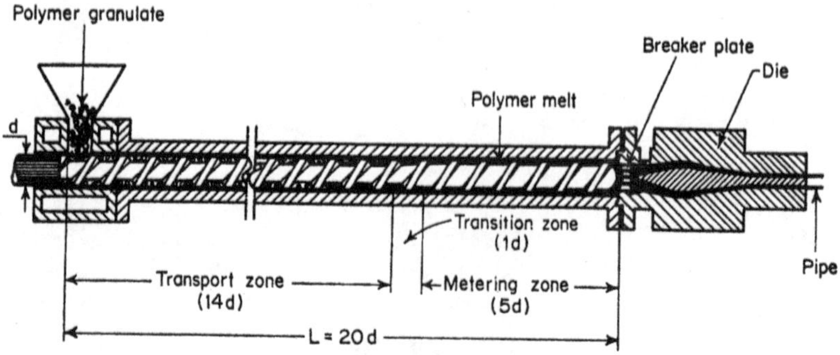

Fig. 26.1 Scheme for a typical single-screw extruder (l/d = 20) shown extruding pipe

The most important component of any extruder is the screw, and it is often found impossible to extrude one material satisfactorily using a screw designed for some other material. Screws are characterized by their length/ diameter ratios (commonly written as l/d ratios) and their compression ratios, namely the ratio of the volume of one flight of the screw at the hopper end to the volume of one flight at the die end. L/d ratios most normally used for single-screw extruders vary from 15 : 1 to 30 : 1 while compression ratios are usually in the range 2 : 1 to 4 : 1. The screw is usually divided into three sections, namely, feed, compression, and metering. The feed section merely conveys the material from under the hopper mouth along to the compression section where the gradually diminishing depth of thread causes volume compression of the melting granules and consequent removal of trapped air which is forced back through the feed section. This ensures an extrudate free from porosity, while another consequence of the volume compression is an increase in the shearing action on the melt, due to relative motion of the screw surfaces with respect to the barrel wall. Improved mixing results, together with the generation of frictional heat, which leads to a more uniform temperature distribution in the molten extrudate. The function of the final section of the screw is to meter the molten polymer through the die at a steady rate and to iron out pulsations. The screw is usually cored for steam heating or water cooling.

Thus far, we have talked about the build-up of back pressures due to the compression ratio of the screw, but, in fact, there would be very little back pressure at normal speeds if the extruder were operating with an open head so that the extrudate could pass through without restriction. The presence of the die alone, of course, sets up an appreciable restriction but it is also usual to have a breaker plate supporting a screen pack consisting of a number of fine or coarse mesh gauzes. In addition to increasing back pressures, the screen pack filters out any contamination which might be present in the raw material. This is particularly important in the extrusion of thin films, where even the smallest of contaminating particles could cause holes or even breaks in the film. For any particular screw speed, a higher back pressure improves the mixing and homogenization (and hence extrusion quality) but slightly reduces extruder output.

The motor employed for driving an extruder should be of more than adequate power for its normal needs. The question of screw speeds is also an important one and variable screw speeds are now considered to be essential. Either variable speed motors or constant speed motors with variable speed equipment such as hydraulic systems, infinitely variable speed gearboxes, step-change gearboxes, etc., are available. Because of the very high back pressures generated in an extruder, it is important to have thrust bearings of robust construction, and the trend towards higher screw speeds makes this even more essential. Overload protection in the form of an automatic cut-out should be fitted. One circumstance which could arise is the failure of a heater at the tip of the extruder when extruding nylon. Since nylon has a sharp melting point, solidification could occur and a blockage arise.

An important piece of equipment at the feed end of the extruder is the feed hopper. It is obviously essential that the flow of granules to the screw of the extruder should be uninterrupted. A common problem in feeding granular material is 'bridging,' which can be avoided by fitting a vibrator or, better still, a screw feed. Another possible cause of blockage at the feed opening of the hopper is melted granules but this is easily avoided by water cooling at the hopper end of the barrel. Special hopper driers are sometimes fitted, especially when extruding hygroscopic materials such as nylon or polyethylene containing carbon black. A typical method of accomplishing predrying is to pass hot air through the granules in the hopper.

We have dealt with some of the requirements for an extruder drive and for feeding the thermoplastic granules and we come now to one of the most important factors governing the quality of the extrudate, namely, temperature control. Heat is supplied to the granules in two ways—by external heaters (which may be oil, steam, or electric heaters) or by frictional heat generated internally by the shearing and compressing action of the screw. The frictional, or adiabatic heating, is appreciable, and in modern high-speed machines can supply the whole of the heating required for steady running, external heating being required only to prevent the machine from stalling at the start of the run when the material is cold.

The most popular type of external heating is electrical because of its compactness, cleanliness, and accuracy of control. Electric resistance heaters are popular because they present few maintenance problems, but induction heating is also widely used and is claimed to give quicker heating with less variation and, also, to facilitate efficient cooling. In extruders running at low to medium speeds (except large machines with screw diameters of about 6-8 inches where frictional heat is again large), external heaters supply a larger proportion of the total heating but in either case temperature controllers are essential and heating and cooling must be carefully balanced. Either air or water cooling may be used, but in either case it should operate automatically. The barrel is usually divided into three or four heating zones, the temperature being lowest at the feed end of the barrel and highest at the head end.

As already mentioned, the screw is also cored for heating or cooling. Where the maximum amount of compounding is required a screw is cooled. This improves the quality of the extrudate but slightly reduces the output.

Correct design of the head is important and should promote streamline flow of the material with no 'dead spots' where material could stagnate and thus decompose due to overheating. This is particularly important in the case of PVC where the degradation point of the material is close to the temperature necessary for adequate flow.

Die design, too, must take account of these factors but to an even more marked degree. Streamlining of the die and die entry is important not only because of polymer degradation but also because turbulent flow can lead to defects in the extrudate. Such defects may be surface effects as, for instance, a roughness or mattness, or a total distortion leading to an effect variously described as knobbly, wavy, or bamboo-like.

A fairly recent development in extruder design has been the use of venting or degassing zones in order to remove any volatile constituents from the extrudate before it emerges from the die. This is achieved by releasing the plastic melt from the state of compression to which it has been subjected, whereupon the water or other volatiles will vaporize and cause the melt to froth. In effect, two screws are used, in series, and separated by the degassing zone. The first has three sections, namely, feed, compression, and discharge, the latter having a shallow thread and normally running full. The second screw has a degassing section, fed by the discharge section of the first screw, followed by a discharge section of its own. The degassing section has a deeper thread than the final section of the first screw and, thus, the polymer suddenly finds itself in an increased volume and hence is at a lower pressure. Frothing occurs as the volatiles are released from the melted polymer, which is again compressed and then fed to the die. The volatile vapors are released either through a vent-hole in the top of the extruder barrel, or through the hollow core of the screw via a hole drilled in the trailing edge of one of the flights in the degassing section. In either case, a vacuum is sometimes applied to assist in extraction of the vapor. Vacuum hoppers are also sometimes used, particularly for dry-blend powders to reduce porosity, in which case the screw must be sealed at the thrust bearing in order to prevent rearward polymer leakage.

Multiscrew extruders, particularly twin-screw extruders are also available and have their own advantages and limitations. In general, multiple-screw machines are more expensive and, because of their more complicated construction, are likely to be less sturdy. A more positive transport of the molten polymer is possible, however, and better mixing is obtained. By virtue of their positive pumping action multiple-screw extruders produce less shear heat and this makes them very suitable for materials that are heat sensitive, have low coefficients of friction or must leave the die at low extrusion temperatures. They find their main application in the extrusion of unplasticized PVC, particularly for large diameter piping produced from powder. Multiple-screw extruders are also used for compounding all sorts of additives into polymers. Multiscrew systems may be intermeshing (either counterrotating or corotating) or nonintermeshing, counterrotating.

FILM EXTRUSION

Film is usually defined as material having a thickness of up to ten thousandths of an inch (10 mil), thicknesses above this being referred to as sheet. There are fundamentally two different methods of extruding film, namely, blow-extrusion and slit die extrusion. The former method produces tubular film, which may be gussetted or lay-flat, while the latter results in flat film.

Blown Film Extrusion

A typical setup for blown film extrusion is shown in Fig. 26.2. In this case the molten polymer from the extruder head enters the die from the side, but entry can also be effected from the bottom of the die. In the die, the melt is made to flow round a mandrel and emerges through a ring-shaped die opening in the form of a tube. The tube is expanded into a bubble of the required diameter (with consequent reduction in thickness) by blowing air through the center of the mandrel. The tubing is usually extruded upwards but can be extruded downwards or even horizontally. The air is contained in the bubble by the pinch-rolls at one end and the die at the other, and it is important that an even pressure of air is maintained in order to ensure uniform thickness and width of film.

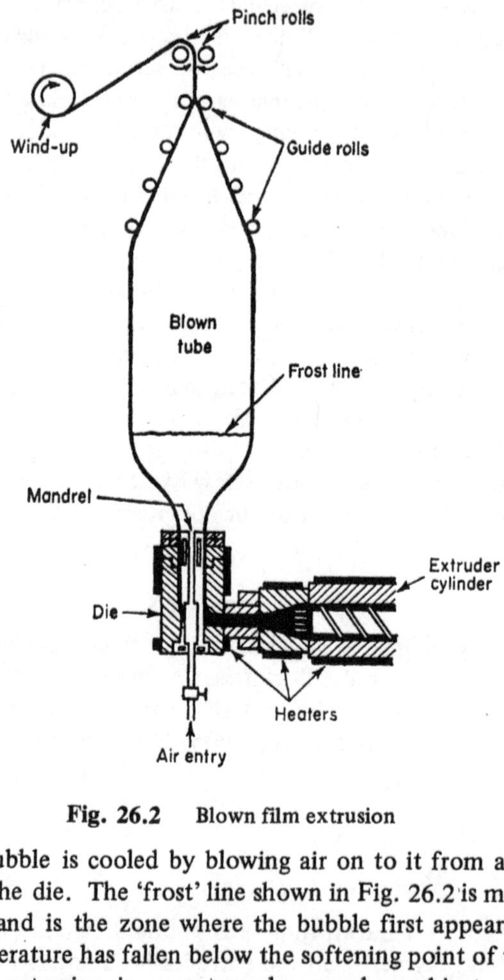

Fig. 26.2 Blown film extrusion

The film bubble is cooled by blowing air on to it from a cooling ring placed round the die. The 'frost' line shown in Fig. 26.2 is more correctly a 'frost' ring, and is the zone where the bubble first appears 'frosty' because the temperature has fallen below the softening point of the polymer.

Blown film extrusion is an extremely complex subject and there are numerous problems associated with the production of good quality film. Among the defects likely to be encountered are variations in gauge, surface defects such as 'fish-eyes.' 'orange peel,' or 'applesauce,' haze, low tensile strength, low impact strength, blocking and wrinkling. The factors affecting film properties are also numerous and include extruder temperatures (barrel and die), extruder output, die design, takeoff speed, rate of cooling of bubble, die ring adjustment, blowup ratio (diameter of bubble/diameter of die) and pinch-roll pressure. With such a large number of film properties and extrusion factors, it is not surprising that it is often necessary to effect a compromise between the various requirements, and it is the balancing of film properties that makes extrusion still very much an art. For instance, good slip and antiblocking properties are obtainable by lowering the temperature of the molten polymer and using a fairly low blowup ratio, whereas toughness and transparency require higher melt temperature and blowup ratios.

Film gauge is affected by a number of factors including takeoff speed, extruder output, barrel and die temperatures, blowup ratio, die ring adjustment and the rate of cooling of the bubble.

Among the surface defects mentioned, 'fish-eyes' are due to imperfect mixing in the extruder or to contamination. Both these factors are controlled by the screen pack. We have already noted that as well as screening-out contaminating particles, a screen pack creates a back pressure which improves homogenization.

It has, of course, been tacitly assumed above that when talking of the influence of extrusion factors on film properties, we were referring to a particular type and grade of polymer. Even when considering a single polymer, such as polyethylene, film quality is greatly affected by such factors as density and molecular weight. High molecular weight polymers give tougher films as will lower density polymers, while high density also means increased stiffness which leads to greater susceptibility to wrinkling during wind-up.

The advantages of blown film extrusion over flat extrusion include the ability to produce film with a more uniform strength in both the machine and transverse directions. In flat film extrusion (particularly at high take-off rates), there is a relatively high orientation of the film in the machine direction and a very low one in the transverse direction. In the blown film extrusion process it is possible, by balancing blow-up ratios against takeoff rate, to achieve physical properties which are very nearly equal in both directions, and this gives a film of maximum toughness. Another advantage of blown film is in bag-making, where the only seal necessary is one across the bottom of the bag, whereas with flat film either one or two longitudinal seals are also necessary. The main disadvantage is the lower clarity of the blown film but this can be improved by more efficient cooling. Blown film is often slit, after extrusion, and is then wound up as flat film. Blown diameters of 7 feet or more have been produced, giving flat film widths of up to 24 feet. This is much wider than anything so far produced by slit-die extrusion. Such large widths of polyethylene film find extensive uses in agriculture, horticulture and building (*see* Chapter 9).

Flat Film Extrusion

In flat film extrusion, the melt is extruded through a slit-die and then either into a water bath or on to a chilled roller. In both cases, the essence of the process is rapid cooling of the extruded film and the cooling is applied within a very short distance of the die lips. Figure 26.3 shows an arrangement using water quenching.

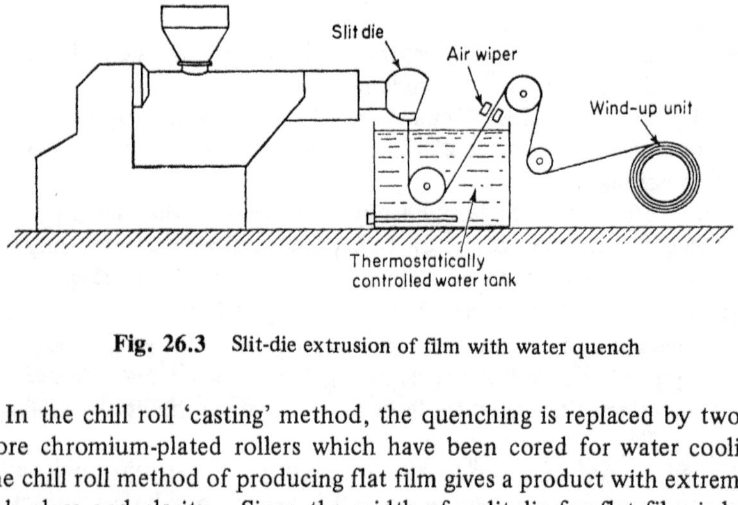

Fig. 26.3 Slit-die extrusion of film with water quench

In the chill roll 'casting' method, the quenching is replaced by two or more chromium-plated rollers which have been cored for water cooling. The chill roll method of producing flat film gives a product with extremely high gloss and clarity. Since the width of a slit-die for flat film is large compared with the diameter of the extruder head, the flow path to the extreme edges of the die is longer than to the center (Fig. 26.4).

Flow compensation is usually obtained, in the case of thin film, by a manifold die, a cross section of which is shown in Fig. 26.5. It consists of a lateral channel (or manifold) of a diameter such that its flow resistance is small compared with that offered by the die lips. The inside surface of the die should be precision machined and well polished since the slightest surface irregularities result in striations or variations in gauge. A manifold is efficient in its task of flow compensation only if the viscosity of the melt

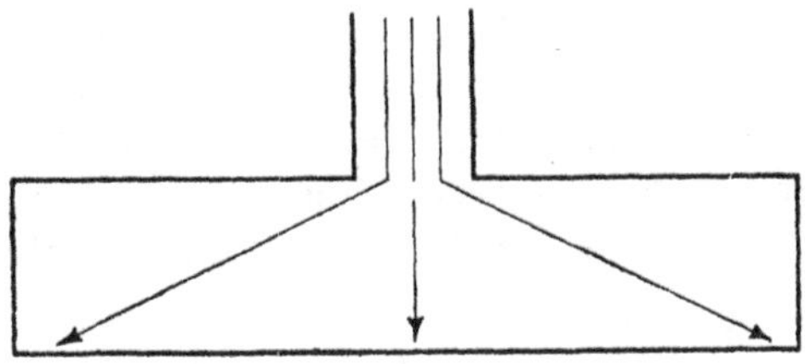

Fig. 26.4 Varying flow paths in simple slit-die

is fairly low, so that higher temperatures are necessary for flat film extrusion. This, in turn, limits the use of the manifold die to materials of good thermal stability. Another consequence of the higher extrusion temperature is that a heavier screen pack is necessary in order to maintain satisfactory back pressures.

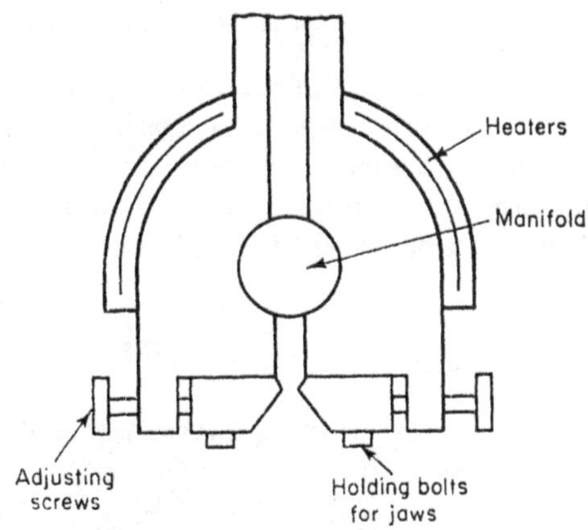

Fig. 26.5 Cross section of manifold-type die for film

SHEET EXTRUSION

Sheet can be extruded in a manner analogous to that outlined for large-width film, namely, by extrusion of a tube, which is then split, continuously, opened out until flat, then passed through polishing rolls. This method, which is usually referred to as the 'indirect' extrusion method has the advantage that it utilizes a simple and cheap die system with good flow characteristics and is therefore suitable for use with thermally unstable polymers such as PVC and polymethyl methacrylate. There is, however, bound to be a residual strain in this sheet and the direct extrusion method is now almost universally used for those polymers which are more thermally stable. As with flat film extrusion, dies giving flow compensation are necessary. The fish-tail die, illustrated in Fig. 26.6 shows one type which was once widely used for sheet extrusion. The fish-tail cavity is so shaped internally as to offer more resistance to flow at the center of the die than at the edge.

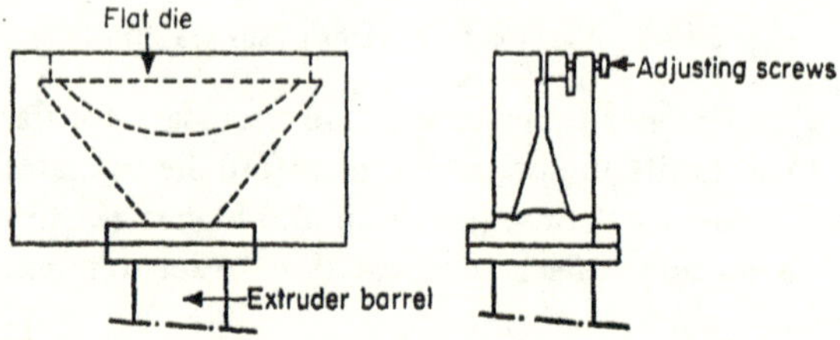

Fig. 26.6 Diagram of fish-tail die for sheet

The fish-tail die is extremely bulky, especially for wide-sheet requirements and the heating of such a large surface becomes a very difficult matter. On the other hand, the simple type of manifold die described for film extrusion is limited, as was mentioned, to low viscosity materials. Greater homogeneity of flow can be obtained by introducing a throttle, as shown in Fig. 26.7.

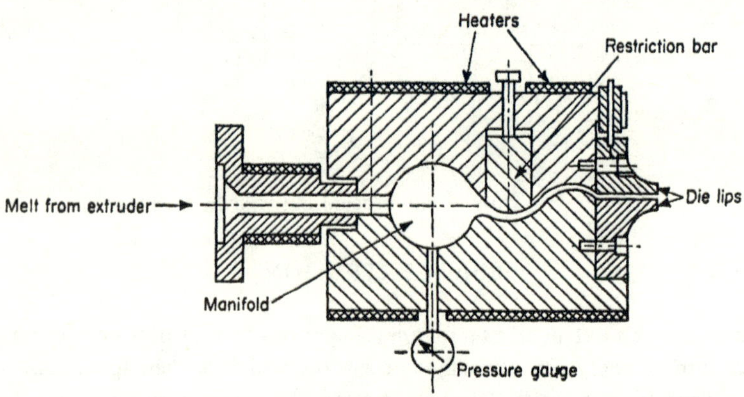

Fig. 26.7

Typical takeoff arrangements for handling extruded sheet are shown in Fig. 26.8. The heated rollers shown are highly polished to give a good surface to the semimolten sheet. This final polishing process cannot be used to cover up defects in the sheet caused by faulty extrusion, and great attention must be paid to eliminating surface defects before the sheet leaves the die. A brief note on sheet faults and their probably causes is given below.

Defect	*Probable Cause*
Continuous lines in direction of extrusion	Die contamination or scored rollers or moisture in the polymer
Continuous lines across the sheet	Jerky operation; maybe too low a temperature
Curved lines across a colored sheet	Color not well enough mixed
Discoloration	Too high a mass temperature or too high a percentage of reground material added
Bad gloss	Roll temperature too low or defective rolls
Dull strip	Die set too narrow at this point

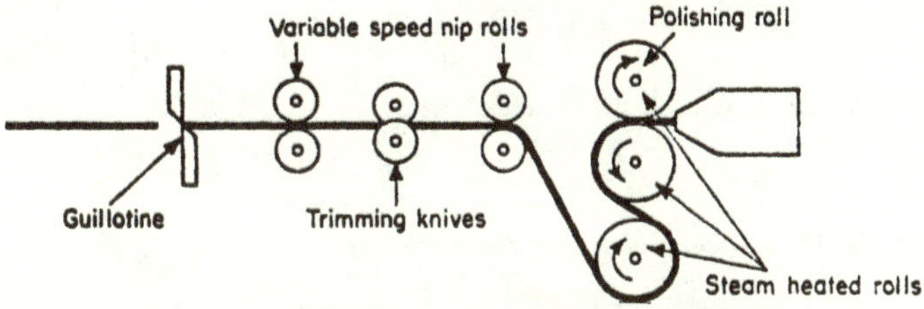

Fig. 26.8 Scheme of process for extruded sheet

WIRE AND CABLE COVERING

The demand for covered wire and cable in the radio and electrical industries is already very large and is steadily increasing, so that this extrusion process is a very important one. A cross-head die is used for the job, a typical design being shown in Fig. 26.9.

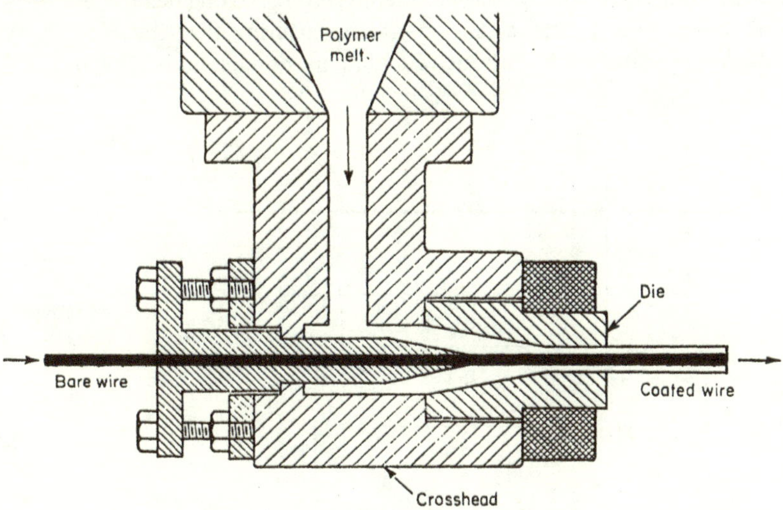

Fig. 26.9 Wire coating cross-head

The polymer melt is forced through the extruder breaker plate and screen pack assembly and then over and round the wire guide carrying the wire, and then out through the die which forms the coating round the wire. This, in essence, deals with the actual covering of the wire but the complete operation requires a complex array of equipment in addition to the extruder, as can be seen from a glance at Fig. 26.10.

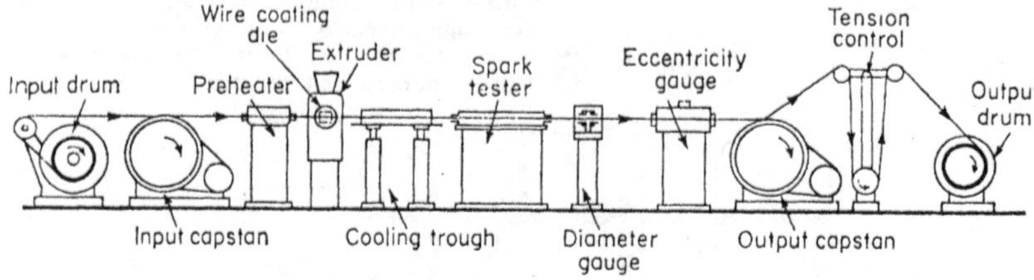

Fig. 26.10 Diagrammatic representation of wire coating layout

The uncoated wire is paid out from a drum and passed through a preheater prior to entering the covering crosshead mounted on the extruder. The preheating is carried out in order to remove any moisture or drawing-die lubricant which may possibly be left on the wire, and which could adversely affect adhesion. In addition to this, the coating adheres more readily to a hot wire.

On leaving the crosshead die, the coated wire is cooled by passing through a water trough. Damage to the insulating coating has to be avoided at all costs, and a neat arrangement for ensuring that the wire travels through the water bath without touching anything solid is shown in Fig. 26.11.

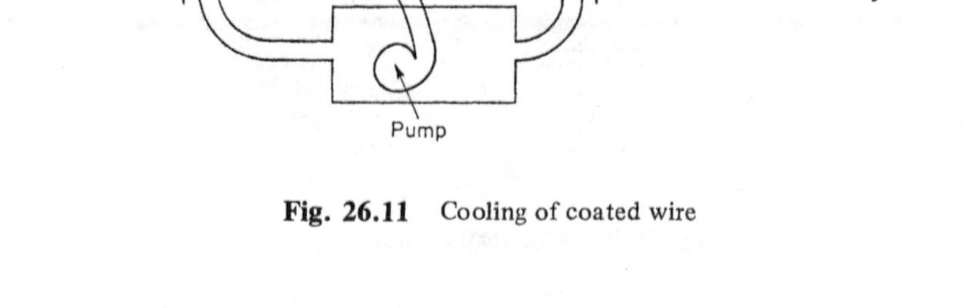

Fig. 26.11 Cooling of coated wire

Various items of equipment for testing the quality of the coating are usually inserted in the setup at this point and may include a spark-tester (to discover the presence of pin-holes or other faults in the insulation), and diameter and eccentricity gauges. The wire is pulled through the coating system by a driven capstan and then passed through a tension controller on to the wind-up drum. The cooling operation is particularly critical in the case of wire and cable covering because the existence of voids or bubbles might well cause a breakdown in the insulation. Cooling should be gradual, particularly where coatings are thicker than about 0.02 inches, and this can be achieved by having more than one water bath held at different temperatures. Variations in coated wire diameter should be avoided, if for no other reason than the fact that polymer is wasted whenever the diameter exceeds the minimum required. Whenever a change of diameter is signaled by the gauge, the speed of the wire through the coating die is adjusted by altering the wind-up capstan speed.

PIPE EXTRUSION

The die used for the extrusion of pipe consists of a die body with a die mandrel and an outer die ring which control the dimensions of the inner and outer diameters respectively. A die mandrel is positioned in the die by a three or four-spoked piece known as a 'spider,' while the outer die ring is held by four adjusting screws which allow centering to be carried out during extrusion. An air channel passes through one of the 'spider' legs, connecting with an air-hole drilled up the center of the mandrel. By sealing the end of the extruded tube with a plug, and supplying compressed air through the air channel described above, the pipe can be pressurized to any required extent.

After the pipe leaves the die, the procedure is calibration (or sizing), cooling, and takeoff. The actual passage of the molten polymer through the pipe die causes a certain amount of molecular orientation. Loss of orientation after leaving the die causes the material to swell. Attached to the die is a water-jacketed pipe which has an inside diameter equal to the outside diameter of the finished pipe. The internal pressure maintained in the pipe causes a slight circumferential stretching. Contact is made with the cooled inner surface of the water-jacketed calibrator. The outer surface of the pipe is thus cooled and set sufficiently to enable it to retain its size while passing through the cooling bath. An interesting variation on the method of calibration described above is to apply a vacuum to the cooled calibrator, so that the molten material is forced against the inner surface by atmospheric air pressure. Takeoff is achieved by means of reeling drums in the case of flexible tubing such as plasticized PVC or low density polyethylene but for the more rigid pipes, a caterpillar type of takeoff is often used.

PROFILE EXTRUSION

Profile extrusion is similar to pipe extrusion except that the mandrel and 'spider' are obviously not necessary. A die plate, in which an orifice has been cut, is bolted to the face of the normal die assembly so that it is then possible to change from one profile to another without the necessity of a major die change. To maintain a smooth flow, the die-plate orifice is blended to match the bore of the die body.

There is a surface drag on the molten polymer as it passes through the die and this has the effect of lessening the flow through the thinner sections of the orifice. To counteract this tendency, it is necessary to alter the shape of the orifice, and there is often a wide difference in the orifice shape and that of the desired extrusion profile. Some examples of sections and the die orifices necessary to produce them are shown in Fig. 26.12

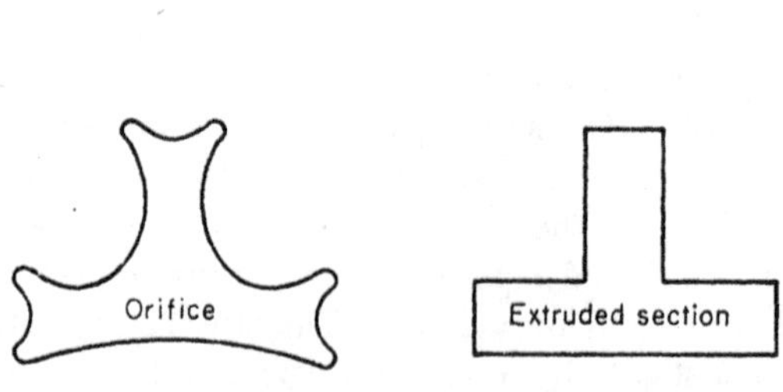

Fig. 26.12 Relationship between extruder die orifice and extruded section

For complex profiles it is impossible to design a die-plate from first principles. It is usual to produce a first approximation in an unfinished form and then to arrive at the correct orifice form by trial and error on the extruder.

EXTRUSION COATING

Many substrates, including paper, fiberboard, cellulose film, aluminum foil, and hessian are today being coated with polymers by the direct extrusion process. By far the commonest polymer used for this process, at present, is low-density polyethylene, the basic process consisting of extruding a thin film of molten polyethylene and pressing it on to the substrate without the use of an adhesive. The equipment used is similar to that employed for the extrusion of flat film, Fig. 26.13 showing a typical extrusion coating setup.

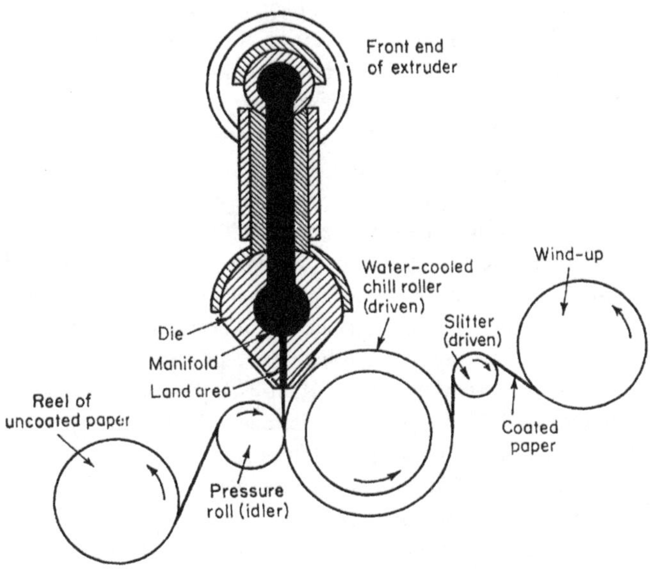

Fig. 26.13 Cross section of paper coating extruder

As it leaves the die lips, the thin molten film is pulled down into the nip between two rolls situated directly below the die. The substrate, which is moving at a speed faster than the extruded film, draws the hot film to the required thickness, while the pressure between the chill-roll and the pressure-roll forces the film on to the substrate. Meanwhile, the molten film is being cooled by the water cooled, chromium-plated chill-roll. The pressure-roll is also metal but is covered with a rubber sleeve, usually neoprene or silicone rubber. After leaving the chill-roll, the coated material is drawn over a driven slitter-roll where the edges are trimmed to remove the uncoated edge and the small bead of polymer which builds up at the edges of the film. After trimming, the coated material is wound up on conventional wind-up equipment.

One or two points in connection with the extrusion coating process may be mentioned here since they have a bearing on the type of equipment used. In the first place, it is not economical to coat narrow substrate widths, since the cost of equipment is high, so that die widths are large with a consequent need for large extruders. Higher temperatures are required for extrusion coating than for unsupported film extrusion, and oil heating is not sufficient in this case. The usual form of heating employed in coating extruders and dies is high-capacity electrical heaters.

The weight of the coating deposited on the substrate is governed by the speed of the chill-roll. For any given extrusion rate, the higher the chill-roll speed, the greater the draw down and, hence, the lower the coating weight. The chill-roll has two other functions, one of which is, of course, to cool the coating to a temperature low enough to enable it to be stripped from the roll. This cooling has to be accomplished in about half a revolution of the roll thus necessitating very efficient water-cooling of the roll. The chill-roll also determines the surface finish of the film. For high gloss, the chill-roll surface must be very highly polished but where dull film surfaces are required, the chill-roll may be etched or sand-blasted.

Adhesion of the coating to the substrate is a matter of prime importance and is affected by several factors. High-melt temperatures are essential, while other factors include coating thickness (thick coatings adhere better than thin ones) and draw-down rate (high draw-down rates have an adverse effect on adhesion). Different substrates also have their own individual problems as in the case of paper and board, where a low moisture content is required for good adhesion. Adhesion of polyethylene to aluminum foil is not an easy matter to achieve and the problems are sometimes overcome by an etching treatment of the aluminum-foil surface. The coating of regenerated cellulose film also presents a problem. If a moisture-proof cellulose film is required, then a single-sided film is used and the coating is carried out on the uncoated side of the cellulose film, the reason being that stripping of the cellulose film's original coating would otherwise occur.

One of the problems encountered in extrusion coating is that of 'beading' at the edges of the hot film where 'necking in' occurs. This local thickening of the film has to be removed during the trimming operation otherwise, when the roll is wound up, the beads at either end support the weight of the roll, leaving a loosely wound and sagging roll in the middle. The extra trimming required means a reduction in the usable web widths, thus increasing the coating costs. In the case of polyethylene the incidence of 'neck in' can be reduced by decreasing the melt index or the density but the risk of voids in the hot film at low coating speeds increases because of imperfect melt flow, so that an optimum combination of these two factors has to be found. The distance between the die lips and the chill-roll should also be at a minimum for minimum 'neck in' while increasing the screw speed also reduces 'neck in.' It can be seen that the balancing of requirements such as minimum 'neck in,' high coating speeds, absence of voids, etc., is a matter of experience.

EXTRUSION OF EXPANDED POLYSTYRENE SHEET

Initial work on the extrusion of expanded polystyrene sheet showed that the normal blowing agents used tended to form solutions in the polystyrene at extrusion temperatures. This led to the formation of large bubbles and hence a coarse cell size. This is undesirable so a nucleating agent, such as a mixture of citric acid and sodium bicarbonate, is used. As mentioned earlier, in Chapter 11, there are two main methods for the production of expanded polystyrene sheet. The first starts with expandable polystyrene beads, i.e., beads that have been impregnated under pressure with a liquefiable gas, such as pentane. The sheet is blow extruded, using a twin-screw extruder, the nucleating agents having been mixed with the beads just prior to extrusion. To avoid premature decomposition of the nucleating agents it is essential that the expandable beads be predried just before the mixing. One suitable method is to blow dried air through the beads for two to three hours. This method of sheet production is particularly suited to low tonnage production of the order of 300–350 tons per annum.

The other method is direct gassing in the extruder. Ordinary crystal polystyrene is premixed with the nucleating agents and loaded into the feed hopper of the extruder. The granules are melted during the first 20 D length of the extruder barrel, the melt temperature being then about 240°C. A liquid blowing agent (one of the fluorinated hydrocarbons) is then injected into the melt, using a high pressure diaphragm metering pump. A mixing section of the extruder screw ensures good dispersion of the blowing agent. The temperature of the melt is then reduced to about 130°C and the melt is forced through the annular die of the blowing head. This method is the most economical for large tonnage sheet manufacture and has the added advantage that the use of fluorinated hydrocarbons as blowing agents eliminates the risk of fire during manufacture and storage.

Whichever method of extrusion is used it will be appreciated that the material leaving the extruder expands considerably with a consequent drop in density from 65 lb/ft^3 to about 5 lb/ft^3 (1040 g/ℓ to about 80 g/ℓ). Thus, a thirteen-fold increase in volume occurs. The sideways component of this expansion would cause corrugation to occur if a slit die were used. Using an annular die the corrugations can be eliminated by blowing the tubular extrudate into a bubble, as with lay-flat polyethylene film. After the bubble has been collapsed during the haul-off operation the lay-flat tube is trimmed on each side to produce two flat sheets that can be reeled up together or separately. This is necessary because the folded edges of the tube are a source of weakness due to the rigidity of polystyrene. Control of sheet orientation is achieved in the usual way by adjusting the blow-up ratio and the haul-off speed. Orientation should be as balanced as possible to avoid splitting of the sheet during any subsequent thermoforming operations.

Although the density of the sheet as it leaves the extruder is about 5 pounds per cubic foot (80 g/ℓ), subsequent operations such as blowing-up, and stretching during haul-off, tend to compress the sheet and so increase the density. In addition, the collapsing of the bubble and subsequent passage of the sheet through nip rollers also increase the density. Normally, therefore, the density of the finished sheet will be of the order of 8–10 lb/ft^3 (128–160 g/ℓ). The density can be reduced by passing the sheet under an infrared heater, at the same time supporting the sheet across its width by a roller. If the expanded sheet is reeled and allowed to stand for a few hours prior to the thermoforming process the cells absorb air and a further reduction in density is obtained when exposed to the thermoforming heaters. This thickening effect is useful in countering the tendency for a thermoformed article to become thin at the corners.

COEXTRUSION

An important development in the extrusion of film is coextrusion, where two or more layers of film are extruded simultaneously to produce a laminate with improved barrier or mechanical properties. Basically the process consists in coupling two or more extruders to a single die head. Tighter control is necessary of extrusion process variables such as melt pressures, screw speeds, melt temperatures and power capacity of the extruder drives. For example, the temperature of each polymer must be closely controlled in order to maintain the proper level of viscosity in the die. The success of the technique lies very largely in the design of the die and the way in which the individually extruded melts are brought together, prior to their extrusion as a multilayer film.

There are three basic methods of carrying out this operation. The first is the multislit die system where the various polymers are each fed to separate manifolds in the die, then pass through separate slits and are combined after they leave the die. In the blow extrusion process the separate films are brought together below the 'freeze line' of both bubbles, i.e., while still molten (Fig. 26.14). Provision can be made in the die for the space between the films to be saturated with a reactive gas, such as ozone, which 'treats' the adhering surfaces and improves the bond strength. This method is particularly useful when combining unlike materials, e.g., nylon and polyethylene.

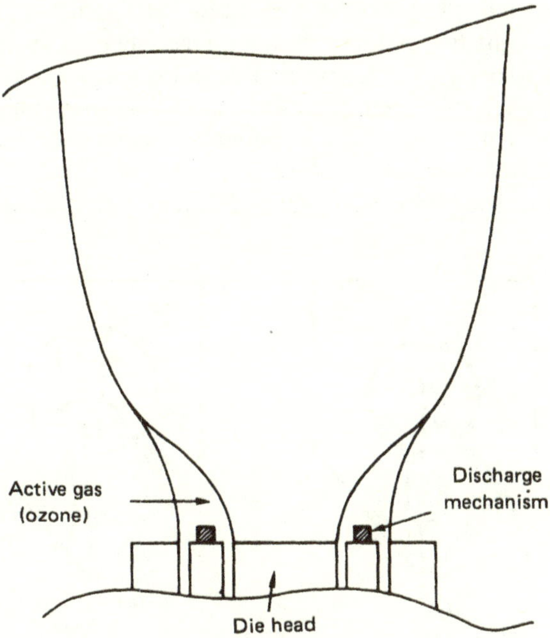

Fig. 26.14 Coaxial extrusion with provision for treating with active gas to improve adhesion

The second method is the multimanifold die. The various polymers are again fed to separate manifolds in the die but here they are combined in the final land prior to leaving the die. Figure 26.15 shows typical setups for two and three film processes.

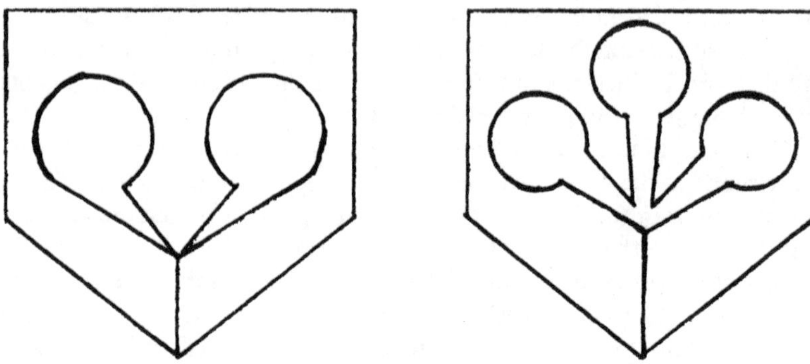

Fig. 26.15 Coextrusion using multiple manifold dies

The third method is the combining adaptor and this is probably the most common method for the coextrusion of flat films. The separate melts are brought together in a common manifold (Fig. 26.16). The flow passages are so designed as to minimize turbulence and avoid mixing of the separate streams. In general, the viscosities of the various polymers should be closely matched when using a combining adaptor. It is also desirable that the viscosities match at about the same temperature to eliminate problems associated with maintaining different melt temperatures in the die.

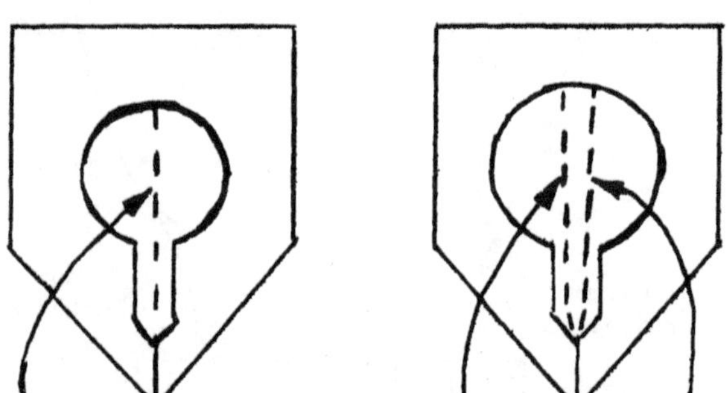

Fig. 26.16 Coextrusion by Laminar Flow Process using single manifold dies

The combining adaptor allows great flexibility in operation. It can extrude two or more layers relatively easily and it can extrude a polymer in a much thinner layer than would be possible if it were extruded as a single film. However, it is not possible to maintain a large temperature difference between layers (which is possible in the multislit process). There are also many combinations of materials that do not flow together and produce a uniform product.

Whichever method is used, coextrusion is an important addition to the various techniques used for producing composite films. The main advantages of coextruded films over conventionally produced laminates are lower cost, a lower tendency towards delamination and a greater flexibility in obtaining a wide range of properties. The possibilities of cost savings arise from the fact that in multilayer extrusion, the finished laminate is produced in one operation. In normal laminating procedures the respective single films have first to be extruded, followed by reeling-up and storage.

Later on, these reels have to be un-reeled and run through the laminating process and this adds to labor costs. Cost savings on material are also possible. The minimum thickness of any particular ply in a conventional laminate is often dictated by its handling characteristics on laminating equipment rather than by the barrier or other properties required. However, very thin plies indeed can be produced by coextrusion so that the optimum thickness can be chosen for the job.

As well as cost savings in the production of the laminated film, there are also possible savings in bag making. Multilayer lay-flat film can be produced by coaxial extrusion. It is then possible to make a bag of laminated construction merely by making one cross seal. Coaxial extrusion, then, provides the only method of producing a bag of laminated construction without side or center seams.

Another method of bonding incompatible films is by the use of 'bridge' materials (or adhesive tie layers) which are compatible with both films. This sort of approach is possible commercially because of the very thin layers that are possible using coextrusion.

The advantage of flexibility in obtaining a wide range of properties depends on the possibilities, already mentioned, of producing very thin component plies. Apart from the obvious factor of cost, there is often a limit to the thickness of a laminate which is set by handleability (stiffness, etc.) or even clarity. In such instances the number of different properties which can be built-in by using different single plies is also limited.

There are also disadvantages to the use of coextrusion. One of these is the difficulty of utilization of the scrap produced during the extrusion process. When extruding a single ply film, a fair proportion of scrap film can be fed into the extruder, and used again since with most plastics there is no danger of chemical degradation. With coextrusion, however, the problem is that a mixture of polymers is produced with no intervening stage, and separation of the plies, even when possible, would not be economic. This disadvantage is, of course, aggravated in the case of short runs. Coextrusion is, therefore, best for long runs with steady running conditions. On the other hand, it is possible to use the trim of certain composite films to produce a blended core, with more specialized polymers as the outer plies. There is, of course, a limit to this type of scrap utilization.

Another disadvantage is that coextrusion cannot be used to produce a laminate with sandwich printing. This is a particular disadvantage in food packaging when a scuff-proof decoration may well be required, but where printing cannot be on the inside of the ply in contact with the food.

Equipment Developments: The most important aspect of equipment development is that of die design, particularly that of circular dies. Early coextrusions were generally two or three layer structures but the requirement for high barrier coextrusions, often based on polymers requiring the use of extrudable adhesive tie layers to promote inter-layer bonding, has necessitated the development of five and seven layer dies. The use of such high performance polymers means that wide differences in rheological behavior are encountered in comparison with those of typical polyethylenes. Viscosities can range from very low (in the case of polyamides, which exhibit almost Newtonian flow behavior) to values several times greater than that of LDPE. The tie layer materials, too, can often represent extremely low viscosity values because of their exposure to die temperatures dictated by the corresponding hot layers.

Many of the high performance polymers now in use require additional streamlining in the die because of their sensitivity to burning. Some, like PVDC and PVC, are generally classified as degradable, while others, like the polyamides, require higher than usual temperatures and so aggravate burning of the polyamide itself and of adjoinging layers. These problems and others, such as increased maintenance complexity, are yielding to programs of development work by a number of companies, and many adjustable and non-adjustable multilayer dies are available.

Another field of development is that of gear pumps, as alternatives to the screw pump which forms the metering stage of a conventional extruder. Gear pumps have been used for many years in low viscosity extrusion processes such as textile melt spinning, but a great deal of development has been necessary to produce gear pumps capable of handling the high viscosity polymers commonly used in coextrusion.

Material Developments: The increasing requirement for improved economy and pack performance has been met by the ability of the coextrusion process to utilize high performance materials. One with high gas and water vapor barrier properties is vinylidene chloride/vinyl chloride copolymer, and this has been used in coextrusion for some time. A more recent development in the field of high barrier materials has been the use of an extrudable high barrier copolymer of ethylene and vinyl alcohol (EVOH). Under dry conditions EVOH is better even than vinylidene chloride copolymers but where containers have to be pasteurized or retort sterilized then vinylidene chloride copolymers are preferable because of their better gas barrier properties under these conditions.

One area of materials development is that of the outer layers, used to provide heat sealability. Linear low density polyethylene (LLDPE) is one material used because it has superior heat sealing properties to those of LDPE by virtue of its very good hot tack properties. LLDPE also has higher tensile strength, puncture resistance and tear resistance. Like LDPE, LLDPE has good printability and surface slip.

Another material development is the use of polycarbonate (PC) in coextrusion. In the UK, for example, a PC/tie layer/LDPE cast film has been produced which is said to have outstanding optical properties, excellent stiffness (for machineability), excellent room temperature impact strength, excellent puncture resistance and good tensile strength (in machine and transverse directions). Resistance to tear initiation is said to be good but resistance to tear propagation is only moderate. Other suggested structures are PC/tie layer/PP and PC/tie layer/LLDPE. The high heat resistance of PC makes it particularly attractive for medical uses, where the need for a sterile package makes high demands in this area. In the U.S.A., for example, a PC/tie layer/EVA coextrusion has been put forward as an alternative to an oriented PET/EVA lamination in chevron pouch production. In thermoforming applications, the addition of a small layer of PC to PP is said to make the film more easy to process on a form/fill/seal machine.

SUGGESTED FURTHER READING

Plastics Forming. (1971) Ed. John D. Beadle (London: Macmillan).

Extruder Screw Design. (1970) R. T. Fenner (London: Iliffe for the Plastics and Rubber Institute).

Processing of Thermoplastic Materials. Ed. E. C. Bernhardt (New York: Reinhold).

Plastic Films. (1989-3rd ed) J. H. Briston (London: Longman Scientific and Technical).

Engineering Principles of Plasticating Extrusion (1970) Z. Tadmor and I. Klein, (New York: Reinhold).

27.

INJECTION MOLDING AND BLOW MOLDING

The injection molding process consists essentially in softening the material in a heated cylinder and injecting it under high pressure into a relatively cold mold where it hardens. The molded article is then ejected from the mold by means of ejector pins, compressed air, a stripper plate, or some other device. The process is a versatile one and can be applied to the production of articles ranging in weight from a fraction of an ounce to several pounds.

The process has undergone great changes since the early days, mainly in the method of injection. Originally, the molten plastic was injected into the mold cavity (or cavities) by a plunger (see Fig. 27.1).

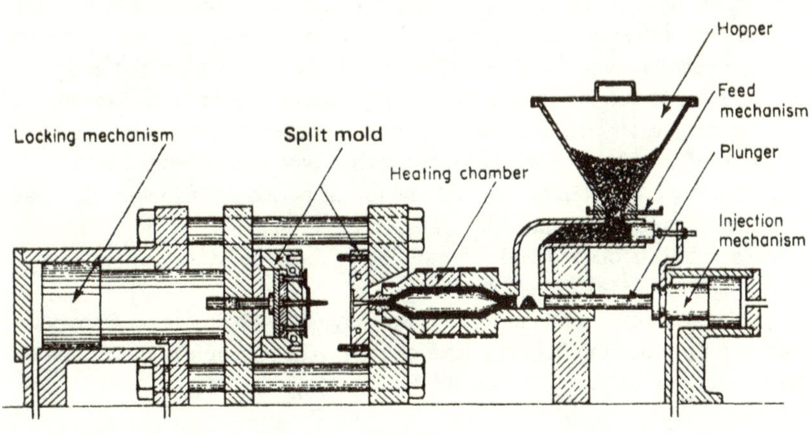

Fig. 27.1 Diagrammatic representation of old-type plunger injection molding machine

In present day machines the plunger has been replaced by a reciprocating screw as shown diagrammatically in Fig. 27.2

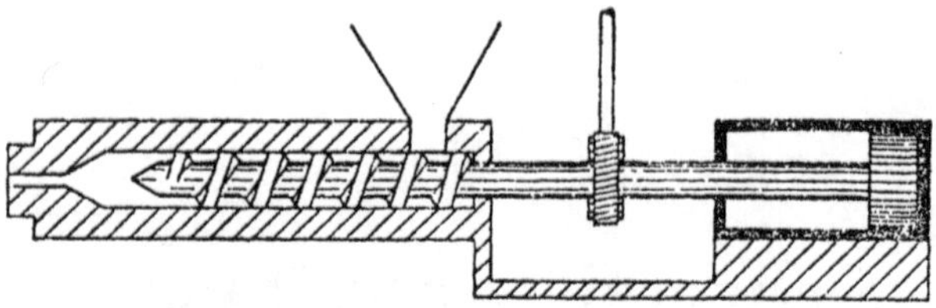

Fig. 27.2 Reciprocating screw injection unit

The sequence of events is as follow:

1. The mold is closed and a locking force is applied.
2. A motor rotates the screw, and the shearing action on the plastic (plus the conductive heat from the external heaters) melts the plastic.
3. The molten plastic passes from the screw flights, through a screw tip non-return valve as the screw continues rotating.
4. The build-up of molten material in front of the screw causes it to move axially backwards.
5. When the required volume of molten plastic has accumulated at the front of the screw, the screw stops rotating and moves forward under pressure to inject the melt into the mold. The melt then flows through the 'sprue' opening in the die into 'runners' which terminate in 'gates' leading to the mold cavity itself. In this operation the screw is performing the function of the plunger or ram in the earlier type of machine.
6. Pressure is maintained on the screw for the period during which the resin in the mold cools and contracts.
7. The mold is opened and the molding is removed.

Another type of injection unit is the two stage unit which has separate components for plasticizing (melting) and injection. The principle of this operation can be seen by reference to Fig. 27.3.

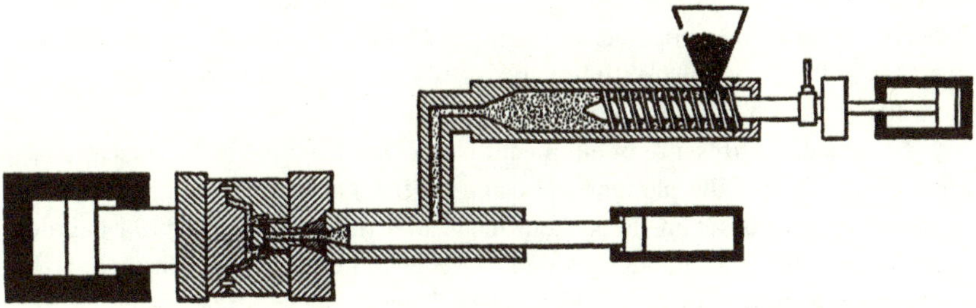

Fig. 27.3 Two-stage Injection Unit

The extruder screw feeds the molten plastic to a separate transfer cylinder from where it is injected into the mold by an injection piston. One advantage is that the extruder screw can continue to turn without interruption during both the injection and hold phases.

Machine Characteristics

Injection molding machines are characterized by their shot capacity, plasticizing capacity, screw diameter, rate of injection, injection pressure, and clamp pressure (or mold locking force).

Shot capacity may be given in terms of the maximum weight which can be injected per shot, in which case it is usually quoted in terms of ounces of polystyrene or cellulose acetate. This has the disadvantage that since the figure is dependent on swept volume of the cylinder during one stroke of the plunger and on the volumetric capacity of the feed mechanism, it is necessary to correct for bulk density and specific gravity when dealing with other materials. A better way of expressing shot capacity is in terms of the volume of material which can be injected by the plunger to a mold at a specific pressure.

One of the factors affecting the output of an injection molding machine is the plasticizing capacity, which is usually defined as the number of pounds per hour of a particular material which can be brought to the temperature required for molding. It was the search for greater plasticizing capacity which led to the development of the reciprocating screw and two stage injection units.

Injection rate is also a very important factor in determining the output of a machine and it is usually expressed as the volume of material discharged per second through the nozzle during a normal injection stroke. This obviously depends on pressure, temperature, the material used, and the size of the smallest aperture in the flow line. With low viscosity materials and fairly large 'gates,' etc., the injection plunger speed may be a limiting factor.

The injection pressure of an injection molding machine is the pressure exerted by the face of the plunger and can vary from about 2000lb/in^2 to 25,000 lb/in^2. Pressure losses in the system mean that pressure in the mold cavities is much less than the plunger pressure. Nevertheless the actual force exerted on the inside surface of the mold is large if the projected area of the mold is large. The clamp pressure is, therefore, an important factor in determining the maximum projected area that can be molded on a particular machine. On large machines, the molds are normally closed by means of large, direct-acting hydraulic rams while on the smaller models a 'toggle' action is usually preferred (Fig 27.4). Direct-acting rams are more easily adjustable for different-sized molds but are slower and need greater pumping capacity. The 'toggle' system has the disadvantage of introducing multiple joints.

Machine Types

Enough has been said to indicate that the range of injection molding machines is extremely wide. Apart from the different sizes, ranging from those with a shot capacity of a few grams to the giants capable of delivering a number of kilograms in one operation, there are reciprocating screw and two-stage units, automatic and semi-automatic types, those with hydraulic or toggle locking, and vertical or horizontal types. Vertical types are used where the molding of inserts is required.

Development of the larger type of injection molding machine has not been due solely to the demand for larger single moldings. Since, for example, a 100 gram machine is slightly lower in cost than two 50 gram machines, there is a tendency to use larger machines with multicavity molds. However, flexibility of operation is a point in favor of single mold, faster cycling machines. Another factor to be considered is where several different colored moldings are called for, and here again it is obviously better to run several smaller machines rather than one large machine using multicavity molds.

Machine Variables

For any particular material, successful molding is determined by mold design and the correct setting of the following injection molding machine variables.

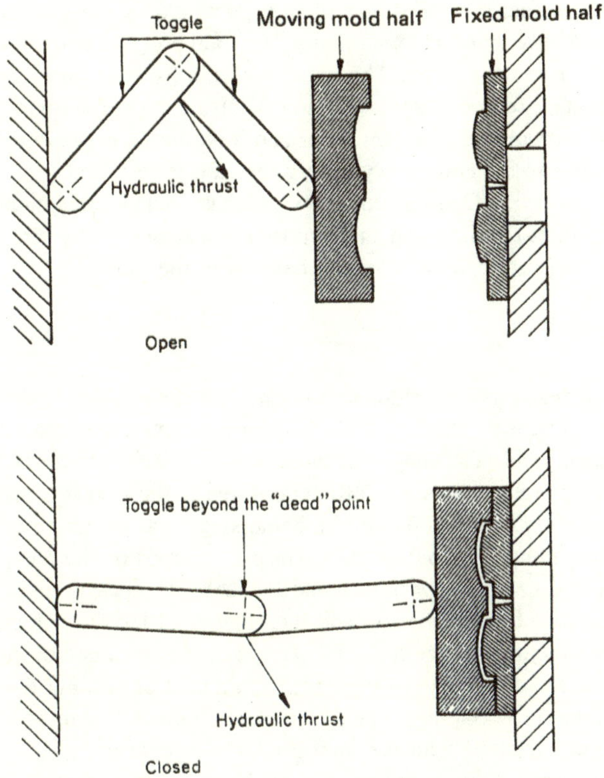

Fig. 27.4 Toggle-action mold closing

Injection Pressure. Injection pressures can vary quite widely and are dependent on mold design and on the size of machine. In general, the aim is use of minimum pressure to produce full shot moldings, free from such defects as surface sink marks or voids. Excess pressure should be avoided as this could lead to 'flashing,' i.e., escape of material from the mold joints.

High injection pressures are vital, however, for the production of thin wall moldings in order to achieve the necessary high injection rates. Molten resin moving through a thin cavity, over a long distance, tends to cool quickly and 'freeze-off' before the mold is completely filled.

Cycle Time. From the standpoint of economics, the aim should be the minimum time for each part of the molding cycle consistent with good quality molding. The question of the injection stroke has already been discussed when it was pointed out that the speed was mainly governed by the viscosity of the material (and hence material temperature), the ram pressure and the minimum aperture size in the flow line. The minimum ram forward time is governed by the fact that if pressure is removed too early from the molding, then surface 'sinking' or internal 'voiding' can occur and it is, therefore, essential that the gate should freeze-off completely before the ram is retracted.

Mold Temperature. Constant mold temperature, below the softening point of the material, is the aim, and this is normally carried out by circulation of a constant temperature fluid through channels in the mold.

Mold Design

Good mold design is a prerequisite for the production of good quality moldings and no amount of juggling with machine controls can compensate for a badly designed mold. The subject of mold design is highly complex but some of the more general points are briefly considered in the next few paragraphs.

Number of Mold Cavities. There can be no single answer to the question of the optimum number of mold cavities, since it depends on the complexity of the molding, the size and type of machine available, the cycle times likely from each possible number of cavities, and the number of moldings required. As far as the last factor is concerned, if a fairly accurate estimate of the various mold costs and cycle times can be made, and a cost of running the machine (without material) is assumed, then a break-even quantity can be calculated which can be compared with the total production required.

Cycle times are not increased pro rata with the number of cavities. This is because in a single-cavity mold the limiting factor is the cooling time of the molding, whereas with higher numbers the plasticizing capacity of the machine tends to be the controlling factor.

Another way of increasing the number of moldings per cycle is by the use of stack molds where two identical sets of cores and cavities are stacked together, back-to-back. Use of a stack mold can almost double the productivity of a machine and result in approximately 20% less energy consumption per molding produced. Because of this stack molds are becoming increasingly popular.

Mold Heating or Cooling

The control of mold temperature is usually achieved by circulating a constant temperature fluid through strategically placed channels. Initial mold design should make provision for these channels, bearing in mind such factors as symmetry of cooling and the effect of the temperature gradient across the mold on the flow of material. Cooling should be more intensive at the mold gate, as the plastic which reaches the furthest extremities has more time to become cooled. This evens up the cooling times and lessens the possibilities of warpage.

With certain materials, e.g. acetals and nylons, the mold is heated in order to achieve the desired degree of crystallinity and so optimize the properties of the molded part.

Mold Venting. If no provision is made in the mold for venting, then the air in the mold is trapped in some corner and is then highly compressed. If the compression is rapid enough, the resultant rise in temperature may be sufficient to cause scorching at this particular point of the molding. With fairly simple shapes it may be possible to arrange the mold so that the point of venting is along the mold parting line. If this is not possible, then trial shots are made and the plastic flow observed. Vents can then be placed in the mold where they will provide for the most efficient release of air. A suitable venting can be achieved by a slot about ¼ inch wide and two thousandths of an inch deep. This allows the escape of air, but not the plastic material because of its great viscosity.

A more modern system is known as 'continuous mold venting'. In essence this consists of venting as much of the periphery of the mold cavities as the runner system will allow. A three mm half-round air channel is provided about 60 to 80 mils from the cavity edge around the whole periphery. This channel is, in turn, connected to the outside atmosphere by relatively large passages. The close-off surface between the cavity and the channel is relieved by 0.1 to 1.0 mils over the whole periphery, the amount of relief being determined by the viscosity of the melt. As the mold is being filled the air passes easily through the continuous vent into the half-round channel.

Removal of Molding. This is essentially a problem of design of the molded piece, together with provision of means of ejection in the mold. It is advisable, for instance, to avoid undercuts which may prevent separation of the molding from the cavity, while for deep moldings it is necessary to provide a slight taper on the walls.

Ejection is carried out by ejector pins, stripper plates, or compressed air. If ejector pins are used, care must be taken that their cross-sectional area is large enough not to push through the molding, and they should be a perfect fit so that no flash is formed. Stripper plates are preferred for flexible moldings such as those from polyethylene.

Shrinkage. When designing a mold, allowance has to be made for the shrinkage of the plastic. This varies with the material, being between 0.002 and 0.008 inch/inch for polystyrene and between 0.015 and 0.025 inches for Nylon 66, and polypropylene. A particular problem arises in the case of polyethylene since the shrinkage rate is different in the direction of flow and across the flow of the material. This can lead to warpage, and if a large, flat area has to be molded it is best to have several gates in order to equalize the flow.

Gating of the Mold. Gating of the mold has a very big effect on the type of molding obtained, one example being briefly mentioned above. In general, the aim is to produce the best possible flow conditions consistent with ease of finishing, while interfering as little as possible with the final appearance of the molding. The type of gate will depend on the type of molding and on the material being molded. The size of gate can also be varied. Small gates are used for preference in order to leave as small a blemish as possible and to provide a smooth flow, but in general, the thicker the section of the molding, the larger is the size of the gate.

The positioning of gates is important, and it is usually advisable for ease of cavity filling and evenness of flow to feed into the thickest section of the cavity and in such a way that the material impinges on the core or cavity wall. Some typical gates are illustrated in Fig. 27.5.

The sprue gate is normally used where the cavity is difficult to fill or when feeding the runners of multiple cavities. Disadvantages of this gate are the expense of trimming, the length of time taken to 'freeze,' and the liability of surface sink marks to form opposite the sprue. On the other hand, there is a low pressure loss across the gate and it is possible to utilize 'after pressure.'

The pin gate is preferable when cavities are easy to fill or when easy molding materials are used. Disadvantages of this sort are, as one might expect, a high pressure loss and an inability to utilize 'after pressure.' It possesses several advantages, however, including ease of trimming, freedom from sink marks opposite the gate, a short 'freeze' time, and a tendency to homogenize the melt by frictional heat.

Hot runners. In normal multicavity molding the sprues and runners are cooled and removed at each cycle, being later reground and remolded. The amount of material needed to fill the sprue and runners at each cycle is all part of the shot capacity of the machine and, hence, reduces the possible size of the molding. For short-cycle molding or for articles of thin section, the hot-runner technique has become popular. Strictly speaking, hot runners are those in which the runners are heated in the mold while insulating the runners from the cooled mold. This is expensive and is technically difficult. More ususal is the technique of insulated runners, widely used with polyolefins. This provides for runners of large diameter so that the outer layer of polymer freezes while the center remains fluid, the frozen polymer acting as an efficient insulant between the 'hot runner' and the cooled mold. In both cases, the effect is the same, and

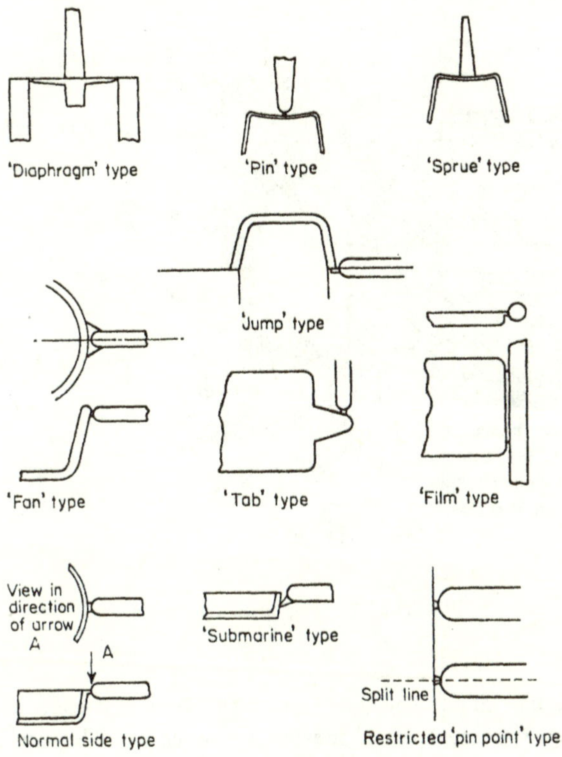

Fig. 27.5 Gating systems

the temperature gradient along the runners ensures that when the mold is opened the moldings part from the runners at the gate area.

A further technique is the 'hot slug' system where the heated nozzle makes direct (though limited) contact with the mold. A hot slug is formed and is insulated from the mold by a frozen layer of polymer, as in the insulated runner technique (Fig. 27.6).

Molding Faults and their Cause and Correction

Enough has probably been said to make it clear that, because of the interdependence of the various machine variables, there is unlikely to be one cause only for any defect in a molding. In addition, many of the possible faults can

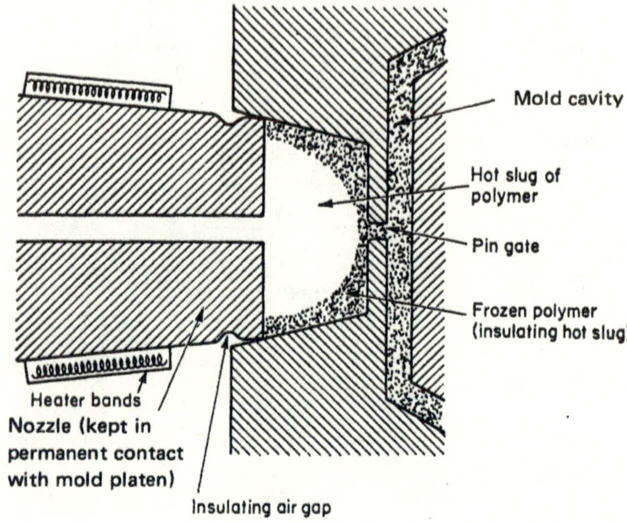

Fig. 27.6 Hot slug injection molding

be aggravated by bad mold design. Table 27.1 gives the most common molding faults, their suggested cause (or causes), and the possible ways of remedying them but, as with most things, experience is the best teacher of all.

Automation and Controls. The injection molding process can be highly automated, from the arrival of resin at the plant to the boxing of the finished parts. At one end, resin can be delivered in rail/road tank cars from where it can be pumped into large silos adjacent to the molding plant. It can then be pumped to coloring stations inside the plant where blending is automatically performed as the material is fed to the individual machine hoppers. At the other end of the line it is often possible for molded parts to be discharged on to conveyors, then automatically inspected, counted, and boxed.

In between, there are such items as fast mold changing (one machine manufacturer claims 45 seconds), robots for the high speed removal of molded components and automatic quality control monitoring.

Computerized controls are now widely available including keyboard setting of the various machine variables and solid state memory storage of optimized settings.

Safety

The pressures and temperatures involved in injection molding make safety an important factor in machine design. Front and rear gates interlock mechanically, electrically, and hydraulically to prevent clamp movement should the gates be left open, or be opened during the operating cycle. In most machines a low pressure safety condition operates during clamp closing. During this part of the cycle, if any obstruction is encountered the slight increase in force above that needed to close the mold signals the clamp to stop.

Injection Molding of Thermosets

Although thermosets are usually processed by compression or transfer molding they can also be injection molded provided great care is taken to ensure proper heating of the molding powder without premature setting in the passage of the mold or even in the cylinder. A screw-injection molding machine is used. Accurate control of heat input is essential. Too much heat causes premature curing of the material with consequent short shots or curing in the cylinder while too little heat makes it difficult to fill the mold (through high viscosity) and this leads to uneconomically long cycle times.

In one process, known as 'jet molding,' the nozzle is heated only while the material is actually flowing. When flow ceases the nozzle is rapidly cooled so that any material remaining in the nozzle does not harden.

Structural Foam Molding

In structural foam molding the injection molding process is modified to produce an article with a foamed interior, integral with substantially unfoamed inner and outer skins.

An inert gas is dispersed throughout the polymer melt, either by introducing the gas directly into the melt or by premixing the polymer with a chemical blowing agent. When the blowing agent is heated during the processing it decomposes and releases an inert gas. The pressure within the injection molding machine barrel is sufficiently high to prevent foaming at that stage. A meausured quantity of the polymer/gas mixture is then discharged under pressure into the mold cavity whereupon expansion takes place and the material fills the mold. The foam tends to collapse at the comparatively cold walls of the mold and a tough skin is formed. Because the mold is filled by expansion of the polymer melt, external pressures can be low so that mold costs are lower and platen areas and molds can be large.

Table 27.1

INJECTION MOLDING FAULTS AND TROUBLE SHOOTING GUIDE

Defect	*Possible Cause*	*Suggested Remedy*
Short moldings	Insufficient feed	Adjust feed setting
	Insufficient pressure	Increase pressure
	Inadequate heating	Increase temperature or lengthen cycle
	Insufficient injection time	Increase injection time
	Cold mold	Increase mold temperature
	Back pressure due to entrapped air	Improve venting of mold
	Unbalanced cavity in a multicavity mold	Check sizes of cavities
Flashing at mold parting lines	Insufficient locking force	Increase locking force
	Injection pressure too high	Reduce injection pressure
	Material too hot	Reduce cylinder temperature
	Mold faces out of line	Re-bed mold faces
	Mold faces contaminated	Clean mold faces
	Flow restricted to one or more cavities (in multicavity mold)	Check and remove restriction
Surface sink marks	Material too hot when gate freezes	Reduce cylinder temperature or enlarge gate
	Insufficient dwell plunger forward time	Increase dwell time
	Insufficient material shot into cavity	Increase feed Increase cylinder temp. Increase mold temp.
	Insufficient pressure	Increase pressure
	Piece ejected too hot	Increase cooling time in the mold
Voids	Material too hot (Gas formation)	Reduce cylinder temp.
	Condensation of moisture on polymer granules	Predry granules
	Condensation of moisture on the mold surface	Increase mold temp.
	Internal shrinkage after casehardening of outer layer	Increase pressure Increase mold temp. Enlarge size of gates Lengthen dwell time

Table 27.1 *(cont'd)*

Defect	*Possible Cause*	*Suggested Remedy*
Weld Lines	Material too cold	Increase cylinder temp.
	Mold too cold	Increase mold temp.
	Injection pressure too low	Increase injection pressure
	Gates wrongly located (including too big a distance from gate to weld joint) or designed	Relocate gates and/or redesign
Distortion of moldings	Ejection of molding at too high a temperature	Increase mold cooling time
	Ejection pin working unevenly	Correct or adjust ejection pins
	Existence of molded-in stresses due to:	
	(1) Material too cold	Increase cylinder temp.
	(2) Bad design	Redesign molding
	(3) Cavity overpacked in vicinity of gates	Check feed setting. Reduce injection pressure and cylinder temperature. Reduce injection time
Crazing and blistering	Excessive surface strain because of cold mold	Increase mold temperature
Surface streaks	Overheating of material	Reduce cylinder temperature
	Moisture in granules	Predry granules
Burn marks	Air trapped in mold cavities	Improve mold venting
Brittleness	Material too cold	Increase cylinder temp.
	Material has degraded	Decrease cylinder temp.
	Contamination with other material	Check the material for contamination

The advantages of structural foams include the following:

(1) High stiffness to weight ratio, due to the foam-sandwich structure.
(2) Lower mold costs which means that smaller runs become economical.
(3) Molded parts exhibit lower warping and are essentially stress-free because filling of the mold is achieved by foaming of the polymer melt and not by high pressure injection.

(4) Sink-marks or depressions are greatly reduced thus allowing the molding-in of ribs, bosses, and intersecting walls. This, together with the possibility of large platen areas allows the injection molding of large parts.

One limitation is the fact that the minimum wall thickness attainable is fairly high (around 5 mm). This is because below about 5 mm the wall consists mainly of the two solid skins with a negligible foam content.

The structural foam technique is applicable to practically any thermoplastic that can be molded or extruded by standard techniques. In practice the main material usage is polystyrene, ABS, polyphenylene oxide, high density polyethylene, and polypropylene. Applications for structural foams include business machine housings, tote boxes, display cases, cutlery and paint brush handles and a wide range of furniture items.

The scope of structural foam moldings can be further widened by reinforcement with such materials as glass fiber, coupled glass, glass beads, carbon fiber, talc and mica. A good example, here, is the molding of a whole cab top for the Jeep using a three per cent glass-filled polycarbonate. The finished molding weighs 36 kg.

BLOW MOLDING

The blow molding process is a versatile one that is not confined to the production of bottles and other containers. Toys, automobile parts and accessories, and engineering components have all been manufactured by blow molding techniques, and many design problems have been solved by their use.

The basis techniques of blow molding were developed by the glass industry. In essence, air is forced under pressure into a sealed molten mass of the plastic, surrounded by a cooled mold of the desired shape. The air pressure causes the molten mass to expand. When the plastic reaches the mold walls it is cooled by contact and re-solidifies. The mold is then opened and the bottle is ejected. The preformed plastics mass which is blown into bottle shape is known as the parison (a term derived from the glass blowing industry). It is generally of tubular form and can be either injection molded or extruded. The two main methods of producing the parison were, not surprisingly, referred to as injection blow molding and extrusion blow molding respectively. A third process was developed during the 1970s and is known as stretch blow molding.

Injection Blow Molding

The basic sequence for injection molding is shown in Fig. 27.7.

The plastics material from which the bottle is to be formed is injection molded around a blowing stick. This effectively constitutes the core of a split mold, the cavity of which is machined to accurately determined dimensions so as to

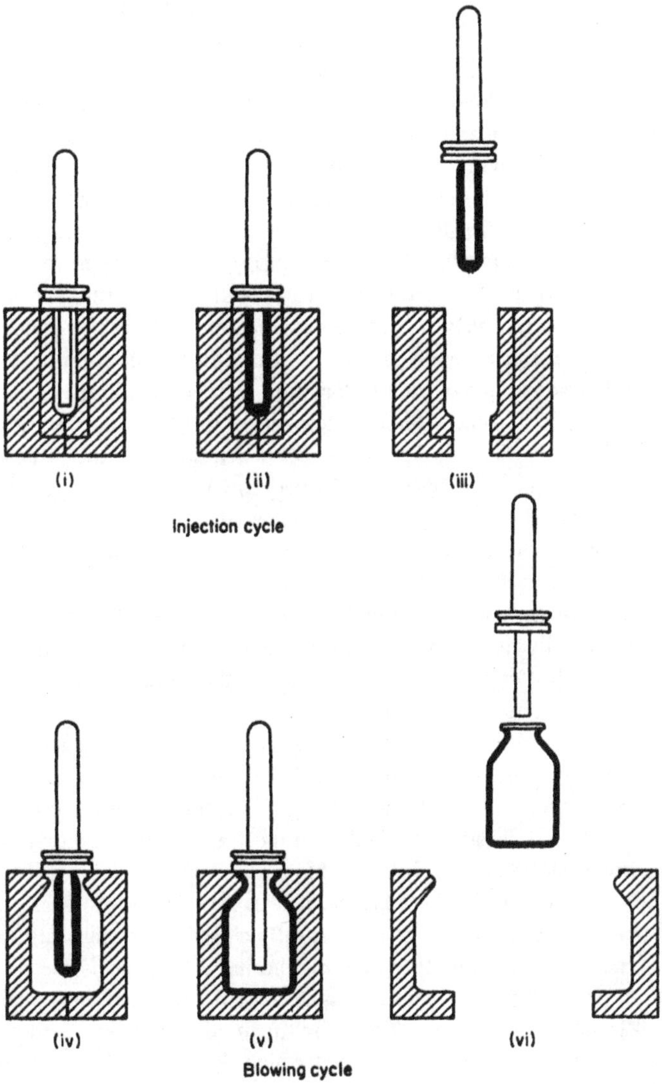

Fig. 27.7 Diagrammatic representation of injection blowing process

give the required thickness profile after blowing to fom the finished bottle. While still molten, the parison and blowing stick are transferred to a second mold (the blowing mold) in which the final bottle shape is obtained by passing compressed air down the blowing stick into the parison. The blown bottle is cooled by the walls of the mold and solidifies, the mold is opened and the bottle ejected.

Originally, the blowing stick and parison were transferred from the injection molding machine to the blowing mold by hand, but a variety of methods for parison transfer were soon developed. Methods for increasing production speeds include the mounting of blowing molds on either side of each injection mold with the parisons being transferred automatically into one or other of the blowing molds alternatively. A further development is the use of multiple mold cavities, thus permitting even greater increases in output speeds.

The chief disadvantage of the two-position method mentioned earlier is that the injection mold and blow mold stations have to stand idle during removal of the molded part. This difficulty has been overcome by the three-position method invented in Italy by Gussoni. Three blowing sticks are spaced radially at equal intervals about a pivot and in a plane between the faces of the mold. The molds are so designed that an injection cavity and a blowing cavity are located at 120° to each other to mate with two of the three parison sticks. The third stick is not enclosed by a mold but serves as an ejection station. In this system, the radially mounted parison sticks can be indexed about their central pivot so that each stick can be brought in turn to the molding station, the blowing station and the ejection station.

One important technical advantage of injection blow molding is the accurate control of wall thickness that is possible. The blowing stick and parison mold can be pre-shaped to give walls of varying thickness so that there is a correct distribution of material after blowing into the final mold, thus avoiding thinning at corners and shoulders. Injection blow molding is also capable of producing a fully finished bottle since the parison is completely enclosed and fully formed in the blowing mold. The latter is fully closed during blowing, whereas in most extrusion blow molding systems the mold itself seals the parison at the bottom on the bottle. The excess material ('flash') necessitates trimming of the bottle subsequent to expulsion from the mold. However, methods of automatic deflashing have been developed which do not interfere with the normal bottle cycle.

The injection blow molding process is more versatile in terms of the number of plastics that can be used. Polystyrene, for example, is not easily blown on extrusion blow molding equipment whereas perfectly good results can be obtained by the injection blow molding process. However, there are also disadvantages to the injection blow molding process. Equipment is usually more expensive (compared with that required for extrusion blow molding) and the process tends to be slower because the injection molding part of the cycle necessitates higher mass temperatures and so the cooling cycle is longer. One contribution to the higher cost of injection blow molding is the fact that two molds are necessary compared with one for extrusion blow molding.

The injection parison mold normally consists of two mold halves for forming the body, a blowing stick, a base and a neck ring. The second (or blowing) mold is similar to an extrusion blowing mold, having two mold halves with an interior cavity corresponding to the outside profile of the finished bottle. The blowing stick has three functions, namely, to act as a core round which the parison is injection molded; to be a vehicle for carrying the parison from the injection mold to the blowing mold, and to provide a means of admitting air to the inside of the parison to expand it to size. The neck of the bottle is produced in its final form at the injection molding stage and is, therefore, dimensionally accurate.

Extrusion Blow Molding

In this method, the parison is in the form of an extruded tube which can be produced continuously or intermittently. In the continuous system, a predetermined length of tube is trapped, at intervals, between the two halves of a split mold. Both ends of the tube are sealed by the closing mold and the trapped portion is inflated by compressed air introduced via a blowing pin or needle. One disadvantage of this method of inflation is that a finishing operation is necessary to produce an acceptable neck because the parison is gripped at both ends.

An alternative is to introduce the compressed air via a mandrel inserted in the bore of the molten tubular parison. Since the hollow blowing mandrel has to enter the open end of the parison there has to be a cutting device independent of the mold pinch-off to sever the parison and provide the open end. One such method, where the extruded parison drops over a blowing tube at the base of the mold is shown in Fig. 27.8.

Blowing can also be effected using a mandrel inserted into the top of the mold. The bottle is then positioned neck-up and the internal diameter of the neck is determined by the external diameter of the mandrel. The cutting action between the mandrel and the top of the mold removes excess material and so produces a finished bottle. The cutting action is not always perfect (particularly as a mold becomes worn) and a patented system, developed by the German company, Bekum, incorporates a sliding sleeve at the mandrel which descends,

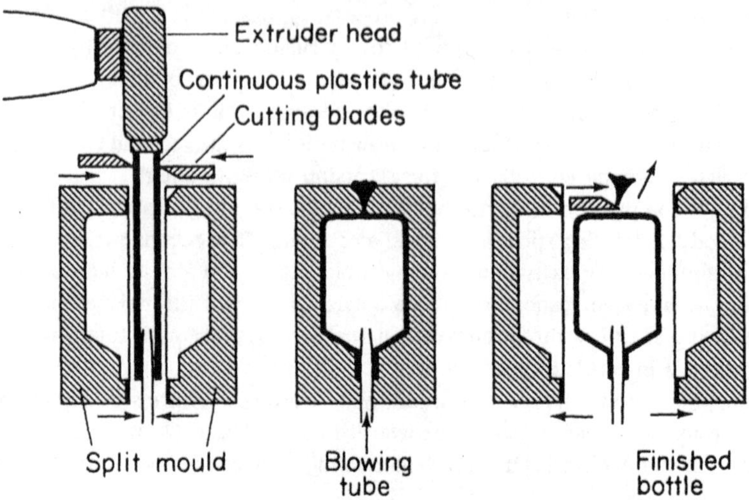

Fig. 27.8 Extrusion Blow Molding Process

after blowing, to form the top rim of the bottle and cut off excess material.

Methods of increasing production speeds include the use of more than one mold, the filled mold being moved away while another is moved into position, ready to receive the next portion of extruded tube. The molds can be reciprocating ones or can be mounted on a rotary table. Each of these methods can be further adapted by having multiple die heads attached to the extruder and feeding them to multiple cavity dies.

In the intermittent process the parison is rapidly extruded after the bottle is removed from the mold and the mold clamping mechanism does not have to move to a blowing station. Blow molding, cooling and part removal all take place under the extrusion head. The stop/start aspect of this method makes the process suitable for polyolefins but not for heat-sensitive materials such as PVC.

For bottles between about ¼ liter and 9½ liter capacity, the reciprocating screw method is used. This is similar to the process described earlier, under the heading of injection molding. After a parison has been extruded, the screw moves backward, accumulating molten resin in front of the screw tip. When the molded bottle has cooled, the mold is opened, the bottle removed and the screw moves quickly forward again. This pushes the accumulated melt through the extrusion head and so forms the next parison. It is possible to extrude up to twelve parisons simultaneously by the use of multiple die heads attached to the extruder.

For large moldings such as large drums and heavy industrial parts, the accumulator-head process is used. If the parison necessary to produce, say, a 50 gallon container, was extruded in the normal manner, it would sag and deform under its own weight long before extrusion was complete, and it would be impossible to produce anything like an even wall thickness. The accumulator-head process consists of an extruder that feeds into a tubular reservoir which is a part of the extrusion head itself. A tubular plunger rapidly extrudes the molten resin from the head annulus with a low, uniform pressure that reduces the overall stresses.

One disadvantage of extrusion blow molding is that when the tubular parison is blown into a bottle shape, the material is thinned more at the mold extremities than elsewhere. The shoulders and base corners of the bottles are weaker, therefore, than is desirable. Thickness control can be achieved, however, by means of a specially designed conical mandrel which is raised or lowered within a mating conical section of the die ring to give a controlled variation in the clearance between the two. The thickness control is programmed to operate with each parison production cycle. Sophisticated models can accommodate up to twelve separate control steps per cycle. Some advantages and disadvantages of injection blowing and extrusion blowing are shown in Table 27.2.

Table 27.2

COMPARISON OF INJECTION BLOWING AND EXTRUSION BLOWING

	Advantage	Disadvantage
Injection Blowing	Greater versatility in terms of polymers that can be used. Good control of neck and wall dimenions. Excellent surface finish. Scrap free; no flash to recycle. Better finish; no pinch-off scars.	Cost Two molds required Relatively slow
Extrusion Blowing	Can be used for very large moldings. Able to produce wide range of sizes and shapes, including bottles with handles, off-set necks, etc. Low cost tooling.	More difficult to control wall thickness (but modern techniques *can* produce excellent results). Trimming operation necessary.

Stretch Blow Molding

Stretch blow molding (or biaxial blow molding) is a process whereby bottles are produced having an appreciable orientation in both the longitudinal and the transverse directions. All blow molding processes produce orientation in the transverse direction, but no longitudinal orientation occurs in extrusion blow molding, and only a moderate amount is produced in injection blow molding. Biaxial orientation improves tensile and impact strength, while in transparent resins there is a reduction in haze. Other advantages include reduced creep and improved gas and water vapor barrier properties. Container weight may also be reduced. The basic principle is that either a length of plastics tube or an injection molded pre-form is mechanically stretched in the axial or longitudinal direction. During this stretching process the temperature is maintained within a narrow range which is specific for each polymer (preferably just above the glass transition temperature Tg. — see Chapter 1.). The stretched parison or preform is then blown to the final desired shape.

When the parison is produced by extrusion, the tube is cut to the required length and the two ends thermoformed, one to give a closed end, and the other to give the neck finish. The axial stretching is achieved in a variety of ways. In one extrusion system the parison is clamped at one end of the mold and then pulled, while in another, the end of the parison is sealed off and a rod, telescoping from a blow-pin, presses against the closed end of the parison and so stretches it. A similar idea is used in the stretching of an injection molded preform where the blow pin elongates to stretch the preform to a pre-determined length before blowing takes place to form the finished bottle.

Whether preforms are made from extruded tube (by thermoforming as described earlier) or by injection molding, there are then two major distinctions that can be made with respect to the subsequent blowing process. Either the preforms are allowed to cool to room temperature and then stored or sent to other locations for subsequent blowing, or they are blown in-line. In the latter process (integrated system) preforms are not allowed to cool to room temperature, but are cooled sufficiently to ensure a clear amorphous preform. An example of an integrated injection stretch blow molding process is that of Nissei. One machine with four stations is used. At the first station the material is melted by heating, injected into the preform mold and cooled rapidly to give a transparent parison. The cooled parison is reheated using an electric heater in order to soften it prior to stretching. At the third station the softened parison is first stretched to about twice its original length in the bottle mold. Compressed air is then introduced into the stretched parison in order to blow it out to shape. Removal of the bottle is effected at the fourth station. The parisons are heated from the inside (by a heated core rod), and the outside (by heating the mold). This allows rapid reheating of the parison without the danger of crystallization of the parison surface with consequent whitening.

Two step systems are exemplified by the Cincinnati Milacron and the Gildemeister Corpoplast. According to the latter company, bottle properties depend on stretching rate and temperature during the stretching operation. Too low a temperature leads to destruction of the molecular structure and stress whitening. At too high a temperature, or during too long a heating cycle, crystallization occurs. The space between these limits is narrowed by raising the stretching rate or orientation, and is further narrowed by a long heating cycle. The Corpoplast process uses short wavelength, open panel infra-red heaters giving rapid heat penetration so that high orientation can be achieved (with consequent improvements in bottle properties.)

Advantages of both two stage and integrated systems seem to be finely balanced. Overall processing costs are much the same when all the ancillary equipment is taken into account. Advantages of the integrated approach are said to be a 10% reduction in energy usage during the reheating process, and the maintenance of the parison temperature at a high level, thus permitting low pressure blowing with accurate dimensional control. Mold changes are faster and less floor space is required.

The two stage approach permits molding of the pre-forms at a central location with subsequent shipping to several bottling plants for blowing and filling. Higher speeds are possible and process parameters can be optimized for bottle resin processing and bottle blowing. Another advantage claimed by manufacturers of two stage processes is that it allows for more frequent quality control checks. Faulty preforms can be spotted immediately after injection molding, and one does not have to wait for the finished bottle to discover the defect. This is claimed to cut the reject rate considerably but, on the other hand, there are allegedly many faults inherent in the preforms than can only by detected in the finished bottle.

Machine Design

1. The Extruder. Since the time taken for the blow molding to cool is the major factor controlling cycle time, the mass temperature is usually kept as low as possible, subject to the production of bottles having good appearance and performance. Good homogeneity of the melt can, therefore, be attained only by the use of high back pressures and high shearing forces, which necessitates the employment of robust extruders with good thrust bearings.

Screws with high L/D ratios (around 20 : 1) are preferred with a short compression stage and a fairly long metering stage (5 to 6D for an L/D ration of 20 : 1).

2. Blow Heads. A typical blowing head is shown in Fig 27.9. In the early heads, the mandrel was supported by a perforated plate or a webbed 'spider' and the material flowed over the mandrel from above. This system gave rise to 'spider' memory effects (flow lines caused by imperfect fusion after flowing past the webbed supports or the perforated plate). Later designs employed the system shown in Fig. 27.10 of a ring-type mandrel which is side fed. Even in this design there tends to be a weld line beyond the mandrel, opposite the point of entry, but this is largely eliminated by restrictions further down the head, causing appreciable back pressures, and the long land length.

3. Molds. Materials in common use for blow molds include aluminum and aluminum alloys, steel, zinc alloys such as Kirksite, cast bronze, and beryllium-copper. For experimental purposes and short runs it is also possible to use metal-filled epoxy resin castings and glass-reinforced thermo-setting resins. Steel is

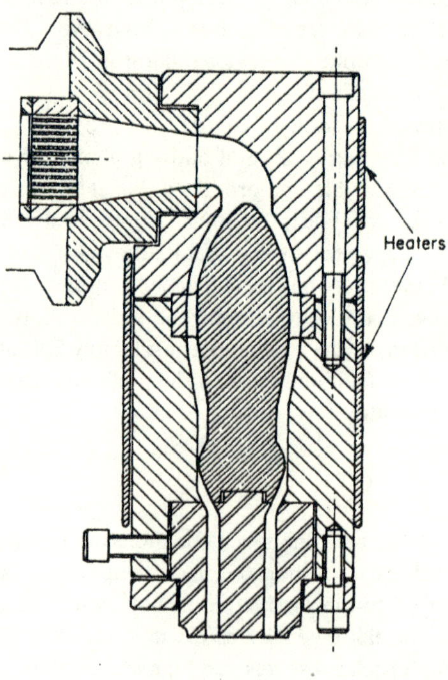

Fig. 27.9 Typical blowing head having a spider supported type mandrel

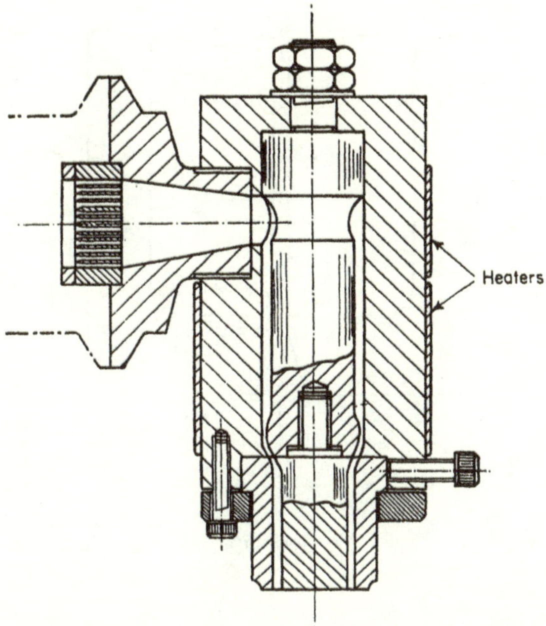

Fig. 27.10 Typical blowing head having an axially supported or ring-type mandrel

expensive and is usually economic only where long production runs are possible. Its use is also dictated, however, where the pinch-off area is large because of its excellent hardness and wear resistance. The heat transfer of steel is lower than those of the other metals mentioned so that the design and location of cooling channels in steel molds is rather more critical.

Design of the pinch-off points of the mold is particularly important because of possible weaknesses in the bottle at these points due to poor welding. If the pinch-off is too thick another difficulty is encountered, namely, that of removal of excess material from base and necks due to too thick a connecting web. Where cast aluminum or zinc alloy molds are used, pinch-offs are made of steel and inserted in the casting.

In general, the pinch-off section has a land length of between 0.020 to 0.100 inch. A typical pinch-off area for high density polyethylene is shown in Fig. 27.11. Vapor honing or fine sand-blasting gives a matte finish to the mold, which is preferable to a polished surface as the latter can lead to air being trapped between the hot plastic material and the mold. Where a matte finish on the mold-

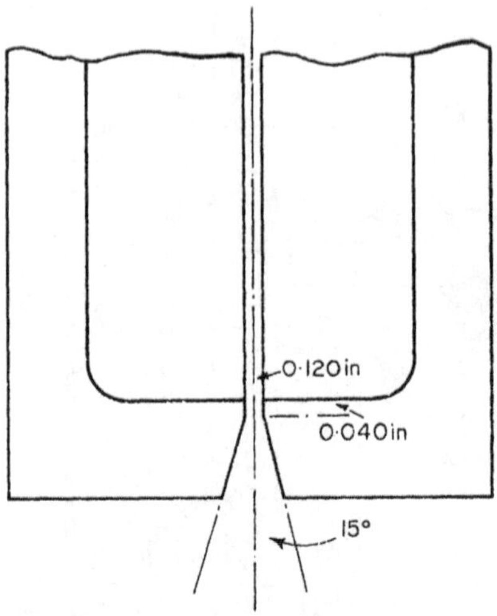

Fig 27.11 Typical pinch-off for high-density polyethylene

ed product is unacceptable, other methods of mold venting must be used. These may take the form of very fine holes or shallow slits. Vents are required where projections occur on the molding, since entrapped air cannot otherwise escape from the corresponding depressions in the mold cavity.

Operating Variables

Mass Temperature. When discussing extruder design, we mentioned that mass temperatures were kept as low as possible in order to reduce cooling time. Too low a temperature prevents complete plasticization which can lead to residual strain in the molding as well as bad surface defects, such as 'melt fracture.' Too high a temperature on the other hand, increases the cycle time and can also give rise to thinning of the parison, especially where a long parison is required. The latter fault depends too, of course, on extrusion rates and on the melt viscosity of the polymer. The effect of parison thinning can be corrected to some extent by using the adjustable mandrel system described earlier. Mass temperature also affects the phenomenon known as 'die swell' (also known as diameter swell). This is measured by the actual relaxed outside diameter of the parison after leaving the extruder die minus the die ring bore diameter, expressed as a percentage of the extruder die diameter. It is caused by the fact that when a molten polymer is extruded through an orifice under pressure, alignment of the molecules occurs along the line of flow. When the extrudate emerges from the orifice, the pressure conditions are relaxed and the molecules tend to resume their original random distribution with a consequent shrinkage in length and swelling in diameter. An increase in mass temperature generally reduces die swell. A reduced rate of extrusion also reduces the amount of swelling. Die swell is affected by material properties such as melt tensile strength and melt viscosity. Increasing melt viscosity tends to increase die swell while a lower melt tensile strength increases sag and draw-down and thus reduces die swell.

Air Pressure. Good reproduction of mold detail and good surface contact with the mold (to increase cooling rate) are both factors which require high blowing pressures, and in practice pressures up to about 80 $lb/in.^2$ are used. Other factors of importance include preventing incoming air from jetting, which could lead to local cooling, and providing air free from moisture and oil from the compressor.

Mold Temperatures. Mold temperatures are always low because of the importance of rapid cooling of the molding.

Mold Closing Rate. Too high a rate of mold closing can cause the container to split on the inside of the neck or shoulder, and it is becoming usual to incorporate damping devices in the machine design.

Blow Molding Developments

There have been a number of developments in the basic processes already described, and two of these are given in the following sections.

Form/Fill/Seal Systems. Machines have been developed that will blow a bottle, fill it and seal it while it is still in the mold. One of the first systems to be developed consisted of an extruder which delivered a vertical parison. This was gripped between two halves of a mold and transferred to a blowing station where a double-channel blowing/filling spigot entered the parison from above. Compressed air was blown down one channel to inflate the parison, after which the product to be filled was passed down the second channel. After filling, the spigot was withdrawn and the bottle sealed by crimping the still hot bottle neck. Filling speeds were low, and later developments concentrated on making the process a continuous one. One such development is the Bottlepack which uses a continuous multi-mold system (with the addition of multiple parison heads, according to the number of molds to be used). The mold halves are mounted on converging caterpillar tracks and move vertically downward. At a point below the parison head of the extruder, the two chains of mold halves meet and trap the tubular parison. This is inflated with air and then filled, using a multiple channel spigot which by this time has closed round the next parison. Meanwhile, the first mold moves to the sealing head where the bottle is sealed. Finally, the caterpillar tracks diverge, the mold opens and the filled bottle is ejected. The operation can be followed by reference to Fig. 27.12.

An important advantage of this system is that at no time does the filling liquid come into contact with the air. This means that aseptic filling is possible since sterile materials filled into the bottles will not be contaminated by airborne bacteria. This is particularly valuable in the case of foods and pharmaceuticals. It is also possible to make very thin wall (and hence cheaper) bottles because they do not exist unsupported (while empty), but remain in the mold until filled and sealed. In normal bottle blowing the minimum bottle weight may well be determined by the performance requirements of the filling line rather than its performance in use. Speeds of 3,000 x ½-liter bottles or 1,000 x 2-liter bottles per hour are claimed. Blowing cycles are usually shorter because some cooling of the bottle occurs by the filled product. This product is, of course, heated by the hot plastic but the temperature rise is small, as can be seen from the following example.

If the weight of a 1-liter bottle is 50g and it is cooled from 180 °C to 50 °C by the product being filled then we can say that:-

Weight of product × temperature rise of product × specific heat of product = Weight of bottle × temperature drop of bottle × specific heat of plastic.

For an aqueous product and a low density poylethylene bottle of 1-liter capacity we have:-

$$1{,}000 \times T \times 1 = 50(180 - 50) \times 0.5$$

$$T = \frac{50 \times 130 \times 0.5}{1{,}000}$$

$$= 3.25^{\circ}C$$

Since the mold also contributes to the cooling of the bottle, the amount of heating of the product is even less.

Ornamentation can be carried out by in-mold embossing or by subsequent labelling. Silk screen printing of the filled bottles is also possible, but the rejection rate is high and may also involve loss of product as well as the bottle.

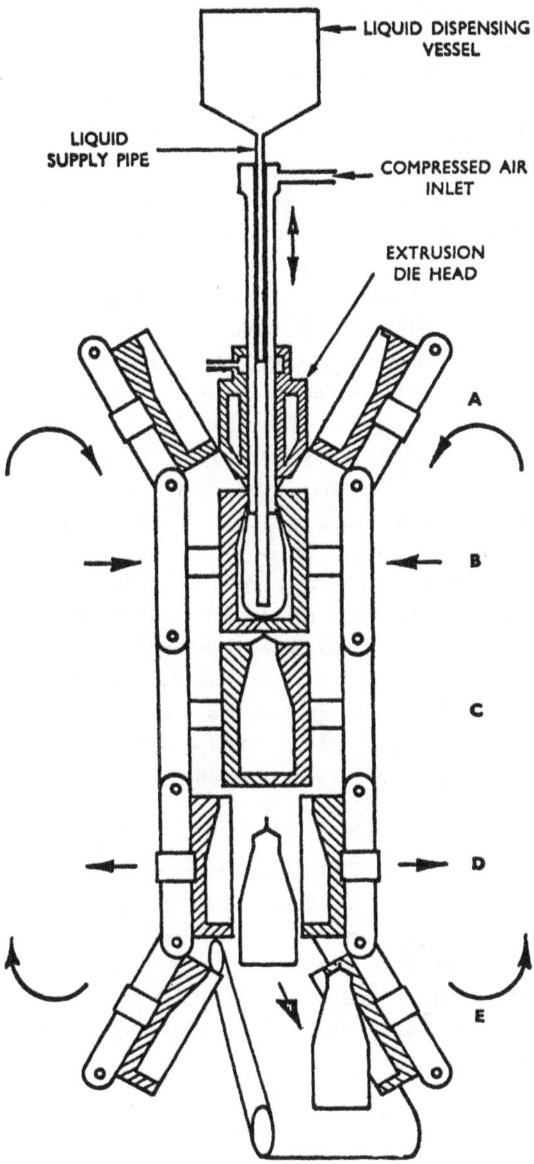

Fig. 27.12 Schematic drawing of the series 4000 Bottlepack Machine

Multilayer Blow Molding. The requirements of many end uses are sufficiently complex, that no one material is likely to be satisfactory except, perhaps, at an uneconomic thickness. Multilayer bottles can provide an answer in such instances. The technique whereby multilayer bottles can be manufactured is based on coextrusion (as described in the previous chapter). A multilayer parison is produced which is then blown to give a multilayer bottle.

A tight control of extrusion process variables such as melt pressures, screw speeds, melt temperatures is necessary. For example, the temperature of each polymer must be closely controlled in order to maintain the proper level of viscosity in the die. The success of the technique lies largely in the design of the die, and the way in which the individually extruded melts are brought together, prior to their extrusion as a multilayer parison. Parison programming is also more critical than in conventional blow molding.

As with films coextrusion provides the facility for producing individual plies of very low gauge and it is hence possible to produce bottles where a thin layer of a relatively expensive high barrier material is sandwiched between an inner and outer layer of low cost material. It has also been suggested that regrind material could be used as the main carrier layer, using polyamide (nylon) as an outer, glossy, scratch resistant surface. Where adhesion of adjacent layers is a problem, then adhesive (tie) layers are used, as discussed in Chapter 26.

Equipment costs are higher than for standard blow molding machines, one estimate being 40–60% more for equipment blowing containers up to 1-liter. On the other hand, coextrusion is claimed to be cheaper than stretch blow molding, which is the other process giving bottles with improved barrier properties. Coextruded bottles are said to be competitive with glass and metal in areas where single material plastics bottles have failed, hitherto.

SUGGESTED FURTHER READING

Injection Mould Design. (1983-3rd ed.). R. G. W. Pye (London: Butterworths).

Injection Moulding of Engineering Plastics. (1988-3rd ed.). Tony Whelan and John Goff (London. Plastics and Rubber Institute).

Plastics in Contact with Food. (1974) J. H. Briston and L. L. Katan (London. Food Trade Press).

28.

THERMOFORMING

In thermoforming, a heat-softened plastic sheet is formed either into or around a mold. The simplest example of this technique is 'bubble blowing' polymethylmethacrylate sheet to form aircraft canopies, but there are now many variations both in basic techniques and in their degree of automation. In general, thermoforming techniques are best-suited for producing moldings of large area, for very thin-walled moldings or where only short runs (including the production of prototypes) are required.

METHODS OF FORMING

Thermoforming may be divided into three main types:

(a) Vacuum Forming. In its simplest form, the method consists of clamping the sheet in a frame attached to the mold box. The sheet is heated until rubbery and a vacuum is drawn between the sheet and the mold. Atmospheric pressure above the sheet forces the sheet on to the mold where it is cooled sufficiently to retain its shape.

(b) Pressure Forming. This is the same as vacuum forming with the exception that a positive air pressure is applied to the sheet from above, which again has the effect of forcing the softened sheet on to the mold. The pressure which can be applied to the sheet is not limited to atmospheric pressure which is, of course, the case with vacuum forming.

(c) Matched Molds Forming. As the name implies, the heated sheet is pressed into shape by trapping it between matched male and female molds.

POLYMERS USED IN THERMOFORMING

Polystyrene

Toughened polystyrene sheet is one of the most widely used materials for thermoforming, being cheap and easy to manipulate. It is fairly rigid and gives good depths of draw. Outdoor weathering is poor, as is surface gloss. One of the largest outlets is for vending and other disposable drinking cups. The material is also used for forming a wide range of tubs, trays, and box inserts. The cigarette pack mentioned in Chapter 11 is a good example of the material's versatility. The wall thickness is around 15 thousandths of an inch (375 microns) with a large depth of draw. Using a 24-impression mold, outputs of 20,000 packages an hour are claimed.

Toughened polystyrene is translucent and cannot be used where good clarity is required. Normally, general purpose (crystal) polystyrene gives too brittle a sheet but if it is biaxially oriented it produces a much superior sheet that can be thermoformed to give crystal clear packages. The strain present in the oriented sheet causes difficulties in thermoforming but it can be pressure formed using special equipment.

Expanded polystyrene sheet can also be thermoformed, matched molds being preferred for deep or complicated draws. The main problem is that the foam structure gives the sheet a very low heat conductivity and it is difficult, therefore, to heat the sheet right through without overheating the surface. Double-sided heating is usually necessary, therefore. Typical applications include trays for prepackaging meat, fruit, and vegetables, and for the transport of delicate electronic components.

Cellulose Acetate

Cellulose acetate is readily thermoformed and has excellent clarity. It gives rigid moldings that are particularly good for blister packaging. It is more expensive than polystyrene but is used where high class formings are required. One disadvantage of cellulose acetate is its dimensional instability under conditions of changing humidity, due to moisture pick-up from the atmosphere. This makes it unsuitable for outdoor applications.

Temperature control during forming is critical because too high a temperature causes surface defects like blistering, while too low a temperature leads to 'milkiness' in the finished forming. Formings from cellulose acetate are easily fabricated either by adhesives or by solvent welding.

Cellulose Acetate/Butyrate

Cellulose acetate/butyrate (CAB) is more expensive than cellulose acetate but has better water resistance which makes it suitable for advertising signs and other outdoor applications. Like cellulose acetate it possesses good rigidity and clarity and is easily fabricated by adhesives and by solvent welding.

PVC

PVC sheet has good clarity and is often used where good chemical resistance is required. Good formings, with good mold detail can be obtained even from very thin gauges of sheet. It is self-extinguishing which makes it particularly useful in applications such as aircraft food trays. Other uses include blister packs, and a wide range of food packaging applications, including margarine tubs and chocolate and biscuit box inserts.

Acrylonitrile/Butadiene/Styrene (ABS)

The thermoforming properties of this material are similar to those of toughened polystyrene (which it resembles chemically) but it is more expensive. It is also tougher so that the price differential can often be reduced by the use of thinner gauges. Thin gauge ABS has been widely used for tubs for the packaging of soft margarine.

Polymethylmethacrylate

This material has excellent optical qualities and resistance to outdoor conditions but it is expensive. Sheet may be prepared by polymerization casting or by extrusion (*see* Chapter 13) and it has found extensive use in the forming of advertising signs and in lighting applications.

Polyolefins

Low- and high-density polyethylene and polypropylene are all available as extruded sheet but are difficult to form because of their combination of high specific heat (leading to slow heating) and low thermal conductivity, which leads to trouble when heating thick sheets. Heating both sides of the sheet at once (sandwich heating) is recommended for the polyolefins in order to overcome these difficulties. Another snag is that because of their crystallinity, the temperatures at which the polyolefins become formable are not far below their softening points, and the sheets have a tendency to sag. High molecular weight grades are recommended, therefore, and molds should be heated slightly to prevent the sheet setting as soon as it contacts the mold.

Thermoforming is easier with high-density polyethylene than with the low-density material but fast heating and accurate control of temperature are still essential. The finished moldings have good rigidity and hardness.

Polypropylene has even greater stiffness and is not subject to environmental stress cracking (as are the polyethylenes). It is also more resistant to high temperatures. The forming properties of polypropylene are similar to those of high-density polyethylene and, in general, the same methods can be used for both.

Applications are usually in areas where the finished formings need to be resistant to boiling water or to moderate sterilization temperatures. At the other end of the scale, polypropylene is not so suitable as high density polyethylene, because the impact strength and other physical properaties decrease with decreasing temperature (below about 0 °C).

Polycarbonate

Polycarbonate can be thermoformed but higher forming temperatures are necessary because of its high softening point. It has very good clarity and a a very high impact strength. It is particularly useful for the manufacture of blisters for hard, sharp articles such as hardware.

Coextrusions

The range of possibilities for thermoforming has been greatly widened by the availability of coextruded materials, particularly the high barrier ones based on PVDC or EVOH. Two products that require high barrier packaging are fats and long life milk products.

Incorporating polycarbonate in the coextrusion gives the added possibilities of hot fill and retortable packaging. Two other possible applications that could benefit from polycarbonate's clarity and heat resistance, are medical blister packs and autoclavable medical trays.

One popular coextruded structure is PP/tie layer/PVDC/tie layer/PP. It has been used to construct a number of different retortable containers. The containers are thermoformed and are able to withstand autoclaving, and subsequent reheating in boiling water, microwave ovens or conventional ovens, up to 110 °C. Gas barrier properties are excellent and are said to be retained for at least one year under unrefrigerated storage conditions. Coextrusions are also used on thermoform/fill/seal machines. In Europe, structures such as PS/tie layer/PVDC/tie layer/PS have been used for thermoform/fill/seal yogurt pots. The high barrier (low oxygen transmission) properties of this structure confer a long shelf life on the product. Other TFFS applications include the unit dose

blister packaging of pills, tablets and capsules. One coextruded structure that has been used successfully in the U.S.A. for the blister of tablets is PP/tie layer/PS/tie layer/PP.

MACHINE VARIABLES

Heating

Temperature control in thermoforming is critical because of the poor thermal conductivity of plastics and because temperature affects the forming characteristics. Too much heat, for example, and the sheet will flow without forming; too little heat, on the other hand, and it may rupture early in the forming process.

Various types of heat source have been used for heating thermoplastics films or thin sheets for thermoforming, including contact heating, infa-red heating, high frequency heating, and heat transfer using hot air or hot fluids. The first two methods are the ones normally used in practice. Whichever method is used, it should meet the following requirements. The film or sheet must be heated to its optimal forming temperature, and this temperature should be even over the whole forming area. Temperature deviations between both sides of the sheet should also be kept to the minimum. The heating period should be as short as possible, but excessive heating of the material must be avoided because of the risk of thermal degradation of the sheet. Energy consumption should also be kept as low as possible. It is desirable to reach the operating temperature as quickly as possible, and so heater capacity should be adjustable.

Contact heating is normally reserved for materials below about 0.38 mm in thickness. It is particularly useful when the sealing flange area of the container has to be of uniform thickness for effective sealing in use. The sheet is clamped at the future sealing flange, and is neither heated nor drawn out in this area. In contact heating the heating platen is often covered with a PTFE separating layer to prevent sticking of the film to the platen. If no PTFE layer is used the heater platen surface is often roughened by sand blasting to allow enclosed air to escape. Another system utilizes heater platens perforated with very small holes (0.5 mm diameter). At the end of the heating cycle, compresed air is injected through the platen, forcing the plastic off the platen and into the mold.

When double-sided heating is necessary (for thicker sheet or for plastics with a low thermal conductivity) or when plug assist is being used, separate heating and forming stations are employed.

Infra-red heaters are of two main types, namely quartz or ceramic. A series of such units is assembled to make up heater panels of various sizes. These can be wired into zones with individual controllers, so allowing more heat to be switched to certain parts of the sheet if required. Double-sided heating is again available when necessary. Infa-red heating acts by absorption of the radiation by the sheet at the surface layers. The generated heat then flows by conduction through the rest of the material. Overheating of the surface can easily occur and effective regulation of the radiation is very important.

Cooling

If cooling is unassisted a very long molding cycle results since all plastics are poor conductors of heat. The molding cycle is made as short as possible by removing the molded sheet from the mold as soon as it is dimensionally stable but the actual cooling time is also reduced, in many cases, by air blasts or water-mist sprays directed onto the sheet. Cooling is also affected by the choice of mold material (*see* below). A certain amount of shrinkage takes place depending on the particular plastic material and on the intensity of cooling. Although shrinkage is basically undesirable it does facilitate the removal of moldings from female molds, particularly those with vertical side walls.

Molds

The construction of molds is governed by several factors, including the number of moldings to be produced (governs the choice of manual, semi-automatic or fully automatic production), the sheet material to be formed, the surface finish required, the thermal conductivity of the mold, the machinability of the mold material, tolerances required, and the allowable cost. For making prototypes, or for small runs, molds may be made from wood or plaster of Paris and so are cheap to produce. For long run production work, however, epoxy resins or metals are preferred. Cast or machined aluminum is the normal material but beryllium copper and brass, chrome-plated for long service life, have also been used. The main advantage of epoxy resins lies in their ease of manufacture and excellent machinability, but they have the drawback of low thermal conductivity. However, if the shape of the mold is not too complex it is usually possible to cool the mold by circulating water within the mounting plate. It should be noted that the low thermal conductivity is sometimes an asset, for prototype work or small runs, because no major warming-up of the mold is necessary and the incidence of chill marks will be greatly reduced.

If the number of items to be produced justifies the extra expense, then metal molds should be used. The main advantages of metal molds are good dimensional stability, good mechanical resistance (long life), excellent thermal conductivity and the excellent surface finish that can be obtained on the formed article. Disadvantages are high material and machining costs.

Male molds are easier to construct (and so are cheaper than female molds) and are generally used for deeper formings. The sides of a mold should be slightly tapered to facilitate removal of the molding, particularly with male molds. With female molds the shrinkage referred to earlier usually leads to ease of release from the mold.

One thing to be avoided in designing thermoforming molds is the existence of deep and sharply angular corners, as they cause severe thinning of the sheet. It is possible to mold undercuts but this necessitates the use of removable sections and so complicates the mold, and lengthens the molding cycle. The vacuum holes drilled through the mold must be kept small otherwise the sheet or film will be drawn into them.

Decoration

Thermoformed articles can be decorated by most of the processes normally used for plastics. Thus they can be labelled, printed, flocked, embossed, or metallized. Printing can be carried out on the finished molding or on the flat sheet, the design being distorted in such a way as to give the required appearance after forming. In the latter case the printed design is distorted in such a way that the appearance after forming is as required. Flexographic, screen or gravure printing may all be employed.

One way to determine the amount of distortion necessary is to form a sheet on which a grid pattern or a system of concentric rings has been printed while in the flat. The amount of distortion that occurs on forming, at various points on the surface, gives an indication of the distortion to be allowed for. Another method is to print a previously formed article, clamp it in the thermoformer again, and heat it up to above its softening point. The sheet returns to its original flat form and the amount of distortion necessary for correct design is determined. In both methods it is important to use a homogenous molding with as little thickness variation as possible and which is free from tension.

Very shallow formings are usually printed without distortion as the sheet is subject to slight stretching only, and this has very little influence on the appearance of the print. With formings that require a high draw ratio, the print design should be such that some slight distortion can be absorbed without affecting the overall appearance, because it is almost impossible to allow for all distortion under such conditions.

Trimming

Trimming equipment can vary from a simple knife to a sophisticated cutting press. For small to medium size moldings a roller press is often used and high production rates can be achieved. For larger moldings a power driven router together with a suitable cutting jig may be more effective. Separation of multi-impression moldings may also be achieved with a combination of slitting knives and a cross-cut guillotine. This method has the merit of being easily adjustable to accommodate a wide range of mold configurations, but can only be used for moldings with a rectangular aperture. If a roller press is used, a separate cutting form is necessary for each mold configuration. Roller presses can be in-line or can operate as separate units.

Automatic Operation

The thermoforming operation can range from manual to fully automatic. Automatic machines are available which will enable the complete cycle to be set in motion at the push of a single button, with the cycle including any or all the features such as drape, plug assist, etc.

A typical sequence of operations for a vacuum forming machine operating on sheet fed is shown below, starting once the sheet has been loaded into the machine.

(1) Push start button.
(2) Heater is brought forward for pre-set time.
(3) Heater retracts.
(4) Air is injected to form bubble (if required).
(5) Drape table rises or plug assist descends.
(6) Vacuum is drawn.
(7) Air blast or water mist spray is applied to surface of molding.
(8) Vacuum and air are turned off. The plug assist (if used) is withdrawn.
(9) Air pressure applied to release forming from mold.
(10) Air pressure off.
(11) Drape table descends.
(12) Operator removes forming and reloads with fresh sheet.

For thinner sheet a continuous reel feed can be incorporated.

THERMOFORMING METHODS

Some of the many methods of thermoforming now available are mentioned below and are illustrated in Figs. 28.1 to 28.7.

Vacuum Forming

Either male or female molds may be used (Fig. 28.1 illustrates the process using a male mold). Vacuum forming into a female mold is probably the simplest of the mold methods but only simple designs are possible because of excessive thinning at bottom and corners of the cavity. With a male mold the thickest section is on top instead of at the side walls.

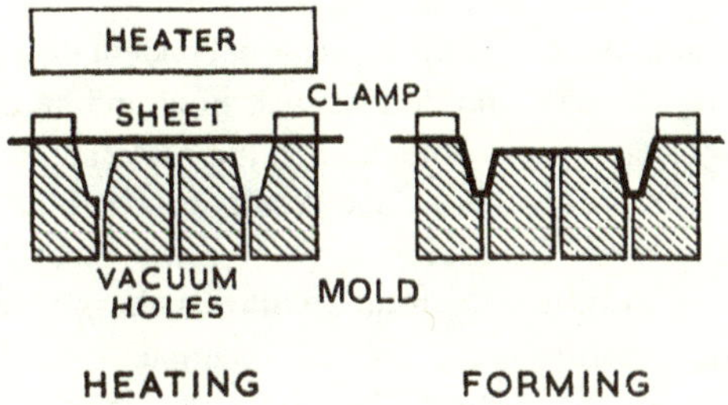

Fig. 28.1 Vacuum forming into male mold

Drape Forming. Drape forming with a male mold is illustrated in Fig. 28.2. The mold is mounted on a piston (not shown) within the vacuum box. The piston rises and pushes the mold into the heat softened sheet, immediately prior to the vacuum being applied. This gives a certain amount of preforming, and so lessens thinning of the sheet at the corners of the mold.

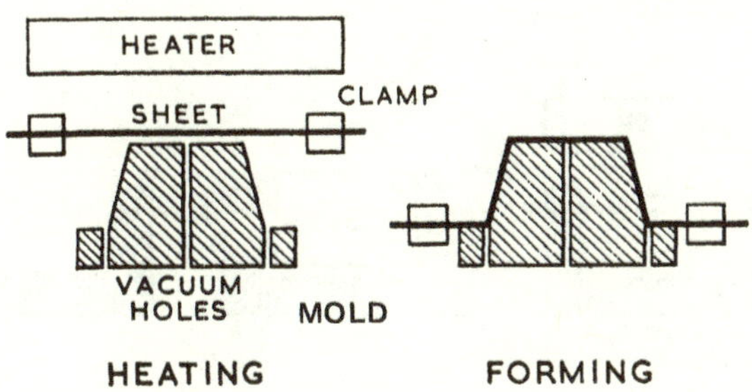

Fig. 28.2. Drape forming into male mold

Plug-assist Vacuum Forming

This method allows even deeper drawing, more even wall thickness and lower cooling cycles. Very close temperature control of the plug and the sheet material is required, however, and complex equipment is necessary (*see* Fig. 28.3). A heated 'air slip' around the plug is often employed to avoid marking the sheet and to give a more even stretch. The function of the plug is to transfer sufficient material towards the lower part of the mold cavity for the formation of, for example, the bottom of the container, and to distribute the remaining material evenly over the side of the molding.

The final distribution of the wall thickness is dependent on a number of variables, including plug material and dimensions, plug and sheet temperatures, and the depth and velocity of pre-stretching. Plug material and plug temperature should be considered toegther. Good results can be obtained, for example, by using a plug made from aluminum or some other heat conducting material, but the plug must then be heated within a sharply defined temperature range. If the temperature is too low, for example, then the high thermal conductivity becomes a disadvantage. The plug chills the sheet excessively and results in a container having a thick base and thin side walls. The advantage of plug materials with low thermal conductivity is that they may be covered and so stand up to more rugged operating conditions.

Another important variable in connection with the plug assist process is that of 'slip-in' of sheet material into the mold. The possibility for slip-in arises from the expansion of the heated sheet. If the sheet is clamped only at the perimeter of the clamping frame then surplus material from the surrounding scrap areas can be drawn into the mold. The slip-in is irregular, or uncontrolled, because of the shape of the scrap areas, particularly in multi-cavity molds. If, however, the clamping edge is directly adjacent to the side walls of the forming, then no slip-in is possible.

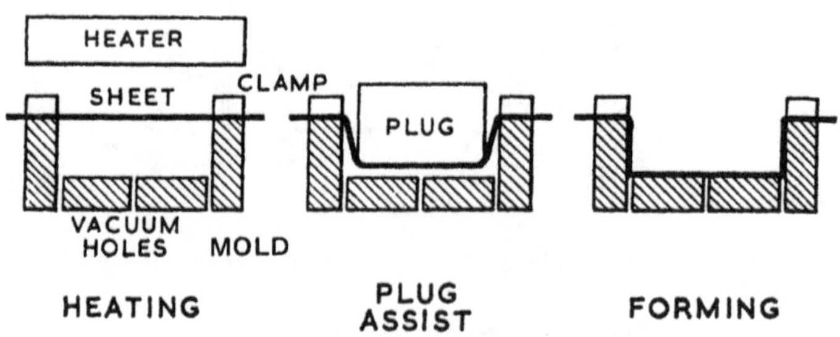

Fig. 28.3 Plug-assist vacuum forming

The third possibility is controlled slip-in. This exists where the clamping edge form an annulus round the mold aperture at a distance of about 3–5 mm. This amount of material is then available for the controlled slip-in of material. If the distance between individual cavities is increased then the annulus width can also be increased, giving more slip-in and a better wall thickness distribution.

Plug and Ring Vacuum Forming

This method uses only simple equipment and is especially suitable for shallow drawing of tough sheets (*see* Fig. 28.4).

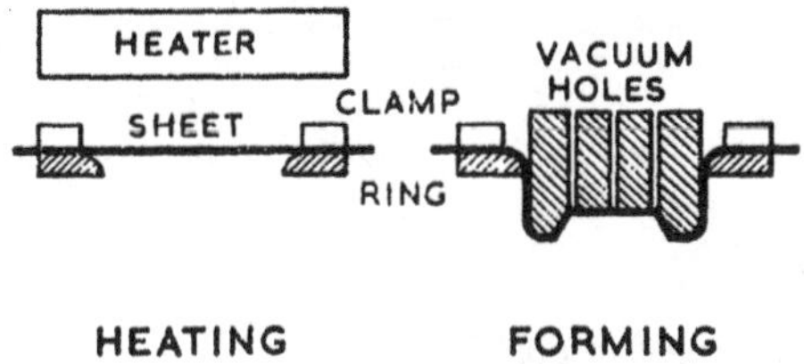

Fig. 28.4 Plug and ring vacuum forming

Snap-back Vacuum Forming

The heated sheet is drawn down into the box by applying a vacuum to the box. The mold is then lowered on to the sheet and the vacuum in the box is released, allowing the sheet to snap back and press tightly against the mold surface. If required, a vacuum may then be applied to the mold as shown in Fig. 28.5. The application of this second vacuum operation allows complex parts with recesses to be formed.

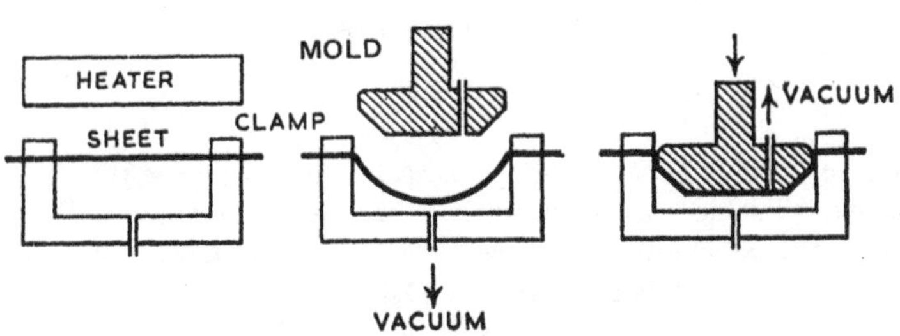

Fig. 28.5 Snap back vacuum forming

Pressure Snap-back Vacuum Forming

This is the reverse of the process described above. Deep drawing is possible but the equipment required is expensive (Fig. 28.6).

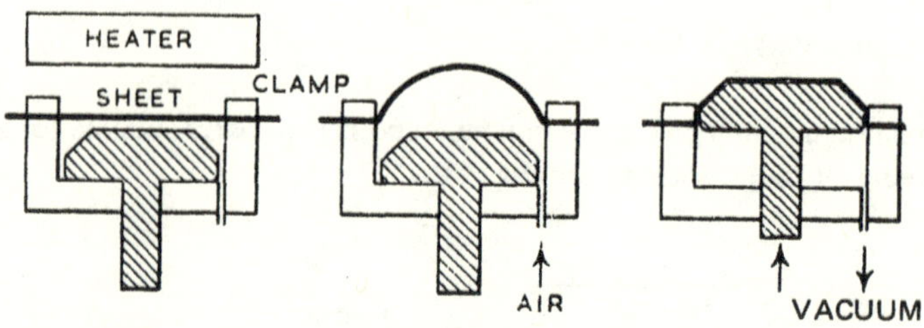

Fig. 28.6 Reverse process of Fig. 28.5 for pressure snap back forming

Blow-back Forming

This method is illustrated in Fig. 28.7, and is used for producing deep drawn parts from heavy, tough sheeting not normally formable by ordinary vacuum forming, since pressures greater than atmospheric can be used.

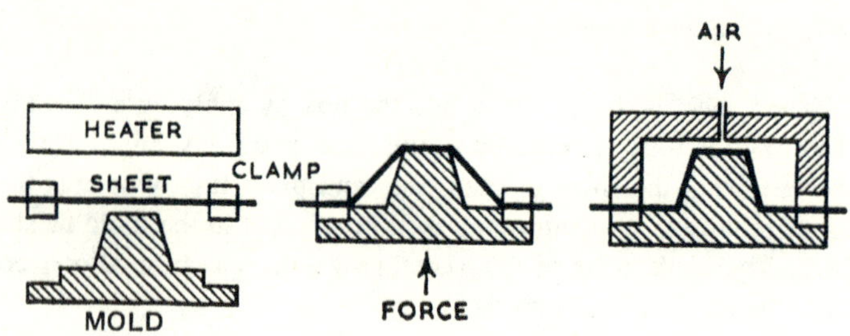

Fig. 28.7 Blow back molding

Bubble Assist

Yet another method is bubble or air assist. In this process the sheet is pre-stretched into a bubble by pumping air into the vacuum box prior to the mold table rising. The possibility of thinning at the sharper or higher parts of male molds in particular is again reduced. If too large a bubble is formed, there is the danger of surplus sheet material when forming. This can be avoided by fitting a photo-electric cell and scanning beam at a suitable height above the sheet so that the air is automatically cut off when the bubble breaks the beam. The air blowing feature can also be used for blowing the finished forming off the mold.

The above notes give some idea of the different techniques available but much work has been carried out to increase the usefulness of thermoforming methods by speeding up the forming process and associated processes such as trimming and, in the case of packages, filling and sealing.

One method of speeding up operations is to use multistation machines of the linear or rotary type. The sheet is mounted on a clamping frame which then moves from a heating station, on to a molding station and then to a trimming station.

In the case of small to medium size moldings, operations can be speeded up still further by the use of multicavity molds.

One of the major recent developments has been the increasing size of articles manufactured by thermo-forming techniques. Examples in the U.S.A. include small car bodies and a caravan body, the latter being formed from sheet 87 inches wide by 192 inches long (2210 × 4877 mm).

BLISTER PACKAGING

A modification of normal thermoforming techniques that is widely used is blister packaging. In a blister pack the product to be packed is supported on a card or plastics backing and then completely covered by a transparent, flanged blister formed from PVC, cellulose acetate, etc. This is glued or stapled to the backing card. The blister may be a simple shape, such as a circle or rectangle, or it may be specially shaped to fit the contours of the object being packaged. An example of the latter is the blister packaging of sponge flan cases. Blister packaging supplies an effective display pack that also gives protection against dust, grease, and mechanical damage. Fig 28.8 shows the stages in producing a typical blister pack.

The product manufacturer wishing to avail himself of the advantages of blister packs can buy-in ready-made blisters (especially if he only requires blisters of simple shape or requires large numbers of a more specialized shape). On the other hand, he can set up his own thermoforming equipment and so make himself independent of outside suppliers. There is a wide range of blister making equipment now available, from manual to fully automatic machines. The latter are only economic if large numbers are required.

Blisters are usually formed on multi-impression molds and can be separated either by slitting knives and a cross-cut guillotine, or by a roller press. If a

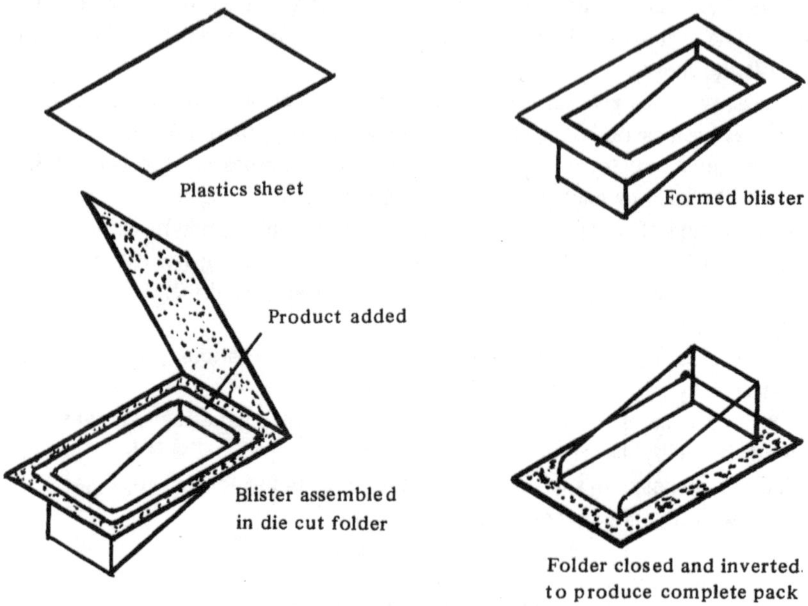

Fig. 28.8 Production stages of a typical blister pack

roller press is used then a separate cutting form is necessary for each mold configuration. With slitting knives and a guillotine it is possible to accommodate a wide range of blister sizes and numbers off. Roller presses can be in-line or can be separate units.

If the blisters are to be heat sealed to the cards (and this is the most common method of sealing) then the card is coated with a heat seal adhesive. Filling is carried out by placing a number of blisters in a jig (with the flange uppermost), the object or objects are placed inside and the cards placed on top. The heated press is brought down and the sealed blisters are removed from the jig. Sealing speeds are increased by the use of rotary tables so that operations can be made continuous. Various degrees of automation are possible, including automatic loading of the heat seal coated boards into the sealing area, and automatic pack discharge with the sealed packs being passed on for further handling to a work table. Loading of the blisters and the objects to be packaged is usually carried out manually, but automatic loading is possible for suitable products of simple shape. The scale of operations can range, therefore, from:

(a) The buying-in of ready formed blisters followed by hand placement of two or four blisters in a simple jig, hand placement of cards and sealing in a small single station blister sealer, to:
(b) Fully automatic operation, starting from a reel of PVC or some other suitable material. The sequence of operations would then be reel feeding of plastics sheet, forming of blisters, cutting out of blisters, transferring of blisters into sealing jigs, loading of blisters, card placement onto blisters, heat sealing of cards to blisters and transferring of completed packs from the end of the line. Speeds of 20 to 25 filled blisters per minute have been claimed for fully automatic lines so that production rates of around two million per annum per machine are possible.

SKIN PACKAGING

This interesting variant of vacuum forming utilizes thin film rather than sheet and the articles to be packaged act as their own molds. The stages in the process are as follows:

(1) The object (or objects) to be packaged are placed on a sheet of porous fiberboard and the assembly placed in the skin packaging equipment.
(2) Film is drawn from a reel and clamped in a frame above the loaded board.
(3) A cutter severs the film from the reel and a radiant heater moves over the film.
(4) The film is heated for a set period.
(5) The heater retracts and the hot film is drawn down over the board and contents.
(6) A vacuum is drawn through the porous board causing the film to be drawn round the articles to be packaged and into intimate contact with the board, forming a permanent bond.
(7) The assembly is moved away and the film is allowed to cool.

It should be noted that the use of a roller press enables the separated cards or blisters to be given rounded corners, whereas slitting and guillotining can only give square cut corners. Rounded corners may be preferred from the marketing point of view and there is also less likelihood of damage to corners if they are rounded.

The fiberboard may be porous enough for the process but it is often considered necessary to perforate it with a large number of micro-perforations. This may be because a denser, more rigid board is required, or because the board is coated to give a better bond with the film. A regular arrangement of perforations can help to guide the operator in placing objects in the right positions on the board. Another method of doing this is to print a regular grid on the board.

Films used for skin packaging should be flexible enough to provide a good contour wrap and have a high melt strength. The latter is essential if pinholing and puncturing at sharp corners are to be avoided. Low density polyethylene and ionomer films are among those that have proved suitable for skin packaging. PVC is often used for heavy objects, especially if they have sharp edges or corners.

Skin packaging can also be used for transit packaging. The skin packed assembly is left uncut and the cards are placed one on another inside an outer case. Suitable shock absorbing dividers are used if the packaged objects need extra protection. This type of packaging has been used for items such as dinner services.

As with blister packaging the equipment available for skin packaging ranges from manual to fully automatic. In fact there are many machines on the market that can carry out either blister forming or skin packaging. The mold in a blister vacuum forming machine is replaced by a drape table, and the same ancillary cutting equipment is used. Production rates for skin packed cards depend very much on the size of the large board and on the numbers of objects that it can accommodate. Cycle times for the skin packaging of an assembly of objects on the fiberboard vary from 1½ to 2 per minute, for a semi-automatic or automatic machine. With 20 objects packaged per cycle this would equate to 30–40 skin packed cards per minute, or three to four million packs per annum. Manual operation would, of course, give much lower figures.

Skin packaging is particularly useful for the packaging and display of multiple items where the presence or absence of the complete set of parts can be seen at a glance.

FORM/FILL/SEAL

The thermoforming process is well-adapted to form/fill/seal operations. In one type of machine, for example, blisters are formed to a profile roughly similar to the article to be packaged. Articles are fed continuously into the blisters and a reel of fiberboard is sealed over them. The individual packages are then trimmed and separated. This type of operation is used for the packaging of high sales volume items such as razor blades, reels of cotton, etc.

The other main area of thermoform/fill/seal is the packaging of a wide range of foodstuffs, including jam, cream, sauces, and cheese. This type of operation is shown schematically in Fig. 28.9. Two reel-fed plastic webs are used, the first being formed into a series of tray-like depressions by heating and drawing a vacuum through the base of appropriately shaped molds. The formed sheet is then indexed under a filling head and the filled compartments are lidded by sealing the second web of material on top. The web of filled and lidded containers is then cut and the individual packs separated. The web used for lidding is often preprinted. A thermoplastics coated paper and aluminum foil may be used for the lidding operation.

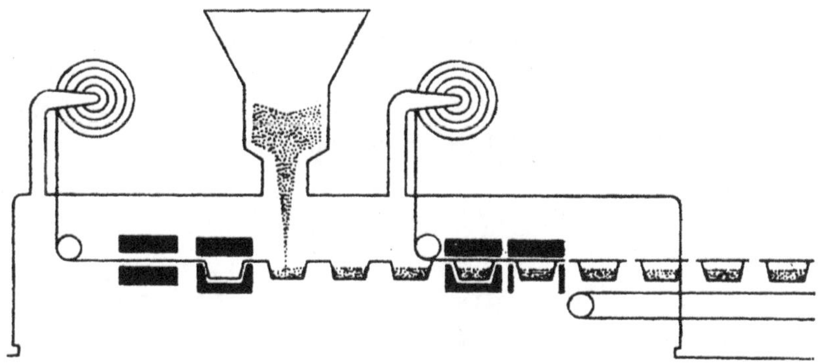

Fig. 28.9 Form/fill/seal thermoforming

A somewhat similar technique is used for the packaging of tablets, pills, or capsules into 'push-through' blister packs. A multicavity mold is used and the pills, etc., are fed automatically into the thermoformed cavities. The coated aluminum foil lidding web is then brought down over the filled plastics web and heat sealed to it. When required, the individual pills, etc., are dispensed by pushing them from the plastics side through the relatively brittle aluminum foil.

Aseptic packaging, as used for long life milk and cream, normally involves the use of hydrogen peroxide for sterilizing both the base and lidding materials. In one type of machine the base web passes through a hydrogen peroxide tank and then remains in a sterile, enclosed cabinet throughout its passage through the machine until it is lidded. The thermoforming is then carried out by means of compressed sterile air, which is filtered to microbiological standards, and plug assist. Filling is carried out in the sterile cabinet, followed by a pre-sealer which forms a seal on either side of the web, thus creatibg an enclosed tube between the base and the lid. A second heat sealing station seals the individual cups as they emerge from the sterile cabinet. The sterilization procedure for the foil/polyester laminate used for the lidding is similar to that adopted for the base. The complete sterilization procedure is fully automatic and is achieved in 30–40 minutes. The machine can cycle up to 30 cups per minute, the limiting factor being the drying time required to exclude all traces of hydrogen peroxide from the pots. Although aluminum foil laminate materials are used for aseptic packaging, gravure printed metallized polyester is widely used for non-aseptic packaging, particularly for preserves and sauces.

SUGGESTED FURTHER READING

Thermoforming Techniques, in J. Agranoff, ed. Moderm Plastics Encyclopedia, 1985/1986, McGraw Hill Inc., New York, 1985

29.

POWDER COATING

Plastic coatings are applied to metallic and non-metallic substrates for a variety of reasons including corrosion protection, electrical insulation, decoration, bonding to other substrates, and prevention of damage to delicate components. Two basic applications originally developed were dipping and spraying either from solution, dispersion or plastisol, but over recent years we have seen the emergence of dry powder techniques. The powder coating process was developed to permit the application of plastic coatings without the necessity to dissolve the polymer in a liquid carrier. This process is akin to the technology of hot melt adhesives, but instead of applying the coating in molten form a powder is used.

The conception of this technology can be traced back to the early 1940's when the precipitation of low-density polythene to produce fine powder was developed. The main substrates which have utilized powder coating technology are metals and textiles. The development of powder coating of metals was essentially for ecological reasons, but for textiles the development arose to fulfill a requirement which was otherwise not achievable with more traditional materials.

The early development of the low-density polythene powder by chemical precipitation provided a base material suitable for producing coatings by preheating an article, and immersing it in a bath of static powder. This coating method was very basic, and little interest was aroused and it was not until some ten years later, in 1953 that the fluidized bed technique was developed in Germany (Knapsack Griesheim). The introduction of this technique was quickly followed by the development of other thermoplastic powders such as nylon and PVC. In the 1950's the thermosetting type of powders were introduced in the USA, and another stone was laid in the foundation of the development of powder coating technology. The basic chemical families developed in this country were the epoxide-type resin, polyester hybrid and polyurethane compounds, and during this period the electrostatic coating technique was developed for processing these types of resinous compounded powders. Even today one of the big drawbacks is the relatively long cure schedules required to achieve acceptable products.

Clearly the early developments in this technology were based on thermoplastics which utilized the fluidized bed techniques, and later the thermosetting powders became more attractive with the development of electrostatic technology. Advances in grinding technology such as freeze and cryogenic processes for thermoplastics led to their wider uses. The super-polyamide, nylon 11, is a difficult material to grind, and initially was cryogenically ground using liquid nitrogen with temperatures below $-130\,^{\circ}$F. This also applies to nylon 12 and because of high basic polymer price, and cost of grinding, the development of markets for this family of polymers has been somewhat restricted. Another limiting factor has been that particle size requirements for metal powder coating, which normally lies between 0–80 and 80–200 microns, makes the economics even less attractive. However, today there is a process for precipitating polyamides into fine powder which has stimulated renewed interest.

Coating of textiles with adhesive powders provides finished pre-coated and heat sealable products which are termed by the garment industry as fusible interlinings. Over recent years these coated fabrics have been laminated to other substrates during the coating process to upgrade or reinforce the total system. This is replacing the more conventional aqueous, solvent and flame laminating techniques as used for wovens, although with the upsurge in non-wovens, we are experiencing a return to paste printing.

The history of fusible interlinigs dates back to the early 1950's, when a company in the UK named Staflex Limited, then still small, developed a coated textile which could be heat fused into garments. These interlinings exert a dimensional control or can even be used to impart controlled stiffness/bulk to a garment part. The early product developed by this company was continuously coated i.e., an all over coating, but this gave a disadvantage to the resultant laminate in that it imparted an undesirable, uncontrolled stiffness. The next development came in 1955 when the random scattering of predetermined size particles of low-density polythene powder was used by this company for coating fabrics. The result was a laminate which still had limitiations in that the polymer used was not completely resistant to the solvents used in the dry cleaning operation. Such a limitation resulted in the interlining only being suitable for use in small area applications. The next breakthrough came in 1964 when once more, through the innovation of Staflex, a plastisol product was introduced with equidistant dots based on plasticized vinyl chloride/acetate copolymer. Copolyamides as described in a previous chapter were made available in 1967, followed by the terpolyamides some two years later, with the introduction in 1969 of the nylon 6/66/12 types in Germany and Switzerland. Since that time little basic process or polymer development has occurred apart from the powder calendar printing, and introduction of thermoplastic polyesters. There have been considerable improvements made in the printing of powders and aqueous pastes. It is very feasible that one day thermosetting polymers, analogous to those used for metal coating, will be realized for textile coating, and the garment producer may no longer require fusible interlinings but will be able to coat his own garment lining immediately prior to attaching it into a particular area of his garment. In actual fact a company in the United Kingdom has been test marketing a urethane type cross-linkable powder which is based upon some novel reactive chemistry.

In more recent times we have seen a move towards the dry laminating of many substrates by film and powder. In particular it has particular advantage over flame lamination which involves the burning-off of a certain amount of foam to use as an adhesive. The burn-off products are toxic and include tolylene di-isocyanate, carbon monoxide and nitrous oxide. Using powder means you have the choice of using the cheaper polyether foam rather than polyester and hence thinner sections of foam up to 2 mil less are possible.

Processing of Powder

The technologies used to produce powders for metals and textiles are very similar, although the amount of compounding for metal coatings is considerably more. A rough demarcation is that metal coatings are in general ground at ambient conditions apart from nylons, and textile coatings are cryogenically ground apart from polyolefins.

Compounding

Before this process is carried out, the ingredients are generally premixed in order to ensure good homogeneity. It has been found advisable to premix particles of comparable size wherever possible. The main types of premixing machines are tumbler, horizontal, conical, high speed blenders, and air jet mixers.

Once all the components are dry blended and homogeneous, they are ready for hot melt compounding. The components for powders in metal coating technology include pigments, extenders, hardeners, flow-aids, and flexibilizers (plasticizers). The two main factors which govern the temperature of compounding are the viscosity of the polymer melt and the reaction temperature of the hardener.

The main compounding techniques used for this particular type of powder technology are all well established procedures in the polymer industry. They are: (i) batch process as typified by the Z-blade mixer, (ii) semicontinuous process which uses heated rolls or Banbury mixers, and (iii) extrusion which is a fully continuous process and can be of the twin screw type.

These compounding processes are not necessary for the production of powders for use in textile coating, although some compounding is required to make certain basic polymers suitable for use under specific conditions. This compounding is largely carried out to incorporate the plasticizer into the polymer matrix, and both cellulose acetate and copolyamide powders have their melting range depressed by incorporating solid plasticizers such as ortho-para toluene sulphonamide, either by the more preferable extrusion technique or simply dry blending.

Grinding

As mentioned in the previous section, it is necessary to compound polymers for use with metals, and obviously the resultant compound must first be granulated to ensure it is in an acceptable condition to be fed into the grinding equipment. The granulating equipment can be a Kibbler or a machine operating on the Masson cutter principle, and it is normal to feed the extruded strands through a cold water bath before cutting.

The basic principle of grinding or milling solids into discrete powders is a process of attrition in which the size reduction is carried out by impacting the solid lumps of resin against a grinding surface, or by reducing the size of the particles simply by impacting themselves with each other, or simply using a knife cutting technique. The impacting can be achieved by means of hammers (Hammer Mills) or by pin discs (Pin Mills). The resins associated with thermosetting powders have low molecular weights and as such, in general are brittle and friable, thus making them relatively easy to grind. However, they must be controlled closely in order to avoid overheating which can lead to premature curing. The thermoplastics have considerably higher molecular weights which can result in general in them being tough and resilient. They are therefore ground by either freeze or cyrogenic methods, whereas precipitation methods are used to produce spherical low particle size powders. Polyolefins in general are relatively easy to grind at ambient temperatures, yet with cryogenics they require considerable amounts of liquid nitrogen before they are at a condition where they can be ground. The technique for PE is normally a knife cutting action which however does produce tails.

The temperatures at which polymers can be ground efficiently vary widely as can be seen from Table 29.1

Table 29.1

GRINDING RANGE FOR RESINS

Material	Gringing Range °F
Cellulose acetate butyrate	0 to −50
Epoxy	+110 to +50
Nylon	0 to −130
Polyethylene	+14 to −180
Polypropylene	+10 to −80
PVC	20 to −50

Obviously the temperature of grinding, depends to a large extent upon the size of particle required. For example, low-density polythene can be ground at ambient temperatures if a mesh size of 35 (500 microns) to 50 (300 microns) is required. To achieve smaller particles it is often necessary to use freezing conditions or some form of chemical precipitation techniques.

There are three main types of refrigerants used with thermoplastic: solid carbon dioxide (dry ice), liquid carbon dioxide, and liquid nitrogen (LIN). The conditions of refrigeration available from these materials are given in Table 29.2.

Table 29.2

CONDITIONS FOR FREEZING THERMOPLASTICS

Refrigerant	Initial condition	Heat absorption (1)
Dry Ice	Solid at −190°F	268
Liquid CO_2	Liquid at 0°F (300 p.s.i.g.)	140
LIN	Liquid at −320°F	166

(1) From initial condition to final gas temperatures of 0°F.

Dry ice is relatively inefficient and requires large amounts of solid carbon dioxide. Even so, it is difficult to maintain thermostatic control, and the large amounts of refrigerants can cause irritation to the eyes of the operators as well as affecting their breathing. Liquid nitrogen is the most favored method because like CO_2 it is essentially nontoxic and may be added in relatively small quantities. The resultant system responds readily to thermostatic control, and at the same time provides an inert gas atmosphere within the grinding chamber, thus reducing the hazards of explosion. The liquid nitrogen is introduced through the spray header (see Fig. 29.1) directly into the feed hopper of the mill. The LIN is stored under normal atmospheric pressure and does not go through a solid state in changing to a gas and so can be metered into the header. There is a greater driving force with the LIN at −320 °F than the dry ice at −109 °F. Embrittlement of the polymer particles occurs very quickly and this results in the particle size distribution being more uniform. However, LIN is relatively expensive and the cost of the refrigerant can add up to 10 percent to the total manufacturing cost. The ratio of LIN to polymer varies, and can be as low as 1.1 up to requiring more than 6 parts of LIN to polymer.

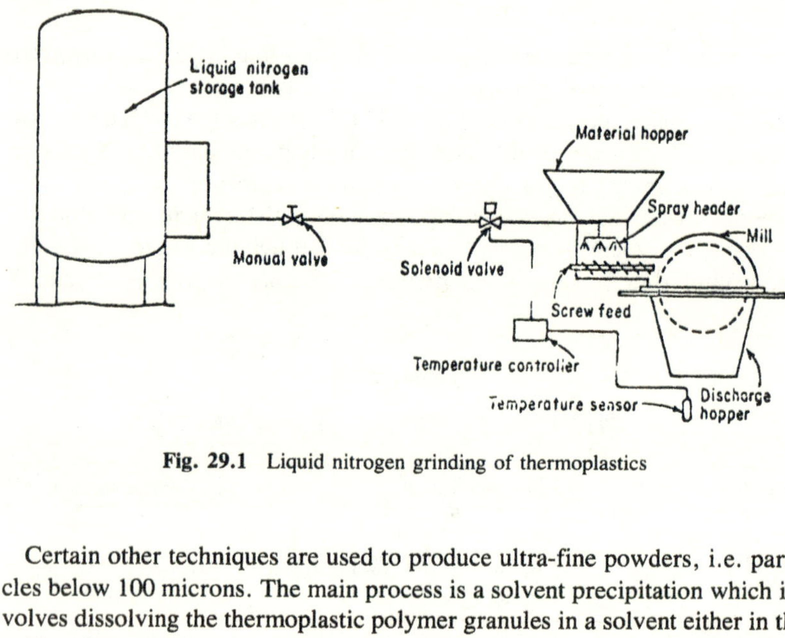

Fig. 29.1 Liquid nitrogen grinding of thermoplastics

Certain other techniques are used to produce ultra-fine powders, i.e. particles below 100 microns. The main process is a solvent precipitation which involves dissolving the thermoplastic polymer granules in a solvent either in the cold or hot, and then by cooling with stirring, and by adding a non-solvent, a powder is precipitated out. This produces a particle with a relatively spherical shape and can be as small as 10 microns. It has one big disadvantage as a process, in that it is a more expensive method than the cryogenic grinding. This technique was first introduced in Europe by Technoplast Spindler (now Plast Labor, Switzerland) and in the USA by USI Chemicals. Both these companies utilize such processes on a commercial scale, and powders are available that are spherical in shape, and are below 100 microns, and can be as small

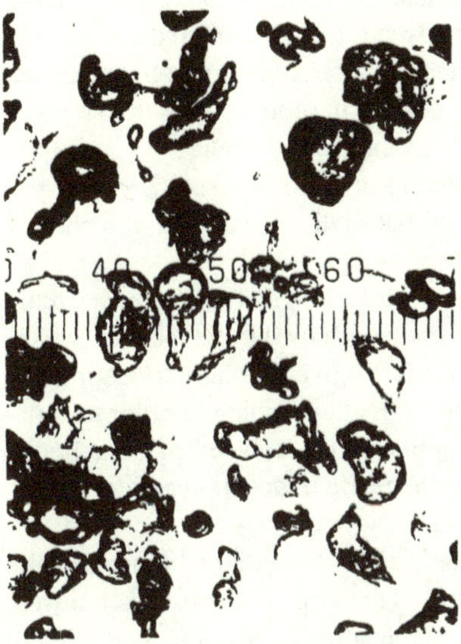

Fig. 29.2 Mechanical grinding

Fig. 29.3. Chemical precipitation

as 20 microns. In Figs. 29.2 and 29.3 it is possible to see the difference of particle shape for powders obtained by mechanical and chemical techniques.

COATING PROCESSES

1. Metals

Before any coating process can be carried out it is essential to remove all forms of oxidation. Descaling is sometimes first necessary and this is usually carried out with mechanical abrasion and/or acid pickling. The acid pickling can be effected using a dilute acid solution and then removing by washing with water. The mechanical method can be carried out by sand blasting, wire brushing, or grinding.

The next step is a degreasing operation which is essential if good adhesion is required between the coating and the substrate. A well established method is a chemical wash with caustic soda, phosphate soaps or synthetic detergents, etc. Solvents can also be used in baths or vapor phase systems and for this purpose trichlorethylene, toluene, acetone, and isopropyl alcohol have all been found to be satisfactory.

The final state is the priming of the cleaned article, and adhesive primers are used that provide satisfactory bonds between the substrate and the powder coating. The primers in general require heating after applying to be effective, and since it is necessary to heat the metal objects in order to pick up the powder for coating this operation also serves to cure the primer.

(a) *Dipping*. This process was developed to coat metals from a liquid system and is mentioned briefly in Chapter 28 but is discussed a little further in this chapter.

Dipping consists in heating the article to be coated in an oven to such a temperature, that when it is dipped into a thermoplastic powder, the powder which adheres to the article melts, flows, and fuses to give a coherent coating. In practice the coated article is replaced in the oven after dipping in order to complete the fusing process. PVC may be applied as a powder or as a liquid (plastisol), but other polymers such as polythene, epoxides, polyamides, polyesters, etc., utilize only powder processes.

When a PVC plastisol is to be used, the article to be dipped must contain sufficient heat to gel the PVC on the metal. The surplus plastisol is allowed to drain off and the coated object is cured by reheating in an oven. This reheating process is necessary to dry and to obtain the maximum physical properties of the polymer. In the case of plastisol coating it is generally referred to as a curing process, and with powders it is sintering. The thickness of the coating is controlled by varying the viscosity of the plasticol and/or the time and temperature of the preheating process.

Dipping by the plastisol technique can be a distinct disadvantage and for this reason many processors prefer to use powders. However, dip coating using powdered polymers is not a straightforward matter of dipping a hot metal article into a box of powder. After a few articles are dipped into a powder reservoir, a number of holes would appear in the powder and subsequent articles would pick up little, or no powder. This limitation is overcome by the fluidized bed technique which is described next and illustrated in Fig. 29.4.

This process is a technique which enables fine powders to be used as if they were liquids. Essentially, as can be seen from the illustration, the equipment consists of two chambers which are separated by a porous ceramic plate. The upper chamber contains the powder and the lower is a plenum chamber fed by an air supply. If sufficient pressure is provided by the air supply each particle becomes bouyant in the ascending air stream, and in effect the powder particle bubbles up and falls back in different places. The overall effect in the fluidized bed is to make it appear like simmering water. As in the other process, the resultant coated product must be reheated in order to sinter the coating and render it homogeneous.

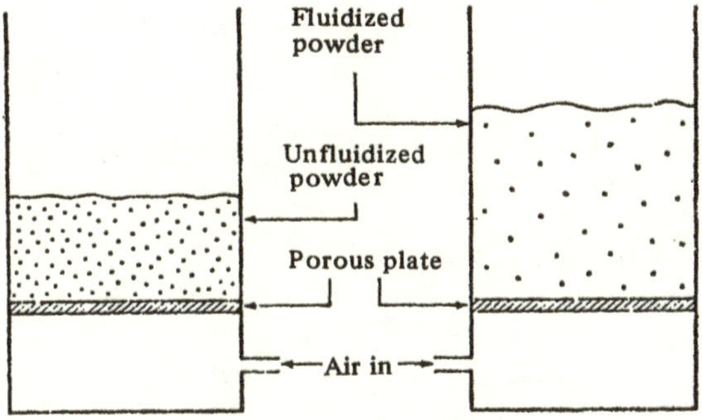

Fig. 29.4 Fluidized bed technique

The particle size of the powder should be in the region 50–250 microns (0.002–0.010 in.) and in shape should be as spherical as possible.

(b) *Electrostatic Powder Spray Coating.* Powder coatings may also be applied by electrostatic spraying or in an electrostatic fluidized bed. This technique does not require the prior preheating of the metal object since the powder clings to the part to be coated by electrostatic attraction. The powder is therefore fed by air through a gun (*see* Fig. 29.5), and at this stage a high voltage, low amperage charge is applied to the nozzle of the powder spray gun. The powder is sprayed at a low velocity and acquires a strong negative charge, and this in turn is attracted to the object to be coated which is earthed. When the powder builds up to a certain thickness, usually about 8 mils, a dielectric barrier is created and further powder is rejected. The particle size of the powder is normally between 10–80 microns (0.0004–0.0032 in.).

It is clear that in electrostatic powder coating techniques there are four processes involved, each of which has an influence on the resultant products. These are: powder charging, transfer of powder from charging region to object, adhesion, and the fusion of the dry powder layer.

Details are given in Table 29.3 which lists details as to the faults occurring with this technique in production, and suggested remedies to overcome such defects.

Mention must also be made here of the electrostatic fluidized bed process which is a combination of the electrostatic spray technique and the more conventional fluidized bed process. This is also referred to as a cloud chamber technique, and the apparatus can consist of a batch of electrostatic elements on the bottom of a shallow bed with a number of auxiliary air jets. The small amount of powder is fludiized by the normal method, and the element imparts

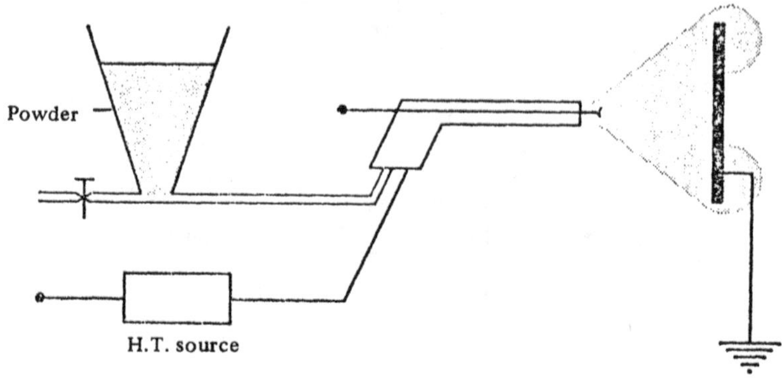

Fig. 29.5 Electrostatic powder coating

a negative charge to the powder with the result that a cloud of charged powder is formed. Any earthed object placed in this cloud is attracted by a layer of electrostatically charged particles rather than the fluidized ones. The big advantage of this process is that it requires only small amounts of powder, but offset against this is the fact that the density of the cloud so formed varies considerably from the top to the bottom of the chamber.

(c) *Flame Spraying*. This method of coating is especially useful for articles that are too large for dip coating. It consists of blowing out finely divided polymer powders through a specially designed burner nozzle, using compressed air as the propelling force. The nozzle is usually flame-heated by means of acetylene or some similar gas, or else it can be electrically heated. The molten polymer particles impinge on the article to be coated where they fuse together and cool, and subsequent sintering is unnecessary. It is difficult to obtain a uniform and homogeneous coating, and a great deal depends on the skill of the operator. One other problem is that at the temperatures necessary to melt the polymers used, degradation is likely to occur.

Before leaving this section mention must be made of another technique of spray coating which is based on a dispersion of the polymer either in aqueous or solvent media. The polymer to be applied is dispersed in an aqueous or solvent base, and then sprayed in a manner similar to that of conventional paint spraying. Sintering is necessary and it is usual to apply one or two subsequent coats with intermediate sintering in order to build up a reasonable coating thickness. This technique of coating is particularly useful with polymers such as PTFE and PTFCE which can best be applied from dispersion, although recently we have seen the introduction of these polymers in fine powder.

Table 29.3

TROUBLESHOOTING FOR ELECTROSTATIC SPRAY COATING
(information supplied by Rilsan Corp. U.S.A.)

Nature of defect	*Origin*	*Remedy*
Bubbles in coating	Any excessive deposit of powder in one spot creating poor air release.	Even out spray pattern or output of gun.
	Overheating.	Reduce fusion temperature.
	Presence of significant water either in the powder or in the compressed air.	Vacuum dry powder. Correct air dehumidifier.
	Release of gas from the object as porous casting.	Preheat to a temperature above fusion to remove occluded gas prior to coating.
	Poor drying of primer.	Assure that all solvent is removed from primer coat prior to fusion.
Yellowing	Overheating of coating during fusion.	Reduce fusion temperature or time. Sometimes film thickness is too thin, allowing primer to show through.
Uneven thickness or over thickness	Wet powder.	Vacuum dry powder and correct air supply.
	Poor spray technique.	Even out spray pattern and projection dwell times.
Orange peel	Powder underfused causing poor flow-out.	Remelt using an additional 20°F fusion temperature.
	Distance between the nozzle and object too short.	Regulate distance from the gun to nozzle and insure uniform and fine powder output.
	Primer too concentrated.	Cut back to normal viscosity and assure smooth and even coat.
Pinholes	Shortage of powder, low coverage.	Increase coverage.
	Poor substrate wetting.	Insure proper removal of grease and/or oil.
Craters	Contamination by surface-active agents.	Reduce fusion temperature to increase melt viscosity. Remove source of contamination through clean up.

Table 29.3 *(cont'd)*

Nature of defect	*Origin*	*Remedy*
Poor adhesion	Object not properly degreased and primed.	Assure adequate surface preparation.
Edge-pull	Poor wetting.	Assure adequate surface preparation and priming.
	Fusion temperature too high.	Reduce fusion temperature.

2. Textiles

There are numerous basic methods of applying a polymer onto a base fabric to produce what are termed fusible interlinings. These include: (i) sintering (random scatter or spray), (ii) paste printing, (iii) powder printing (engraved roll or rotary screen technique), (iv) preformed plastic net, (v) hot melt printing, (vi) extrusion, and (vii) spraying from solution or aqueous based systems.

Only the first three techniques are used widely, and these are dealt with in detail although brief mention is made of certain of the other techniques.

(a) *Sintering.* This is a method of applying powders by sprinkle coating in a random pattern using vibratory screens or brush rollers. The range of powder particles is from 100–800 microns. The finer powders are used mainly for coating fabrics with a fairly tight construction (close weave), e.g. nonwoven, cotton shirt, and blouse fabrics. In practice it would be normal to use powders in the range of 100–300 microns. Heavier interlinings such as canvasses, made originally from horse hair, etc., used in the construction of men's jackets and top coats would require 300–800 microns. The disadvantage of this method is the irregular application of particles provided by the random technique. Another disadvantage is that if there is a considerable proportion of small particles they tend to fall into the interstices of the interlining, providing no contribution to adhesion, but at the same time giving a disadvantage of a stiff hand. A schematic diagram of this technique is given in Fig. 29.6.

On sprinkle coating the powder onto the substrate, it is then essential to heat the powder particles in order that they may become molten and thereby adhere to the fabric interlining. The coated fabric can then be calendered to modify the hand of the fabric, or can be simply passed over chill rolls for subsequent rolling up and inspection.

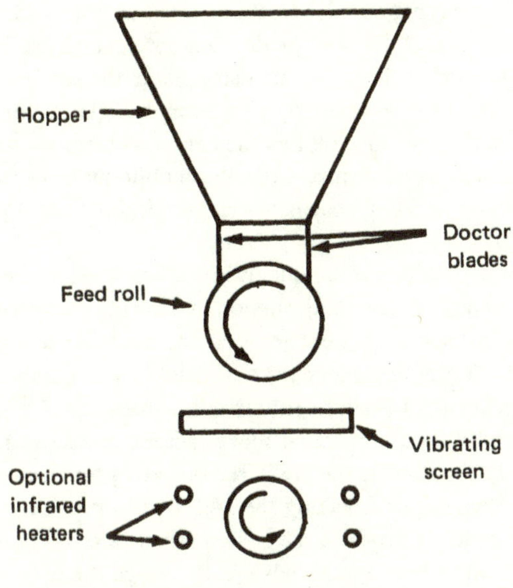

Fig. 29.6 Sinter coatingn equipment

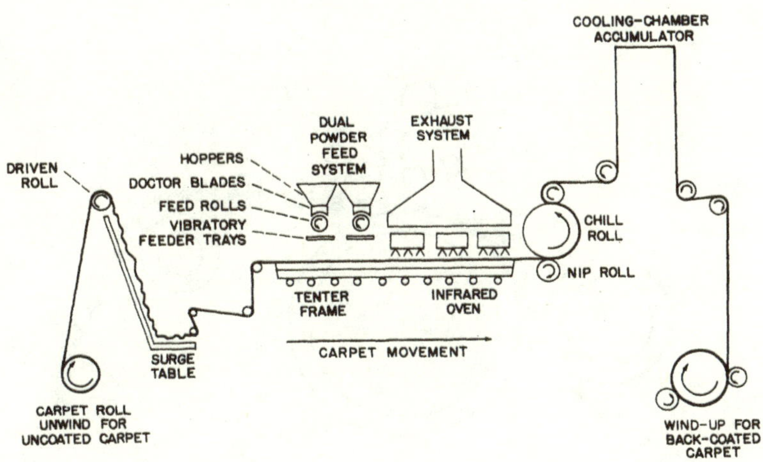

Fig. 29.7 Typical carpet coating line

This method of applying powder to a substrate has been widely used to coat other substrates such as wood and metal. One interesting use for this coating process has been found in the carpet industry where the sinter coating of polyethylene powder has been used to apply backing material or for tuft retention. Coatings of this nature have meant that the carpet so formed could be molded, and thus have found extensive use in the automobile industry (see Fig. 29.7). Where more defined molding shapes are required, the tendency is to move to polyester powders.

(b) *Paste Printing*. In this technique, powder below 80 microns is first converted into an aqueous or solvent dispersion. A suitable thickener such as polyvinyl alcohol and a dispersing agent are used and a solids content of 50 percent is recommended in order to achieve the best results. The pastes so formed may be printed by means of a gravure roller, silk screen, or if a regular pattern of dots is required then a perforated nickel screen is necessary. For fusible interlinings, the last method is the preferred one and provides a process which has the distinct advantage of placing the adhesive in precisely located spots. After the screen printing process, the coated fabric is passed under heaters in order to drive off the volatile materials. At this stage the dots, while they are loosely attached to the fabric, are little friable dots and must be further heated in order to coalesce them, and render them homogeneous and better keyed.

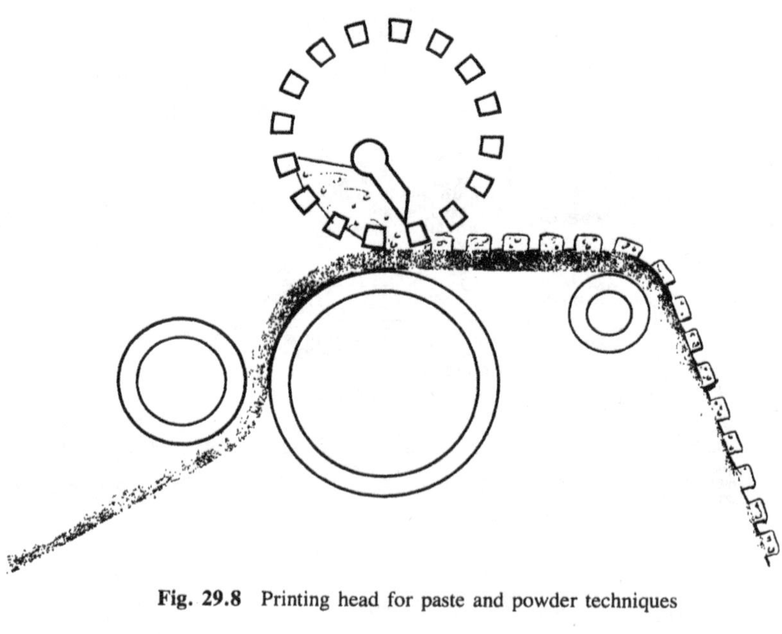

Fig. 29.8 Printing head for paste and powder techniques

The fabric can then either be calendered or simply passed over cooling rolls in order to solidify the molten dots (see Fig. 29.8).

The geometry of the pattern of the dots available is normally 9 up to 30 mesh. For example, a 14 mesh screen infers that there are 13 - 14 dots per inch measured on the diagonal of a square one inch in measurement, as indicated below.

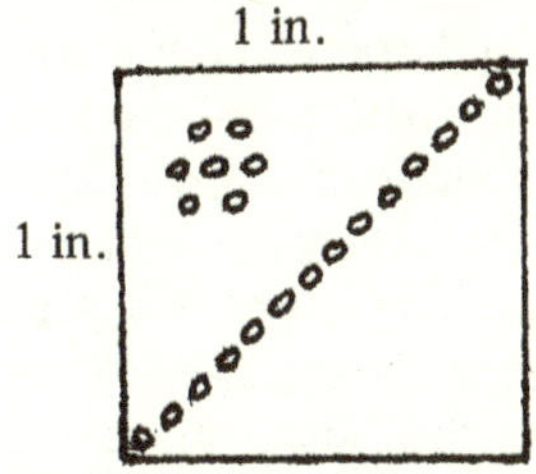

(c) *Powder Printing.* There are two methods of printing directly from powder; one is based on an engraved roll and the other uses virtually the same equipment as for paste except dry powder is used. The former technique of using the engraved roll is more widely used by the textile industry and combines the advantages of the high quality of paste printing with the relative ease of handling, which has become associated with the sintering technique previously described.

In this process, thermoplastic powders are filled into the recesses of a rotating engraved roll by means of a doctor knife system (see Fig., 29.9).

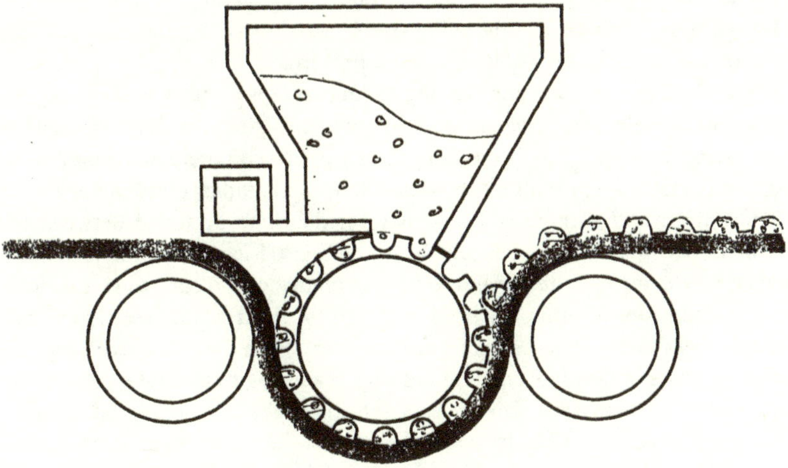

Fig. 29.9 Powder dot printing by means of engraved roll

The engravings are generally in the range of 9 up to 30 mesh and the application of the doctor blade ensures that powder collects only in the depression. The engraved coating roll is heated to the point where the powder in the depressions is just beginning to agglomerate. At this stage the powder is evacuated from the depressions by transferring it to the heated moving web. The hot roll heats the moving web at the widest possible angle of contact to temperatures well above the melting point of the powders. It is obvious that the powder dot must be pulled out at the right moment, and therefore the machine operators must have a certain amount of skills. If the powder flows too freely it trickles out of the gravure holes before touching the web, but if it is agglomerated too much it simply clogs up the printing roller. Once the powder has agglomerated, the hot web is pressed against the engraved roll by means of heating rolls. The agglomerated powder dots then adhere to the web and are then pulled out of the depressions in the form of loosely sintered particles. The coated web is then passed under a sintering zone where the dots melt and all the separate particles fuse together to form nicely rounded spheres which are sitting proudly on the surface of the web. The coated web is then calendered, if required, or simply cooled ready for batching.

The powder size for this process, as well as the powder technique using a rotary system, is either 0–200 microns or 50–200 microns. It is apparent that the particle size distribution not only plays an important part in determining the ability of the powder dot to be pulled from the engraved roll, but it is very important for controlling the amount of adhesive which is applied to the web. In practice, during the coating operation, it is possible to add additional fine particles, say 0–80 microns which can modify the coating weight.

For garment fusibles, an engraving size of 9–17 mesh is generally used but for shirt collars, 25 and 28/30 mesh are preferred.

In certain areas, the powder printing technique uses a rotary screen, largely because of the relative cheapness of the equipment. However, limitations have been established with this process in that it is not commercially possible to produce coated fabrics with mesh sizes above 17. Another disadvantage with this technique is that there is a tendency for debris to be found between individual dots. Machines are available which can be used both for paste and powder all on one machine. Reports have been made of recent work in which the screen powder printing has been modified by electrostatic techniques, but nothing of commercial acceptance has been seen. With the ever increasing use and acceptance of non-wovens, the industry is turning more to the paste technique where it is possible to coat heat sensitive fabrics to better advantage.

(d) *Preformed Net.* This net was developed largely as a result of a joint effort by Staflex Limited and Smith & Nephew Limited in the U.K. Later Hercules marketed the product in the USA under the trade name of Delnet. Essentially the process consists of embossing high-density polythene film and stretching the film in two directions, which leaves the structure of bosses connected by filaments. This net, in turn, is laminated to a web, and on heating the coated web the filaments shrink back into the bosses. This leaves a web coated with equidistant spaced dots which are thermofusible. The big disadvantage with this process is that there is a limiting factor as to which polymers are capable of being embossed and stretched. Only high-density polyethylene with very specific physical characteristics has proved suitable, and for mono structure net there are reports of polyamides and other polymers coextruded with the HDPE.

A table listing the processes available, particle size of powder, polymers most widely used, and manufacturers of coating machines is given in Table 29.4.

Miscellaneous Coating Techniques

Extrusion coating of fusibles was one of the original processes used to produce fusible interlinings, but it suffers from the severe limitation in that it provides an overall coating which makes the hand very stiff and boardy. However, low-density polyethylene film is still used today to laminate to certain cotton basic interlinings in order to produce fusibles for shirts.

The coating of fabrics by spraying the adhesives from solutions was used in the early development of these materials, but suffers from severe limitations, especially from the ecological standpoint. The early polyamides were alcohol soluble and were coated by a spray technique, and added to the disadvantages mentioned previously, the resultant coating had poor wash fastness. These polymers were quickly replaced by the terpolyamides.

In recent years considerable interest has been given to the possibility of using polymer granules rather than powders. Obviously to avoid a grinding cost of 10–20 percent would be a big saving in the cost of the final fusible interlining. Melting of the polymer granules, and directly printing the molten adhesive onto the fabric would not only save the cost of grinding, but theoretically it could lead to higher speeds in the coating operation. Unfortunately, processes which have been demonstrated so far, suffer from two main disadvantages: (i) poor print definition with smearing and (ii) excessive polymer degradation. These disadvantages are pertinent to the polyamides and polyesters which have to be cryogenically ground, and would realize the biggest cost saving if they could be hot melt printed. There are some companies in the USA who are coating polyamides by the hot melt process. Polythene and ethylene-vinyl acetate do not present such problems in printing, but interest in their use for this technique is not so great, presumably because of the relatively low cost of grinding.

Table 29.4

POWDER REQUIREMENTS FOR FUSIBLE COATING PROCESSES

Process	*Powder size Microns*	*Polymers*	*Manufacturers*
(1) Sprinkle (also termed scatter, spray sintering)	100–800	Polythene Polyamides Polyesters E.V.A.	Caratsch AG Saladin AG Eugen Knobel AG Schaetti & Co. Zappa
(2) Paste (a) aqueous (b) Plastisol	 0–30 0–80	 Polythene Polyamides PVC	Caratsch AG Saladin AG Eugen Knobel AG Schaetti & Co. Zappa Zimmer Stork Brabant Schaetti
(3) Powder (a) Calender Intaglio (b) Screen Printing (c) Electrostatic	 0–200 50–200 0–200 0–200	 as in (1) as in (1) as in (1)	 Caratsch Saladin Eugen Knobel AG Schaetti & Co. No commercial suppliers–under development.
(4) Preformed Net		High-density Polythene	Patented process Staflex Ltd. Smith & Nephew Hercules Corp.

Principal Polymers for Powder Technology

(a) *Metals.* The main materials used for metal coating include low and high-density polythene, PVC, nylon, epoxy, polyester, PTFE, and PTFCE.

The main systems for the thermosetting powders are:

1) Epoxies (epoxy resins + specific hardeners).
2) Epoxy-polyesters or hybrids (epoxy resins + carboxyl polyester resins).
3) Polyesters or pure polyesters (carboxyl-polyester resins +triglycidyl isocyanates/TGIC/hardener).

(4) Polyurethanes (hydroxyl-polyester resins + cycloaromatic or cycloaliphatic isocyanate hardeners).

These are all well covered in the relevant chapters so no further elaboration is given. However, for interest two formulations are provided as a guide as to what is required in order to produce a satisfactory powder coating compound.

A typical decorative compound would be:

Material	Parts by Weight
Epikote 1004 (1)	6.0
Titanium dioxide	27.5
Cadmium red	3.0
Polyvinyl butyral (2) (Butvar D510)	5.0
Aerosil 2941/380 (3)	0.5
Dicyandiamide	3.0

A typical polyester powder coating compound would be:

Material	Parts by Weight
Crylcoat 311 (4)	100.0
Araldite GT 7004 (5)	100.0
Modaflow	1.0
Titanium dioxide	200.0
Coloring agent	Trace
Irganox 1010 (5) (antioxidant)	0.10

Key:

(1)	Epikote 1004	Shell Chemical Company UK
(2)	Butvar D510	Monsanto Chemicals
(3)	Aerosil	DeGussa AG
(4)	Crylcoat 311	UCB, Belgium
(5)	Araldite GT 7004	
	Irganox 1010	CIBA Geigy

(b) *Textiles*. The polymers used for fusible interlinings at present are all thermoplastics, and include low and high-density polythene, plasticized PVC, copolyamides, copolyesters, plasticized cellulose acetate, ethylene-vinyl acetate copolymers and hydrolyzed ethylene-vinyl acetate. Like the metal coatings polymers, these are all discussed in greater detail, but Table 29.5 lists some of the main characteristics of some of the fusible polymer systems.

Uses for Powder Coated Articles

Let us first look at the requirements for industrial and indoor coatings.

a) Excellent decorative aspect, perfect flow-out, attractive color or special surface effects like matt, satin, textures or wrinkle appearances.
b) Good corrosion resistance which highly depends on the surface treatment.

In Europe much of the use for powder coating has been in the field of decorative finishes. It was soon established that single coatings of powder could provide tough decorative finishes without the hazards associated with solvent based systems. One of the major fields for powder coating is in the use of coating garden furniture. Powder coated furniture has many attractions and the fact that it provides a very smooth surface, which is both dirt free and detergent resistant, makes such a method eminently suitable for this particular application. Markets are also growing steadily in the appliance, building, and electrical industries.

The weather resistant decorative coatings are mainly used in outdoor architecture for parts requiring outstanding weathering performance, and are applied in thin films 50–100 microns. The most popular powder material for this application is the polyester-triglycidyl isocyanurate.

Economically, powder coatings can be compared with the cost of two coats normally required with conventional solvent systems. The application of powder to substrates is also cheaper and provides a lower reject rate, although in general the initial outlay on capital expenditure is higher. The use of polyvinylidene difluoride in solvent systems competes with the polyester-TGIC.

It would appear that at present, the use of powders for coating applications is more widespread in Europe than in the U.S.A. However, we are seeing increasing uses of powders in the U.K. for more technical applications. Particular ones already reported are the coatings obtained with epoxide powders for large diameter pipes for gas and oil transmission. The huge potential of powders for coating car bodies has yet to be realized, however, material and process technologies are still evolving and the automobile industry has already shown some directions. A disadvantage with these powder coatings is their relatively long cure cycle, which can be from 8–20 mins, at temperatures as low as 165° up to 200°C.

Table 29.5

CHARACTERISTICS OF MAIN FUSIBLE POLYMER SYSTEMS

Polymer	*Melting range* °*C*	*Melt Index* *19°C*	*Fusing temperature* °*C*	*Resistance to*	
				Washing	*Drycleaning*
L.D. polythene	100–120	2–200	120–150	Fair	Fair
H.D. polythene	125–136	6–50	155–175	Excellent	Good
Hydrolyzed ethylene-vinyl acetate	95–110	50–100	120–150	Good/fair	Good
PVC plastisol	100–120	Med.	140–160	Good/fair	Good
Polyester	110–135	20–80	140–175	Good	Fair
Copolyamide	110–120	50–100	Steam and/or 130–160	Good/fair	Very good
Terpolyamide	90–120	20–50	Steam and/or 130–160	Good/fair	Very good
Plasticized copolyamide	90–120	Med.	Steam and/or 130–160	Fair	Good

Fusible interlinings offer several advantages over conventional interlinings, which are placed into a garment by the conventional technique of sewing. Apart from easier, higher production rate and less expensive manufacture, it is possible to impart special qualities of controlled stiffness, shape retention, stability, or resilience into selective areas of a garment in order to improve its fit, aesthetics, or performance. Other advantages include the stabilization of the outer fabric with improved hand retention in wear. Disadvantages include the fact that once the interlining has been positioned and fused any error is extremely difficult to rectify. Very few problems occur in manufacture providing the correct fusing conditions are used in conjunction with modern presses.

The uses of fusible interlinings are too numerous to mention here, but suffice it to say that a fusible can be found in almost any garment where sewing can be involved apart from seaming. Major uses are in the chest area of ladies and men's jackets, shirt and blouse collars, cuffs and the hemming of garment parts, and household draperies.

Recently there has been an increase in the use of heat sealable powders for dry laminating, particularly in the field of lamination for the automotive field. Such applications are door panels, parcel trays, sun visors and seating.

SUGGESTED FURTHER READING

Powder Processing of Thermosetting Plastics. (1975) R. H. Chambers. Paper included in Developments with Thermosetting Plastics Editied by A. Whelan and J.A. Brydson (London: Applied Science Publishers).

The Technology of Powder Coatings. (1976) S. T. Harris (London: Portcullis Press).

Cryogenic Grinding pf Plastics. (1972 Feb) Richard H. Snyder (New York: Modern Plastics).

Developments in Thermosetting Powder Coatings. (1974) DR. S. Kut (London: Proceedings 4th International Powder Symposium organized by "Surface Coatings").

Flame Spraying of Powder Coatings. (1974) Professor H. Nagasaki (London: Procedings 4th International Powder Symposium organized by "Surface Coatings").

Rilsan Coating Techniques. No. 853/76 Technical Bulletin (U.S.A.: Rilsan Corp.).

Fluidised Bed Coating with Vestamid. Technical Bulletin (Germany: Chemische Werke Hulls AG).

Coating with Microthene Powdered Polythene. Technical Bulletin (U.S.A.: USI Chemicals).

Coating Fabrics with Microthene. Technical Bulletin (U.S.A.: USI Chemicals).
Coating Equipment for Coethylene Dispersions, Pastes and Ground Material. Technical Bulletin (Switzerland: Plast-Labor SA).
Chemical Bases of Adhesives for Fusible Interlinings. (1971) K. H. Stukenbrock, Clothing Institute Journal, Vol. XIX No. 3.
Powder Coatings in Europe. (1989) E. Bodnar, European Coatings Journal (W. Germany).
Heat Fusible Powders. (1989). Dritex International Ltd. (UK).

30.

MISCELLANEOUS PROCESSING TECHNIQUES

CALENDERING

Calendering is the process by which continuous sheet (either rubber or thermoplastic) is made by passing the heat softened material between two or more rolls. Calenders were originally developed for processing rubber, but they are now widely employed for producing thermoplastics (mainly flexible PVC) sheet and coatings. Although the subject of calendering is included here under miscellaneous processes, its importance should not be underestimated.

In essence, the calendering process consists of feeding a plastics mass into the nip between two rolls where it is squeezed into a film which passes round the remaining rolls. It then emerges as continuous film, the thickness of which is governed by the gap between the last pair of rolls. The surface of the sheet is determined on the last roll and may be glossy, matt, or embossed. After leaving the calender, the sheet is cooled by passing it over cooling rolls and then through a β-ray thickness gauge before being wound up. If fabric or paper is fed through the last two rolls, the film can be pressed into the surface of the web and the calender thus becomes a coating machine. A four roll calender coater is shown in Fig. 30.1. This method of coating is generally restricted to heavy coating weights and is often used for applying high viscosity vinyls to paper or fabrics. The thickness of coating is controlled by the gap between the last two rolls.

The plastics mix fed to the calender may be a simple hot melt as in the case of, say, polyethylene, but in the case of PVC a great deal of effort is put into the precalendering processes such as blending, mixing, gelling, and straining. In addition to the polymer there may be inert mineral fillers (added to reduce the cost and modify the physical properties), pigments, lubricants (to assist in processing), stabilizers and plasticizers. The dry components (with the exception of the pigment) are loaded into a ribbon blender and agitated sufficiently to produce an even dispersion. If plasticizers are to be added, they are sprayed into the powder mix during the initial period of blending. When correct dispersion has been achieved the blend is removed via a valve at the base of the mixing chamber and weighed-out into batch tins. If a pigment is required it is added at this stage to each separate batch. The batch tins are then emptied into a primary mixer, such as a Banbury, for about five to ten minutes at 120–160°C. The combination of heat and the 'kneading' action of the Banbury causes partial gelation of the mix. This partly gelled material is then passed to a two-roll mill and made to form a sheet round the front roll. This may be fed directly to the calender, but for good quality sheet or for thin sheet (say about 0.1 mm) a straining process is added to remove any coarse particles. A typical strainer consists of a single screw extruder with a filter screen immediately in front of the screw. The screen consists of a fine stainless steel woven wire mesh backed up by a medium and coarse mesh for mechanical support, together with a thick perforated strainer plate.

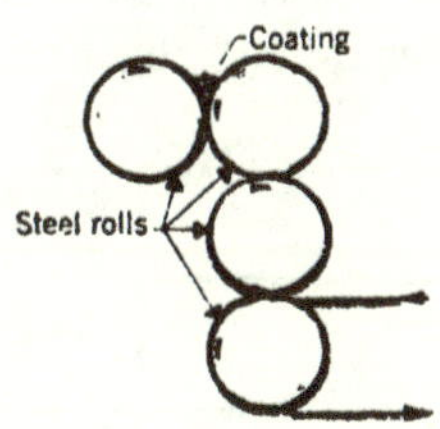

Fig. 30.1 Four-roll calender coater

Types of Calenders

Calenders may consist of two, three, four, or five hollow rolls (or bowls) arranged for steam heating or water cooling, and are characterized by the number of rolls and their arrangement as, for example, I, Z, or inverted L types. Four-roll inverted L and Z types are the most usual, the Z types having the advantages of lower loss of heat in the sheet (because of shorter travel) and of being easier to construct. The latter factor arises from the fact that less attention has to be paid to compensation for roll-bending since there are no more than two rolls in any vertical line (as opposed to three in a four-roll inverted L).

Other variations in calender design are dictated by the fact that very high forces have to be exerted on the rolls in order to squeeze the plastic material to a thin film. These high forces cause the rolls to bend, thus giving sheet which is thicker in the middle than at the edges. This can be counteracted in a number of ways, three of which have been used commercially:

(*a*) The use of 'crowned' rolls, i.e., rolls having a greater diameter at the middle than at the edges.
(*b*) Crossing the rolls slightly, thus increasing the nip opening at either end of the rolls.
(*c*) Roll bending. In this method, a bending moment is applied to the end of each roll by having a second bearing on each roll neck, which is then loaded by a hydraulic cylinder.

As mentioned earlier, thickness of the sheet leaving the calender is measured using a β-ray gauge (Fig. 30.2). This gives an average reading of thickness over an area of sheet. Radiation from an isotope such as Thallium 204 is passed through the sheet and then collected in an ionization chamber. The radiation received by the chamber is inversely proportional to the weight per unit area of the material being measured.

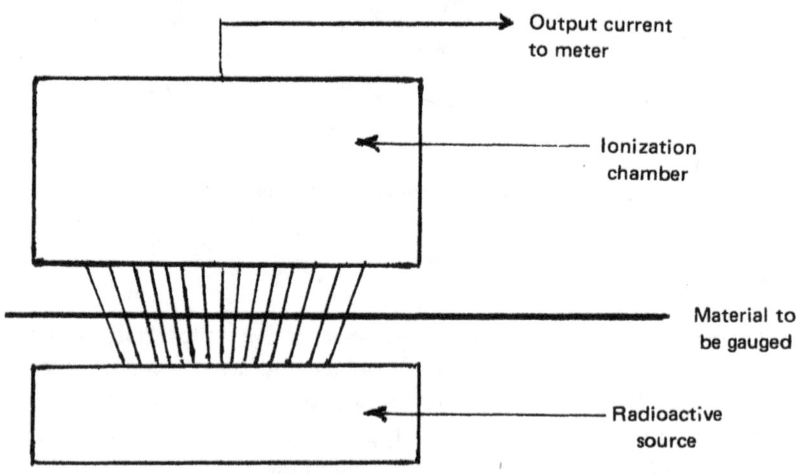

Fig. 30.2 Beta-ray gauge

Calender Requirements

As will be seen by a glance at the illustration of a typical machine, calenders are of massive construction. They call for high temperatures, with little variation across the rolls, and high pressures, again with low tolerances.

The required uniformity in temperature can be achieved by leading steam through concentrically drilled holes beneath the surface of the rolls and by circulating hot oil through the bearings in order to prevent heat losses at the end of the rolls through transfer to the bearings.

The high forces are necessary to squeeze the plastic mass into a very thin film and any unevenness in the forces along the roll is reflected in variations in the film thickness. One cause of pressure fluctuations is too much clearance in the bearings, which can be solved either by pulling back the rolls against one side of the bearing or by using tapered roll bearings.

Lubrication is usually of the 'flood' type, a system in which copious supplies of oil are circulated by gravity or pressure through sealed bearings. After passage through the bearings the oil passes to a sump where it is cooled and then pumped through filters on its way back to a header tank for recirculation. The system is shown schematically in Fig. 30.3.

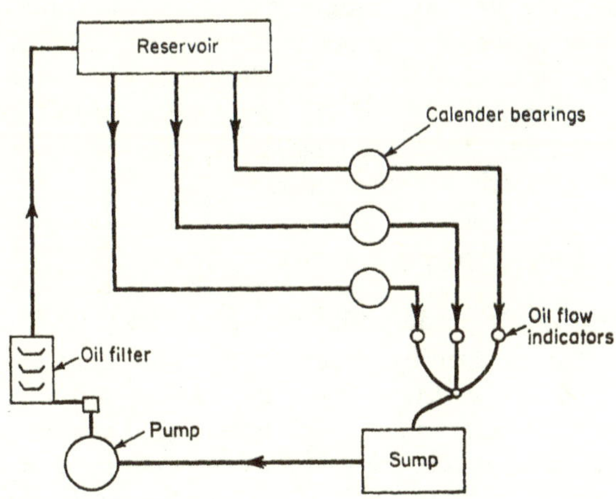

Fig. 30.3 Flood lubrication of calender

Calendering Faults

1. Bank marks: shiny patches on the surface of the sheet.
2. Cold marks (crow's feet): these are marks, thus, 〉〉〉〉〉〉〉, caused by the stock being too cold.

3. Blistering: this is due to too high a temperature or too large a bank of rolls and is prone to occur mainly in thicker sheet.
4. Pin-holes: these are caused by foreign matter or by the presence of unplasticized polymer particles.
5. Water marking: usually due to lubrication contamination.

Luvitherm Process

Because calendering is a capital intensive process, calenders tend to cater for wide width film (around 1.8 meters wide) because the running costs per unit width are proportionately less. However, such large machines tend to be used only for plasticized PVC because unplasticized PVC has a much higher melt viscosity , which makes it harder to handle on rolls of this width.

A special calendering process known as the Luvitherm process was developed, therefore, to produce unplasticized PVC films. The PVC is flash-heated to a temperature of 220°C while in contact with a specially designed aluminum roll. The hot film so produced is normally oriented immediately after calendering in the high temperature zone. The calender unit is narrower than the large units described earlier (usually less than one meter) and the output is significantly lower. Special grades of PVC are used with special stabilizers, and the compounding stage can be an extruder compounder feeding directly to the calender.

Advantages of Calendering

Calendering usually produces film having a better gauge (thickness) uniformity than that obtained by extrusion. One reason for this is the great care that is paid to the engineering of the calender bowls (as mentioned earlier. The final gauge is dependent on the gap between the final two bowls, whereas in extrusion the gauge is dependent more on draw-down speeds (in the case of slit-die film) or blowup ratios (in the case of tubular film). In addition, in an extruder cross-head die there may be a range of path lengths which lead to preferential flow with consequent variations in gauge. In the tubular die process there are complicating factors such as the structures supporting the mandrel in the die. This also affects the melt flow, leading to gauge variations.

Another advantage of calendering is that better mixing is obtained. Because the amount of energy expended in a calendering line is so much greater than in an extrusion line, calendered film is less dependent on feedstock uniformity.

The main advantage of extrusion methods where PVC is concerned is that shorter runs are possible at economic rates.

COATING (WEBS)

One of the principal methods of coating substrates with a polymer, namely, extrusion coating has already been dealt with in the chapter on Extrusion while another method was briefly described above under Calendering. The coating process is important, not only as a means of modifying film properties but also as a means of applying adhesive in the adhesive lamination of two substrates.

Coating processes can be divided into two main classes. In the first, an excess of the coating material is applied to the web and the surplus is removed by means of a rigid knife, smoothing bar or air-jet knife (air doctor). In the second class of processes, a predetermined amount is applied to the web by means of a calender, rollers, brushes, or by curtain coating. Extrusion coating also falls into this class.

Excess Application Systems

In knife coating, the thickness of the coating is controlled by a sharply profiled bar (or knife), the coating having previously been applied either by passing the web over a roll running partly immersed in the coating material or by running the coating material on to the face of the web, just in front of the knife. This technique is used extensively for coating fabrics with PVC, where it is often referred to as spreading. The PVC is prepared in the form of a paste and this is applied to the fabric by means of a knife. In general more than one layer is applied, each layer being gelated by means of heat before the next layer is added.

Three methods of providing a base, over which the knife can spread the paste, are generally used. In the floating knife system, the fabric alone provides the required base while in other systems a rubber blanket or a backing roll may also be present (Fig. 30.4).

Gelation of the applied film is usually carried out by radiation or convection in a tunnel-type oven, followed by treatment with an embossing or polishing roll to give the required finish.

Where low to medium viscosity coatings are to be applied, the use of an air knife is to be preferred. The excess of coating material applied is com-

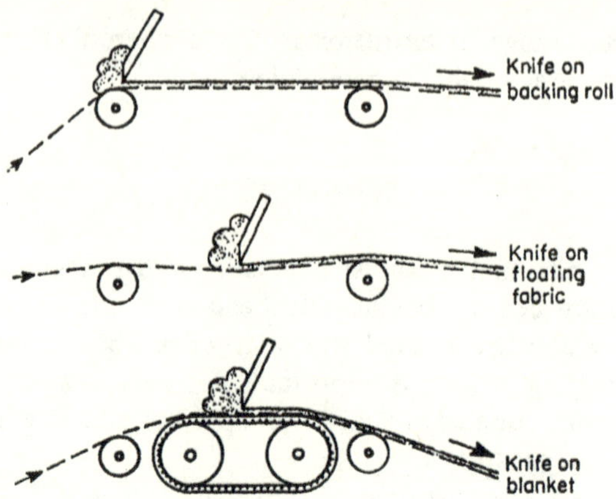

Fig. 30.4 Spreading

paratively small (about 1 : 1 to 1.6 : 1), and the main determining factor for coating weight is the air knife pressure. Other factors include air stream angle, web speed, and the physical properties of the coating. The air knife method is widely used for coating a variety of webs, including plastics films. In the past its use has been more or less restricted to the application of water-based coatings because of the difficulties in handling large volumes of air mixed with flammable solvent vapors, and surface drying of the coating by the air knife. New equipment and techniques are leading to improvements in this respect. An air knife coating system is shown in Fig. 30.5.

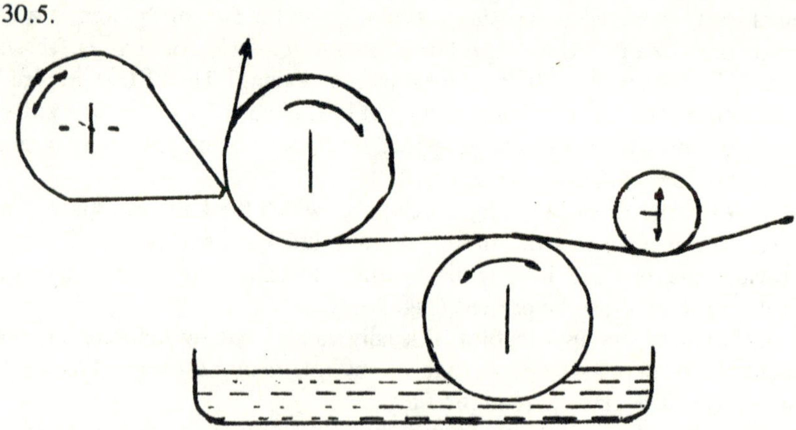

Fig. 30.5 Air doctor coating

The smoothing bar coating process employs either a smooth or a wire-wound bar, which is usually rotated slowly against the direction of web travel. The smooth bar is normally used for the china clay coating of fiber-board but the wire-wound one can be used for hot melt coatings. It is difficult to achieve an even coating weight with this method and it is liable to produce occasional scratch marks on the coating.

Predetermined Systems

Common members of this class are reverse roll coaters. They are capable of metering and applying uniform films of a coating material to a wide variety of surfaces, using a wide range of coating solids and viscosities. The simplest forms of roll coaters, such as the single roll and two roll reverse coaters, have only a limited metering action, however, and are not recommended when precise coating control is required. In a one-roll reverse coater, the roll rotates in a bath containing the coating and the web passes over the top of the roll in the opposite direction to the rotation of the roll. The two roll coater merely adds a rubber backing roll run at the line speed (Fig.30.6).

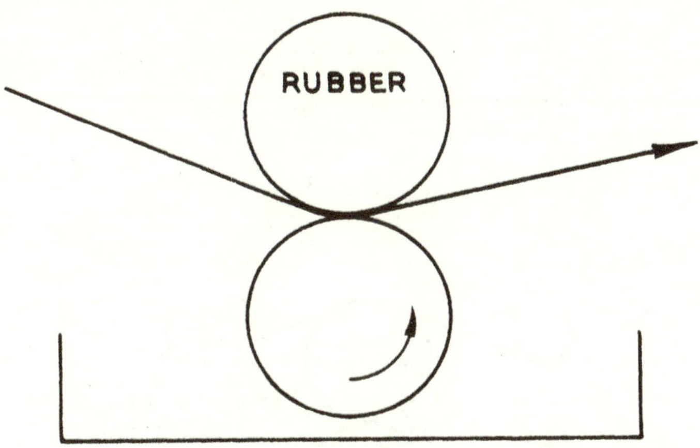

Fig. 30.6 Two-roll reverse-roll coater

In three or four roll coaters the extra rolls are used to give more precise metering of the coating (Fig. 30.7). Coating weight is controlled primarily by the metering roll gap and the rate of rotation of the applicator and metering rolls in relation to the web speed. In systems where the web passes over the applicator roll without the assistance of a backing roll, the process is referred to as 'kiss' coating. Uniform coating then depends on maintaining complete contact of the web with the applicator roll and good control of web tension. These are usually achieved by the use of hold-down idler rolls on either side of the applicator roll (Fig. 30.8).

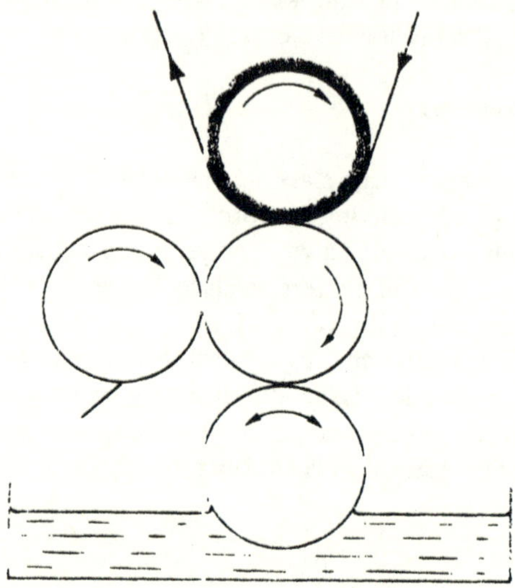

Fig. 30.7 Four-roll reverse coater

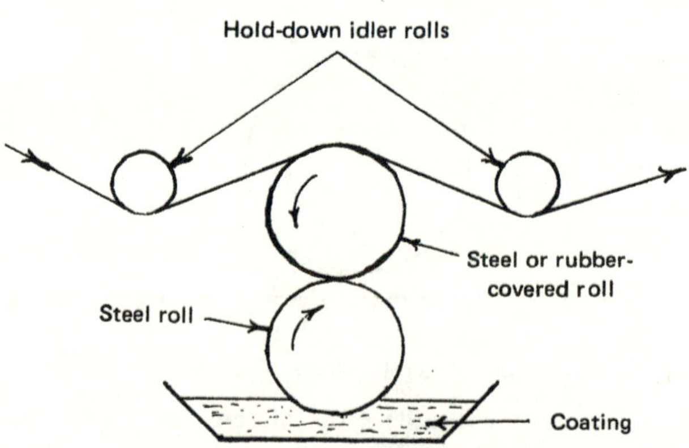

Fig. 30.8 Two-roll kiss coater

Nip roll coaters are another method of applying a predetermined amount of coating. They are often used for the application of thin coatings, difficult to apply by normal reverse-roll methods. The coating is applied to the web as it passes between two coating rolls, one of which is usually rubber covered (Fig. 30.9).

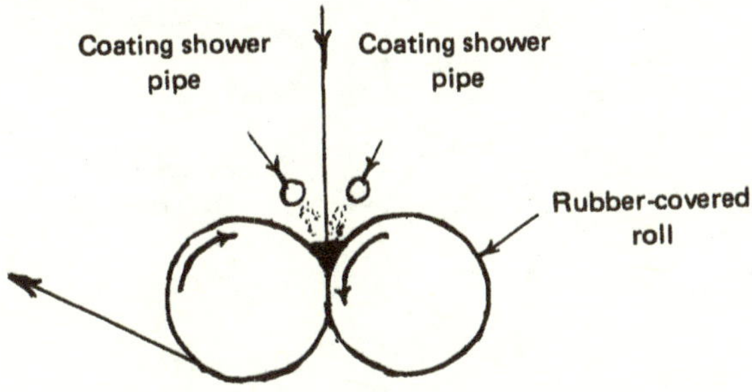

Fig. 30.9 Two-roll nip coater

The nip between the two rolls is flooded with the coating and the amount that is actually applied to the web is controlled by the hardness of the rubber rolls, the nip pressure, the web speed, the coating viscosity, and the characteristics of the web material.

Next, there are gravure coaters which employ an engraved cylinder, running in a trough of coating material, to meter the transfer of coating to the web. The engraved roll is doctored with a thin blade to ensure that an even amount of coating is left in the engraved cells. Direct gravure consist of two rolls only (Fig. 30.10). The engraved roll transfers the coating directly to the web at the nip formed by the engraved roll and a rubber backing roll. This has the disadvantage that unless there is a good flow-out of the coatings, the engraved pattern is transferred to the web. This is eliminated by the use of a three-roll system as in the off-set gravure process, shown in Fig. 30.11. Coating weights are generally low when using the gravure process but precise control is possible. One disadvantage is that the engraved cylinder must be changed when a change in coating weight is required.

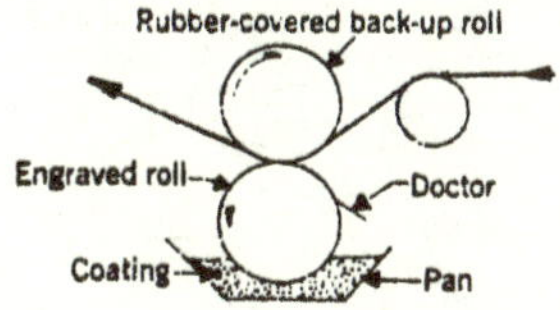

Fig. 30.10 Direct gravure coater

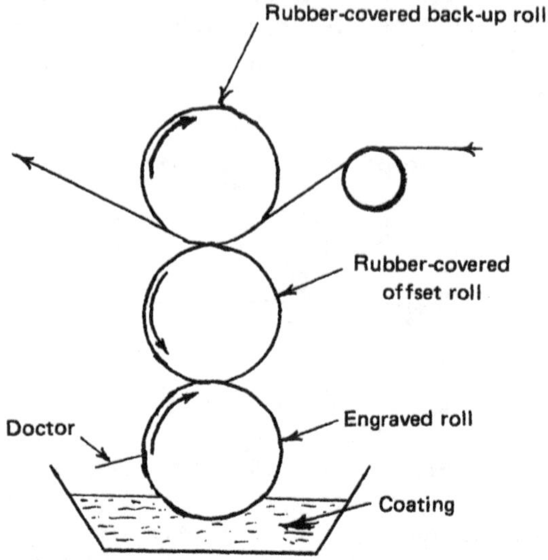

Fig. 30.11 Offset gravure coater

Finally, there is curtain coating, a typical example being shown in Fig. 30.12. A laminar film of the coating material is extruded through a slotted orifice or over a weir. Curtain coating is suitable for moderately thin, evenly flowing materials, such as lacquers, hot melts, and some latices. It gives good coverage of irregular surfaces and is particularly suitable for carton blanks. Coating thickness is governed by the orifice dimensions or the flow rate over the weir, the web speed and the coating formulation. The thinness of the falling curtain is limited by the need for it to be continuous and this also limits the speed of the coating line. Curtain coating is widely used for hot melt coatings, particularly with hot wax blends, many of which are now specially formulated for this process.

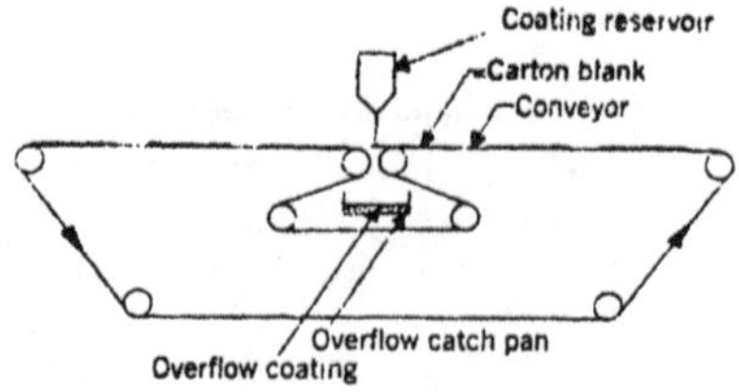

Fig. 30.12 Curtain cooler

SLUSH MOLDING

This process was developed for manufacturing hollow articles from PVC plastisols (and, sometimes, low molecular weight polyethylenes). Either a split or one-piece metal mold may be used. This is heated, and a measured amount of plastisol poured in. The plastisol is dispersed evenly over the inner surface of the mold by rotation. Gelling and, finally, fusion of the plastisol occurs and the mold is then cooled. The finished product is removed either by splitting the mold or, in the case of a one-piece mold, by collapsing the object by applying a vacuum.

Automatic machines have been applied to this process using a slightly different method. The molds are filled to the top with the plastisol and then carried by a conveyor belt through an oven where the plastisol gels to a thickness of about 1/16 inch on the inside wall. The excess plastisol is poured out and the molds are transferred to another oven where curing is completed.

Slush molding has been used extensively for making dolls, balls, and flexible toys of all sorts.

MELT SPINNING OF FIBERS

In this process the polymer is heated and the resultant viscous melt is pumped through a plate containing a large number of fine holes, known as a spinnerette. Solidification of the extruded fibers is carried out by passing them through a transverse current of air. Special pumps have been developed to operate in the temperature range necessary to produce a manageable melt (230-315°C). Another practical detail is the provision of an inert atmosphere in the melting chamber, prior to the pump. A schematic diagram of a typical set-up is shown in Fig. 30.13.

Fibers produced by the melt spinning process are oriented by cold-drawing from two to six times their original length. This brings the linear molecules closer together and increases the attractive forces between them. Cold-drawing is most important since it increases the tensile strength to many times its initial value.

FILM FIBERS AND TAPES

The production of fibers by melt spinning is relatively expensive and

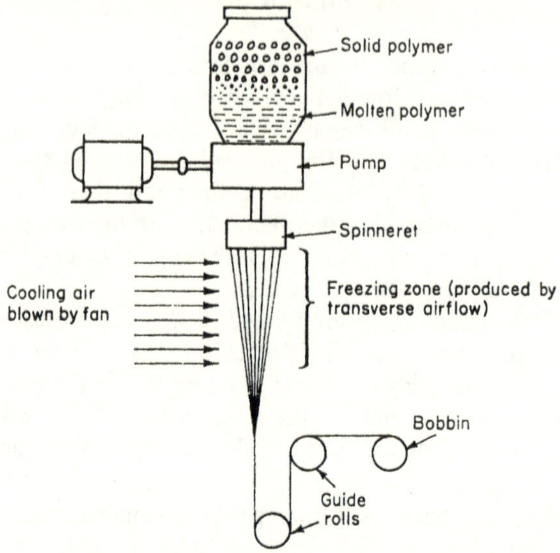

Fig. 30.13 Melt spinning of fibers

finds outlets mainly in the field of textiles. A much more varied field of applications is opening up for film fibers and tapes. The earliest outlets were for coarse fibrillated yarns used in the making of ropes and twines. This was quickly followed by the manufacture of woven sacks from stretched film tapes. Later applications include primary carpet backing, sail cloth, artificial grass, and artificial seaweed.

Before studying the subject in more detail it is helpful to define some of the terms used so that they can be referred to again without further explanation.

Film Tape

Water bath, chill roll cast, or blown film is split into narrow tapes (usually 2 or 3.5 mm wide and 50 or 28 mm thick, respectively) and stretched under heat. Stretch ratios are of the order of 7 : 1. Stretched film tape or film yarn are equivalent terms.

Twisted Film Fibers

Water bath, chill roll cast, or blown film is split into wider tapes than the film tapes and again stretched under heat. Additionally, the tapes are then split by mechanical means into a coarse, fiber-like form. Stretch ratios are somewhat higher, being of the order of 10 : 1. Fibrillated film is an equivalent term.

Fine Denier Fibers

These result when the above process is taken still further, until a fine, soft, wool-like mass of fibers is obtained. Super fibrillated film is an equivalent term.

Denier

This is the weight, in grams, of 9000 m of the fiber in the form of a continuous filament. The SI unit is the tex, which is the weight in grams of 1000 m of the fiber.

Tenacity

This is a measure of tensile strength and in SI units is measured in mN/tex.

Process and Equipment

The complete process for producing stretched film tape consists of the following stages:

(*a*) Production of the film.
(*b*) Slitting of the film to tapes of predetermined widths.
(*c*) Heating and orientation of the tapes.
(*d*) Winding of the stretched tapes on to a suitable spool.

The process may be made continuous or it may be split into two stages such that the film is wound on to a roll after extrusion. The roll is subsequently unwound, slit, stretched, and wound on to spools. Running the process in two stages has the advantage that the ultimate line speed is not governed by the extruder speed so that both extruder and stretcher can be run at their optimum speeds. There is a wide range of equipment commercially available for each step but the make-up of an optimum complete line depends on a number of factors including size of plant, equipment available, cost of equipment, previous technology, and the precise products to be produced.

Film Extrusion

Blow extrusion and both slit die extrusion methods can all be used satisfactorily for the production of the basic film. There are, however, advantages and disadvantages to each of the three methods.

Slit die film systems are generally reckoned to give higher line speeds since the film can be more efficiently cooled. Better control of film thickness is also possible. One disadvantage is the greater amount of scrap generated, due to the edge thickening that occurs in flat film extrusion. When the film tapes are to be used for the manufacture of ropes, however, this is not so important since the thicker tapes can be used as part of the rope's central core. Chill roll casting has a tighter control of film thickness than has water-bath quenching but it is less efficient for thick film extrusion because of differential cooling of the two sides of the film. Apart from the higher variation in film thickness, the water quench method has one other disadvantage, namely, water carry-over. This means that devices for the removal of excess water have to be incorporated. One possible method is to include vacuum/heat driers in the line.

Blow extrusion has the advantage of lower cost but it is important to realize that blowing of film always produces some lateral orientation of the film and this is never completely eliminated during the subsequent longitudinal orientation. The final film tapes have, therefore, a reduced tendency toward self-splitting. The strength is also somewhat less and the film has a softer hand than cast films. This softer feel is desirable in tapes intended for weaving and fibrillation, and especially for fabrics from fibrillated film because of the enhanced consumer appeal. The reduced self-splitting characteristics are also an advantage in weaving and for fibrillation. the latter is because reduced splittability leads to a more controlled splitting, at least when a pin roller unit is used, and to less post-fibrillation.

Other advantages of blow extrusion include lower extrusion temperatures (giving higher tensile strength) and the elimination of edge trimming.

Slitting and Stretching

The film is slit into tapes of the desired width, either by rotary knives or by a slitter bar in which are set razor-like blades. During the slitting operation the film should be kept under controlled tension in order to give accurate cutting. This is best achieved by placing the slitter bar close to a roll or drag bar. After slitting, the tapes are led into a pull-roll stand (or godet unit) and through a stretching oven. The godet unit may consist of three, five, or seven power driven rolls, together with a nip roll covered with neoprene or urethane rubber to prevent slippage. The stretching oven is usually about 9 to 12 feet long and is fitted with accurate temperature control (with special attention to uniformity of heating). Circulating hot air ovens are the most usual but infrared heaters are also used and hot-plates, which heat by contact, have been developed. At the other end of whatever heating unit is employed there is a second godet unit operating at a higher speed than the first. The ratio of the speed of the second godet unit to that of the first is the draw ratio (7 : 1 for weaving tapes and 10 : 1 for tapes for fibrillating). Where tapes are to be used for woven carpet backing the weaving process may generate enough heat to cause tape shrinkage, unless the tapes are previously heat set. The process of relaxation and heat setting is carried out by running the tapes through a second oven followed by a third godet unit.

In later developments the film is stretched first and slit afterwards. The blown or cast film is stretched on a unit consisting of preheated rolls, stretching rolls, and annealing rolls. It is then wound up as wide stretched film which can later be slit into tapes. In this process the stretching is controlled over a very short distance so that only a slight reduction in the film width occurs. It is possible, therefore, to produce a larger number of tapes from a given film width than with conventional methods based on slitting before stretching. Advantages of hot-roll stretched polypropylene film include:

(*a*) High output,
(*b*) High tenacity, and
(*c*) Low splitting tendency.

A disadvantage of this process is that output is limited by the high initial film thickness required. This is because the film thickness after stretching is directly proportional to the stretch ratio used. With a stretch ratio of 10 : 1, for example, the initial film thickness would be approximately 400-500 mm and adequate cooling and preheating of such films is difficult.

Winding

The winding of film tapes is a more difficult process than the winding of monofilaments, whether natural or synthetic. This is because of the added complication of possible twisting which can reduce the tensile strength by 15-20 percent. Another critical factor is tension which must be kept constant. Constant tension winders are available but are expensive. A cheaper alternative is the use of constant torque winders but these require manual adjustment by the machine operator as the diameter of the wound bobbin increases.

Fibrillation

Film tapes can be fibrillated by mechanical splitting in a controlled manner to give a range of fine denier fibers. The three basic systems for fibrillating stretched film tapes are:

(*a*) Pin-roller system,
(*b*) Castellated die system, and
(*c*) Embossed roller system.

Pin-Roller System

The equipment consists of a roller which revolves at a peripheral speed higher than the linear speed of the film and which can incorporate up to approximately 30,000 needles on a series of bars; an assembly of rollers for adjusting the contact area of the tapes; and draw off rollers for adjusting the tension. As they revolve the needles split the tapes, forming a regular or irregular network pattern. The size of this network can be adjusted by altering the roller speed and the contact arc. The fibrous material produced by this method is hairy and bulky, and comes nearest in appearance to a natural fiber. The fibers have a rectangular cross section which gives good coverage in, for example, carpet applications.

The tow, as it comes from the fibrillator, can be either tufted or woven directly into a carpet, or it can be collected, cut, carded, and spun into a yarn. The particular use to which it is put depends on the nature of the network structure which can be varied by altering the fibrillation conditions.

Castellated Die System

In this system the film is extruded, not through the usual flat film die but three flat dies, each having 35 profiled, castellated slots through which a total of 105 tapes are extruded. The profiles of each separate slot are such as to produce 25 ribs in each tape. Stretching of these ribbed tapes causes them to shear, so forming 25 strand fibers.

Roll-Embossed Film Process

This process is similar to the castellated die system inasmuch as it gives a profiled film. On stretching this produces continuous multifilaments as before. The method by which the embossing is produced differs, however. A flat film is fed into the nip of two rolls, one of which is profiled with a fine pattern of grooves around the circumference. The other roll is plain and serves as a pressure roll. The film may be either above or below its crystalline melting point; in other words it can be run through the embossing system straight from the extruder, or after it has been reeled and stored.

It will be seen that the main difference between the two embossed film systems is that the castellated die system profiles the film while it is in the die, whereas the roll-embossed film process profiles it outside the die. The former is subject to die blocking when powdered additives to the polymer are used. This includes color masterbatches.

Texturization

Texturized yarns are often required because of the higher bulk they offer, compared with that obtained with conventional polypropylene film fibers. A relatively new technique is now available, however, based on the use of laminates instead of single-layer film. Bicomponent films can be made into highly crimped yarns provided that the two layers in the laminate film are chosen from materials with a sufficiently great difference in their stress-strain behavior.

When yarns are textured in this way they can have an elongation of between 100 and 400 percent depending on the grades and on the processing conditions used. Crimping can be carried out immediately after stretching and fibrillation, or it can be applied to previously wound uncrimped yarns, after heating.

WELDING

In spite of the various techniques available for producing complicated moldings, it is often necessary to join two or more components to produce the particular setup required. One common method of achieving this union is by welding and several processes are available, some examples of which are described below. The heat sealing of film to form large sheets, or to close plastic bags and sacks, is also a welding operation and is discussed here.

Hot Gas Welding

This process, which bears a superficial resemblance to acetylene torch welding of metals, is particularly useful for producing items of chemical plant, such as tanks and ducting, from thermoplastic sheets. The sheets to be joined are first bevelled and cleaned. A welding rod of the same composition as the sheets is obtained. Sheets are placed side by side so that the two bevelled edges form a V-shaped channel. The tip of the filler rod is placed in the channel and both it and the adjacent area of sheet are heated with a stream of hot gas produced by a flame or electrically heated welding-gun. Compressed air is suitable for PVC but for polyethylene an inert gas such as nitrogen must be used. Welding guns are also available which incorporate a welding rod feed (Fig. 30.14).

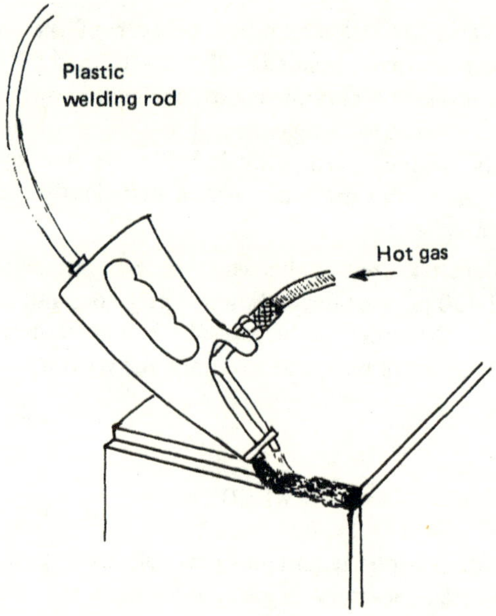

Fig. 30.14 Hot gas welding

Heated Tool Welding

In this process, a heated tool (usually of metal) is used to bring the plastic surfaces to the required temperature. An example of this type of welding is the polyfusion process for joining plastic pipes by means of injection molded couplings. The polyfusion tool is shaped so that one side of it fits into the coupling while the other fits over the pipe. The tool is heated in a flame and then used to soften the inside wall of the coupling and the outside wall of the pipe. The pipe and couplings are fitted together and allowed to cool.

Friction Welding

A particularly good example of friction welding is spin-welding which may be used for joining two rods or two injection molded halves of a bottle. One piece is fixed in a chuck which is then rotated at high speed. The matching piece is held against it until frictional heat causes the polymer to flow. The chuck is stopped, and the two pieces allowed to cool under pressure. There are limitations to this method inasmuch as the cross sections of the two pieces to be welded must be either circular or ring-shaped. The advantage of spin-welding when making a bottle is that the two portions can be injection molded and thus have completely even wall thickness. The technique has been used for aerosol bottles where an even wall thickness is important because of the high internal pressures when filled.

In spin-welding, any excess melt is forced from between the two mating surfaces in the form of a thin flash. This should be kept to a minimum, especially in the case of hollow containers where internal flash cannot be trimmed. If necessary, the joint can be designed so that the flash is retained within pockets or undercuts, or is forced to the outside diameter where it can be trimmed. Spin-welding is usually confined to the joining of similar materials as using dissimilar materials normally leads to low-strength joints.

Ultrasonic Welding

The process of ultrasonic welding depends on the vibratory energy of an ultrasonic head, or horn, being transmitted through two pieces of plastic in contact. At the interface between the two pieces this mechanical energy is converted into heat and an almost instantaneous weld is produced. The principle of the method, briefly, is that the vibratory energy from the horn radiates outward from the horn tip and is transmitted through the plastic as a wave motion. At the interface of the two components there may possibly be some reflection of energy but this should be small because the joint is under pressure due to the force applied at the horn. Most of the energy, therefore, passes through to the lower piece of plastic. The energy is then reflected by the anvil on which the plastic is resting and sets the mating plastics surfaces vibrating against each other. Because of the high frequencies involved, the two surfaces are literally hammered together and the mechanical energy of motion is transformed into heat which melts the mating surfaces of the plastics. The frequency of vibration is around 20,000 cycles/second (cps). A power supply is used to convert the standard 60 cps current into the 20,000 cps current required and this is then fed into a transducer which transforms electrical energy into mechanical vibration. A diagrammatic representation of the process is shown in Fig. 30.15.

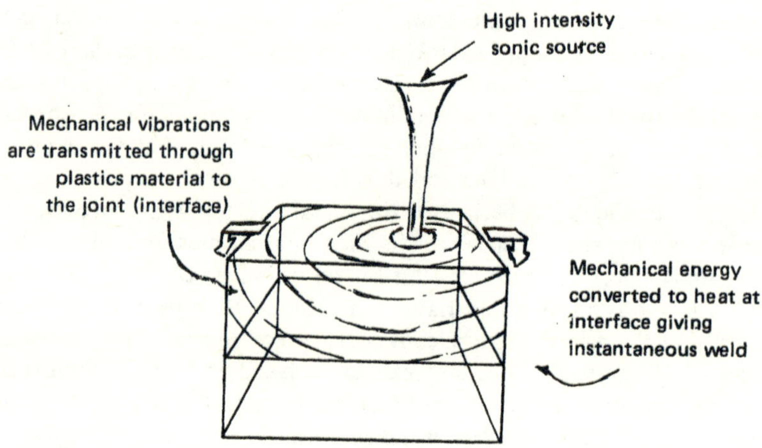

Fig. 30.15 Diagrammatic representation of the ultrasonic welding process

Ultrasonic welding is normally restricted to the joining of thermoplastics of the same basic family, although it has proved possible to weld ABS to an acrylic. This was done in the case of an acrylic lens welded to an ABS retaining ring in a car headlight reflector. Members of the styrene family of thermoplastics, such as basic and high impact polystyrenes, styrene acrylonitrile, and ABS give particularly good results with ultrasonic welding, as does polycarbonate. In general, rigid thermoplastics having a high modulus of elasticity and a low melting point are the ones which respond best to ultrasonic welding. The high modulus of elasticity is required to transmit the impressed vibrations while low melting points give rapid fusion of the surfaces. Materials with these characteristics give good welds even when the mating surfaces are remote from the horn tip. This is known as 'far field' sealing. Welds of up to 8 inches away from the horn have been made in polystyrene. Less rigid materials, such as polypropylene and high-density polyethylene are unable to transmit the impressed vibrations over very large areas and good seals are made only in the immediate vicinity of the horn tip. Such seals are known as 'near field' seals. Table 30.1 shows the results obtained by ultrasonic sealing with a range of plastic materials.

The presence of fillers or reinforcements also affects the weld. Some fillers, such as talc or asbestos, reduce the efficiency of the process, but glass fibers can improve the ability of the material to weld ultrasonically.

Another factor affecting the efficiency of ultrasonic welding is the design of the joint. The important consideration is that a small area of contact is desirable in order to concentrate the impressed energy and so allow rapid development of a melt. Fig. 30.16 shows various joint designs with comments on their likely performance using ultrasonic welding.

Horn design is also important and must be tailored to the specific application. When contact has to be made with contoured surfaces then the face of the horn must be shaped in conformity, in order to prevent marking of the surface.

Unlike spin welding the cross section of the mating surfaces need not be circular. Advantages over solvent or adhesive joints include lack of toxicity of flammable solvents, cleanliness, and speed. Joining times usually range from 0.5–4 s but can be as low as 0.1 s.

Heat Sealing of Film

So far, we have considered only the fabrication of fairly thick sections of plastic. When it comes to the joining of two pieces of film or flexible sheet, the only method (of the ones so far discussed) that is still applicable is ultrasonic welding.

There are many other methods, however, which are used for the joining of film and the choice is dictated by the chemical nature of the plastic, the type of seal required, and economics. By far the largest proportion of the flexible sheet and film market is held by two plastics, namely, PVC and low-density polyethylene, so it is convenient to consider methods of sealing those. On the whole, completely different methods are used for poly-

Table 30.1

ULTRASONIC WELDING CHARACTERISTICS OF SOME THERMOPLASTICS

	Ultrasonic Welding Characteristics	
Material	Near Field	Far Field
ABS	Excellent	Good
Cellulose Acetate	Poor	Poor
Nylons	Good	Good
Polyacetals	Good	Good
Polycarbonates	Excellent	Excellent
Polyethylene		
Low-density	Good/Poor	Poor
High-density	Good	Good/Fair
Polypropylene	Fair	Poor
Polysulfone	Good	Good
Polystyrene		
General Purpose	Excellent	Excellent
High Impact	Excellent	Good/Poor
PVC		
Rigid	Fair	Poor
Flexible	Poor	Poor

ethylene and PVC. PVC is difficult to heat seal by direct application of heat because it tends to degrade thermally at temperatures fairly close to its melting point although stabilizers have been developed which help in this respect. On the other hand, PVC is a material with a relatively high power factor so that appreciable heat is generated when a high frequency current is passed through it. PVC sheet, then, is generally welded by high frequency current heating. In general, frequencies of around 50–80 Mc/s are used with outputs of up to about 25 KW. For very thin materials it is essential to use the higher end of the frequency range and to lower the voltage in order to avoid 'flash.' This occurs when the high frequency voltage is high enough to break down the insulation resistance of the material. In other words, since a given current produces a given amount of heat, it is safer to obtain this current by using a high frequency but a low voltage in order to operate below the flashpoint.

One advantage of high frequency welding is that shaped welds are easily produced by using appropriately shaped electrodes. A typical application is the encapsulation of expensive and delicate parts for automobile and aero engines. These are placed in a vacuum formed sheet of transparent, rigid PVC and a corrosion inhibitor added. A similar molded sheet is added on top and a weld made which is shaped to the component, thus giving a transparent, rigid, hermetically sealed pack which provides perfect protection.

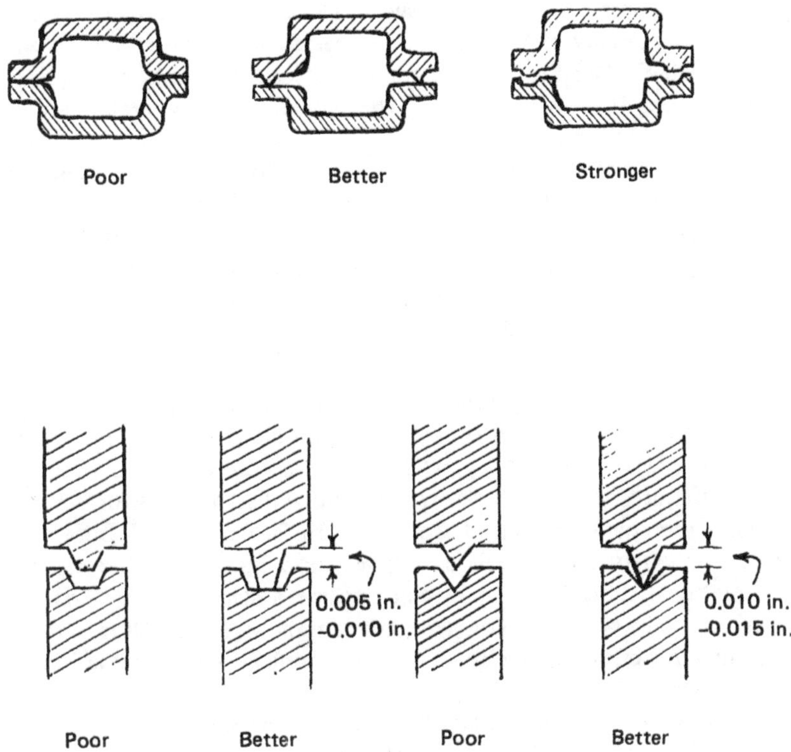

Fig. 30.16 Ultrasonic weld designs

Another advantage of high frequency welding is the ease with which the weld can be made through a liquid. This is made use of in the manufacture of shampoo sachets where a completely filled cushion-type sachet is required.

The power factor of low-density polyethylene is much lower and high frequency welding cannot be used. The application of heated metal jaws to two layers of film is not suitable, either, because polyethylene sticks to hot metal surfaces. The heated jaws have to be coated, therefore, and PTFE or a silicone rubber are commonly used. Another snag with the straightforward use of heated jaws is that the seals are obviously weak until the polyethylene has cooled and if the seal is removed from the jaws in a molten condition, then there is a danger of rupturing the seal. One method of overcoming this is the 'impulse sealer.' The sealing jaws are of very light construction, usually of resistance wire covered with PTFE, and are electrically heated. Because of their small heat capacity, they cool quickly when the current is switched off. The sequence in, say, sealing a polythene bag, is as follows:

The bag is placed between the jaws and the jaws are closed. The jaws are heated (the current is usually activated by the pressure of the jaws when closed) for a certain time, the current is switched off but the bag is retained by the jaws for a few seconds, to cool. The jaws are opened and the bag is then removed. The heating and cooling time are usually preset with a sequence timer and a lamp lights up when the heating cycle is in operation.

Another method, widely used in automatic bag making, is to use a hot wire as a knife so that a seal is made on the trailing edge of one bag and the leading edge of the next bag, while at the same time separating them.

One difficulty when sealing heavy duty polyethylene sacks for, say, fertilizers, is the contamination of the mating surfaces by dust. This makes the production of strong seals by conventional methods rather difficult and a method has been developed which utilizes a blast of hot air along the top of the sack, while the edges are held firmly together. Dust particles are completely encapsulated by molten polyethylene and do not affect the heat seal strength.

Finally, both polyethylene and PVC films can be welded with the application of infrared radiation. The efficiency depends on the temperature achieved at the junction of the two pieces of film and this is, in turn, dependent on the amount of radiation absorbed by the material, part being reflected and part passing through. Maximum absorption for most plastics occurs at a wavelength of three microns and this corresponds to a surface temperature of 700°C.

In the case of thin films, the rate of temperature rise and the final temperature are dependent on the backing material on which lies the welding area. An appreciable amount of the incident radiation passes through the film and heats the backing pad with consequent heating of the plastic film thus giving a better weld. Best results are obtained with compressed lamp black.

Heat Sealing of Closures

Although the power factor of polyethylene is too low to enable the method of high frequency heating to be used directly for heat sealing, an adaptation has been used for welding a disc-type closure onto a blown polyethylene bottle. The aperture to the bottle is circular and has a ridge which fits into an annular space in the closure. Inside this annular space a metal ring is fitted. The completed assembly is then passed through a high frequency field, the metal ring heats up and melts the plastic in the vicinity, thus forming a completely tamper-proof seal.

SINTER CASTING

Sinter casting processes utilize powdered plastics rather than granules. There are basically two main methods.

Engel Process

In this process an open-ended mold is completely filled with powder and is then placed in an oven for a predetermined time. The heat from the oven causes the powder to melt, from the mold wall inwards. The mold is removed from the oven and the excess powder discharged after which the mold is replaced in the oven to complete the fusion. The wall thickness and final weight of the molding are determined by the length of the heating cycle and this is determined by experiment.

The process is simple and cheap (both capital and mold costs are low) but only open articles can be manufactured. Applications include rowing dinghies, tanks, and open end drums.

Rotational Casting

In this process the mold is loaded with a measured quantity of powder and is then rotated about two different axes inside an oven. The powder melts and flows over the inside of the mold. When all the powder has melted and is adhering to the inside surface of the mold the cooling cycle is commenced. Rotation of the mold is continued until the molding can be removed without distortion occurring. A multistation machine is shown. in Fig. 30.17.

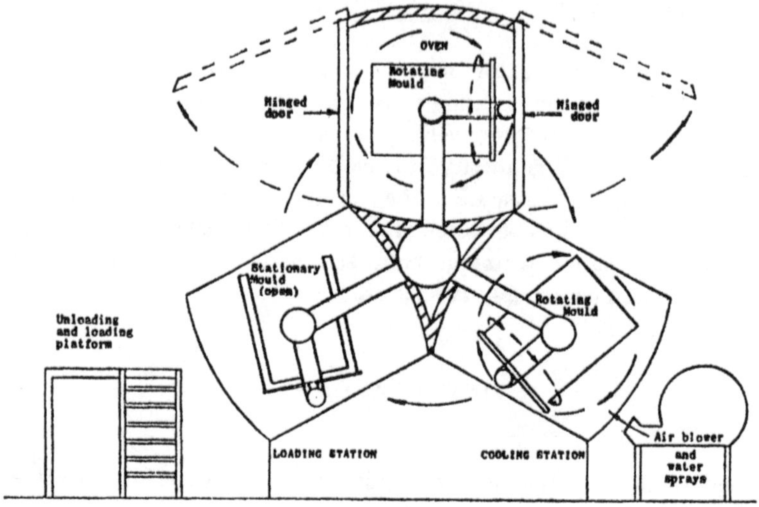

Fig. 30.17 Rotational casting

Molds for rotational casting can be of light construction as high pressures are not used. The molds are rotated at fairly low speeds, just sufficient to ensure good flow of the molten polymer. Both open and closed hollow products can be made by this process and because of the low pressures used, molded-in stresses are absent.

SOLID PHASE FORMING

This heading refers to a group of thermoplastic processing techniques where the material is shaped while it is in the solid, rather than the melt. phase. The phrase 'solid phase forming' is a more accurate description of these processes than 'cold forming,' which is often used, because the plastic is heated, albeit to a temperature below its melting point.

One of the advantages of solid phase forming, in general, is a reduction in molding cycle times. With injection molding or thermoforming, for example, the plastic is heated to well above its melting point in order that it should flow easily. It is then forced into or around a cold mold where it is held until sufficiently cool to be removed from the mold without distortion. Plastics are poor conductors of heat so that this part of the cycle is often time consuming, especially where thick sections are involved. In solid phase forming the heat content of the molding is much less and cooling times are greatly reduced. Equipment costs may also be less.

Physical properties of the finished molding can also be improved using solid phase forming techniques. The inherent strength of the molding may be increased because of the relatively cold working involved, while there are also less strains in the molding because of the absence of orientation, which is always present in melt forming techniques due to material flow.

Another factor is that higher molecular weight materials can be used because molecular weight choice is not constrained by the need for low viscosities to aid flow through the mold. Many physical properties are improved with increasing molecular weight, so that solid phase forming can produce stronger moldings.

The four main processes are Press Forging, Rubber Pad Forming, Stretch Forming, and Solid Phase Pressure Forming.

Press Forging

In pressure forging a solid billet is squeezed between a pair of mated molds. One variation on this technique is shown in Fig. 30.18 where two shaped punches mate within a common die ring, rather than directly with each other. Advantages of press forging include simply-made, inexpensive tools and ease of removal of the finished molding.

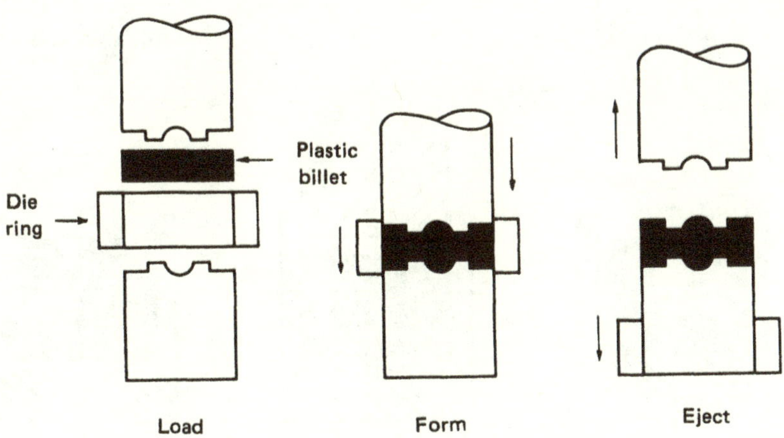

Fig. 30.18 The press forging process

In the system shown, the die ring is placed round the lower punch and the preheated billet placed inside the ring. The upper punch is brought down into the ring and pressure is then applied via a simple compression press. The time taken to deform the billet varies from about ten to twenty seconds and the upper punch is then raised and the die ring is pulled down over the lower punch. The forged part is then easily removed.

The pressure required for forging rises rapidly as the part thickness decreases and too high a pressure produces poor quality moldings. Optimum pressures are of the order of two to four tons per square inch of projected area and this more or less limits the minimum thickness of press forgings to about 0.125 in.

The technique of press forging is particularly suitable for the production of pump impellers, gear wheels, large diameter valve handwheels, industrial trolley wheels, and pipe flanges.

Rubber Pad Forming

Matched die stamping, as practiced with metals is one of the fastest of all forming operations but when tried with plastics spring-back usually

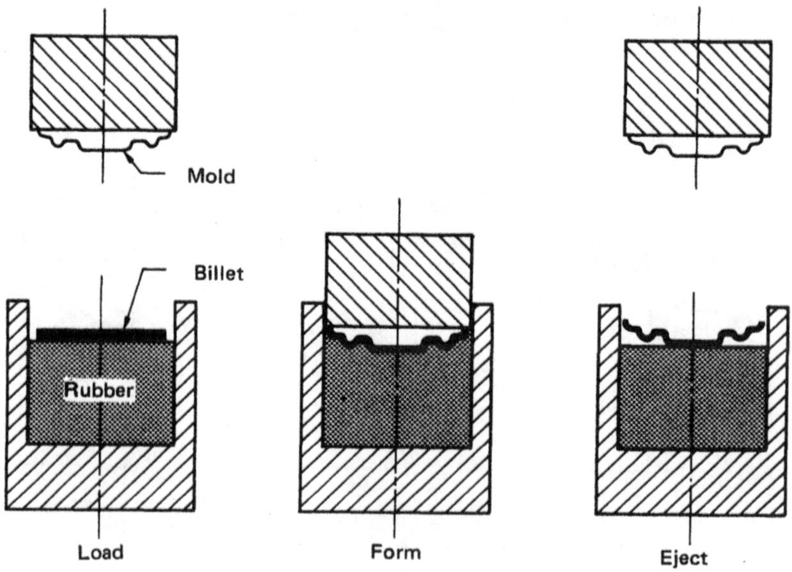

Fig. 30.19 The rubber pad forming process

occurs thus preventing good mold reproduction. However, a similar technique known as rubber pad forming can be adapted to plastics with excellent results. In this process a solid block of rubber is substituted for one half of the matched die sets (Fig. 30.19).

The rubber block is placed within a metal cylinder, closed at the bottom end, and placed on the lower platen of the press. To the upper platen is mounted a steel plate which can be in the form of either a male or female mold. Plastics sheet is then placed on the rubber and the press closed using pressures of around 1000 pounds per square inch of rubber surface. The mold pushes the sheet into the rubber which, being easily deformed, wraps the plastics sheet around the mold. Dwell times are of the order of 10 to 15 seconds (to eliminate spring-back) as compared with practically zero in the case of metals. Best results are obtained by preheating the sheet to within about 10°C of its melting point. Parts already made commercially by this process include a snow shovel, trays, and display items such as emblems. In general, wall sections are fairly thick (above 0.1 inch) and draws are shallow (about two inches maximum).

Stretch Forming

In this process a preheated disc of plastic is placed in a split mold, the disc fitting into a recess at the top of the mold. The disc is then clamped into position and a punch descends, stretching the plastics disc into the shape of the mold. Although the disc is heated the temperature is kept below the crystalline melting point of the plastic, so that the cooling cycle is low. Forming under these conditions also improves the clarity of the finished article, in the case of polypropylene, giving results superior to those obtained by injection molding. The process is particularly suited to open end containers such as quart oil cans and is shown diagrammatically in Fig. 30.20.

Solid Phase Pressure Forming (SPPF)

This process was developed mainly for polypropylene. Normal pressure forming (*see* Chapter 28) is a well-established process for polystyrene, PVC, etc., but polypropylene is at a disadvantage in this area because polypropylene sheet is difficult to control at conventional pressure forming temperatures and cooling times are longer than for polystyrene and PVC. The SPPF process effectively overcomes these processing difficulties by carrying out the forming at temperatures below the crystalline melting point of

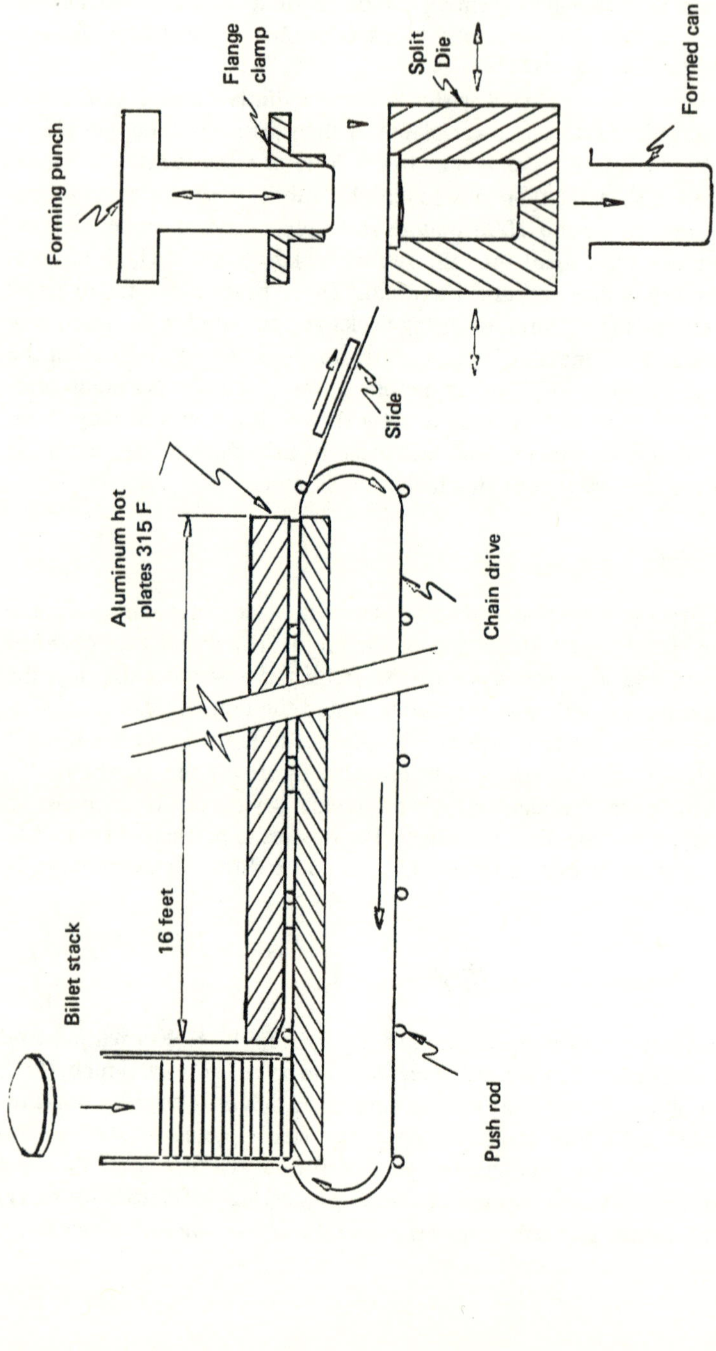

Fig. 30.20 Stretch-forming machine–design rate is 900 quart cans per hour

polypropylene. At such temperatures cooling times are reduced and overall cycle times similar to those of PVC and high impact polystyrene can be attained. The sheet is heated by infra-red heaters in the normal way; the heat input being adjusted to bring the sheet to a uniform temperature of 150 °C. The heated sheet is first stretched into the mold cavity, using a heated plug. At the lowest point of travel, cold air is forced into the plug chamber under high pressure, thus pushing the sheet against the cooled inner wall of the mold, and finishing the forming operation at high speed. Both the strength and clarity of PP containers are improved by this process.

One other advantage of the lower forming temperature is that the PP containers are free from the odor and taint that are sometimes present at normal forming temperatures. This odor/taint has often debarred PP thermoformed containers from food contact applications, in spite of their suitability from other points of view (notably their high softening point).

Only minor modifications to standard pressure forming equipment are required. Tool design is slightly more complex than for polystyrene and hence tools are more expensive but are still much cheaper than for injection molding.

ORNAMENTATION

Although not a processing technique in the same sense as calendering, melt spinning, rotational casting, etc., ornamentation is an important aspect of plastics and is best dealt with here. Ornamentation includes printing, labelling, embossing and vacuum metallization. At its simplest, ornamentation may be required only for identification, but in these days of sophisticated marketing techniques it often has to act as a selling aid as well. In addition, foodstuff containers have to carry instructions for serving or for recipes based on the contents.

'Free' Ornamentation

By 'free' ornamentation, is meant ornamentation which adds little or nothing to the manufacturing cost of the article, once the mold has been made. One method of ornamentation which does not involve any post molding process is self coloring, and this is easily achieved with plastics. A wide range of colors is available with most plastics and this method of ornamentation has many attractions.

Self coloring gives what might be called background ornamentation, but it is also possible to obtain 'free' printing. This is achieved by cutting into the mold surface so that the required message appears as embossing on the finished container. There are limitations, here, on the minimum size of printing obtainable, and the method is not suitable for lengthy texts, such as explanations as to directions for use. For brand names and simple descriptions of the contents, however, embossing can be most effective. Another method of producing an embossed surface on the container is by the use of inserts in the mold. This enables the pattern or printing to be changed without the necessity of producing a completely new mold. It is particularly useful where changes have to be made after short production runs, e.g., a change of grade. In addition to embossing, the technique of texturing the mold surface can be adapted to produce graining and other textures on the outside of the container, and this is yet another way of producing 'free' ornamentation.

Labelling

Labelling is a generally applicable technique and most standard methods can be used. In the case of very inert plastics, such as polyethylene and polypropylene, it may be necessary to use pressure sensitive labels. These are coated with a permanently tacky adhesive, which is protected before use by a backing paper. The latter is coated with a release coating to permit easy stripping of the label.

One method of labelling containers which overcomes the problem of adhesion is a plastic film label, in the form of a band, placed round the container and then shrunk on by heat. The films normally used for shrink-on labelling are oriented PVC, polyethylene and polypropylene.

Printing

Printing is a widely used means of ornamenting plastics and several methods are available. Some of these are applicable to injection molded or thermoformed items, blow molded bottles and films, but others are more limited. For instance, silk screen printing can be applied to most plastic items, but gravure printing is usually limited to film (although it is possible to gravure print special heat transfer labels and apply them to bottles, etc., in such a way that the decoration becomes part of the pack—see Heat transfer labelling.).

It is necessary to pretreat certain plastics before printing, in order to obtain satisfactory adhesion. Polypropylene and the polyethylenes are examples of plastics that need pretreatment because of their inert, nonpolar structure, which inhibits any chemical or mechanical bonding between the printing ink and the plastics surface. Pretreatment processes are aimed, therefore, at producing a surface to which the ink can key and they are usually based on oxidation processes. One process is flame treatment, which is widely used for the pretreatment of blow molded and injection molded articles. It consists of exposing the plastics surface to a suitable oxidizing flame for a short period of time, usually in the range 0.2–3.0 sec. The flame treatment changes the plastics surface in such a way that it becomes wettable and allows a strong adhesive bond to be formed between the surface and the ink. A variety of flame pretreatment machines are available, including linear conveyors where the article is taken past burners placed on one or both sides of the conveyor belt. In some cases the article may be rotated as it passes the flames. Where cylindrical bottles are to be pretreated it is possible to use a ring burner, and drop the bottles through the flame, and then through a water cooled chute. Another process is based on electrical pretreatment and is particularly useful for films, where the thin sections involved make it difficult to use flame processes. The pretreatment is effected by means of a corona discharge between two electrodes. The equipment consists of a high voltage AC generator, an output transformer and two electrodes. One electrode is a metal blade fixed in an insulated holder while the other is an earthed roller made from steel covered with a dielectric, such as polyester film. The metal blade electrode is slightly narrower in width than the film to be treated, in order to prevent direct discharges to the roller. The film passes between the two electrodes, where its surface is oxidized by the ozone generated by the intense ionization of the air.

Screen printing. This is essentially a stencilling process through a screen, usually made from nylon or stainless steel, and having about 200–300 perforations per inch (8–12 per mm.). Screens are prepared by photographic methods and are made porous only in the areas where decoration is required. The screen is supported on a frame which holds the ink supply and keeps the screen taut. A rubber squeegee is drawn across the screen and the ink is forced through the porous areas of the screen on to the plastic surface (Fig. 30.21). The screen is moved out of contact with the molding and the printed area is stoved to dry the ink film. The 'lay down' of ink is heavy; the print is even and sharply defined, and either a matt or gloss finish can be obtained. The process is widely used for the decoration of blown bottles, injection moldings and thermoformed articles at speeds in the range 100–4,000 items per hour. Fast drying inks which allow drying of the ink film to be carried out by cold or hot air or by rapid flaming, have been developed during the past few years.

Advantages of silk screen printing include low cost of equipment and screens, short changeover time and the easy training of operators. It is also an economically attractive process for short runs, while also being adaptable to longer ones, using more sophisticated equipment.

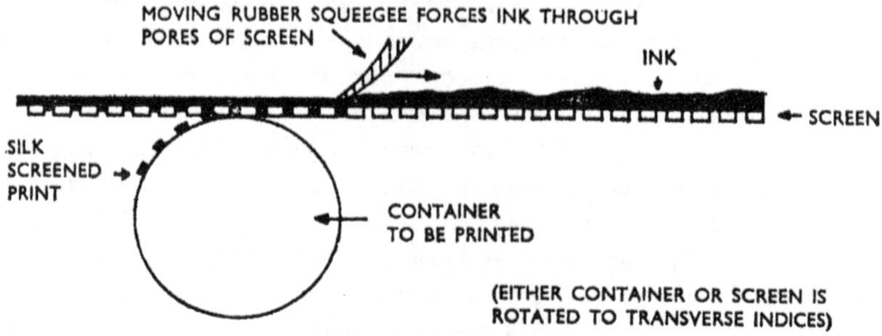

Fig. 30.21 Silk Screen Printing Process

The main disadvantage of silk screening is that, for multicolor printing, it is usually necessary to use a separate screen for laying down each color, with suitable drying equipment between each station. If the colors in the final design do not overlap, however, it is possible to use a split screen to give a multicolor effect in one pass.

Dry offset printing. Another name for this process is offset letterpress. It is used when multi-color printing of cylindrical, conical or flat moldings is required. The principle of the process can be seen from Fig. 30.22, which shows a cylindrical printing system. The image is in relief and may consist of type matter, etched zinc or copper blocks, or various rubber or plastic alternatives. A collection of such items for printing from the flat is known as a forme. A cylindrical replica of the whole forme is known as a stereo. Stereos are not easily made and the modern trend is to prepare a thin flat plate and wrap it around a plain cylinder.

Tone reproduction is achieved by breaking the picture into a large number of small dots. These are equally spaced but the size varies with the tone. Light tones are achieved by using very small dots whereas in dark tones the dots are larger and merge into one another. The rubber blanket cylinder is an important part of the dry offset process because it picks up the image from the raised plate and transfers it to the article being decorated. Accurate reproduction of the image depends on a faithful reproduction of each dot and this depends on accurate control of the pressure between the relief plate and the rubber blanket. Excess pressure between plate and blanket leads to an increase in dot size because the rubber then picks up ink from the edges of the raised image.

Where multi-color printing is required, the rubber blanket cylinder is made to contact a number of relief plate cylinders each of which carries part of the design and is in contact with the inking roller of the appropriate color (see Fig.

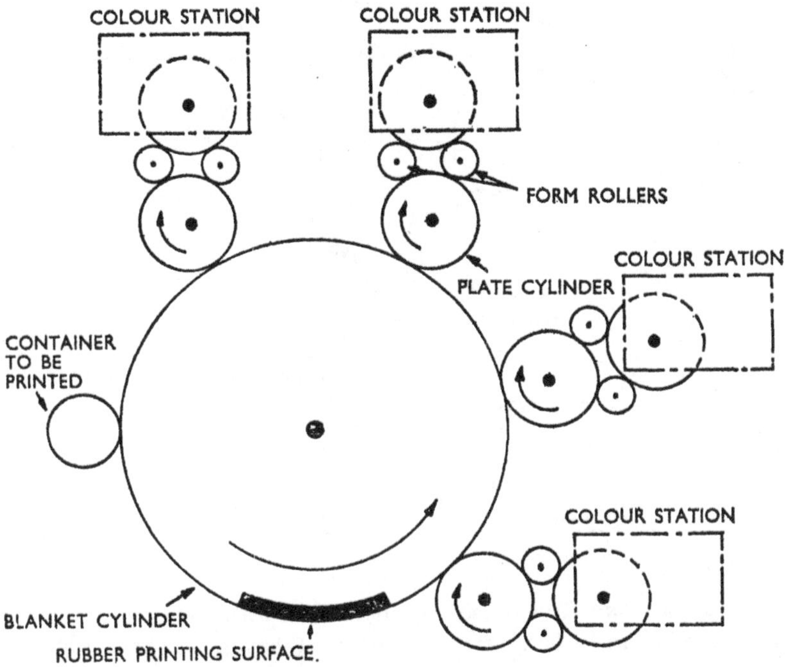

Fig. 30.22 Dry Offest Printing Process (multi-color)

30.22). As with silk screening printing, the moldings have to be stoved before filling or packing. Printing speeds vary in the range 500–6,000 items per hour.

Flexographic printing. Flexographic printing is the process most widely used for the decoration of packaging films. The principle of operation is shown in Fig. 30.23. Ink from the ink reservoir is transferred to a so-called anilox roller, via a rubber inking roller. The anilox roller is of engraved stainless steel and holds ink in the recesses of its design, thus acting as a metering device to the flexible rubber plate. The latter is mounted on the plate cylinder by an adhesive and has raised characters on its surface. The film is kept in firm contact with the rubber plate by means of the steel impression cylinder. The inks used are based on fast drying solvents.

Flexographic machines can run at speeds of 60–300 meters per minute, but oven drying is essential to achieve the highest rates. Up to six color printing can be achieved. Very fine details cannot be achieved but the rubber plates are relatively cheap, changeover times are short and printing speeds are high.

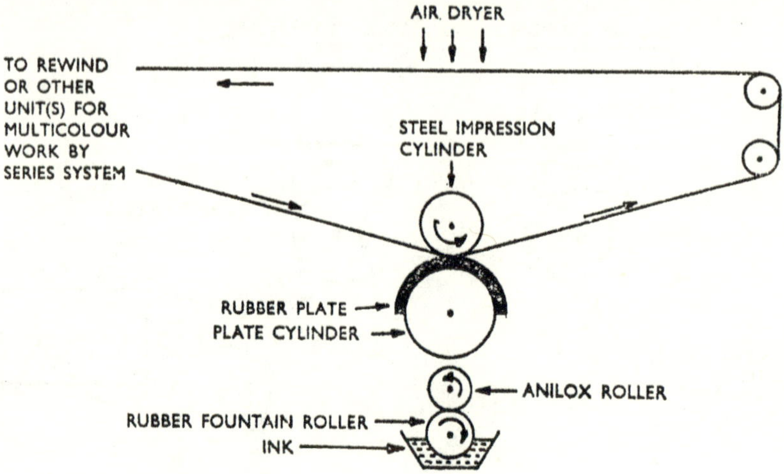

Fig. 30.23 Flexographic Printing Process

Gravure printing. This is another process used for film and sheet. It is expensive and normally used for the decoration of vinyl sheeting but is also employed for high quality packaging films. The process consists of revolving an etched metal roller in an ink reservoir. The ink is held in the recesses of the etched design, excess ink being removed by a doctor blade. The film is pressed against the etched roll by a rubber impression roller and the image thus transferred. With multi-color printing, drying between each color is essential.

The main advantage of gravure printing is that it can produce a high quality, multi-color, fine detailed print, which cannot be achieved by flexographic methods. Its main disadvantage is high cost due to the etched metal rollers. Printing speeds, too, are lower in comparision with flexographic methods, being in the range 20–130 meters per minute.

Heat transfer labelling. Basically, this type of operation consists of the automatic and instantaneous heat transfer of a gravure print, from a special release coated carrier paper, to the container surface. The carrier web is then rewound. The print cools rapidly on transfer and has good adhesion to the molding. However, it is usual to pass the decorated containers through an oven directly after transfer, in order to improve the surface gloss and ensure maximum adhesion. Speeds of 30–100 articles per minute are possible.

The advantages include greater freedom of design and the fact that a multi-colored print is achieved in one pass. In addition, the process is completely automatic and can be run by one unskilled operator. The reject rate is also lower than with screen or offset processes.

The main disadvantage is the high cost of equipment and of the gravure printed labels. If very large quantities are to be printed, then the heat transfer process becomes economically viable in comparison with offset and screen printing.

In-mold decoration. This is a process which has been widely used for the decoration of thermosets and is now applied to the decoration of injection molded thermoplastics, particularly in the fields of tableware, kitchenware and toiletry products.

A reverse printed transparent film (some 75–125 microns thick) is placed in the mold, before the molding process starts. The decoration thus becomes permanently embedded in the molding. The film must be of roughly the same refractive index as the molding. The printing process used for the printing of the films can be gravure, flexographic or screen, according to the effect desired. The film can be held in place in the mold by gravity, mechanically or by inducing a static electrical charge. One advantage of the process is that freedom of design is increased because of the number of different printing effects which can be used. Other advantages include elimination of postmolding decoration processes, with savings in floor space, labor and equipment, and protection of the decoration from chemical or physical attack.

The main disadvantage is high cost, due to increased cycle times (it takes about 10 seconds to insert the printed film in the mold) and the fact that it is more difficult to reprocess reject moldings due to the presence of the decoration.

Hot foil stamping. This is an expensive method of decoration, used for cosmetic containers and for small items, such as dials, knobs and closures, where high quality and definition are needed. It consists of the fusion and release of a special coating from a heat resistant carrier tape, and its transfer onto the molding to be decorated. There are two effects possible with this process, namely, an inlaid or embossed effect and the top stamped effect. The stamping foil consists of a carrier tape (polyester, cellulose film or glassine paper), a release layer, which is a wax melting at a predetermined temperature when brought into contact with a heated die, a metal or pigmented film coating, and an adhesive layer, which acts as a bond between the imprint and the molded plastic. There are three process variables, namely, die temperature, contact pressure and dwell time.

Machines range from hand operated units to fully automatic ones, with outputs from 500–5,000 prints per hour. In spite of the high cost there are many advantages; for instance the print is dry, and the article can be handled immediately after printing. Also no pretreatment is required, even with the polyolefins. The print has good opacity. The finish can range from gloss to matt and a wide color range is available. In this respect, it should be noted that some decorative effects, such as metallic finishes—gold and aluminum, for instance—cannot be achieved by other techniques. Fine quality printing with good accuracy is obtainable.

Vacuum metallizing. The fundamental principle of this process is the heating of aluminum, in a vacuum, to a high enough temperature to vaporize it. This causes minute particles to be ejected from the surface in all directions. These will condense onto the moldings or film present in the vacuum chamber to give a bright, decorative surface.

In practice, temperatures in the range 1,500 to 1,800 °C are necessary. One way of achieving this is to hang short pieces of aluminum wire onto a tungsten filament which is then heated electrically. This is only suitable for short heating cycles, and so it is used for metallizing plastics moldings on a batch system. One of the earliest uses of vacuum metallization was for decorating control knobs on items such as radio sets.

The interesting markets, these days, lie in the realm of film metallizing. For the longer cycles that are needed here, the aluminum wire is fed onto a block of metal, usually tantalum. The tantalum is heated by holding it in a carbon crucible across which is arced a high voltage current.

Although vacuum metallizing of films is often carried out for purely decorative reasons, it is increasingly being used to improve the barrier properties of the base film. Such films are sometimes used as competitors to aluminum foil.

A wide variety of plastic films are suitable for metallizing, including polyester, polystyrene (both normal and expanded), polypropylene, unplasticized PVC and cellulose films. There are many decorative uses for metallized films including Christmas trees and garlands (usually uPVC, because of its non-flammable properties). Lacquered, metallized polyester film is also slit into very narrow strips and used as decorative textile threads.

Electrical uses for metallized polyester film are also increasing. Its use in capacitors, for example, allows a great reduction in size over capacitors made from paper and aluminum foil. High fidelity electrostatic loudspeakers are also in service based on metallized polyester. The requirements for this application are high tensile strength, absence of plasticizer, high electric strength, chemical inertness, freedom from pinholes and dimensional stability over long periods of time, and a wide range of climatic conditions.

Trends in metallizing include the use of metals other than aluminum and the use of non-metallic materials. Silver, gold and zinc are already in use but consideration is also being given to the use of chromium, nickel/chrome alloys and stainless steel, because of their acid and alkali-resistant properties.

Among the non-metallica materials being considered, alumina and silica show promise as transparent, high barrier coatings, while stannic oxide would be transparent and have anti-static properties.

SUGGESTED FURTHER READING

Plastics Forming. (1971) Ed. John D. Beadle (London: Macmillan).
Plastics Films. (3rd ed., 1989) John H. Briston (London: Longman Scientific and Technical).

PART IV TESTING

31.

PHYSICAL AND CHEMICAL TESTING OF PLASTICS

The testing of plastics is generally carried out for the same reasons as apply to the testing of other materials, namely, to determine their suitability for a particular application, for quality control purposes or to obtain a better understanding of their behavior under various conditions. It is also necessary for the manufacturer of a new material to be able to measure its performance with relation to other materials and thus be in a position to assess the market likely to be available to him.

Physical Testing

The physical testing of plastics must be standardized if the comparison of physical data from two or more different sources is to have any meaning. The results of physical tests carried out on a material often depend on the ambient conditions of temperature and humidity, the size, shape, and method of preparation of the test pieces, and the techniques of measurement employed. Much of the early measurement of plastics properties utilized methods of test which had been developed for testing metals, textiles or wood. These were often unsatisfactory because of the difference in behavior of the newer materials and it was not until some understanding had been obtained of the nature of plastics, and of their behavior under conditions of stress at various temperatures, that real progress was made in standardizing test methods.

The main organizations dealing with the standardization of test methods for plastics are, in Great Britain, the British Standards Institution (B.S.I.); in America, the American Society for Testing Materials (A.S.T.M.); and in

Germany, the Deutsche Normenausschuss (D.N.A.). The latter organization issues its standards under the initials D.I.N. (Deutsche Industrie Normen).

References to some of the most important of the standard methods evolved by the above organizations will be made in the text but it cannot be emphasized too strongly that for information on any particular test method, the appropriate publication should be consulted.

Unfortunately, the provision of a standard test method does not mean that there is likely to be universal agreement on the interpretation of the test results, and there are still many controversial points connected with physical testing. In addition to the interpretation of the results of various test methods, difficulties are often encountered when the results of tests are used to assess the likely performance of a molded or extruded article in practice.

In general, the physical testing of plastics can be classified as follows:

(*a*) Dimensional
(*b*) Thermal
(*c*) Mechanical
(*d*) Electrical
(*e*) Optical
(*f*) Transmission (permeation)

Before discussing the physical testing of plastics in more detail, however, the following points, which affect testing results, should be noted.

1. CONDITIONING

Many of the properties of plastics vary according to the ambient temperature or humidity or both and it is essential, therefore, to condition samples before testing and carry out tests under standard conditions. The effect of temperature is obvious in the case of thermoplastics, properties such as yield stress, for example, decreasing as the temperature rises towards the softening point of the plastic.

The effect of humidity can be a factor with certain plastics such as nylon and cellulose acetate. In conditions of high humidity, water is absorbed, with consequent impairment of physical and electrical properties, although these are restored to their original values when excess moisture is removed. Conditioning procedures are laid down in ASTM. D.618-61.

2. RATE OF TESTING

Because of the long-chain structure of thermoplastics, their mechanical properties are time dependent. This is well illustrated by their behavior under tensile stress. At very slow speeds the molecules can readily disentangle, and the measured tensile strength depends mainly on the magnitude of the weakest intermolecular forces. At faster speeds there is little time for disentanglement or slipping and the breaking point does not occur until the strongest intermolecular forces have been overcome. It is difficult, therefore, to compare tensile test data from different sources unless the rate of clamp separation is standardized.

3. METHOD OF FABRICATION OF TEST SPECIMEN

The method of fabricating test specimens can sometimes have an effect on the magnitude of the test properties. Orientation during fabrication and rate of cooling (affecting spherulite size in crystalline polymers) can both vary according to the type of fabrication used and both can affect the test results. In connection with the above factors, and where relevant, a standard specification for testing a particular material usually lays down the method of preparing and conditioning a sample before test. The number of test specimens is also stated in the relevant material specification but it may sometimes be desirable to alter this if a statistical approach is to be be made.

(a) Dimensional Tests

One of the commonest dimensional tests is the measurement of density or specific gravity, and it is usually based on Archimedes principle, i.e., the apparent loss in weight on immersion in water. Though basically simple, the preparation of the sample, in the case of thermoplastics, has to be carefully controlled because the density of a polymer depends on its thermal history. An annealing process is, therefore, carried out in order that all specimens undergo the same heat treatment.

Density is usually quoted in grams per cubic centimeter and is thus numerically the same as specific gravity, which is defined as the weight of a given sample of material compared with the weight of an equal volume of water at the same temperature.

For film samples a different approach is necessary. One method utilizes a density gradient column as described in ASTM D. 1505-60T or BS.2782: Part 5: 1970. A density gradient column consists of a mixture of two fluids of different densities, the proportions of which change uniformly from the top to the bottom of the column.

The liquids are chosen for their densities but must also be without any effect on the film samples. Ethanol-water and methanol-water mixtures are often used, especially for measuring the density of polyethylene films. A density gradient column is prepared as follows. Both component liquids are degassed by boiling and are then mixed in the proportions required to give compositions with the densities desired for the two extremes of the column. These are then placed in separate flasks. The flask containing the lighter mixture is fitted with a capillary tube outlet which runs to the bottom of the column. This flask is then stirred and connected to the flask with the heavier mixture. The density of the liquid in the first flask thus increases linearly with time and so the flask is gradually filled with a liquid having a linear change of density.

A series of calibrated glass floats covering the range of densities required is then placed in the column. Film samples are cut and then immersed for half a minute in, e.g., ethanol or methanol, and then placed in the column by means of tweezers, taking care not to introduce air bubbles. After three hours measurements are taken of the heights of the floating samples and the calibrated markers against their respective densities. The heights of the markers are then plotted against their respective densities and the best fitting curve is drawn through the points. The densities of the film samples are then read off against this curve. Finished samples are removed at intervals using a slowly moving wire gauze basket.

BS.2782: Part 5: 1970 gives an alternative method, using a series of liquids of different densities made up in stoppered glass vessels, kept at a constant temperature of 23°C. Film samples, prepared as before, are placed in each of the vessels and observations made as to whether the sample sinks or floats. The densest material in which the film sinks and the least dense in which it floats is observed. The density of the film then lies between the densities of the two liquids.

Other dimensional tests include Bulk Density (of granular molding material) ASTM.D.792-50, and Apparent Powder Density, B.S.771: 1959.

(b) Thermal Tests

The important thermal properties of plastics are thermal conductivity, coefficient of linear expansion, specific heat, softening point, heat distortion temperature, and mold shrinkage.

Both softening-point tests and heat-distortion tests are designed to give the plastic user some idea of safe temperatures in use but care must be taken when using the results of such tests because of the possibility of creep under continuous loads and, in the case of some polymers, an accelerated breakdown of the material due to heat. When the latter possibility exists, stabilizers are added by the plastics manufacturer and the efficiency of various stabilizing systems is usually measured in terms of 'oven life.' The test is a simple one which involves placing samples of the plastic material in an air oven at 135°C and examining them at intervals.

Some of the more important softening-point or heat-distortion tests are given below:

Vicat Softening Point. This test is defined in B.S. 2782: Pt 1: 1956 and DIN. 57302. Basically, it is a penetration-type test, and results are expressed as the temperature at which an indentation of one mm occurs when the specimen is subjected to pressure by a standard needle (1 mm^2 surface area) under a standard load (usually 1 kg). The temperature is raised steadily during the test at a rate of 50°C/h.

Deflection Softening Point. This test is also described in B.S. 2782: Pt 1. The test specimen is clamped at one end and loaded at the other with a 20-kilogram weight. The assembly is then placed in an oil-bath and the temperature raised at a rate of 1°C/min. The temperatures at which angular deflections of 30° and 60°C occur are noted and reported.

Heat-Distortion Temperature. This is an American standard method (ASTM.D.648-56) that gives some idea of the immediate tendency to deform under very low conditions of stress. Samples 5 in. long, ½ in. wide and 1/8 in. and ½ in. thick are supported ½ in. from each end and loaded at the center to give fiber stresses of either 66 or 264 $lb/in.^2$. The temperature is then slowly increased (2°C/min) until the center of the sample is deflected to the extent of 0.01 in. The temperature at which this deflection occurs at the two different loadings is then reported.

This test has certain limitations and the data obtained may be used to predict the behavior of plastics at elevated temperatures only in applications in which the factors of fiber stress, temperature, method of loading, and duration of stress are similar to those used in this test.

Cold-Bend and Cold-Flex Temperatures. Among the tests concerned with the effects of temperature on the behavior of plastics, B.S. 2782: Pt 1: 1956 describes the two mentioned above. The former consists of winding a standard test piece round a mandrel 5 mm in diameter at a rate of 1 rev/min. Tests are carried out at various temperatures, differing by 5°C. The

Cold-Bend Temperature is defined as the lowest temperature at which all the samples pass this test.

The Cold-Flex Temperature (also known as the Clash and Berg test) measures the temperature at which a standard angular deflecttion (200°) is obtained when subjected to a standard torque.

Flammability

Any assessment of the burning characteristics of a plastic involves the measurement of two distinct properties, namely, the temperature at which it ignites and the rate at which it burns away after ignition. The flame resistance of plastics in general is good and a large quantity of heat is usually necessary for ignition. On an average, their rate of burning is also slow. However, there is still a wide range of behavior from that of cellulose nitrate, which ignites readily and burns with a fierce flame, to that of some of the thermosets which are very difficult to ignite and which char rather than burn.

As with other materials the presence of oxygen is necessary for combustion to occur but this may be present in the molecule, as is the case with cellulose nitrate. Generally, however, the rate of burning depends on the amount of ambient oxygen. Experiments show that polypropylene, for example, ceases to burn when the oxygen content drops to 16 percent, unless the temperature of the flame is very high.

With the ever-increasing use of plastics in high temperature environments or where there is the occasional risk of flame, the question of flammability of plastics has become extremely important.

The various degrees of resistance to flame can be classified as follows:

(i) Nonburning – the ability of a material to resist combustion.
(ii) Fire resistance – the ability of a material to stop the penetration of flame.
(iii) Flame retardancy – the ability of a material to retard the penetration of flame.
(iv) Self-extinguishing – this means that the material ceases to burn as soon as the source of the flame is removed.

One of the applications for plastics where the degree of fire resistance is important is in constructional work. In the U.S.A. a flame spread rating system has been developed on the basis of the Underwriters' Laboratory Inc., Steiner 'Tunnel Test.' This test was designed to measure the spread of fire along materials exposed in the interiors of buildings. The burning chamber is a long, ventilated, horizontal tunnel in which there are controlled gas flames. The test is carried out on specimens of the material 20 in. wide and 25 feet long. The rate of flame spread is based on a scale in which asbestos board is zero and red oak is 100. Based on this scale, the ASTM.E.84 standard considers any material with a flame spread of 0 to 25, to be noncombustible; from 25-50, fire-retardant; from 50-75, slow burning; and from 75-200, combustible. Anything over 200 is highly combustible.

Another standard is ASTM.D.635-63 which uses specimens measuring ½ in. × 5 in. × 'thickness as normally supplied.' For thermoplastics this thickness is usually in the range of 0.05 in. to 0.25 in. Each specimen has a line scribed one inch from each end. It is then clamped at one end on a ring stand so that the longitudinal axis is horizontal, and the transverse axis is inclined 45° to the horizontal. A piece of 20-mesh Bunsen burner gauze is clamped horizontally 3/8 in. below the specimen. A Bunsen burner, placed so that the flame contacts the end of the specimen, is held there for 30 s and then removed. If the specimen fails to ignite, the burner is replaced for another 30 s. The extent of burning is measured along the lower edge of the specimen.

A specimen that doesn't ignite is classed as 'nonburning by this test.' If it does burn, it is timed until it stops or until the four-inch mark is reached. A specimen which burns up to the four-inch mark is classed as 'burning by this test' and the rate of burning reported as in./min.

If the specimen does not continue burning to the four-inch mark, it is classed as 'self-extinguished by this test,' and the length of the burned portion is reported as the material's 'extent of burning.'

(c) Mechanical Tests

The mechanical properties of plastics vary over a wide range from soft and flexible to hard and brittle, with many intermediate combinations. It is, therefore, difficult in some cases to lay down standard tests for the measurement of a particular property which is suitable for all plastics. Some of the more important tests are described below.

Tensile Properties. Two figures may be quoted for the tensile strength, namely, ultimate tensile strength and yield strength. The relation between these two, and their relative importance, can best be seen by reference to a stress-strain curve. Other properties which can be measured from a stress strain curve are elongation (at yield and break) and the modulus of elasticity. The latter is given by the ratio of stress divided by strain and is a measure of the stiffness. It should be noted that the term 'strain' is used here in the strict sense of the increase in length per unit length or, more usually, as a percentage. Stress/strain curves of polymers have been divided into five classes by T. S. Carswell and H. K. Nason (*Modern Plastics,* June 1944) and an example of each is shown below in Fig. 31.1.

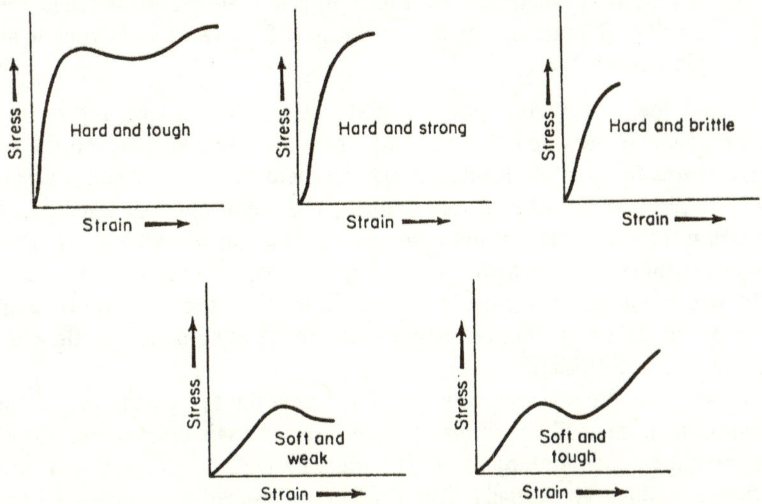

Fig. 31.1 Typical stress/strain curves obtained with polymers

The curves shown in Fig. 31.1 represent the following types of polymer properties:

(i) *Hard and tough.* Such materials have a high modulus of elasticity, yield stress, ultimate tensile strength and elongation at break. Typical materials showing such a curve are cellulose acetate and certain of the polyamides.

(ii) *Hard and strong.* This type of material has a high modulus of elasticity, yield stress, and ultimate tensile strength and a moderate elongation at break. Typical polymers giving this type of curve are rigid PVC formulations and polystyrene copolymers.

(iii) *Hard and brittle.* The hard, brittle type of material is characterized by a high modulus of elasticity and ultimate tensile strength. There is no well-defined yield point and the elongation at break is low. Typical examples are polystyrene, polymethyl methacrylate and certain phenol formaldehyde polymers.

(iv) *Soft and weak.* Such materials have a low modulus of elasticity, yield stress, and ultimate tensile strength. The elongation at break is moderate. This type of behavior is characteristic of soft polymer gels.

(v) *Soft and tough.* These materials have a low modulus of elasticity and yield point. The elongation at break is high and the ultimate tensile strength is usually much higher than the yield stress. Examples of materials giving this type of curve include rubber and plasticized PVC.

During the straight line portion of the stress/strain curve, the strain is proportional to the applied stress and when the stress is removed the test piece returns to its initial length. After the yield point, the strain is greater than it would be for a linear relationship, and the nonproportional part of the strain is irreversible. In other words, a molded article which has passed the yield point never returns to its original shape. This means that the yield stress is usually a more practical figure than the ultimate breaking stress except in cases such as those classed above as hard and brittle where the yield point is ill-defined.

As we have already seen, the speed at which the test-pieces are stressed is important, especially for thermoplastics, since their mechanical properties are time-dependent by virtue of their long-chain structure. Tensile testing machines are usually fitted with multispeed gear boxes so that several speeds are available on the one instrument. Most machines are also fitted with charts, and the stress/strain characteristics of the material under test are recorded automatically. Computer systems are also available which can be used in conjunction with tensile testers to give print-outs and visual display unit read-outs.

One effect of time dependence of mechanical properties is that where long term stresses are concerned a plastics article may rupture or deform by an unacceptable amount at a constant stress well below its short term yield stress. This continuous deformation with time, under constant load, is known as creep. It is also encountered with metals though to a lesser extent. Creep can be studied by applying a constant load to a standard test specimen and measuring the strain at intervals over long periods of time. The resultant figures for strain are plotted against time when a curve of the type shown in Fig. 31.2 is obtained. When the load is first applied there is an immediate elastic deformation, OA, which can be calculated from Young's Modulus for the material.

During the initial stage the strain increases at a diminishing rate (section A–B on the curve) and this is followed by secondary creep (from B–C) where the strain increases at a constant rate. The final stage is known as tertiary creep and here the strain increases at an accelerating rate. This stage is often accompanied by crazing or cracking of the test specimen.

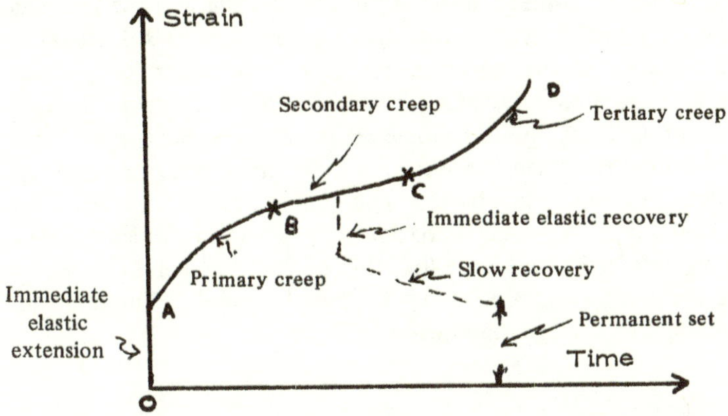

Fig. 31.2 Typical creep curve for thermoplastics

In general, thermosets have lower creep values than have thermoplastics, because of their network structure. Thermoplastics exhibit varying resistance to creep. Those materials with high intermolecular forces (e.g., hydrogen bonding) exhibit low creep, as do those thermoplastics that are below their glass transition points under the conditions of test.

In practice, it often happens that the load is applied intermittently and for limited periods. When the load is removed there is an immediate elastic recovery followed by a further slow recovery. The test specimen does not return to its original dimensions, there being a residual strain or permanent set (Fig. 31.2).

For meaningful results which can be used in practice it is necessary to obtain a series of curves at different stress levels and at different test temperatures. Any increase in temperature increases the magnitude of the creep because of the reduction in the attractive forces between molecules.

Flexural Strength. Flexural strength is sometimes referred to as cross-breaking strength or the modulus of rupture. It is the maximum stress developed in the surface of a prescribed shaped test-piece when supported near the end and loaded at the center until failure occurs. Flexural strength is quoted as lb/in.^{2} or kg/cm.^{2} Fuller details of the test are given in ASTM.D.790-63 or B.S.2782: Pt 3.

Impact Strength. The measurement of impact strength is probably the most controversial topic in the subject of plastics testing and many criticisms have been leveled against the methods used for its measurement.

Unfortunately, no one has yet put forward a universally accepted substitute, so that tests such as the Izod or Charpy continue to be used in spite of their disadvantages. Indeed, although it may be true that the test data obtained are of doubtful quantitative significance, they are still useful as providing a basis for estimating the relative shock resistance (or, perhaps more correctly, the relative resistance to stress concentration) of the various plastics. Impact testing has become standardized round the Izod test, in which a notched specimen is clamped in a vice and is struck by a swinging pendulum. The energy lost by the pendulum in breaking the test-piece is registered on a dial. A popular variation is the Charpy in which the test-piece rests on two anvils (one on either side of the notch) and the pendulum strikes the test-piece immediately behind the notch. The different methods are illustrated diagrammatically in Fig. 31.3.

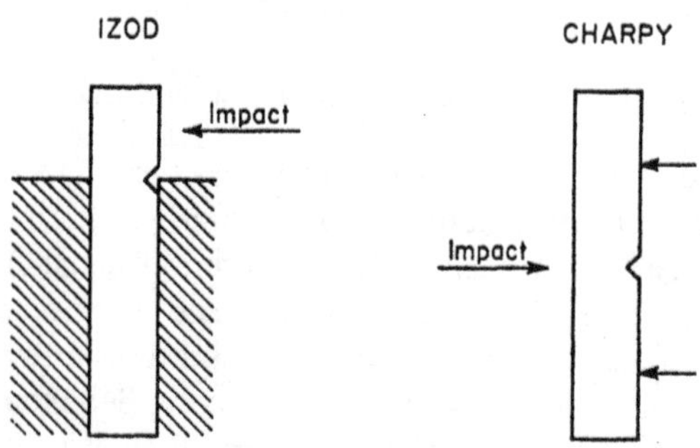

Fig. 31.3 Diagrammatic representation of Izod and Charpy impact tests

For thermosetting materials, the test specimens are compression molded, while thermoplastic materials may be either injection molded or compression molded. The method of preparing thermoplastic specimens must be stated when test results are published, since this has an effect on the magnitude of the result.

One fault of the Izod test is that the energy lost by the pendulum consists, not of the energy needed to break the specimen, but of a complex figure consisting of the energy necessary to produce the first crack, the energy necessary to shear off the top half of the specimen, and the kinetic energy in the broken-off piece which is manifested in its flight through the air after impact. Only the first figure represents the true impact strength and it has been maintained by some researchers that the remaining two factors can contribute to errors as high as 100 percent of the true impact strength.

Another criticism of the Izod test is that the results are influenced by the 'notch sensitivity' of the material under test. It will be seen that, although the Charpy test overcomes the first objections to some extent, it is still liable to errors due to 'notch sensitivity.'

A low Izod value may indicate notch sensitivity but may straightforwardly be due to an intrinsically low impact strength. One way of determining whether a material is indeed notch sensitive is to carry out a series of impact tests with pieces having different notch tip radii. The Izod values are then plotted on a logarithmic scale against the notch tip radius on a linear scale. The steeper the slope of the curve, the greater is the notch sensitivity of the material, and this gives a clear warning to the designer to avoid sharp radii and other points of stress concentration in the molding.

Another factor to be considered when looking at Izod test figures for any two materials, with a view to assessing their relative usefulness for molding, is the question of stresses in the final article. Two polymers may have similar Izod impact strength and yet behave very differently in practice. This can be due to differences in flow. The one with the highest flow gives a more shock-resistant article, especially if the molding has thin walls or has complex flow paths.

An alternative method of measuring impact strength, and one which simulates more closely the knocks likely to be encountered in practice, is the Falling Weight Impact test (BS.2782). There are many variations in respect of the test pieces used but basically the test consists of dropping a known weight from a known height on to the test piece. The estimated level at which 50 percent of the samples are likely to fail is normally reported in this test.

Test pieces which have often been used for this test include compression molded sheets and injection molded shallow dishes. The sheets are usually mounted on a circular support and the point of impact is applied at the center. Injection molded dishes (having their 'gate' at the center of the dish) are inverted and located so that the point of impact is 20 millimeters away from the center, the area near the gate being one of the recognized weak spots of an injection molding.

Finally, mention should be made of the impact strength of film. This is usually measured by the 'falling dart' method, in which a dart with a hemispherical end-piece is dropped from a known height onto a center of a circular piece of film. Adjustment of the weight of the dart can be made by means of detachable weights which fit over the shaft of the dart. The weight which causes fracture of 50 percent of the specimens is determined, and this, multiplied by the height from which the dart is dropped, is known as the impact strength.

Hardness. This is a complex property to measure and expresses the resistance to deformation. It is usually measured by some modification of the methods used in the testing of metals, the relevant standards being ASTM.D.785-62: ASTM.D. 676-55T: and B.S. 903 (for rubber). The basis of these tests is measurement of the indentation caused by the application of a steel ball or diamond cone under a standard load. Different sizes of penetrators are used for the different ranges of hardness, and there are given letters which are quoted at the same time as the hardness number. Standard methods are Rockwell Hardness and Shore D for rigid materials; B.S. softness number and Shore A for plasticized materials, rubber, and rubber-like materials.

(d) Electrical Tests

Several electrical tests are in use, the properties measured being related to the particular end-uses in view but the most important are described briefly below. Full details of British Standard electrical tests are to be found in B.S. 2782: Pt 2: 1957. When American tests are also of interest, they are mentioned below.

Arc Resistance. This is sometimes known as resistance to tracking and is defined as the ability of the material to withstand the discharge, across its surface, of a specified voltage. Details of the American test are given in ASTM.D. 495-61.

Volume Resistivity. In this test a potential gradient is applied to two electrodes separated by the material under test. The volume resistivity is then defined as the ratio of the potential gradient applied to the electrode, to the current flowing in this material. The electrical resistance, R, along the length l, of a conductor having a cross-sectional area, a, (normal to l and to the direction of the current) can be expressed as $R = \frac{xl}{a}$, where x is the volume resistivity, i.e., $x = \frac{aR}{l}$.

If l and a are made equal to unity then the volume resistivity may be regarded as the electrical resistance between the opposite faces of a cube of unit dimensions (*see* ASTM.D. 257-61).

Power Factor. When an alternating current is applied to two conductors separated by a dielectric (in this case a plastics material) the dielectric undergoes alternating disturbances and energy is dissipated as heat. The power factor of a material used as the dielectric is the ratio of the energy losses in the capacitor to the total energy passing through the material.

A high power factor, therefore, means that a great deal of heat is liberated and the material is a poor insulator for high-frequency conductors.

Thermosetting materials usually have high power factors, as have highly polar thermoplastics such as PVC. The heat generated in PVC by the application of a high-frequency alternating current is sufficient to melt the PVC and this is the basis of high-frequency welding techniques for PVC sheeting. Polyethylene and polystyrene, on the other hand, have very low power factors and are, therefore, used in radio-frequency insulation. The American tests are covered in ASTM.D. 150-64T.

Dielectric Constant. This property is sometimes referred to as 'permittivity' and is the ratio of the capacity of a condenser using the material under test to that of a similar condenser with air as dielectric. This test is also covered by ASTM.D. 150-64T.

Dielectric Strength. This is a measure of the applied voltage necessary to cause breakdown of the material under test and is specified as the maximum voltage which the material speciman can withstand for one minute, divided by the thickness of the speciman (measured in thousandths of an inch). Details of the American test are given in ASTM.D. 149-64.

(e) Optical Tests

These include measurements of gloss (important for film and especially for surface coatings), haze, light transmission, 'see-through' clarity (all of which are important for film), and refractive index (important for plastics used for lenses or light diffusers).

Gloss. The property of a film or coating which we know as 'gloss' is a measure of the ability of the film to reflect incident light specularly (i.e., as in a mirror – angle of incidence = angle of reflection). A high gloss will, therefore, produce a sharp image of any light source and will hence give rise to a pleasing sparkle on the film or coating.

Gloss is measured by a device that measures the percentage of the light, incident at an angle (usually 45°) to the surface of the film, that is reflected at the same angle.

The essentials of one standard test (BS. 2782: Part 5) can be seen by reference to Fig. 31.4. The glossmeter is first calibrated by using an optically flat, highly polished, black glass plate of refractive index 1.52. This has a gloss value assigned to it of 53 units. A light trap with matt black interior is first placed in position and the lamp switched on. The measuring device is then set at 0. The reference standard is next placed in position and the sensitivity control adjusted to give the correct reading for the standard. Finally, the test speciman is placed over the light trap and the new reading noted.

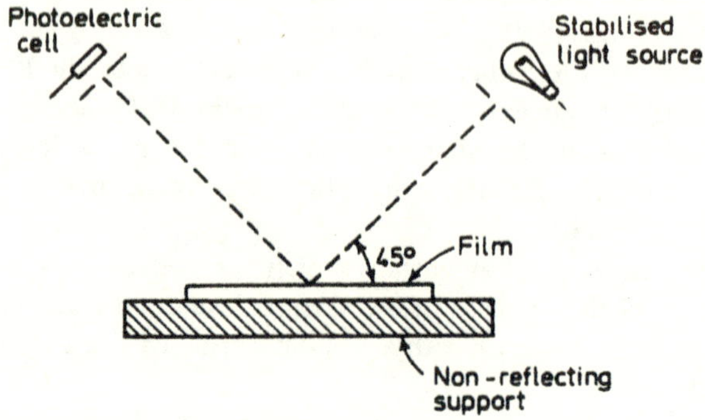

Fig. 31.4 Glossmeter

Haze. The term 'haze' when applied to a film is a measure of its 'milkiness' and is important in the case of packaging films. One cause of haze is surface imperfections of the film so that the amount of haze is not necessarily proportional to film thickness. The appearance of haze is caused by light being scattered by the surface imperfections or by inhomogeneities in the film. These latter can be caused by voids, large crystallites, incompletely dissolved additives or cross-linked material.

Haze is measured by a technique specified in ASTM-D.1003 and in BS.2782: Part 5: 1970. It is defined in these standards as the percentage of transmitted light which, in passing through a specimen, deviates by more than 2½ ° on average, from an incident parallel beam by forward scattering from both surfaces, and from within the specimen.

A beam of light is passed through the film sample and onto the highly reflective internal surface of an integrating sphere in such a manner that all the transmitted light is collected into a photo-electric cell. Next, the sphere is moved through a small angle so that the path of the light beam now falls on an extention of the sphere comprising a matt black surface that absorbs all the light transmitted straight forward along the path of the beam. Any light that is appreciably scattered in a forward direction is still gathered by the highly reflective internal surface of the sphere and into the photocell. The ratio of the second photocell reading to the first is a measure of the haze.

Light Transmission. This is an indication of the proportion of incident light passed by the film, whether or not the light is scattered on passing through. No account is taken of the quality of the transmitted light, and it is possible that a blurred image might be obtained even when looking through a film with a high light transmission value. Such a film might be suitable for a tight overwrap, but not where the film is acting essentially as a window.

Light transmission is measured by means of a photoelectric cell. The intensity of a light source is measured by the cell, both with and without the interposition of the film sample. The light transmission is the ratio of the light intensity measured with the film, to that obtained without it, and is expressed as a percentage.

'See-through' Clarity. A better guide to the quality of an image when viewed through a film is given by the property known as 'see-through' clarity. This indicates the degree of distortion of an object when seen through the film. In one test the optical definition of a standard well-illuminated wire mesh grid viewed through the test film is compared with a set of eight standard photographs. These cover the range of clarities normally encountered in the material under investigation, and the test specimen is given a number corresponding to the standard it most closely resembles. Another way of assessing 'see-through' clarity is based on the assumption that definition of an object is controlled by small-angle scatter. Fig. 31.5 shows the basis of the method.

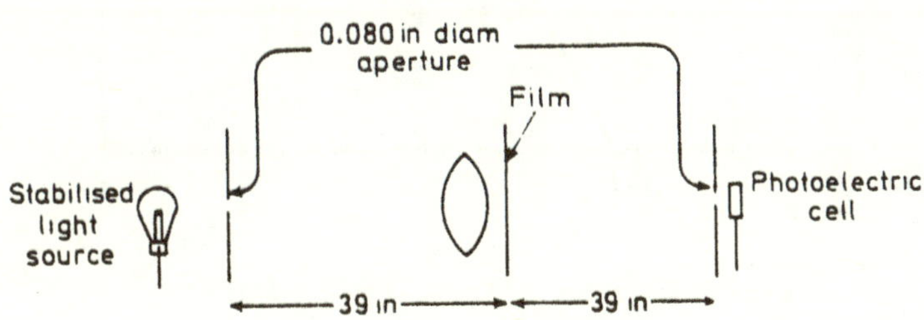

Fig. 31.5 'See-through' apparatus

A stablized light source serves to illuminate a small (2.03mm or 0.080in diameter) orifice which then acts as the optical object. A large lens focuses this image onto a similar orifice placed in front of an electric cell. The distance between the orifices is 1.98m (78in) and the lens is placed midway between them. When a test speciman of film is placed behind the lens, as shown, only light which is transmitted by the film with a deviation of less than 4 min will enter the photocell. The photocell reading, expressed as a percentage of the intensity of the incident beam, is the 'see-through' clarity. This figure is, of course, a measure not only of the distortion, but also of the light transmission, and the 'see-through' clarity, and the light transmission (determined as described earlier) must be compared in order to assess the effect of deviation alone.

Refractive Index. This can be measured by any of the standard methods used for determining the refractive index of glass. One simple method, using a block or slab of material, is outlined below and is illustrated in Fig. 31.6.

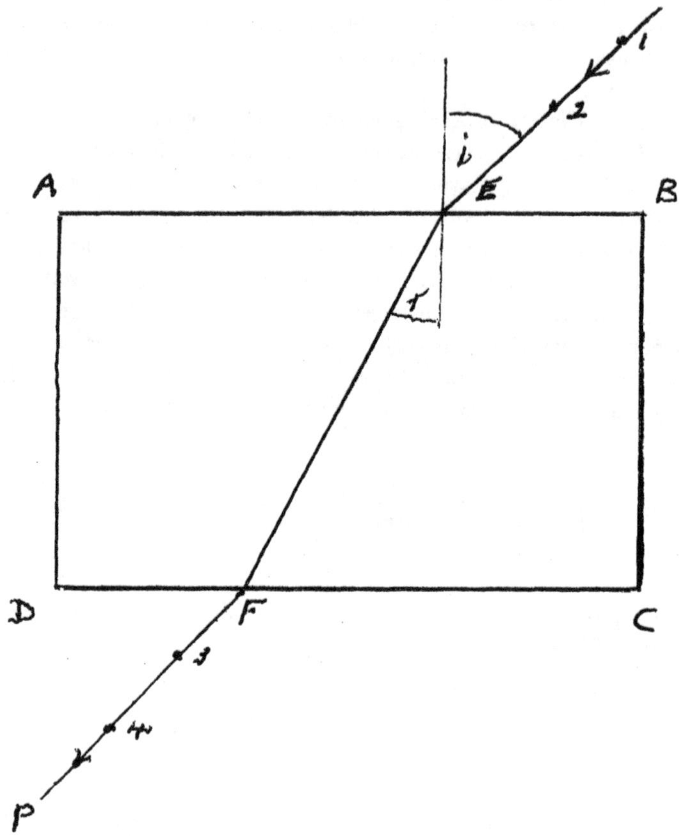

Fig. 31.6 Measurement of Refractive Index

AB and CD are the parallel faces of a block of the material whose refractive index is to be measured. Two pins (1,2) are used to indicate the incident ray of light. The position of the emergent ray of light is found by looking along the direction PF, which is then marked by two more pins (3,4). The four pins will then appear to be in the same straight line. The block is then removed and the points E and F are joined by a straight line. The angles of incidence and refraction (i and r) are measured and the refractive index is given by the ratio sin i/sin r.

(f) Transmission Tests (permeation)

There are basically two mechanisms by which a gas or vapor is able to pass from one side of a plastics film to the other. If the film is porous then the gas or vapor can flow through the holes. This is not likely to be the case except where extremely thin films are concerned. When porosity is not present then gases and vapors can still pass through the film by a process of solution (or absorption) at one surface of the film, followed by diffusion through the film and evaporation at the other surface.

A full treatment of the permeability of gases and vapors through plastics, and its measurement, is beyond the scope of this book but a brief summary is given here. Where permanent gases are concerned, and under condtions of constant temperature, and a constant partial pressure differential, a steady state will be achieved after a certain time has elapsed. The quantity of gas, Q, passing through the film is directly proportional to the difference in gas pressure on either side of the film, and inversely proportional to the thickness of the film. In addtion, it is directly proportional to the time during which the permeation has been occurring, and to the exposed area.

We can, therefore, write:

$$Q \propto \frac{AT(p_1-p_2)}{x}$$

Where Q = quantity of gas that passes through the film;
A = surface area in contact with the gas;
t = time;
(P_1-p_2) = partial pressure differential;
x = thickness of plastic

The forgoing can also be expressed as an equation, thus:

$$Q = \frac{P\ At(p_1-p_2)}{x}$$

where P is a constant for a specific combination of gas and plastic at a given temperature. P is variously known as the permeability factor (or 'P-factor'), permeability coefficient or permeability constant. It should be emphasized that the above equation applies only to steady state conditions.

The position with regard to vapors, including water vapor, is less clear. For one thing, the steady state condition is reached more slowly, while there may also be chemical interactions between the permeant and the plastics film.

The practical importance of permeability can hardly be overemphasized. One of the prime functions of a packaging film, very often, is to act as a barrier to gases and vapors. Biscuits, for example, need to be kept dry, while conversely, cigarets and tobacco need to be protected from moisture loss. Similarly with the permanent gases, fresh produce needs to be able to lose carbon dioxide and pick up oxygen, while fatty foods may go rancid if oxygen is not kept out. Many foods are packed in a vacuum, and a good barrier is essential if this condition is to be maintained. The measurement of permeability is, therefore, important and standard methods are available for both water vapor and the permanent gases.

Water Vapor Permeability. Tests for water vapor permeability are described in BS.2782: Part 5: 1970. Methods 513A and 513B are carried out by sealing test specimens with wax over the mouth of metal dishes containing a desiccant. The dishes are weighed initially and then placed in a temperature and humidity controlled cabinet. Weighings are carried out at regular intervals and the gains in weight are measured. In Method 513A, the temperature of the cabinet is maintained at 25 °C ± 0.5 °C and the relative humidity at 75% ± 2%. These conditions are taken as representative of 'temperate' climates. For 'tropical' conditions the tests are performed at 38 °C ± 0.5 °C and 90% ± 2% relative humidity.

Where a quick indication of the potential barrier properties of a film is required, the film can be clamped in an instrument where one side of the film is exposed to a high humidity atmosphere, while the other side is in contact with dry air. As water vapor diffuses in, it is detected by an infra-red absorption cell or by a resistor, the value of which is affected by changes in relative humidity. The instrument then measures the time required for a particular change in relative humidity as detected by the cell or resistor.

Gas Permeability. Measurement of gas permeability is carried out under controlled conditions of pressure, as well as of temperature and relative humidity. In essence, the usual tests consist in making the film a partition between a test cell and an evacuated manometer. The pressure across the film is usually one atmosphere. As the gas passes through the film specimen the mercury in the capillary leg of the manometer is depressed. After a constant transmission rate is achieved, a plot of mercury height against time gives a straight line. The slope of this line can be used to calculate the gas transmission. Gas permeability measurements are described in ASTM.D.1434 and BS.2782: Part 5: 1970.

The foregoing method is known as the pressure increase method. Another common method is that of concentration increase. Two gases are used, a reference gas and the test gas. A partial pressure difference across the barrier material with respect to the test gas is created, without a difference in total pressure. Because the method of measuring the concentration of the test gas can be specific to that gas even in the presence of other gases or vapors, equipment can be developed in which the relative humidity of both the test and the reference gases can be controlled. This is not possible where equipment has to be evacuated, and is very important when determinations are carried out on barrier materials having a hygroscopic base. Various methods of measuring the concentration change are used including chemical analysis, gas chromatography and radioactive tracer techniques.

Odor Permeability. There are no standard tests for the measurement of odor permeability. One method which can be used to compare the efficiency of several films as odor barriers is to make up pouches with each film. The pouches can be filled with some odorous material and then placed in separate clean glass bottles, sealed by crimping with aluminum foil. The minimum time for an odor to be apparent in the bottle can be measured, and will give a rough ranking list of the test films. This ranking list may vary somewhat according to the type of odorous material used so that the substance to be packed should be used as the test material if possible.

Measurement of Flow

The flow properties of any plastic material, whether thermosetting or thermoplastic, are of extremely great importance to the manipulator. They are, unfortunately, also difficult to measure. In the case of thermosets, the Cup Flow Test (B.S. 771) is probably the best known. This is carried out in a thimble-shaped mold and enough molding powder is introduced to produce a complete molding plus a flash of 2 to 2½g. The powder is placed into the mold (which is kept at a temperature of 163 °C $\pm$ 1 °C) by means of a scoop, and a presure of 10 $\pm$ 0.5 tons applied to the mold. The number of seconds which elapse between the first indication of pressure on the gage and the moment that the flash ceases to move is taken as the 'time of flow.' A great deal of experience is necessary in assessing the results of the test and relating them to the flow properties of the molding powder used.

When dealing with thermoplastics, the point of interest is usually the behavior of the molten polymer when caused to pass through an orifice under pressure. The Melt Flow Indexer has already been mentioned (Chapter 9). This was developed by Imperial Chemical Industries Limited and has since been modified and accepted as a British Standard (B.S. 2782: Pt 1) while a similar method has also been adopted as an ASTM procedure. A schematic diagram of the Melt Indexer is shown in Fig. 31.7.

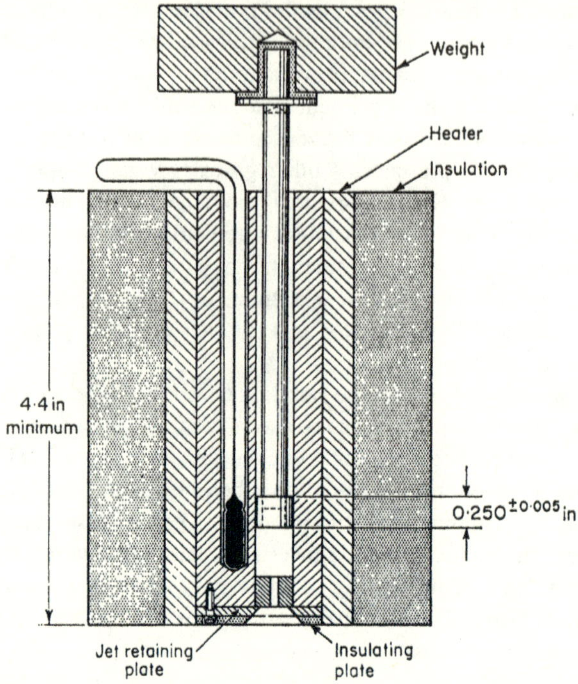

Fig. 31.7 Apparatus for determining Melt Flow Index

The apparatus is operated at 190°C for low- and high-density polyethylene, and 230°C for polypropylene. The extruder piston operates under dead-weight conditions and the pressure is quite low (43.2 lb/in.^{2}). The Melt Flow Index is the weight of material (in grams) extruded in ten minutes. Although the results are very useful as a method of grading, there are limitations due to the fact that the flow of molten polymers is non-Newtonian. In consequence, it is impossible to forecast the flow behavior under the high shear conditions obtained in commercial injection molding machines and extruders from the results of tests using the very much lower shear forces in the Melt Indexer. Other equipment is available, however, which can simulate the conditions existing in an extruder or injection molding machine.

Environmental Stress Cracking

Before turning to the section on Chemical Analysis, mention should be made of a test to determine the susceptibility of a sample of polyethylene to environmental stress cracking (*see* Chapter 9). The test was devised by the Bell Laboratories in the U.S.A. and was later adopted as a tentative standard by the ASTM. In essence it consists of taking a strip of polyethylene, 1½ in. long × ½ in. wide × 1/8 in. thick and scoring a line along the central ¾ in of the strip to a depth of from 0.020 in. to 0.025 in. The test strips are then bent along their long axes through 180°, with the scored line on the outside. The strips are put in a holder which maintains them in the bent position and the whole assembly is inserted into a test tube filled with the product under test. The test tube is kept at a temperature of 50°C and the time to failure is noted.

The test can be used for evaluating different grades of polyethylene (by using a standard environment) or for the investigation of the stress cracking properties of different chemical environments (by using a polyethylene grade with a known reaction to a standard environment).

Chemical Analysis

Polymer materials are complex in nature, and the commercial materials often contain various additives such as stabilizers, plasticizers, and fillers which sometimes appreciably alter the physical properties of the product. All this makes it very difficult to devise some simple scheme for their identification. The general chemical inertness of most plastics, which makes them so attractive to the user, also presents difficulties when attempting to analyze them.

Using the fairly simple tests described in this chapter it should be possible to obtain a reasonably accurate idea of the nature of the plastic under examination, although conclusive results can not be obtained in every instance. It is found that experience plays a very important part in this kind of investigation and a careful note should be made of any relevant information available, such as the method of fabrication employed and the end-use to which the material was put.

Only a few years ago it used to be possible for an expert to determine not only the nature but often the previous history of a polymer merely by the use of a penknife and a match. The extremely wide range of materials now on the market means that this kind of examination is no longer a substitute for chemical analysis, but as a preliminary exercise, it is, nevertheless, valuable, and some of these nonchemical tests are described.

A preliminary visual examination should always be made and a note taken of the color of the material. This can, in some instances, act as an initial 'sorting-out' operation. Most plastics can be obtained in a wide range of color effects except for polytetrafluorethylene which has a bluey-white mottled appearance. Among the thermosets, a light pastel shade rules out the possibility of its being phenol-formaldehyde, as moldings in this material are invariably dark in color. A completely transparent molding might be polystyrene or styrene copolymer, an acrylic such as polymethylmethacrylate or a polycarbonate. A visual examination will usually tell which method of processing has been employed, i.e., compression molding, injection molding, vacuum-forming, etc., and certain deductions might also be made from this observation. An injection-molded article, for instance, would not be cellulose nitrate since this material is not amenable to injection molding because of fire risk. The vacuum-formed article is very likely to be cellulose acetate, PVC, or high-impact polystyrene.

In addition to their obviously different behavior towards heat, thermosets and thermoplastics also differ in their behavior when cut with a sharp knife. A thermoplastic cuts cleanly and may have quite a 'soft' cut, while a thermosetting material tends to chip when cut. Certain plastics have other distinctive properties which give a very good guide in subsequent testing. Polystyrene, for instance, has a characteristic metallic 'ring' when dropped on a hard surface, polyvinyl alcohol is water-soluble, polypropylene and the polyethylenes have a density less than unity and hence float on water, and PTFE and the polyethylenes have a waxy feel.

As an illustration of how even the little information gleaned from the above tests can prove informative, we can take the case of a screw-top bottle closure. If this is black or dark brown in color it is most likely made from phenol formaldehyde (especially if it is of simple design, i.e., designed for a cheap product) but an examination could be made for a 'gate' mark to check that it is not an injection molding. This can also be checked by the application of heat and observing the behavior when cut. If white or colored and a thermoset, it is most likely urea-formaldehyde, as there is very little advantage to be gained from using melamine formaldehyde at a greater cost. If the cap is made from a thermoplastic material, it can be checked to see if it is polystyrene by dropping on a hard surface. If there is no metallic ring, it could be cellulose acetate, polypropylene, or polyethylene and the behavior when dropped in water divides cellulose acetate from the other two. A further subdivision may be attempted by noting the reaction when the material is scratched with a fingernail. Polypropylene normally has a high gloss and does not scratch, high-density polyethylene is also glossy but scratches to a certain degree, while low-density polyethylene is not usually glossy and scratches easily. Rigidity also acts as a guide between high-density and low-density polyethylenes. It is thus possible to obtain a fairly close idea, in this particular case, from a knowledge of the materials normally used for various end-uses, plus a few simple tests. It must be admitted that the case taken was simpler than many, nevertheless a knowledge of the plastics used for particular end-uses is a help and can be gained by experience.

Certain polymers, such as PVC, PTFE, and polychloroprene can be recognized by the results of a simple technique, well-known to chemists as the Beilstein Test. A freshly cleaned piece of copper wire is heated in a bunsen flame and placed in contact with the material under test, some of which will then adhere to the wire. The wire is returned to the flame, where the production of a green-colored flame is an indication that a halogen (i.e., fluorine, chlorine, bromine, or iodine) is present. The test should be taken only as an initial guide, since many compounding ingredients also contain halogens so that misleading results can sometimes be obtained.

Much information can be obtained by heating the material on a clean spatula or penknife. The following information should be noted: ease of ignition, whether self-extinguishing or not, the nature and color of the flame and the odor of any volatiles. These may all be an indication of the nature of the polymer. Some results of flame tests are given in Table 31.1.

A simple method for identifying the more common plastics is given in Table 31.2, although it is again emphasized that experience is often needed before a definite interpretation of some of these tests can be given. Once an indication has been obtained, confirmatory tests can be carried out. Some common confirmatory tests are given below:

CONFIRMATORY TESTS (20 MILLIGRAM SAMPLES ARE SUITABLE FOR THE FOLLOWING TESTS)

(a) Casein

The sample is dissolved in concentrated nitric acid by boiling for about five minutes. After cooling, excess ammonium hydroxide is added. Development of an orange coloration usually indicates a protein containing a phenyl group, but this reaction may also occur with phenolics.

(b) Cellulose, Cellulose Esters and Ethers

The sample is shaken with 1 cm^3 of water containing two drops of β-

naphthol in chloroform (10 percent w/v). At least 2 cm^3 concentrated sulfuric acid is then carefully added down the side of the test-tube so as to form a lower layer. A violet ring is obtained at the interface and, on shaking, a purple solution is obtained. Cellulose nitrate does not give a positive reaction.

(c) Cellulose Nitrate

The sample is first dissolved in 1 cm^3 of acetone, and a solution of diphenylamine in concentrated sulfuric acid (5 percent w/v) is then added a drop at a time. A dark blue coloration is first produced but the addition of excess diphenylamine solution destroys this to give a brown coloration.

Table 31.1

CHARACTERISTIC FLAMES PRODUCED BY BURNING POLYMERS

(i) No flame is produced	*A*. Sample deforms slowly	1. CAUTION–FUMES MAY BE TOXIC Sample not attacked by nitric acid/hydrochloric acid mixture FLUOROCARBON 2. Hydrochloric acid odor POLY-VINYLIDENE CHLORIDE or PVC or COPOLYMER
	B. Retains shape Odor of formaldehyde	1. No after-odor UREA FORMALDEHYDE 2. Strong fish-like odor MELAMINE FORMALDEHYDE 3. Odor of phenol only PHENOL FORMALDEHYDE (Inorganic Filler)
(ii) Burns but extinguishes itself when flame is removed	*A*. Yellow flame, ignites with difficulty	1. Acid odor, green spurts PVC 2. Odor of phenol mixed with that of burning wool PHENOL-FORMALDEHYDE (Organic Filler) 3. Sparks from flame. Acetic acid odor resembling vinegar CELLULOSE ACETATE (Flame Resistant Grade)
	B. Blue flame with yellow tip. Melts and drips. Odor of burning wool or hair. Slowly soluble in hydrochloric acid NYLON	

Table 31.1 *(cont'd)*

	A. Burns slowly and self-extinguishing. Black smoke and sharp acrid odor similar to P/F. Sample may be light colored EPOXY	
	B. Burns rapidly, intense white flame CELLULOSE NITRATE	
	C. Slow burning. Clear smokeless blue flame. Odor of formaldehyde ACETAL	
(iii) Continues to burn after removal from flame. Note color of flame during first second of burning	*D.* Blue flame, yellow at top	1. Fruit odor. ACRYLIC 2. Melts and drips. Drippings may burn. (*a*) Odor of rancid butter CELLULOSE ACETATE/ BUTYRATE (*b*) Pungent odor CELLULOSE PROPIONATE 3. Melts and drips. Drippings burn. Paraffin odor similar to burning candle. Floats on water (*a*) POLYETHYLENE (Low-density flexible, high-density more rigid) (*b*) POLYPROPYLENE (Rigidity similar to HD/PE but more difficult to scratch with fingernail)
	E. Yellow flame	1. Melts and drips. Drippings may burn. Acetic odor CELLULOSE ACETATE 2. Dense smoke. Flakes of soot in air (*a*) Odor of illuminating gas POLYSTYRENE (*b*) Acetic odor POLYVINYL ACETATE 3. Flame blue at edges (*a*) Melts and drips. Drippings may burn. Odor burnt sugar ETHYL CELLULOSE (*b*) Cracks and breaks POLYESTER (Nonfire resistant)

Table 31.2

SIMPLE TESTS FOR THE IDENTIFICATION OF THE MORE COMMON PLASTICS

<table>
<tr><td colspan="10">Cut thin sliver from edge of sample</td></tr>
<tr><td colspan="3">Powdery chips formed–indicates thermosetting resin.
HOLD SAMPLE IN LIGHTED MATCH, AND SMELL RESULTANT VAPORS</td><td colspan="7">Smooth sliver obtained indicates thermoplastic. Confirm by application of a hot metal wire or rod to the sample. Melting indicates thermoplastic
DROP SAMPLE ON TO HARD SURFACE</td></tr>
<tr><td rowspan="3">'Carbolic,' smell. Sample usually black or brown in color. Burns but extinguishes itself when flame removed = PHENOL-FORMALDEHYDE</td><td colspan="2" rowspan="3">Fishy smell–sample usually brightly colored (or white)
UREA-FORMALDEHYDE or MELAMINE-FORMALDEHYDE
TRY SCUFFING THE SAMPLE WITH FINGERNAIL</td><td colspan="2" rowspan="3">'Metallic' ring indicates presence of Styrene polymer
BURN A SMALL SAMPLE AND BLOW OUT FLAME SMELL RESULTANT SMOKE</td><td colspan="5">Absence of 'metallic' ring precludes polystyrene (unless foamed) but may be high-impact polystyrene (containing butadiene rubber)
PLACE SAMPLE IN WATER (+ A FEW DROPS OF DETERGENT)</td></tr>
<tr><td rowspan="2">If sample floats, indicates a polyolefin (unless it is foamed polystyrene–easily identifiable)
(a) SCRATCH SAMPLE WITH FINGERNAIL
(b) BURN USING LIGHTED MATCH</td><td colspan="4">If sample sinks: not a polyolefin.
BURN SMALL SAMPLE. NOTE EASE OF IGNITION AND TYPE OF FLAME</td></tr>
<tr><td>Burns with a clear flame. BLOW OUT FLAME AND RESULTANT VAPOR</td><td colspan="3">Difficult to ignite and is self-extinguishing
NOTE COLOR OF FLAME</td></tr>
<tr><td>Self-extinguishing, black smoke. Sharp acrid odor Sample lighter in color than P.F. = EPOXIDE</td><td>Scuffing = UREA FORMALDEHYDE</td><td>No scuffing = MELAMINE FORMALDEHYDE</td><td>Characteristic smell of styrene = POLYSTYRENE</td><td rowspan="2">Smell of styrene and butter smell = STYRENE-ACRYLONITRILE
As above and smell of rubber = ABS copolymer</td><td>Glossy surface, does not scratch. Burns with smell of paraffin wax = POLYPROPYLENE</td><td>Fruity odor = ACRYLIC (PROBABLY POLYMETHYLMETHACRYLATE)</td><td>Flame has greenish tinge. REMOVE SAMPLE FROM FLAME AND SMELL VAPOR</td><td>Flame is yellow REMOVE SAMPLE FROM FLAME. SMELL VAPOR</td><td rowspan="2">Little flame. Decomposition of material but no charring. A cellular structure is developed before decomposition = POLYCARBONATE</td></tr>
<tr><td></td><td colspan="2">The above test is not easy to interpret without experience, and definite conclusions can be drawn only from chemical tests</td><td></td><td>Glossy surface, slight scratching. Burns and drips like sealing wax = HD/PE</td><td>Smell as for burning paper = CELLULOSE ACETATE OR CELLULOSE PROPIONATE (NO SIMPLE CONFIRMATORY TEST AVAILABLE)</td><td>Acrid smell. Material soft and flexible = PLASTICIZED PVC (IF FILM, COULD BE VINYLIDENE CHLORIDE COPOLYMER</td><td>Smell of formaldehyde = POLYACETAL</td></tr>
<tr><td></td><td></td><td></td><td></td><td></td><td>Not particularly glossy, scratches easily. Burns with smell of paraffin wax = LD/PE</td><td></td><td>Acrid smell. Material nonflexible (and glossy if in form of molding = RIGID PVC OR PVC/PVA COPOLYMER</td><td>Indefinite smell. Sample has slippery feel
PRESS COLD METAL POINT TO HEATED SURFACE THEN WITHDRAW IT</td><td></td></tr>
<tr><td></td><td></td><td></td><td></td><td></td><td></td><td>Smell of rancid butter = CELLULOSE ACETATE BUTYRATE</td><td></td><td>Threads easily formed = NYLON</td><td></td></tr>
</table>

(d) Nylon

The sample is burnt and a filter paper, moistened with a saturated solution of *o*-nitrobenzaldehyde in 2N sodium hydroxide solution, is placed in the fumes. Nylon 66 gives a deep mauve coloration. Other nylons give no color.

(e) Polyethylene Terephthalate

The sample is burnt and a filter paper moistened with a saturated solution of *o*-nitrobenzaldehyde in 2N sodium hydroxide solution, is placed in the fumes. A green-blue color is produced.

(f) Epoxide (based on diphenylolpropane)

The sample is dissolved in 1 cm^3 of concentrated sulfuric acid and 1 cm^3 of concentrated nitric acid is then added. This solution is then poured into 100 cm^3 of N sodium hydroxide solution. A bright orange-red coloration is produced.

(g) Phenol-Formaldehyde

A reagent (known as Millon's reagent) is first prepared by dissolving 10 g of mercury in 10 g of fuming nitric acid, without heating, and is then diluted with 20 cm^3 of water. Any precipitate which forms is filtered off and discarded.

The sample is then boiled with 1 cm^3 of this reagent for 2 minutes. A red coloration is produced. This coloration is also given by some proteins.

(h) Urea and Melamine Formaldehydes

Reflux the sample with dilute sulfuric or phosphoric acid, and divide the solution in two parts.

(i) Add few drops of a saturated solution of xanthydrol. A white precipitate indicates urea formaldehyde.
(ii) Add a saturated aqueous solution of picric acid. A bulky yellow precipitate indicates melamine formaldehyde.

(i) Phthalates and Phthalic Acid

The sample is heated with 0.1 g of pure phenol and a drop of concentrated sulfuric acid until the melt turns orange or brown in color. When cooled the mass is triturated with 2 cm^3 of water and is then filtered. The filtrate is made alkaline with 10 percent aqueous sodium hydroxide, when the characteristic red color of phenolphthalein is produced.

If in the test above, the phenol is replaced by resorcinol a green fluorescence is produced.

(j) Polyvinyl Acetate

An iodine reagent is prepared by dissolving 0.1 g of iodine and 1 g of potassium iodide in a mixture of 10 cm^3 water and 10 cm^3 ethyl alcohol and then making up to 100 cm^3 with 2N hydrochloric acid.

When the sample is covered with 0.5 cm^3 of the above reagent, the solid becomes stained a deep red. This effect is enhanced if 10 cm^3 of water is added.

(k) Polyvinyl Alcohol

One drop of the iodine reagent described above is added to 1 cm^3 of a 0.5 percent aqueous solution of the sample, when a blue color should result. If no coloration is observed, then one drop of a saturated aqueous borax solution should be added to the mixture, and the blue color should then appear.

Polyvinyl alcohol containing more than 10 percent acetate groups gives a red coloration in this test.

(l) Polyvinyl Chloride and Vinyl Chloride Copolymers

The sample is dissolved in 5 cm^3 pyridine by boiling and, while still hot, 0.5 cm^3 of a solution of sodium hydroxide in methyl alcohol (2 percent w/v) is added. A brown coloration is produced and eventually a brown precipitate is formed.

(m) Polyvinylidene Chloride and Vinylidene Chloride Copolymers

The sample is heated on a water-bath with 5 cm^3 of morpholine. After about two minutes, the solvent turns dark reddish brown, while after about five minutes both solvent and polymer turn almost black.

(n) Shellac

This polymer is readily recognized from the characteristic odor of sealing wax obtained on burning. It is soluble in dilute alkalis such as ammonium hydroxide. When heated to 150°C the polymer is converted to an alcohol-insoluble, horn-like mass in 20 to 60 min.

(o) Silicone

The white ash remaining after burning the sample is soluble in hydrofluoric acid and leaves no residue when the solution is evaporated. The presence of silica as a filler will, of course, invalidate this test.

Should any doubt exist after carrying out the above tests, further chemical tests may be applied which will determine the presence or absence of certain elements such as nitrogen, the halogens, or sulfur. The basis of this scheme is the Lassaigne Sodium Fusion Test. Initially, 0.1 to 0.5 g of the polymer are decomposed by heating with metallic sodium in an ignition tube. When the reaction has subsided, and while the tube is still red-hot, the tube is plunged into about 10 cm^3 of distilled water contained in a small mortar. Care is obviously needed in carrying out this operation, and the ignition tube should be kept with the open end facing away from the operator at all times. The ignition tube and contents are carefully ground under water with a pestle and the mixture filtered.

The filtrate is divided into five portions which are then examined as follows:

1. Nitrogen. A portion of the test solution is placed in a boiling tube with several drops of freshly prepared ferrous sulfate solution and brought to boiling. After cooling, the contents are acidified with dilute sulfuric acid and a drop of ferric chloride solution is added. The formation of a blue precipitate (Prussian Blue) indicates the presence of nitrogen in the polymer under examination.

2. Sulfur. A few drops of freshly prepared sodium nitroprusside solution are added to the test portion contained in a boiling tube. The formation of a violet color indicates the presence of sulfur.

3. Chlorine or Bromine. A portion of the test solution is placed in a boiling tube and acidified with dilute nitric acid and boiled for 2 to 3 minutes. A few drops of silver nitrate solution are then added. A white precipitate, soluble in ammonia, indicates the presence of chlorine while a yellow precipitate, insoluble in ammonia, indicates the presence of bromine.

Fluorine does not give a precipitate.

4. Fluorine. An aqueous solution of 0.1 percent zirconium nitrate and 0.3 percent alizarin S is prepared, and this is used to impregnate a filter paper, which is then dried. The filter paper is then moistened with 50 percent acetic acid. The test sample is neutralized with dilute hydrochloric acid and one drop placed on the moistened filter paper. The red color of the filter paper turns yellow if fluorine is present.

5. Phosphorus. The test solution is acidified with dilute nitric acid and an excess of ammonium molybdate is added. The formation of a yellow crystalline precipitate indicates the presence of phosphorus.

If the results from all the tests described above are correlated, it is usually possible to make a positive identification. As a guide to using the results of Lassaigne Test, Table 31.3 lists the elements present in many of the polymers commercially available.

Table 31.3

ELEMENTS PRESENT IN CERTAIN POLYMERS

Elements	*Present in*
Chlorine	Chlorinated rubber; rubber hydrochloride; PVC; polychloroprene; polyvinylidene choride; PTCFE; VC/VA copolymers; chlorinated polyether.
Nitrogen	Cellulose nitrate; polyamides; polyurethanes; polyacrylonitrile; epoxides; urea- and melamine-formaldehyde; phenol-formaldehyde (due to the presence of hexamine).
Sulfur	Rubbers, polysulfides; polysulfones.
Phosphorus	Due to the presence of additives such as phosphate plasticizers (which must be extracted).
Chlorine and nitrogen	Vinyl chloride–acrylonitrile.
Chlorine, nitrogen, phosphorus	Phosphonitrilic chloride.
Chlorine and sulfur	Chlorsulfonated polyethylene (Hypalon).
Fluorine	Fluorocarbons.
Nitrogen, phosphorus, sulfur.	Casein.

SUGGESTED FURTHER READING

Electrical Properties of Polymeric Materials. (1973) M. E. Baird (London: The Plastics and Rubber Institute).

Flexural Testing of Plastics (1969) R. D. Heap and R. H. Norman (London: The Plastics and Rubber Institute).
Impact Tests and Service Performance of Thermoplastics. (1971) P. I. Vincent (London: The Plastics and Rubber Institute).
Handbook of Plastics Test Methods. (2nd ed. 1989) Ed. R. P. Brown (London: Longman).
Mechanical Testing and Plastics (2nd ed., 1984) S. Turner (London: The Plastics and Rubber Institute).

INDEX

A

F

P

Q

R

S

T

U

www.ingramcontent.com/pod-product-compliance
Lightning Source LLC
LaVergne TN
LVHW102006210926
848483LV00009B/101
* 9 7 8 0 8 2 0 6 0 3 4 4 5 *